Latin America

TEXTS IN REGIONAL GEOGRAPHY

A Guilford Series

Edited by James L. Newman, Syracuse University

LATIN AMERICA: REGIONS AND PEOPLE
Robert B. Kent

AFRICA SOUTH OF THE SAHARA: A GEOGRAPHICAL INTERPRETATION
SECOND EDITION
Robert Stock

THE EUROPEANS: A GEOGRAPHY OF PEOPLE, CULTURE, AND ENVIRONMENT
Robert C. Ostergren and John G. Rice

Latin America

Regions and People

ROBERT B. KENT

THE GUILFORD PRESS
New York London

A Division of Guilford Publications, Inc.
72 Spring Street, New York, NY 10012
www.guilford.com

Printed in the United States of America

This book is printed on acid-free paper.

Last digit is print number: 9 8 7 6 5 4 3 2 1

Library of Congress Cataloging-in-Publication Data

Kent, Robert B.
Latin America : regions and people / Robert B. Kent.
p. cm. — (Texts in regional geography)
Includes bibliographical references and index.
ISBN 1-57230-909-1 (pbk.)—ISBN 1-59385-269-X (hardcover)
1. Latin America—Geography. 2. Latin America—Social conditions. 1. Latin America—Economic conditions. I. Title. II. Series.
F1408.9.K46 2006
918—dc22

2005027886

Cover photos, left to right:

Paulista Avenue, São Paulo, Brazil, 2004
Todos Santos Lake, Los Lagos Region, Chile, 1993
The portal of a colonial residence, Potosí, Bolivia, 1997
The Altiplano, Department of Oruro, Bolivia, 1980
A restaurateur, Cliza, Bolivia, 1991

To all Latin Americans,
and especially to my wife, Marialena,
and to our children, Robert and Anika

Preface

It's an old story. At first, a subject catches your interest, then it fascinates you, and then, suddenly, without warning, you are consumed by it. So it was with me and Latin America. My family lived in California about 15 miles north of the United States–Mexico border when I was in junior high school. I began studying Spanish then. Who knew any French or German speakers? My family went to Tijuana once or twice, and later in college I made a few trips with surfing buddies to Rosarito Beach and Enseñada—more memorable perhaps for our interactions with Mexican traffic police than with the surf.

Some time in the early 1970s my passing curiosity about the region metamorphosed into something more urgent. I only remember the vaguest outlines of my first extended trip to Latin America, hitchhiking to Calexico and then crossing the border and taking the train from Mexicali to Guadalajara. I do remember fondly the Mexican teacher and his son who befriended me on the train. They were returning home after a summer working as dishwashers in California and invited me to their home in Guadalajara. I stayed with them for several days before heading off to Lake Chapala and Colima . . . eventually back to California.

With that trip began a lifelong fascination for me with Latin America. It has led me back there to travel, to work, and to live there again and again. On one of these sojourns, to Bolivia in 1980, I met Marialena Torrico Canaviri. We married later that year. Together we have lived in and traveled through much of Latin America, at first by ourselves, and then later with our children, Robert and Anika.

This book provides the reader with an overview of the human geography of Latin America, from both thematic and regional perspectives. On the one hand, the book utilizes a cultural–historical approach to the macrolevel processes and patterns that define Latin America as a culture region. On the other hand, specific places matter to geographers; the unique characteristics of individual places and regions are central to understanding a region's human geography. As such, the book also identifies and examines in detail the principal geographical subregions that constitute Latin America. Therefore, from an organizational standpoint, the book is divided into alternating chapters: one that examines a key theme in human geography, and another that examines in detail one of the nine principal geographical subregions.

Many individuals helped me to bring this book to completion. Early drafts of many of the book's chapters were completed during a sabbatical leave at the Universitat Jaume I in Castellón, Spain, during 1998–1999. Drs. Vicent Ortells and Eugenio Burriel provided critical support during that year that facilitated work on the book. In addition, I am indebted to Risa Patarasuk, Emily Aronson, Fred Boateng, Heather

Bickenheuser, and Badri Lakkur, all of whom worked as research assistants on this project. A handful of cartographers contributed to the production of the maps. Thanks are due to Elvin Delgado, Francisco Jimenez, Ernesto Carreras, Cesar Piovanetti, Fred Boateng, and Keith Pitts, who drafted maps for this text. Special thanks are due to Iraida Galdon for her cartographic expertise. She produced many of the maps and diagrams herself and also helped bring greater order and consistency to maps produced by different cartographers over several years. The organizational skills of Kevin Butler, GIS research manager for the Laboratory for Cartographic and Spatial Analysis in the Department of Geography and Planning at the University of Akron, are also acknowledged and appreciated. Bob Barrett provided invaluable assistance in reviewing and editing the manuscript in the final stages of the book's preparation. Laura Monroe reviewed an earlier draft of the book, and her advice and editing are also greatly appreciated. Kristal Hawkins, of The Guilford Press, has been supportive and patient throughout the book's preparation. Thanks are due to Jim Newman, Series Editor of Guilford's Texts in Regional Geography, for encouraging me to take on this project and providing me with guidance as I undertook my work. Also, the comments and suggestions of three anonymous reviewers were very useful and helped me to improve the manuscript in a number of ways.

I am also indebted to a number of my teachers, colleagues, friends, and family who have contributed to my enthusiasm for and knowledge of Latin America. Dennis Dingemans, David Robinson, Rolf Wesche, Ray Bromley, Francois Belisle, David Hess, Olav Sibille, and Luis Sanchez inspired and encouraged me over many years. Further, I wish to acknowledge the support and encouragement of my colleagues in the Department of Geography and Planning at the University of Akron. They have created and sustained a collegial and stimulating departmental environment there for many decades. My friends John Markoff, Rob Puddicome, Gary Bradski, and Hank Bennett have prodded, encouraged, and commiserated with me on this and other projects since we became friends in high school and college in California. Finally, I owe a tremendous debt to my wife, Marialena, who accompanied me for extended periods of residence in Peru and Argentina, and has traveled with me from Tijuana to Cape Horn. Her support and deep personal knowledge of the region have been invaluable.

Finally, I hope that through this book I am able to share some of the excitement and satisfaction that the study of Latin America and its people has brought me over the last 30 years. It is also my hope that it will encourage readers, including Latin Americans, to learn more about the region and its peoples by traveling there and experiencing Latin America themselves firsthand. As Eva Crane, the inspiring editor of *Bee World*, the journal in which I published my first scholarly article, wrote:

> One learns so much, even from a short visit to a new country, that cannot be understood by reading books or even by correspondence with people in it. One also learns in a tantalizing way, a little of the countries just beyond, and of the new interests and problems they present. Beyond British Columbia is Alaska, and beyond Mexico is Guatemala, and then Costa Rica, and Peru. . . . If one only had the time and the means . . .

Contents

1	Introduction	1
2	Latin America as a Culture Region	8
3	The Environment	37
4	The Hispanic Caribbean	55
5	Peopling Latin America to 1820	67
6	The Mesa Central of Mexico	91
7	Early Cities and Urban Development	113
8	The Andes	130
9	Migration, Population Change, and Race	154
10	The Central Valley of Chile	179
11	The Central American Highlands	187
12	Land and People since the Conquest	206
13	Brazil's Atlantic Coastal Plain	236
14	Contemporary Cities and Urban Patterns	246
15	The Southern Brazilian Highlands	271
16	Latin American Development in Perspective	292
17	The Pampa	308
18	Latin America in the World Economy	322
19	The Amazon Basin	348
20	The Latin American Diaspora	365
21	Conclusion	390
	Chapter Notes and Additional Sources	396
	Index	408
	About the Author	422

1 Introduction

Latin America is a vast region spanning parts of North America, almost all of South America, and much of the West Indies. It encompasses 19 countries as well as Puerto Rico, a commonwealth territory of the United States, and, arguably, even parts of the southwestern United States. It is characterized by a long history of shared cultural values and traditions, most notably language and religion. While its human geography is distinguished by a large measure of cultural and historical homogeneity, its physical geography is diverse and includes deserts, steppes, rain forests, fertile plains, rugged highlands, and alpine glaciers. Any region, but especially one as immense and complex as Latin America, defies easy description and understanding.

This book is a primer on the human geography of Latin America. On the one hand, it aims to provide the reader with an overview of Latin America's human geography, and, on the other hand, its regional complexity. First, it defines and examines Latin America as a culture region, identifying factors that have contributed to the varied human geography and landscapes of the region. Second, the book uses a regional perspective to identify and explore the distinct environments that are the component parts of this enormous world region.

Approaches to Understanding the Geography of Latin America

Over the past 30 years, geographers have taken a range of approaches to providing an overview of the region's geography. A strictly regional approach to the study of Latin America characterizes some of these efforts. Individual countries are the principal focus in Preston James and Clarence Minkel's *Latin America* 5th ed. (1986), while in *Latin America: A Regional Geography* 3rd ed. (1973) Gilbert Butland divides the area into five macroregions, and begins the treatment of each with a brief regional summary, followed by a country-by-country analysis. A slightly different approach is taken by Harold Blakemore and Clifford Smith in *Latin America: Geographical Perspectives* 2nd ed. (1983). Here Mexico and Brazil are examined individually, while in other cases macroregions that include several countries, like Central America, the Central Andes, or the Southern Cone, form the basis for discussion.

Other geographers have combined the regional approach with thematic treatments of key geographical themes, such as physical geography, population and settlement, agriculture, economic geography, urban systems, and economic development. Robert West and John Augelli's *Middle America: Its Lands and Peoples* 3rd ed. (1989) is a masterful example of this approach using macroregions as the basis for its regional analysis. *South America* 4th ed. (1991) by Arthur Morris takes a similar tack, but uses countries as the basis for the book's regional studies. Most recently, in *Latin America and the Caribbean: A Systematic and Regional Study* 4th ed. (2002) Brian Blouet and Olwyn Blouet have pursued a similar approach.

Fewer geographers have approached the study of Latin America from a strictly thematic perspective. David Clawson utilizes this approach in *Latin America: Lands and Peoples* 2nd ed. (2000), as do Cesar Caviedes and Greg Knapp in *South America* (1995). More frequently, thematic or systematic approaches to the geographic study of Latin America have been directed at more narrowly defined topics and have not sought to present a comprehensive overview of the region. A focus on economic and social development has been common in these regional thematic treatments. Among these books are Alan Gilbert's *Latin American Development: Geographical Perspectives* (1974), Rosemary Bromley and Ray Bromley's *South American Development: A Geographical Introduction* 2nd ed. (1988), David Preston's *Latin American Development: Geographical Perspectives* 2nd ed. (1996), and Robert Gwynne and Cristobal Kay's *Latin America Transformed: Globalization and Modernity* (1999). Other systematic geographic studies of the region are less common, but do include Jonathan Barton's *A Political Geography of Latin America* (1997), and some might argue for inclusion of Alan Gilbert's *The Latin American City* (1994).

The Organization of This Book

The focus of this book alternates between thematic chapters that examine Latin America in its entirety and regional chapters that spotlight the significance, environmental conditions, historical geography, economic development, and contemporary conditions of the principal geographic regions that comprise Latin America.

Thematic Perspectives

By and large, the book's thematic chapters follow a cultural–historical approach to the study of contemporary Latin America. This perspective on geographical study holds that in order to understand the contemporary landscape, one must understand the culture and historical geography that have shaped it. Thus, after an introductory chapter that defines Latin America as a culture region and a subsequent chapter that outlines the region's environmental context, the thematic chapters are organized in a roughly chronological manner that tracks the region's development. These chapters focus on the pre-Hispanic cultures as well as the historical geography of the region, then proceed to chronicle the population history, urban development, agriculture and land use, social and economic development, economic geography, and the contemporary patterns of the Latin American diaspora.

Regional Perspectives

The geographic concept of a region is one that helps to organize our understanding of places and environments. Essentially, *regions* are areas that share geographic characteristics. A wide range of criteria may be utilized to define geographic regions. These are often related to the physical environment, culture, or economy, but any number of criteria may be used depending on the reasons for identifying them.

Often a single factor is used to define a region. This may be done because this variable is central to the analysis being undertaken or because it serves as a surrogate for a wide range of other variables. For instance, predominant language use might be mapped as a surrogate for defining broad cultural regions. Thus, the dominant use of English, French, Portuguese, and Spanish in the Western Hemisphere might be mapped to define the distinct regional cultures associated with each.

In other cases, geographers use a more extensive set of criteria, or multiple factors, to more accurately reflect the cultural or physical homogeneity of distinct geographic regions. In the United States one of the most widely recognized regions is "the South." The criteria used to identify the South are myriad and can include elements of the physical environment, climate, soils, and natural vegetation, as well as a range of characteristics intimately linked to the region's culture: history, social conditions, attitudes, foods, and dialect.

A key tenet of this book is that place matters. It is argued here that familiarity with Latin America's principal regions is essential to an appreciation and understanding of the region as a whole. This background assists in the comprehension of the social, economic, and environmental linkages that have shaped the region in the past and will continue to do so in the future. While physical geography often forms the basis of the initial identification of these regions, and their names reflect this fact, they are multiple-factor regions representing the congruence of both environmental and cultural variables. Typically, each region is distinguished by both historical and contemporary importance, as well as a sizeable population.

Nine principal regions are identified: the Hispanic Caribbean, the Mesa Central of Mexico, the Andes, the Central Valley of Chile, the Central American Highlands, Brazil's Atlantic Coastal Plain, the Southern Brazilian Highlands, the Pampa, and the Amazon Basin (Figure 1.1). The order of their treatment in the text roughly mirrors the chronology of their integration into the Latin American realm. Each regional chapter identifies the area's significance, examines the principal characteristics of the physical environment, reviews its historical geography and economic development, and ends with an appraisal of its contemporary social and economic geography.

This book also identifies and examines notable peripheral regions in Latin America. More often than not, these are well-recognized physiographic or environmental regions that appear on almost all continental-scale maps of Latin America. Typically, these peripheral regions are remote and inaccessible from the population and economic heartlands of the principal regions. These regions are often impoverished, sparsely populated, and characterized by a limited resource base. The economy of these areas is usually dependent on the exploitation of a limited number of natural resources and extractive industries (most often mining, forestry, or fishing), or extensive agricultural pursuits, like livestock raising. Limited economic opportunities and high birthrates often make these peripheral regions prime sources for migrants to the cities and metropolitan areas of the principal regions, where they go to seek brighter futures. The Sertão of northeastern Brazil is an excellent example of a peripheral region that has long supplied large numbers of migrants to the country's principal regions, the Atlantic Coastal Plain and the Paraná Plateau, and more specifically São Paulo.

The peripheral regions described here are the Mesa del Norte of Mexico, the Caribbean Coastal Lowlands of Central America, the Llanos, the Guiana Highlands, the Gran Chaco, the Sertão, and Patagonia (Figure 1.2). Each regional treatment includes a map of the region and provides an overview of the region's environmental characteristics and contemporary conditions. A short overview of each peripheral region is included in the chapter of the principal region with which it is most closely linked. Thus, the treatment of the

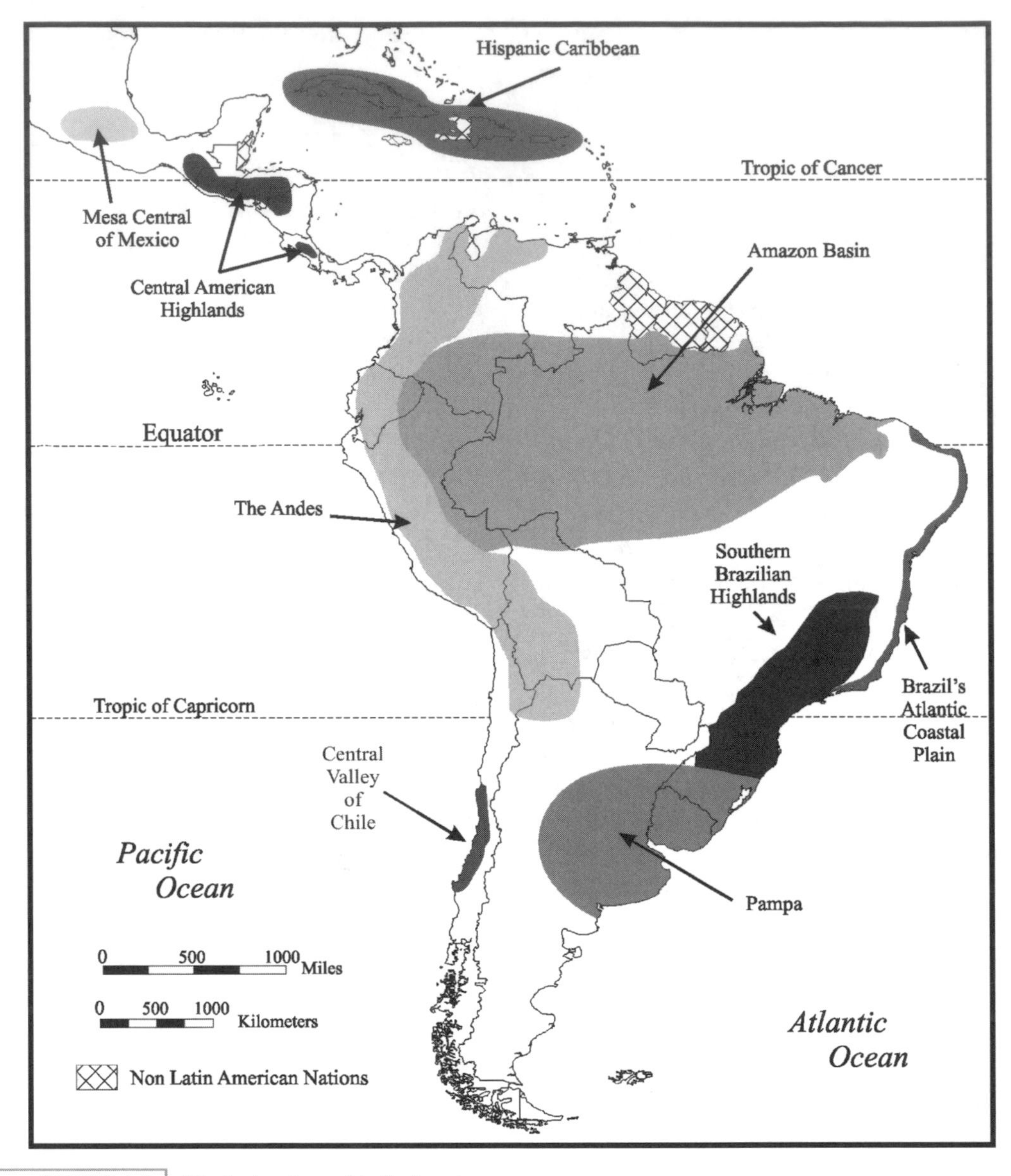

FIGURE 1.1. Principal regions of Latin America.

Mesa del Norte of Mexico is included in the regional chapter on the Mesa Central of Mexico.

It is important to note that the peripheral regions discussed in this book do not represent an exhaustive survey of peripheral regions in Latin America. Many others can be identified. The Borderlands of the United States and Mexico, the Southern Highlands of Mexico, the Yucatan Peninsula, the Chocó region of Colombia's Pacific coast, the Yungas of Bolivia's Andean piedmont, the vast Pantanal wetlands of southwestern Brazil, the Cuyo in western Argentina, and Chile's Sur Chico and Sur Grande are but some examples.

The identification of regions and their boundaries can be controversial. Indeed, it is often the subject of heated debate among both professional geographers and the public at large. The regions identified in this book represent one geographer's interpretation of their extent, characteristics, and boundaries. Other geographers and scholars, as well as Latin Americans in general, may view Latin America's geographic subdivisions through substantially different lenses and may contest the validity of particular regional identifications or the boundaries of those regions. Furthermore, the identification of regional boundaries is not a precise science and is further complicated here because the scale of the analysis is continental.

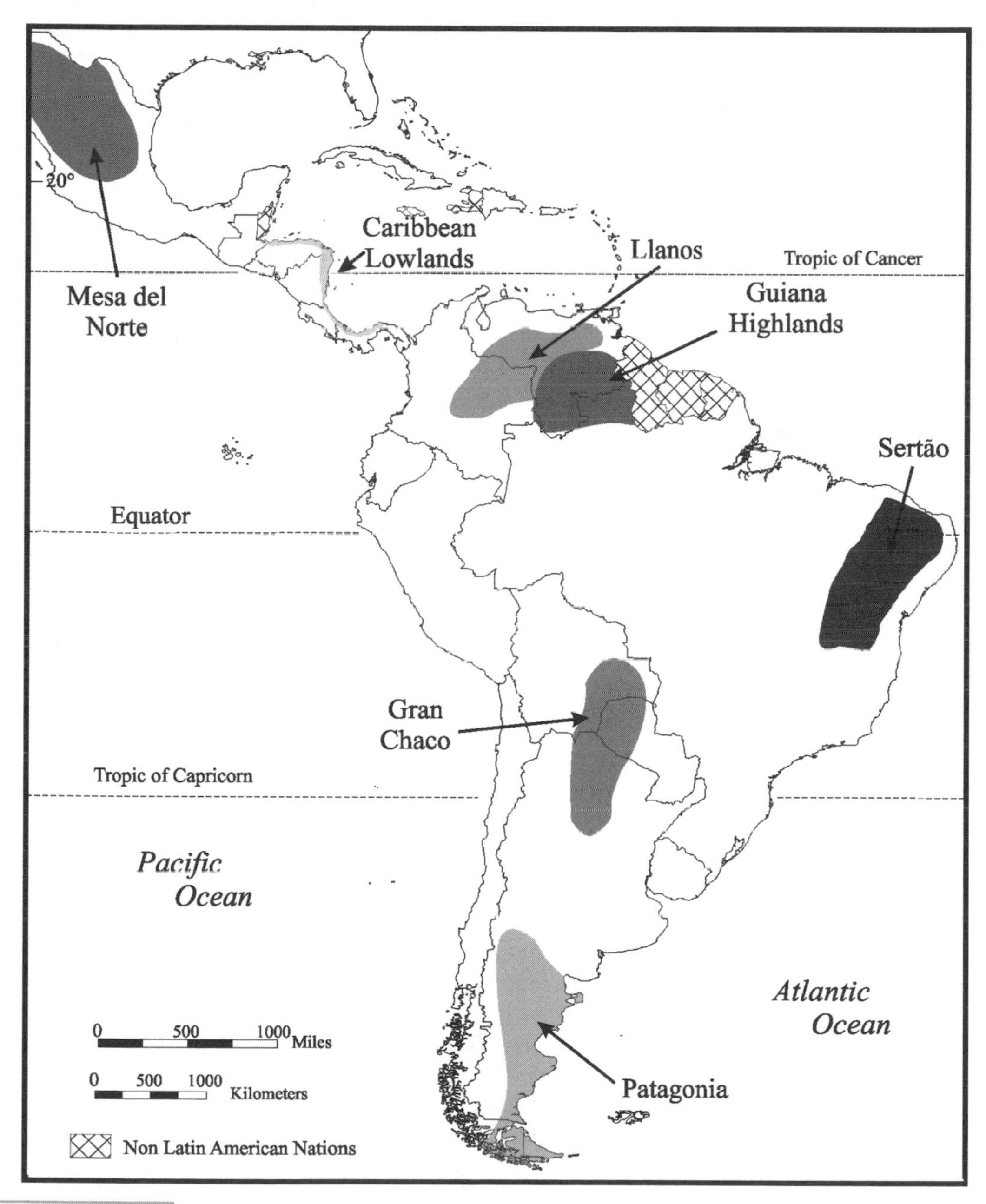

FIGURE 1.2. Secondary regions of Latin America.

A Note on Sources and Resources

The source material drawn on for this book includes books and journal articles from the scholarly literature as well as statistical data from government sources and international organizations. In addition, newspapers, notably *The New York Times*, and other news outlets have provided contemporary perspectives on current and evolving conditions in Latin America. Although much of this material is in English, sources in Spanish and to a lesser degree in Portuguese have also been utilized.

In addition to the sources noted above, background information, especially for the chapters focusing on Latin America's principal geographic regions, has been drawn from several key reference sources. Noteworthy among these are the *Encyclopaedia Britannica*, the *Encarta Multimedia Encyclopedia*, the *Columbia World Gazetteer*, and the *Cambridge Encyclopedia of Latin America*. The climatic data presented in the climographs in the regional chapters comes from *www.worldclimate.com*. Country-specific geographical studies as well as a range of atlases published in Spanish and Portuguese have also provided supporting documentation and background information.

Few comprehensive travel guides for the Latin American region existed before 1990. The notable exception is the venerable *South American Handbook* published for over 70 years. However, recently there has been a veritable explosion of country-specific travel guides. Many of these are strictly tourist guides with little to recommend them to the educated traveler or scholar. However, some, especially those published by Moon Publications and Lonely Planet Publications, are carefully researched and field-checked regularly. These provide timely and useful information about country conditions as well as about remote areas for which there is often little readily accessible contemporary information. Several of these publications have been utilized as reference resources in this book.

The World Wide Web, which has revolutionized research in the last decade, has been employed frequently in the preparation of this book. Many academic resources are available through the Web, especially articles from scholarly journals and statistical databases produced by government institutions. A wide range of institutions, including museums, think tanks, and research institutes, and many individual scholars maintain websites with a wealth of pertinent information. Nevertheless, the integrity of such web resources must be assessed carefully. Many are of questionable origin, while others are clearly advocacy organs for social causes, political viewpoints, or business endeavors.

Finally, the author's travel, residence, and research in Latin America over the last three decades provide the basis for many of the observations and comments about the region.

Bibliographic information is provided in two places in this book. Each chapter is followed by a short list of further reading in English, typically books, book chapters, and journal articles. These are intended to provide the reader with additional resources to further explore the principal themes developed in each chapter. More detailed bibliographic information pertaining specifically to the material covered in each chapter is provided at the end of the book in the section "Chapter Notes and Additional References." Besides referencing the sources of information and statistical data provided in the corresponding chapter, these lists include a variety of additional resources for the interested reader, including non-English-language resources, principally in Spanish. In neither case, however, are these lists intended to be exhaustive; instead, they are intended to provide a starting place for further reading and research.

Further Reading

Augelli, J. P. (1962). The rimland–mainland concept of culture areas in Middle America. *Annals of the Association of American Geographers, 52*(2), 119–129.

Barton, J. R. (1997). *A political geography of Latin America*. London: Routledge.

Blakemore, H., and Smith, C. T. (Eds.). (1983). *Latin America: Geographical perspectives* (2nd ed.). London: Methuen.

Blouet, B. W., and Blouet, O. M. (Eds.). (2002). *Latin America and the Caribbean: A systematic and regional survey* (4th ed.). New York: Wiley.

Bromley, R. D. F., and Bromley, R. (1988). *South American development: A geographical introduction* (2nd ed.). Cambridge, U.K.: Cambridge University Press.

Butland, G. J. (1973). *Latin America: A regional geography* (3rd ed.). London: Longmans.

Caviedes, C., and Knapp, G. (1995). *South America*. Englewood Cliffs, NJ: Prentice-Hall.

Clawson, D. L. (2000). *Latin America: Lands and peoples* (2nd ed.). New York: McGraw-Hill.

Gilbert, A. (1974). *Latin American development: A geographical perspective*. Harmondsworth, UK: Penguin Books.

Gilbert, A. (1994). *The Latin American city*. London: Latin American Bureau.

Gwynne, R. N., and Kay, C. (Eds.). (1999). *Latin America transformed: Globalization and modernity*. London: Arnold.

James, P. E., and Minkel, C. W. (1986). *Latin America* (5th ed.). New York: Wiley.

Morris, A. S. (1991). *South America* (4th ed.). Seven Oaks, Kent, UK: Hodder & Stoughton.

Nostrand, R. L. (1970). The Hispanic–American borderland: Delimitation of an American culture region. *Annals of the Association of American Geographers, 60*(4), 638–661.

Preston, D. (Ed.). (1996). *Latin American development: Geographical perspectives* (2nd ed.). London: Longmans.

West, R. C., and Augelli, J. P. (1989). *Middle America: Its lands and peoples* (3rd ed.). Englewood Cliffs, NJ: Prentice-Hall.

2 Latin America as a Culture Region

In its simplest form, we might describe *culture* as the ability of humans to learn—but more specifically to accumulate knowledge and experience in one generation and pass it on to subsequent generations. A more formal definition is that culture is the "totality of socially transmitted behavior patterns, arts, beliefs, institutions, and all other products of human work and thought characteristic of a community or population."

Cultures—or human lifeways—manifest themselves on a variety of geographical scales. Often the concept is employed to define or to characterize broad areas of cultural affinity, or culture realms, on a global scale—for instance, the West, the Arab world, or Latin America. Such global culture regions, indeed all culture regions, are defined by one or a combination of common traits: history and traditions, language, or religion.

Latin America as a Culture Region

Latin America is one of the world's most coherent and easily identified culture regions. It generally corresponds with South America, Central America, and the Caribbean and is distinguished by a series of common characteristics that provide a foundation of cultural unity for the nearly 500 million people who live there. Language is perhaps the single most important trait. Closely related Romance languages, Spanish and Portuguese, are spoken by about 350 and 150 million people, respectively. A dominant religion, Catholicism, provides a framework of common beliefs, customs, and even religious architecture that binds the people of Latin America together. Although Protestant denominations, especially Evangelical Christian ones, have experienced significant growth in Latin America during the last 30 years, the historical dominance of the Catholic Church has not been seriously threatened.

Common political and social institutions also create cultural continuity throughout the region. For instance, Latin American legal systems are modeled on a system of civil law practiced in Spain and Portugal and based on the Napoleonic Code, and are quite distinct from the closely related systems of common law in force in Great Britain, the United States, Canada, Australia, and many other former British colonies. Throughout the region, governance in both a political and an administrative sense is highly centralized, a product of the colonial experience and bureaucratic systems operative in Spain and Portugal. Also, informal social institutions as well as social attitudes and values contribute to the definition of Latin America and demonstrate the common cultural roots of the region's peoples.

Shared social values and attitudes manifest themselves in many ways. *Compadrazco*, coparenthood, is a traditional form of ritual parenthood throughout most of the region. At the time of their child's baptism in the Catholic religion parents ask acquaintances, friends, or relatives, preferably of higher socioeconomic status, to stand as *compadres* and *comadres* for their children. While the *compadre* and the *comadre* serve as the child's godparents, their relationship with the child's birth parents is also a mechanism to extend the social and economic network of the parents.

The concepts of *machismo*, the apparent dominance of men in Latin American family life, and of *marianismo*, the outward passiveness of women in those same Latin American families, are stereotypical behavior patterns that do bear some resemblance to reality (Vignette 2.1). In business and public life a similar type of ritualized behavior, *clientelism*, based on socioeconomic and political dominance, on the one hand, and subservience, on the other, is played out between individuals and businesses. This paternalistic system is reminiscent of feudal social relationships; loyalty to powerful individuals and families by their clients and subordinates is rewarded and disloyalty is punished. Historically in Latin America such relationships characterized those between owners of haciendas and plan-

VIGNETTE 2.1. *MACHISMO*

Although people in the United States, Canada, and Europe may not know much about Latin America, many are at least vaguely familiar with the concept of *machismo*. The prototypical *macho* male is proud, aggressive, and fearless, and also controlling and highly protective of female family members, girlfriends, and wife. He is sexually powerful—demonstrated by his history of having numerous girlfriends, fathering many children, and if married, having a mistress. The culture of machismo permeates not only personal relationships, but in some cases the legal system. In Brazil, for instance, men who have killed their wives or their wives' lovers frequently claim that they killed "in defense of honor." The logic of this defense, often accepted by Brazilian juries, is that the damage to the male's dignity and manhood, when added to his emotional trauma, is a sufficient justification for murder. This defense traces its origins to Portuguese colonial law that exonerated a husband who caught his wife in an adulterous act and killed her. Such occurrences are not rare. A survey in São Paulo state in the early 1980s revealed that nearly 800 women a year were murdered by spouses and lovers and that almost none of the perpetrators were punished.

In Mexico, the case of Claudia Rodriguez illustrates the double standard often applied to women in a macho culture. According to witnesses, a drunken male attacked and assaulted her. When Rodriguez threatened him with a handgun, he attacked her again and she shot and killed him. She languished in jail for over a year awaiting trial. A district judge denied an appeal for her release in 1996, arguing that, because her attacker was drunk and Rodriguez was sober, it was Rodriquez who was responsible for the attack, not the drunken attacker. The case enraged Mexican feminists and widespread protest about Rodriguez's fate led to her eventual release in 1997.

These examples demonstrate the most extreme face of macho behavior and attitudes. In all fairness, it must be noted that many Latin males reject machismo and macho behavior as an anachronism. Indeed, today a significant percentage of Latin American marriages, especially among the middle class, are predicated on the same concepts of personal equality and the sharing of responsibilities that characterize many marriages in the United States, Canada, and Western Europe.

tations, *los patrones*, and their slaves or workers, *los peones*. A passive fatalism is often said to characterize Latin Americans' attitudes, especially in regard to economic success, as opposed to the activist attitudes the Protestant ethic is said to imbue in Western Europeans and North Americans.

Although differences in culture may be considered fascinating curiosities, U.S. companies and business personnel are finding that a solid understanding of Latin American culture helps them penetrate Latin American markets and do business there. The removal of trade barriers between the United States and Mexico by the North American Free Trade Agreement (NAFTA) has underscored this fact. U.S. companies find that a whole new business environment exists south of the Rio Grande—and that it is not enough just to speak Spanish to achieve success in this environment.

Daily scheduling, decision making, and even office etiquette are distinct. Mexican executives and managers prefer long lunch breaks and work late into the evening, while their U.S. counterparts tend to like shorter lunch breaks and end the workday by 5:00 or 6:00 P.M. Decision making is highly centralized in most Mexican businesses with only the highest level executives participating in policy discussions, while the decision-making process in U.S. firms is often more broadly based, involving at least some input from second-tier executives and managers. Mexicans believe that a more genteel style characterizes their business communications and practices, which are often laced with social pleasantries and tend to avoid direct confrontation. Americans, on the other hand, often perceive these practices as feints, deceptive pretense, and an unwillingness to address problems head-on.

It is cultural cohesiveness, not geography, that defines the Latin American region. So, while many assume that everything south of the Rio Grande River is Latin America, that is not the case. In South America, the former British, Dutch, and French Guianas—now called Guyana, Surinam, and French Guiana, which sit along the continent's northern coast—must be excluded from Latin America. As former colonies of Great Britain, the Netherlands, and France, respectively, each of these small political units has a cultural environment that bears no similarity whatsoever to Latin America. At present, Guyana and Suriname are independent countries, while French Guiana is an Overseas Department of France. On the Central American isthmus a similar situation defines the nation of Belize, an English-speaking former colony of Great Britain. In the Caribbean Basin only three among the myriad of political units found there are part of Latin America: Cuba, the Dominican Republic, and Puerto Rico. Collectively these three represent well over half of the population and land area of the Caribbean Basin (Figure 2.1).

Regional Terminology

While the term "Latin America" is most commonly used to define the culture region of American nations with Spanish and Portuguese colonial histories, some might reasonably question its use. Why "Latin" America? The use of the term "Latin" implies a broader sweep than simply Spain and Portugal. The Latin nations include France, Italy, and Romania, that is, all countries where Romance, or Latin-based, languages are spoken. Yet in the Americas the term "Latin America" excludes countries or political entities with French colonial histories like Haiti, French Guyana, Martinique, and Guadelope. The origin of the term is somewhat obscure but it appears to have first been used to describe the region by the French and Spanish American exiles living in Paris in the 1850s. It came into common usage in English at the beginning of the 20th century.

Although rarely used, perhaps the most correct term when referring to the region is

FIGURE 2.1. Latin American nations.

"Iberoamerica." It refers specifically to those areas of the Americas that trace their origins to the countries of Europe's Iberian Peninsula: Spain and Portugal. Culturally, Iberoamerica can be further subdivided into Hispanic America, those areas with direct historical and cultural links with Spain, and Luso-America, those areas with direct historical and cultural links with Portugal—in a word, Brazil.

A myriad of other terms are often used to describe Latin America and/or its constituent parts, although these are largely not cultural in origin and frequently refer to broad geographical regions (Figure 2.2). "South America," for instance, refers to the continent of the same name—and is strictly a geographical definition of a continental landmass. Similarly the terms "the Caribbean" and "the West Indies," often used interchangeably, refer to the plethora of islands that occupy the Caribbean Sea and the Gulf of Mexico. "North America," although usually associated with the United States and

Canada, also encompasses part of Latin America: Mexico.

"Central America" is widely used to describe the Isthmus of Central America, but the term is used in at least three distinct ways. First, it refers to the region's structural geology and physical geography and describes the isthmus, or land bridge, that connects North America with South America. Geographically, that land bridge begins at the Isthmus of Tehuantepec in southern Mexico and extends to the northwestern corner of the South American continent. Second, in its most common usage, it is used in the context of the region's political geography, where it refers to those nations that in their entirety comprise the isthmus: Guatemala, Belize, Honduras, El Salvador, Nicaragua, Costa Rica, and Panama. Third, the term describes Central America as a cultural–historical region dating from the early 1800s. At the end of Spanish colonial rule in the 1820s, almost all of those areas that had comprised the Captaincy General of Guatemala, a Spanish administrative unit that in-

FIGURE 2.2. Latin American geographical regions.

cluded the Central American isthmus and part of southern Mexico, united to form the short-lived nation of the United Provinces of Central America (1823–1840). This country was comprised of the present-day republics of Guatemala, El Salvador, Honduras, Nicaragua, and Costa Rica. Residents of these nations still demonstrate an affinity toward this historical region; written postal addresses from these nations frequently end with the words "Centro América" or the simple abbreviation "C.A." Thus, Belize, the former British colony on the Caribbean, and Panama, then part of the Viceroyalty of New Granada and later part of Colombia, are not considered Central American nations from a historical–cultural perspective.

Although not widely used in the popular literature, geographers have often employed the term "Middle America" to describe the region *between* North America proper and South America proper. In geographical terms, the region includes Mexico, all of the nations of the Central American isthmus, and the West Indies. The region is characteristically divided into two distinct areas, the Mainland and the Rimland. The Mainland is characterized by a Hispanic culture that blends indigenous and European influences. The Rimland encompasses the West Indies, as well as a narrow coastal area running from the northeastern corner of the Yucatan Peninsula along Central America's east coast until it merges with South America. The culture and population of the Rimland reflect a mix of European and African influences. Historically, Hispanic influences have dominated in Cuba, the Dominican Republic, and Puerto Rico, whereas the colonial histories of the remainder of the Rimland are linked largely to northern European nations: France, Great Britain, and the Netherlands. During the last half of the 20th century, proximity to the United States has ensured significant cultural, economic, and political influence from the United States in the Rimland and to a somewhat lesser extent in the Mainland.

Culture, Subcultures, and Scale

On a macroscale, Latin America is a coherent culture region. However, as one begins to examine the region in detail, it becomes apparent that Latin American culture is not a monolithic phenomenon and that a range of subcultures can be distinguished within the region. Distinguishing between these distinct Latin American cultures is critical to understanding the human geography of Latin America.

Language is one of the most basic characteristics of cultural differentiation. While the use of different languages generally demarcates major cultural differences between human groups, more subtle aspects of language use can also identify important cultural differences. Such variation works at many levels, as the worldwide use of English demonstrates. North American English speakers can readily identify "foreign" native English speakers by their accent, intonation, and vocabulary—and the more widely traveled can often identify these variations more specifically, for instance, as Scottish, English, or Australian. At the same time, North American English speakers can identify a range of regional variations in English language use in North America—for example, New England, Canadian, Southern, Midwestern, and New York City accents. A careful listener may also be able to distinguish variations within the latter, demonstrating how hints about education and social class can be inferred from language use as well.

Such is the case in Latin America too. A useful approach to understanding the subdivisions of the Latin American culture region is through language use. Spanish is the mother tongue for two-thirds, or about 350 million, of the region's inhabitants. It is the primary language of the home, commerce, and government in all nations of the region except Brazil.

Yet there are considerable regional variations in Spanish American culture that are manifested in the use of language. One example occurs in Spanish-speaking countries where there is a pronounced distinction between the lowlands and the highlands, as in the Andean nations of South America—especially Colombia, Ecuador, and Peru. Differences in culture between the highlands and the coastal lowlands are reflected in regionally distinctive accents, *serrano* and *costeño* , respectively. *Costeños* speak rapidly, often "eating" syllables, and ignore *s*'s or soften them tremendously. *Serranos*, on the other hand, speak slowly and enunciate syllables clearly. Argentines, especially those from Buenos Aires, who are known as *Porteños* , meaning "people from the port," are readily identified within Spanish America. They use an archaic form of the familiar personal pronoun for you, *vos*, instead of *tu*; drop the final *s* in most words; and employ a distinctive pronunciation for the Spanish letter *ll*. Spanish speakers of Mexican origin can be identified by their strong intonation, which often gives their speech a kind of lilting singsong character.

But these differences are not just about accents and intonation—vocabulary use also underscores cultural variation within Spanish-speaking Latin America. Miscommunication and a lack of understanding do occur, even between Spanish speakers. In the early 1990s, the U.S. corporation American Express began promoting its credit cards in Latin America and targeted Argentina for its first advertising campaign. Its advertising agency took the company's highly successful slogan from its U.S. advertising program, "Membership has its privileges," and translated it into Spanish as "*Ser miembro tiene sus privilegios*" or "To be a member has its privileges." It was very successful in Argentina. With the creation of NAFTA in 1994 and increased accessibility to the Mexican market, American Express asked its ad agency to initiate the same campaign in Mexico. To the chagrin of American Express and its advertising agency, Mexicans whispered, snickered, and laughed at the ads because in Mexico the word *miembro*, or "member," refers to the male sex organ. A quick modification of the campaign's slogan using a more regionally sensitive Mexican Spanish vocabulary salvaged the advertising campaign.

Brazilian Portuguese reveals geographical variations in dialect that reflect regional variations in Brazilian culture. The country is divided into two major dialect regions: the north and the south. Two major divisions characterize the north: an Amazonian dialect characteristic of the area encompassing the Amazon River and its major tributaries and a Northeastern dialect used in the nation's early colonial coastal states and their backlands, the Sertão.

Southern Brazilian speech is divided into four regional groupings. First, the Bahian dialect, which centers on the state of Bahia but includes considerable areas of neighboring states, owes much of its character to the strong African influence found in this region. Second, Fluminense is associated with a small geographical region centering on the city of Rio de Janeiro and including the states of Rio, Espirito Santo, and eastern Minas Gerais. Yet this limited geographical distribution belies its significance: the Fluminense dialect, and more particularly the variant associated with the city of Rio de Janeiro, known as *Carioca*, sets the linguistic standard for educated speakers across a broad spectrum of the media, the professions, and geographical regions in Brazil. Third is the Mineiro dialect, which is spoken only in a small area of the Brazilian highlands in the state of Minas Gerais. Spoken by more Brazilians than any other variant, the Southern dialect spans a vast region of southern Brazil. It centers on the large and populous state of São Paulo and extends north into Minas Gerais, west into Goiás and Mato Grosso, and south and southwest to Brazil's borders with Uruguay and Paraguay.

Indigenous Cultures and Languages

Although Spanish is absolutely dominant in most of the Hispanic American republics, indigenous language use is common in several of these countries. In Paraguay, Guaraní, a member of the indigenous Tupi–Guaraní language family, is spoken by over 90 percent of the population. In no other Western Hemisphere nation does such a large proportion of the population speak an indigenous language. This is particularly notable since Paraguay's indigenous groups now account for only a miniscule part of the nation's population—perhaps 1 percent. In Paraguay, as in other Spanish colonies, the *conquistadores* (conquerors) and subsequent colonists interbred with native women. Their offspring, known as *mestizos*, represented a mixture of European and indigenous gene pools. Eventually mestizos became the dominant group. Other exchanges occurred, including the adoption of indigenous farming practices and foods, but the most remarkable was the enduring use of the natives' language, Guaraní. As a consequence Guaraní is the primary language of most Paraguayans. It is spoken in the home, on the street, and in many business situations. However, almost all Paraguayans are fully bilingual in Guaraní and Spanish.

The cultural context of indigenous language use in both Guatemala and Bolivia is different from that in Paraguay. In Paraguay, Guarani is widely spoken by all social classes, from the streets of the nation's capital to remote rural villages, and no social stigma is associated with its use. In Guatemala and Bolivia, on the other hand, indigenous language use is characteristic of marginalized rural residents or recent rural migrants to urban centers. In mainstream society non-Spanish speakers are disdained. Social and economic success in these societies is associated with the acquisition of strong Spanish language skills.

In the cases of Guatemala and Bolivia, several million of each nation's population speak an indigenous language. In Guatemala, indigenous language speakers may account for as much as one-third of the nation's population of about 10 million. Most speak one of the 23 languages that belong to the Maya language family. These speakers are concentrated in the western and northern highlands and in the rain forests of the southern Petén in Guatemala.

The relative proportion of indigenous language speakers in Bolivia is even greater. Close to 50 percent of the country's population of 7.5 million speak either Quechua or Aymara (Figure 2.3). Quechua, the language of the Inca, is more widely spoken, with most speakers concentrated in the departments of Cochabamba, Potosi, and Sucre in the central Andean highlands. In some of the country's departments the proportion of Quechua speakers exceeds two-thirds of the population.

Quechua is not just spoken in Bolivia. It is the most widely spoken indigenous language in South America. Its use is concentrated in the Andes and is roughly coincident with the extent of the former Inca Empire. At the extremes of its distribution in Argentina and Colombia, the number of speakers in each country is probably less than 100,000. However, the number of Quechua speakers in Bolivia, Peru, and Ecuador likely exceeds 10 million.

Aymara is spoken by fewer Bolivians and is geographically more concentrated than Quechua. Aymara speakers are located in the far-western areas of Bolivia around the shores of Lake Titicaca and along Bolivia's border with Peru. In Bolivia there are about 1 million Aymara speakers, about the same number found just across the border in Peru.

Mapping the combined distributions of Quechua, Aymara, and bilingual Quechua–Aymara speakers shows a wide swath of indigenous language use across Bolivia's Andean highlands. Few indigenous language speakers are found in the eastern lowlands or in prov-

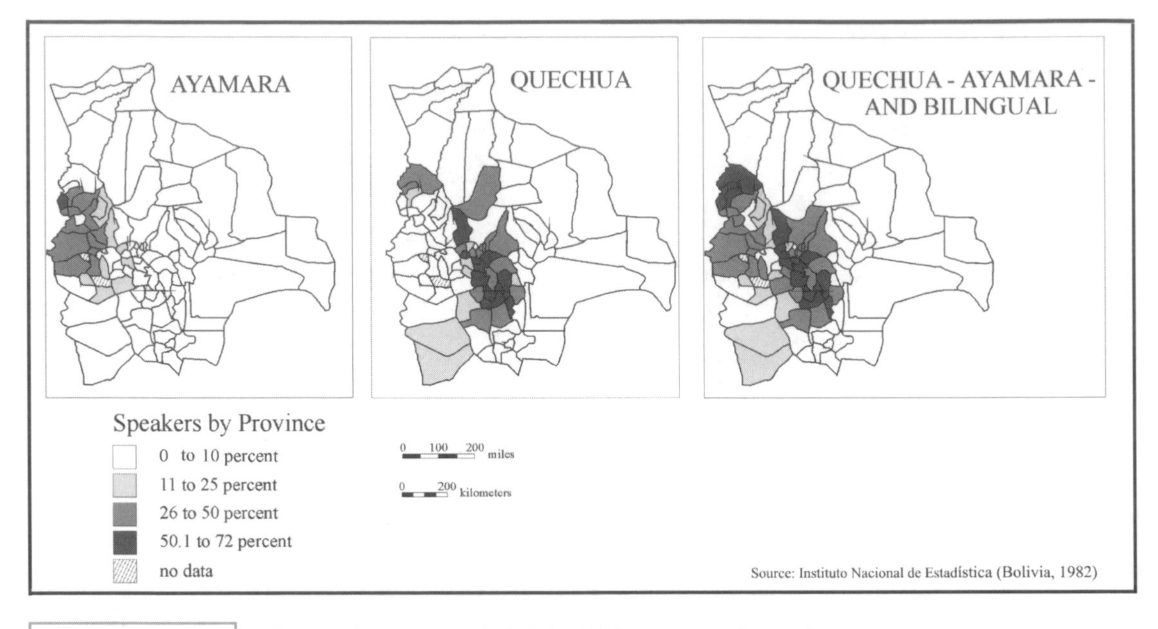

FIGURE 2.3. Indigenous language use in Bolivia, 1976, percentage by province.

inces of the southern highlands near the Argentine border.

Other areas of indigenous language use and cultural survival are found in Latin America (Figure 2.4). In Mexico several million Maya speakers concentrate on the Yucatan Peninsula and in the highlands of the southern state of Chiapas, while Nahuatl speakers, the descendants of the Aztecs, can be found in the country's central highlands. In southern Chile, nearly 1 million Araucanians, or Mapuche, maintain their native language as well as many cultural survivals from pre-Hispanic times.

Indigenous language survivals have modified spoken Spanish and Portuguese in a wide range of geographical regions in Latin America. In Brazil, for instance, it is estimated that as many as 10,000 words from the Tupi language family, including names of plants, animals, people, and places, have been incorporated into the lexicon of Brazilian Portuguese. In Peru and other Andean countries, a range of Quechua words have been incorporated into regional Spanish. These include *charqui* for jerked meat, or "jerky" as it is called in North American English, instead of the standard Spanish word *cecina*. The use of *huaco* for landslide instead of the Spanish word *derumbe* is another example. In Mexico, where there still are a large numbers of native speakers of indigenous languages, many words of Nahuatl (Aztec) origin have been adopted into Mexican Spanish. Even words from indigenous languages that are now extinct have worked their way into regional Spanish lexicons—and beyond! "Hurricane," or *huracán* in Spanish, is a survival of the language of the Arawak peoples who once peopled much of the Caribbean. "Barbecue," or *barbacoa* in Spanish, is a survival of the language of the Taíno people, who were annihilated by war and disease early in the colonial period. They had occupied the island of Hispaniola, and used the word to describe the heated rock slabs they used for cooking.

Mesoamerica

Mesoamerica is the largest, most cohesive, and most distinctive indigenous culture region in

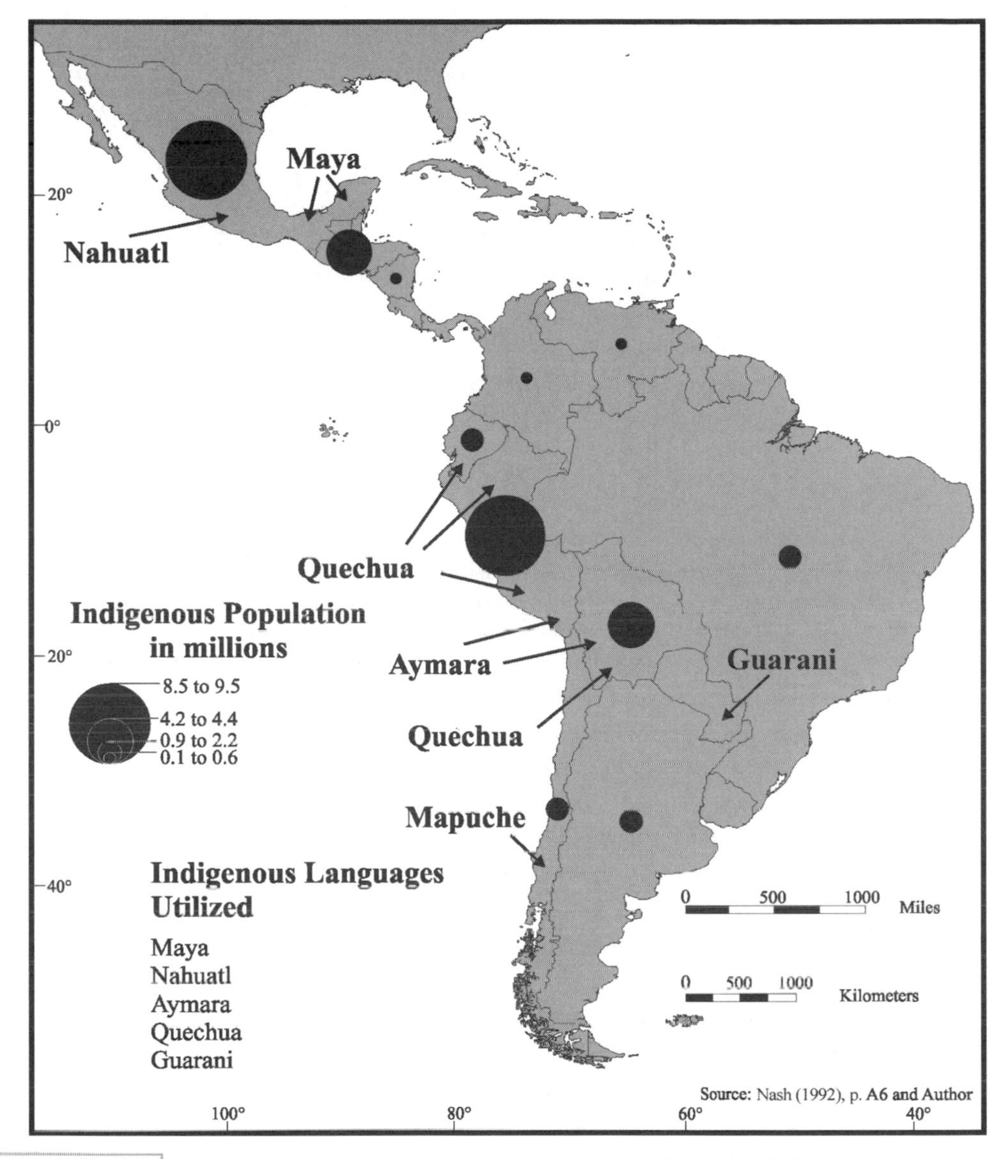

FIGURE 2.4. Indigenous population and language use in Latin America, circa 1990.

Latin America (Figure 2.5). Geographically, Mesoamerica extends from the northern margins of Mexico's *Mesa Central*, or Central Plateau, southward to include all of central and southern Mexico, including the Yucatan Peninsula, Guatemala, Belize, most of El Salvador, the western portions of Honduras and Nicaragua, and Costa Rica's Nicoya Peninsula. At the time of the Spanish Conquest in the early 1500s, the Mesoamerican culture region included a wide range of advanced indigenous civilizations dominated by the Aztecs of central Mexico and the Maya in the northern Yucatan. Other important groups included Tarascans, Zapotecs, and Mixtecs. Although these indigenous peoples were never united in a single state, they shared a range of characteristics that gave their area cultural unity. Shared characteristics included land terracing, irrigation systems, a calendar, the construction of permanent and specialized religious structures, ball courts, and the cultivation of corn, beans, and squash as the principal agricultural plant complex. A common set of domesticated plants grew in their gardens and fields.

The most enduring of these characteristics, and those that readily identify the area as Mesoamerican today, are those related to agricultural and food habits. Corn, beans, and squash remain the primary food crops grown by peasants (*campesinos*) and smallholders throughout this region. The dominant regional cuisine reflects this crop mix as well. Corn and corn-based products—for instance *tortillas* and *tamales*—are consumed in large quantities, as are black beans, squash, and hot chili peppers (Vignette 2.2).

Regionally specific vocabulary usage also serves to distinguish Mesoamerica as a distinctive culture region within Spanish America. In Mesoamerica many words whose origins are from indigenous languages have been incorporated into the region's Spanish speech. Words of Nahuatl (i.e., Aztec) origin are particularly notable, and when used in Spanish-speaking countries of South America are unknown to most of the population. Peanuts are not *manis*, as they are in South America, they are *cacahuetes*; turkeys are not *pavos*, they are *guahalotes*; and avocados are not *paltas*, they are *aguacates*. However, several words of Nahuatl origin have been incorporated into standard Spanish, used even in Spain, and derivatives of some also have been adopted into standard English, notably *tomato*, *chocolate*, and *coyote*.

African Cultural Influences

The descendants of African peoples comprise an important element in Latin America's population mix. While the Spanish and the Portuguese initially used the indigenous inhabitants of their America territories to satisfy the need for cheap, abundant labor for farms, plantations, and mines, they quickly turned to imported slave labor from Africa as a more dependable and economical source of supply. The trade in African slaves began in the late 1500s and continued for over 300 years. During that period millions of Africans were brought to both Spanish and Portuguese colonies.

The presence of Africans contributed to

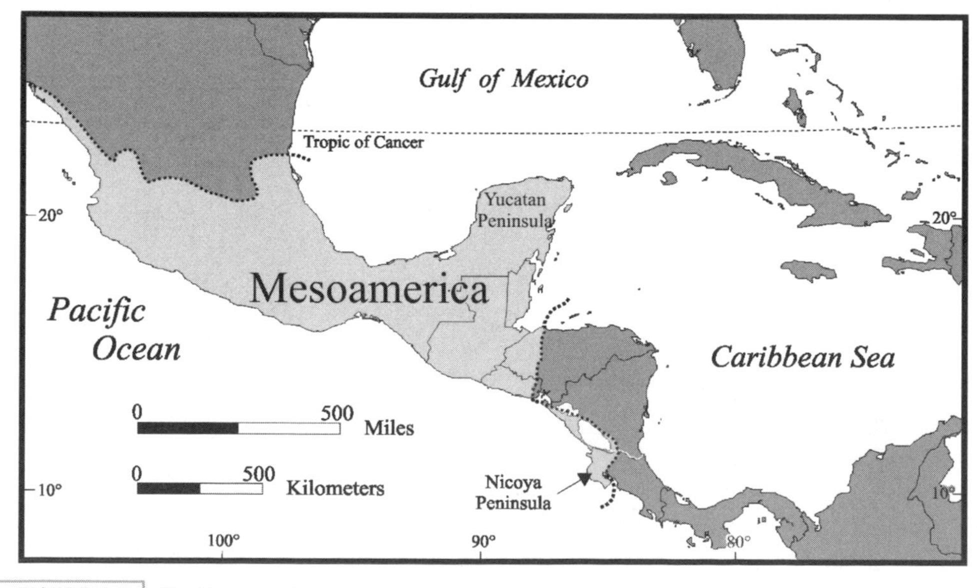

FIGURE 2.5. The Mesoamerican region.

VIGNETTE 2.2. TORTILLAS

No food item is more closely identified with the Mesoamerican culture region than the tortilla. Tortillas are produced from ground corn soaked in lime and mixed with salt and fat (Figure 2.6). During pre-Hispanic times, tortillas were a staple foodstuff for Mesoamerican peoples. Despite the introduction of wheat by the Spanish and a widespread acceptance of bread and other wheat-based products, tortillas continue to be a mainstay in the diet of the vast majority of Mexicans and Central Americans. Traditionally, the corn for tortillas is ground into corn flour on a low stone slab called a *metate*. The flour is mixed with water and salt and turned into a dough, or *masa*, which is patted by hand into small, round, flat cakes, tortillas, which are cooked on a flat stone, a *comal*. Typically, Mexican tortillas might be 4–5 inches in diameter and 1/8-inch thick, yet considerable regional variation in tortillas does exist. In Tuxtla the traditional tortilla is small, about 2.5 inches in diameter, and crisp, while in Oaxaca traditional tortillas measure nearly 8 inches in diameter. A Salvadoran variation is the *pupusa*, corn tortillas that are 3–4 inches in diameter and about ¼-inch thick and stuffed with meat, cheese, or beans. On the southern margins of Mesoamerica, on the Nicoya Peninsula in northwest Costa Rica, the traditional tortilla is a giant 10 inches in diameter.

However, the effects of culture change and modernization have modified Mesoamerica's dietary habits as well as the traditional corn tortilla. The wheat-flour tortilla, *tortilla de harina de trigo*, found throughout Mexico, but most frequently in northern states like Sonora, evolved as a result of the introduction of wheat into the New World during the colonial period. During the last half of the 20th century the production of tortillas has been increasingly mechanized. Most Mexicans living in cities and small towns buy their tortillas ready-made from specialized tortilla shops, *tortillerias*, where the corn is ground, the ingredients mixed, the dough stamped, and the tortillas cooked in a semi-industrial process. Others may purchase ready-made tortillas packaged in plastic from neighborhood stores and supermarkets.

The diffusion of the tortilla, at least as a specialty dietary item, has taken it far beyond the limits of Mesoamerica and into Anglo-America. The presence of a population of Mexican origin in the southwestern United States has assured the presence of the tortilla in the regional cuisine there. However, since the 1980s, the eating of Mexican-style cuisine has been adopted across a broad social and geographic spectrum in the U.S. population. Tortillas are readily available in grocery stores throughout the country, even in areas with comparatively few Hispanics, such as Cleveland, Ohio, or Syracuse, New York. American consumers' tastes are even demanding that a broad range of tortilla types be offered for sale, including low-calorie varieties.

a significant modification of the dominant Iberian culture, Lusitanian or Hispanic, in areas where they became numerically significant. The largest numbers of African slaves arrived in the Caribbean—some 4.7 million. Brazil followed, with 3.5 million, while the Spanish American mainland only received about 600,000. African cultural traditions and practices, including elements of material culture, religion and belief systems, language, and food habits contributed to a complex mosaic of cultural variety in Latin America. Their impact on modern Latin American culture is especially profound in the Caribbean and Brazil.

The immense geographical and human scale of the Brazilian experience and the integral role Africans have had in it provide ample

FIGURE 2.6. A Mexican meal in the Yucatan. The ubiquitous tortillas are wrapped in a napkin in the basket with a bowl of black beans beside them, 1987.

examples of these African cultural influences. Indeed, in Brazil, the peculiarities of the Portuguese slave trade and the Brazilian slave economy did not result in the complete deculturalization of African slaves—as was nearly the case in the United States. As a consequence, it is possible to identify distinct cultural influences from Africa. The Africans who arrived in Brazil belonged to three broad cultural groups. These are Hausa/West Africans, primarily Yoruba and Dahoman, now concentrated in Bahia and Maranhão; Moslems from the Guinea region, focused largely in Bahia as well; and Bantu peoples from the Congo, Angola, and Mozambique, now located in Rio de Janiero and Minas Gerais.

African cultural influences in Brazil span a spectrum. A wide range of Brazilian dishes trace their origins to Africa. *Feijada*, a dish made of black beans, rice, meat, a leafy green vegetable, and manioc flour, is so widely consumed throughout the country that some consider it the most typical national dish. Other African foods are consumed widely in the northeast, especially in Bahia, where African dishes like *vatapá*, *acarajé*, and *acaça* are common. Sometimes known as "Bahian dress," the turbans, colorful fabrics, lace-trimmed dresses, and multiple bracelets and necklaces often worn by women along Brazil's northeastern coast are of African origin. The internationally known Brazilian dance, the samba, owes many of its rhythms and movements to African influences, while other, lesser known dances like the *maractu* are directly traceable to African cultural traditions. The latter, for instance, was danced at the coronation ceremonies of African tribal chiefs in the Congo region. Some musical instruments, agricultural crops, and elements of Brazilian placer mining techniques also exhibit African cultural influences.

From the sacred to the profane, African cultural influences are ingrained as permanent components of Brazilian society. Brazilian Portuguese reflects the influence of Africans in its vocabulary, and perhaps more importantly in its melodic inflections and syntax. It has been suggested that the African nannies who cared for the children of their white masters "impregnated Brazil from within" by exposing young whites to African lullabies, stories and folktales, folk medicines, and beliefs. The Christian pre-Lenten festival Carnival has been transformed by African Brazilians and owes much of its distinctiveness to them. Even the practice of the Roman Catholic religion in the country has been modified by African culture, resulting in a syncretic Catholicism. Yet African religious practices did survive independently as well, and millions of Brazilians actively participate in these belief systems—*candomblé* in Bahia, *xango* in Pernambuco, and *macumba* in Rio de Janeiro.

While African cultural influences are probably strongest and most visible in Brazilian culture, they are also present throughout African Caribbean societies. In the Hispanic Caribbean, Cuba and Puerto Rico specifically, the spiritual practice of *santeria* is a manifestation of African cultural traditions. Such traditions are considerably more restricted on the mainland of Hispanic America, although in coastal Caribbean areas of Central America, Colombia, and Venezuela they can be observed.

Non-Iberian European Cultural Influences

During most of the colonial period (1521–1821), as well as in the early decades of the independence period, the cultural dynamics of Latin America changed little. Iberian, African, and indigenous cultural influences were the key components of the evolving culture, although their individual effects varied considerably on a regional basis. Independence, however, freed the new republics from Iberian colonial policies that severely restricted immigration of non-Iberians. Some countries, notably Brazil, Argentina, Uruguay, and Chile, actively promoted immigration through aggressive policies and in some cases even subsidized it. Between 1880 and 1920, Brazil received over 4 million immigrants and Argentina 2 million, largely from southern and northern Europe. These Europeans modified the cultural characteristics of the countries and regions where they settled.

Italians were the largest group, locating primarily in Brazil's southern states of Rio Grande do Sul, Santa Catarina, Paraná, and São Paulo, and in Argentina. Their influence on Argentine culture was dramatic. Before the massive influx of Italians during this period Argentina was a *Hispanic-criollo* country, strongly influenced by traditional Spanish values and populated by a small number of the descendants of early colonists and a large population of mestizos. But since the Italians began to settle, Argentina has become a largely European nation in terms of its population, material culture, and values. The Italians transplanted many elements of the café culture characteristic of Mediterranean Europe into Argentina. Coffee bars and cafés with outdoor seating are common elements in the urban landscape of the country's cities and towns, and espresso machines are an almost omnipresent feature in restaurants and cafás. Italian cuisine—for example, spaghetti, gnocchi, and ravioli—is as ubiquitous as beefsteak on Argentine tables. In Buenos Aires a local dialect, *Lunfardo,* developed as a result of the hybridization of Spanish with Italian vocabulary and speech patterns, and has influenced the Spanish spoken throughout the country. Italian surnames like "Galtieri," "Benedetti," and "Sabatini" are as common in Argentina as those of Spanish origin.

But beyond the influence of immigration, Argentines of the late 19th and 20th centuries embraced Europe enthusiastically. Argentina welcomed British investment and business, and embraced French culture and the scientific and military expertise of the Germans. Religious, elite, and public architecture in Argentina also followed European patterns. Some parts of Buenos Aires, especially districts like wealthy El Retiro, bear an architectural similarity to European, particularly French, cities. While only a few Spanish colonial-style churches can be found in Argentina, many churches strongly resemble the gothic styles characteristic of Europe (Figure 2.7). Perhaps not surprisingly, many Argentines tend to view themselves as Europeans, not as Latin Americans—an attitude many other Latin Americans interpret as arrogant and aloof.

European immigrants dramatically impacted the business environment in those areas where they settled, even though their numbers were often small. Immigrants often found that hereditary tradition-bound elites controlled the productive agricultural land, and that their access to government jobs and even military careers was severely limited. As a consequence many turned their energies to business, commerce, and manufacturing, bringing with them a dynamism and spirit of entrepreneurship that was sorely lacking in traditional Hispanic and Lusitanian societies. In Argentina, Chile, and southern Brazil, European immigrants, including appreciable numbers of Germans, transformed the business and industrial cultures (Vignette 2.3).

FIGURE 2.7. A gothic cathedral in Mar de Plata, Argentina, 1994.

Their contributions contributed to the long-term economic development and prosperity of these regions, which distinguished them from the more traditional Iberian areas of Latin America.

Minority Cultures and Enclaves

While Europeans have exerted the strongest cultural impact on Latin America since the independence period, immigration by other groups has also modified the region's culture and cultural landscape in significant ways.

Asians have been important, particularly the Japanese in southern Brazil. Beginning in 1908 and continuing through the early decades of the 20th century, Japanese immigrants arrived in modest numbers in São Paulo state, but their numbers reached a total of nearly 200,000 by the mid-1930s. Many arrived as contract laborers, but eventually moved into played-out coffee plantation lands and established themselves as small farmers. Producing for the large urban market of São Paulo, they invigorated the local agricultural economy while providing a rich diversity of agricultural products to urban consumers. Today, there are nearly 1 million Brazilians of Japanese descent. Most are still located in São Paulo, where they constitute the largest concentration of ethnic Japanese anywhere in the world outside of Japan.

Small numbers of Japanese, about 20,000, also immigrated to Peru as contract laborers. Eventually they occupied the same type of economic niche as in São Paulo; they became fruit and vegetable farmers producing for Lima's urban market. As in Brazil, many second- and third-generation immigrants gravitated toward urban centers and professional occupations. In Peru, Alberto Fujimori, a second-generation Japanese Peruvian, became the nation's president (1990–2001). More limited Japanese immigration occurred in Paraguay, Bolivia, and the Dominican Republic during the 1950s and 1960s, as governments there sought agricultural colonists to settle vacant lands and to energize local agricultural economies.

Chinese immigration has also been a factor in a few nations. This is especially true in Peru, where the Chinese arrived initially as railroad-building contract laborers in the late 1800s. Although their numbers were never great and their settlement is largely limited to Lima and other coastal cities, they did add at least two dishes to the national cuisine, *arroz chafu*, a kind of hybrid Chinese Peruvian rice-and-vegetable dish served throughout the country in small restaurants and cafés, and *lomo saltado*, a stir-fried dish usually including at least beef, tomatos, and onions. Modest Chi-

VIGNETTE 2.3. GERMAN SETTLEMENT IN SOUTH AMERICA

German immigration and settlement were encouraged during the mid- to late 19th century in both Chile and Brazil. Both governments wanted immigrant settlers because they believed that the Europeans would inject an element of economic dynamism in their economies and could occupy marginal areas beyond the fringe of permanent settlement.

In Chile, German immigrants settled mostly in the country's southern Lake Region between Valdivia and Puerto Montt between 1846 and 1880. Although few in number, the isolation of the region and a lack of other large population groups allowed the settlers to leave a permanent cultural imprint on the area (Figure 2.8). The largest numbers, close to 4,000, came from Hesse, Westfalia, Wurtenberg, and Saxony and settled in the vicinity of Valdivia and Osorno. A second concentration of perhaps 3,000 settlers hailing from Moravia, Bohemia, and Silesia settled on the lake plain of Lake Llanquihue just north of Puerto Montt. A final flow of immigrants between 1895 and 1896 brought about 1,000 settlers from the region around Berlin to the northern portion of Chiloe Island. During the first half of the 20th century a clearly defined German–*criollo* ecumene encompassed the region, running some 100 miles northward from Puerto Montt to Valdivia. Evidence of the German cultural heritage of this region is still clearly apparent today—both rural and urban architecture bear strong evidence of German influences, German surnames are common, and *kuchen* and *strudel* are available as desserts in many restaurants.

In Brazil, large numbers of German settlers occupied the country's two southernmost states, Santa Catarina and Rio Grande do Sul, with the earliest settlement being in the vicinity of the town of São Leopoldo in the 1820s (Figure 2.9). In subsequent decades, additional settlement expanded along the Jacui River Valley until an area nearly 100 miles long and 25 miles wide was colonized almost exclusively by Germans. Conservative estimates suggest that at the end of the 19th century between 150,000 and 200,000 German nationals or those of German descent lived in southern Brazil. Considerable immigration to that region continued until the beginning of World War I.

The cultural landscape in these rural settlement areas had a distinctly German flavor—for example, in farm architecture, the layout of rural street-villages, or *strassendorfers*, and agricultural practices. As in Chile, German settlers here lived in comparative isolation from the remainder of Brazilian society—linguistic and social acculturation did not begin to take place until well into the 20th century.

German settlement was not restricted to rural areas: many towns had sizable German populations. These immigrants contributed to the development of an artisan middle class that was often totally lacking or extremely small in most other Brazilian towns and cities. Two of the largest of these towns, Joinville and Blumenau, boasted almost exclusively German populations of between 15,000 and 20,000 in the late 1880s.

As late as 1920 in Blumenau over two-thirds of the population spoke German as their mother tongue. The town supported two German-language newspapers. Blumenau's German heritage makes the city a major tourist attraction today—and its Oktoberfest is reported to be the largest in the world after that of Munich. Today the state of Santa Catarina, with nearly 5 million residents, boasts a standard of living close to that of Western industrial countries. Its literacy rate is 90 percent, 96 percent of households have electricity, the infant mortality rate is similar to that of the United States, and its per capita annual income is higher than that of any country in Latin America.

FIGURE 2.8. A German settler's homestead on the margins of Lake Llanquihue in southern Chile, 1993.

nese immigration has also occurred in Central America, Mexico, and Cuba. In Central America the Chinese presence is evidenced largely by the frequent occurrence of Chinese restaurants in most national capitals and port cities. In Mexico, scattered Chinese immigration occurred in the northwestern states of Baja California Norte, Sonora, and Sinaloa during the late 1800s and the early decades of the 20th century. Chinese entrepreneurs and laborers played a significant role in the agricultural development of the Mexicali Valley on the United States–Mexico border in the early 20th century—farming nearly 75,000 acres in 1919. Mexicali, the valley's principal town, was home to several thousand Chinese in the 1920s. A distinct Chinese commercial and residential district developed, La Chinesca, which survives to this day. Although Mexicali's "Chinatown" is not as clearly defined as it once was, nearly 10,000 Chinese Mexicans live in Mexicali and its environs today. Other Chinatowns are found in Lima, Peru, and Havana, Cuba.

Like immigration from Asia, immigration from the Middle East has been limited both geographically and numerically. It has been locally significant in some countries, notably Argentina and to a lesser extent Chile, where these immigrants worked initially as merchants and entrepreneurs and later as intellectuals and professionals. In Argentina, for instance, Levantine immigrants, peoples from the nations bordering the eastern margins of the Mediterranean Sea, carved out a profitable niche for themselves as traveling merchants serving a market ignored by city-based retail-

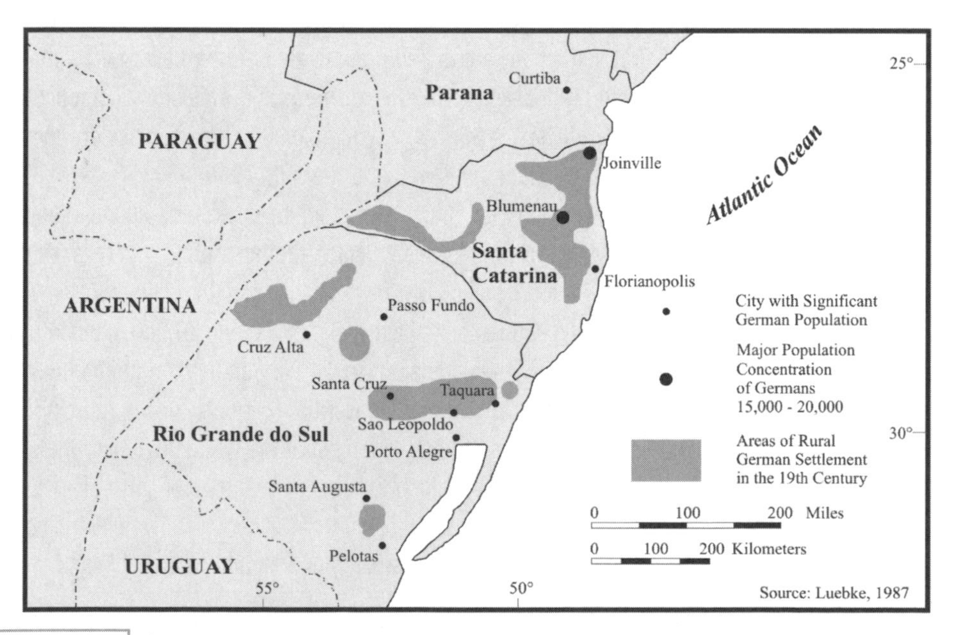

FIGURE 2.9. German settlement in southern Brazil in the 19th century.

ers. The immigrants found willing customers on the vast expanse of the Argentine Pampa by traveling by train from one railroad station to the next, and then walking from homestead to homestead offering goods for sale. In some provinces in western Argentina, like San Juan, the number of Christian Arabs, Syrians, and Lebanese reached sufficient numbers to support the construction of local clubs. Argentina's past president, Carlos Saul Menem, hails from western Argentina and is a descendant of early Arab immigrants to the country.

Jewish immigration into Argentina began in the early 1900s and included agricultural as well as urban migrants. Argentina's Jewish population is the largest concentration of Jews in Latin America, with some 200,000 today, most residing in Buenos Aires.

Cultural enclaves, geographic regions that are occupied by ethnically distinct peoples, are few in Latin America and have had only locally significant impacts. Mennonites are perhaps the most notable of the enclave cultures, with small concentrations in Mexico, Bolivia, and Paraguay. Seeking freedom from what they perceived to be increasingly invasive governments in the United States and Canada, and responding to the wishes of some Latin American governments to settle new or marginal lands, Mennonites established agricultural colonies in northern Mexico, the tropical lowlands of Bolivia near Santa Cruz, and the Chaco region of Paraguay. Mennonite settlement in Mexico dates from the early decades of the 20th century, while their settlement in Bolivia and Paraguay occurred during the 1950s and 1960s. Some Mennonite settlements are large: the Cuauhtemoc colony in the state of Chihuahua, Mexico, numbers over 50,000, and in eastern Bolivia and northwestern Paraguay the Mennonite populations number approximately 30,000. Mennonite communities have remained apart from their host societies. Their contributions have consisted largely of stimulating local agricultural economies and markets with increased agriculture production and a wider range of high-quality food products, often fresh milk and cheese, for local consumers.

Much of the influence of the United States in Panama during the 20th century has been manifested in economic and geopolitical terms. For example, after the initiation of its construction in 1904 the Panama Canal became an economic and military enclave of the United States, bisecting the country of Panama and casting a long shadow over that nation. The annual fees paid by the United States for use of the Panama Canal provided critical funds for the government's operation, and the money spent by the thousands of Canal Zone residents provided an important impetus to local businesses in Panama City. While the *balboa* is the official currency of Panama, its value is pegged to the U.S. dollar and U.S. currency is used for all financial transactions. Even Panama's coinage, which is minted by the U.S. Mint, is the exact size, shape, metallic content, and value as U.S. coins. These coins even work in U.S. vending machines! The United States has intervened politically and militarily in Panama's internal affairs, most recently in December 1989, when it invaded Panama and deposed and arrested the country's military strongman, Manuel Noriega. The long-term presence of nearly 15,000 U.S. citizens who lived in the Canal Zone did have some influence on local culture and values. English, for instance, is widely understood as a second language among Panamanians. The signing of the Panama Canal Treaty in 1977 and the takeover of the canal by Panama in 1999 has led to some lessening of U.S. influence in the country. However, in a country that uses U.S. dollars as its currency, and where root beer and pancake syrup are readily available on many grocery store shelves, it is unlikely to disappear.

In the last 20 years, however, another type of enclave with the potential to significantly affect local culture and life ways has become increasingly common: the tourist enclave. Nowhere in Latin America have such enclaves been more important than in Mexico. Here, port towns and coastal fishing villages like Manzanillo, Mazatlán, Puerto Escondido, and Cozumel have been transformed by foreign tourism into bustling centers with an international clientele and flavor. Since foreign tourism now is one of Mexico's primary sources of foreign exchange, the Mexican government itself has developed new tourist enclaves like Cancún and Ixtapa. While the immediate goal of these endeavors is to increase tourist traffic and revenues, these enclaves are also slowly changing local cultures and values as foreign models of dress, behavior, and taste are adopted and adapted by local residents.

While the impact of these tourist enclaves in Mexico is still geographically restricted, their potential to affect regional and national culture in smaller Latin American nations like those of Central America or the Caribbean is great. In the Dominican Republic the massive tourist complex of Puerto Plata, on the nation's north coast, has direct jet connections with Miami and other international destinations and receives tens of thousands of visitors each year. With the demise of its former patron, the Soviet Union, Cuba has turned increasingly to tourism as a source of jobs and foreign exchange earnings. Tourism's impact on Cuban culture, although still limited, has included a rise in sexual tourism and prostitution. The economy also became increasingly "dollarized" until 2004. Then, in response to a tightening of the U.S. economic embargo, the Cuban government outlawed the use of the dollar and replaced it with the euro. In the last 20 years, the Pacific coast of Costa Rica's Nicoya Peninsula has been the focus of a minor construction boom as hotels and tourist complexes serving an international clientele have been built.

The influence of North American and Western European tourists on Latin American culture manifests itself in peculiar ways. Mexican restaurant cuisine and etiquette demonstrate clear influences from the United States. North American–style pancakes are found on the menus of a wide range of restaurants, not just those serving resort populations. But they are almost never found anywhere else in Latin America. Similarly, the custom of tipping waiters and waitresses is confined to elite and upscale restaurants in Latin American capitals and major urban centers. Yet throughout much of Mexico tipping is now customary even in workaday cafés and restaurants with only the most modest pretensions. In some countries the "dress code" has also been modified by the presence of North American and Western European tourists. In the late 1960s the wearing of shorts by men or women outside the narrow confines of beaches and swimming pools was not tolerated by Latin Americans. Whistles and catcalls greeted the occasional offender. Today Latin Americans tolerate the shorts worn by tourists, and in some areas they themselves have adopted the custom and wear them in public as well.

Cultural change also has come to Latin America as its residents have migrated in increasing numbers to the United States. A generation ago, once migrants arrived in the United States, contact with family and friends in their country of origin was often limited to letters, the occasional long-distance phone call, and the rare trip back home after years of living in the United States. However, by 2000, significant reductions in the cost of telephone calls, as well as improvements in communication technologies like e-mail, the Internet, and more recently video conferencing, meant more frequent contact with home. Travel costs have also declined, and as a consequence increasing numbers of immigrants are transnational, traveling back and forth between the United States and their country of origin with regularity. These are people with roots in both worlds. As such, they facilitate the flow of North American ideas and influences back to Latin America and vice versa (Vignette 2.4).

But immigrants need not return to their homelands to be transnational. The Domini-

VIGNETTE 2.4. RETURN MIGRATION AND CULTURAL CHANGE

Cultural change comes in all kinds of ways. Sometimes it is sinister and unwelcome. In El Salvador, a country racked by a long violent civil war during the 1980s, a new threat to civil society is the rapid proliferation of street gangs modeled on, and often linked to, street gangs in Los Angeles and other U.S. cities. Tens of thousands of Salvadorans fled the violence of the country's civil war and immigrated, both legally and illegally, to the United States in the early 1980s. Growing up in tough inner-city neighborhoods in Los Angeles, some El Salvadoran children joined street gangs. When the civil war ended, the United States deported many of the illegal immigrants-some of them gang members. In El Salvador, these people have formed new gangs or established branches of those gangs to which they belonged in Los Angeles. An ineffective police force and an inefficient justice system provide these gangs with considerable opportunities, as do their contacts in the United States, and trafficking in drugs, weapons, and stolen vehicles offers lucrative income sources for them. United States-inspired gang graffiti now increasingly adorns adobe walls in urban slums and villages, as young tattooed men in oversized baggy pants and shirts saunter through the streets.

In the Dominican Republic a different kind of youthful reverse migration is occurring. In an effort to avoid the negative social consequences of urban life in the United States, Dominican immigrants, largely from New York City, have been sending their children back to the Dominican Republic to live with relatives and finish their high school educations there. In the late 1990s as many as 10,000 "Dominican Yorks," as they are known in the Dominican Republic, enrolled in the nation's schools. Many, only semifluent in Spanish and unaccustomed to local customs and food, find the experience of living in their parents' homeland difficult. At the same time, many Dominicans view them with distrust and suspicion. Some schools will not allow older transfer students to enroll, noting unruly behavior and an unwillingness to accept local rules as significant problems. Their peers also complain that the return migrants exhibit behaviors stereotypical of New Yorkers: they are loud, pushy, and excessively self-confident. The local parents of their peers worry that association with the Dominican Yorks will expose their children to drugs and encourage the kind of rebellious behavior they associate with teenagers in the United States. For their part, the Dominican Yorks say they are all too happy to return to the United States just as soon as they graduate from high school. Quipped one 17-year-old student originally from the Brooklyn, "The day after I graduate, I'm outta here."

can Republic recognizes dual citizenship and permits overseas Dominicans to vote in the nation's presidential elections, as many in the United States, Mexico, and other countries did in 2004. Colombia also allows its citizens to hold dual citizenship. This creates curious and potentially troubling situations in which emigrants who have become naturalized citizens in the United States may vote in elections in the United States as well as those in their native country. Similar legislation is being considered in Mexico, but with as many as 10 million Mexicans in the United States potentially eligible to vote this is a complicated and potentially volatile issue.

The Media, Mass Culture, and Other Cultural Influences

Latin American culture is constantly evolving. Today the influence of the mass media is nearly omnipresent. Latin America has a long and venerable history of newspaper publication; some of the daily newspapers now in circulation have been published for over 100 years. Nevertheless, the rate of newspaper readership in most Latin American countries is low. On the other hand, radio and television ownership and use are widespread throughout the region. The cultural values and messages projected by these media penetrate into some of its remotest corners. The influence of North American popular music may be the most ubiquitous—in remote jungle settlements or on the packed buses of the national capital the strains of "golden oldies" or the latest "heavy metal rock" and rap music from the United States are blared out in a regular fashion to Latino listeners. The cinema, mostly movies from Hollywood, is dubbed or subtitled into Spanish or Portuguese. In Brazil, the region's largest single market, slightly less than 40 percent of all cinema presentations were produced in the country during the final decades of the 20th century. In Spanish America, Mexico and Argentina have the largest and strongest local film industries, with more limited contributions from a handful of other countries. North American television shows invariably constitute a major segment of the local television programming on Latin American stations. For many Latin Americans, television shows like *Bay Watch*, *America's Most Wanted*, and *The Simpsons* provide most of their understanding of the culture and values of their northern neighbors. Little wonder that some Latinos believe that most North Americans carry firearms, cheat on their spouses, and get divorced with regularity.

Rates of radio and television ownership in Latin America can be used to measure the extent of penetration by the mass media. They show surprising variation among the region's nations. Radios are inexpensive and so are within the financial reach of the vast majority of the population, even the poor. In the early 1980s, for instance, a study of peasant families in the Cochabamba region of Bolivia showed that 90 percent owned radios. Indeed, Bolivia, with one of the lowest per capita incomes in Latin America, has the second highest rate of radio ownership. It is tied with Uruguay at about 57,000 per 100,000 population, falling just behind Argentina (Figure 2.10). Somewhat surprisingly, Brazil and Mexico post only intermediate rates of radio ownership—35,000 and 22,000 per 100,000, respectively. Guatemala has the lowest rate, only 5,000 per 100,000, just one-third the rate of the next highest country, Paraguay.

Widespread television ownership, announced by the staggering number of television antennas evident in Latin America's cities, presents a different pattern than might be expected. Affluent countries tend to have the highest rates of television ownership, while more impoverished nations have lower rates. But the correspondence is not exact. The rate of television ownership in Mexico, perhaps because it is a major manufacturer of televisions, is by far the greatest in the region. With a rate of nearly 60,000 per 100,000, its rate is almost three times larger than that of the next four highest rated countries, Cuba, Argentina, Brazil, and Uruguay, whose rates range from around 22,000 down to 19,000 per 100,000. At the other end of the spectrum, Nicaragua, Guatemala, and Honduras have rates that range between 3,000 and 5,000 per 100,000, while Bolivia brings up the rear with only 700 televisions per 100,000 population (Figure 2.11).

By 2000 new technologies, especially personal computers, cell phones, and the Internet, became increasingly available in Latin America (Figure 2.12). These technologies revolutionized communication for many of the region's residents, especially those in urban centers.

Marketing

The rate of both radio and television ownership has been increasing steadily over the last

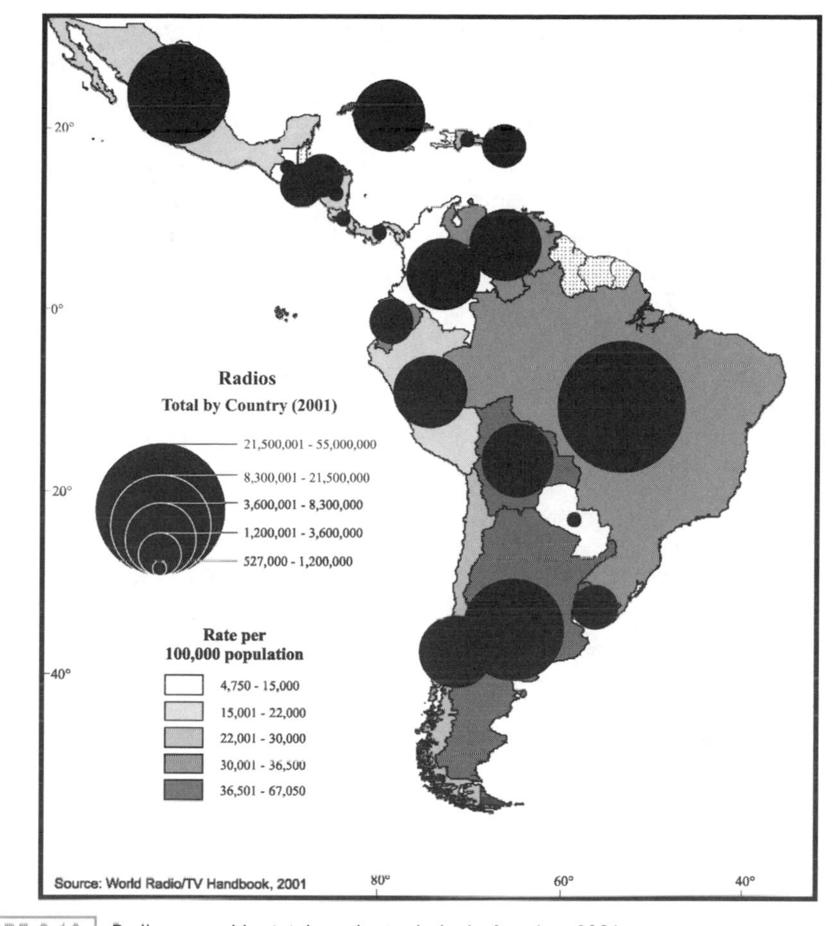

FIGURE 2.10. Radio ownership, totals and rate, in Latin America, 2001.

three decades. Today the ownership of either a radio or a television is standard in nearly all Latin American households.

North American and European companies have marketed their products in Latin America since the 1800s. But beginning in the early decades of the 1900s the visibility of these products and brands became widespread as marketing and brand-name recognition became key elements in modern business strategies. Brand names and the product labels associated with them made their entry into Latin America in the period between 1920 and 1960. Consumer durables like motor vehicles and tires, appliances like refrigerators and radios, and some basic commodities such as petroleum products represented the initial influx. Brand names like Ford, Chevrolet, Goodyear, Firestone, Zenith, Motorola, as well as Esso, Texaco, Shell, and Mobil, became increasingly visible as car dealerships, appliance stores, gas stations, and other businesses included their logos on their stores and in advertising. The public too became accustomed to these, sometimes even incorporating product names into local vocabularies—thus in some countries refrigerators are not only known by the generic term in Spanish, *nevera* or *refrigeradora*, but also as *Frigidaire*, the brand name for the Westinghouse refrigerator line.

Consumer goods also began to penetrate Latin America in these decades. Signs for carbonated beverages or soft drinks from the United States became increasingly common, with Coca Cola and Pepsi leading the pack (Figure 2.13). Brand names also began to affect local and regional vocabularies. In Peru *gillette* is used for "razor" instead of the traditional *hoja de afeitar*, and in both Peru and Bolivia the preferred term for "oatmeal" is *cuaquer* from the brand name Quaker Oats instead of *avena*, the correct term in Spanish.

In the second half of the 20th century, a broader range of business activities and companies began to make inroads in Latin American economies. In the largest urban areas, Latin Americans began to become familiar with the logos and practices of major financial services companies. Citibank, First Bank of Boston, Aetna, and Prudential became part of the streetscape in downtown business districts and in advertising on a city's skyline (Figure 2.14).

The last decade has witnessed dramatic changes in the commercial landscape of urban areas. Franchise businesses with home bases in the United States, Canada, and Europe have entered Latin America in a big way. Fast-food restaurant franchises may well be the most numerous and most visible of these commercial imports. While franchises of U.S. chains like Pizza Hut and Kentucky Fried Chicken (now KFC) appeared in Mexico as early as the 1960s and McDonald's hamburgers and fries became available in many Latin American capitals during the 1980s, the 1990s

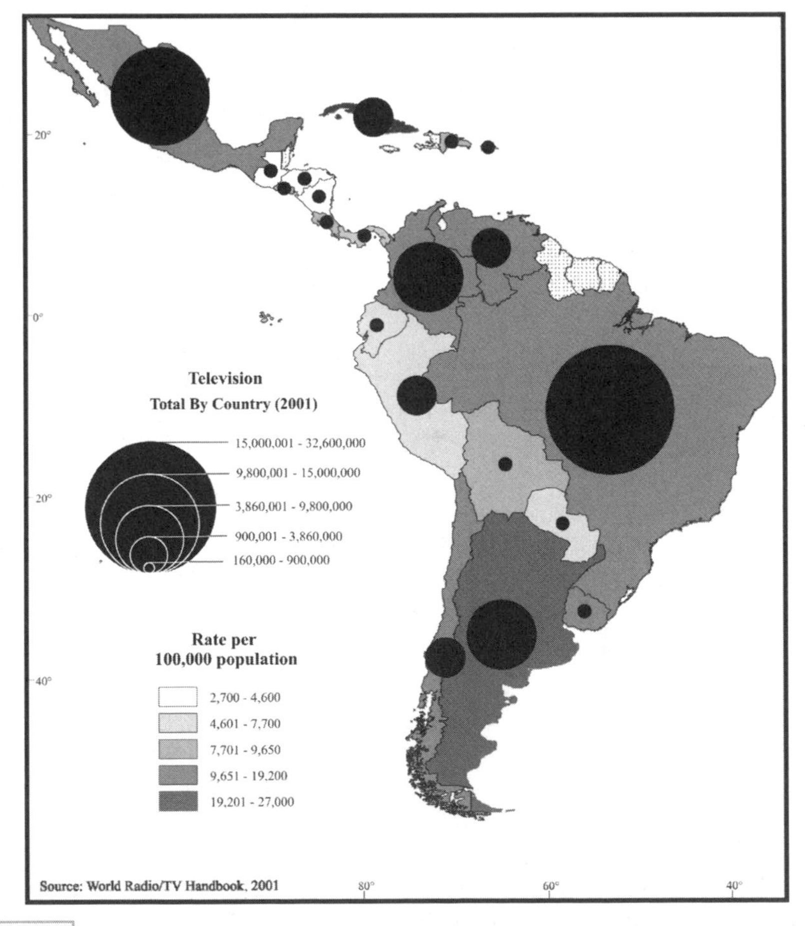

FIGURE 2.11. Television ownership, totals and rate, in Latin America, 2001.

FIGURE 2.12. An Internet café in Tegucigalpa, Honduras, 2002.

saw a veritable explosion in the number of fast-food franchises operating in the region.

The tide of neoliberal economic reform that has swept Latin America since the mid-1980s encouraged the entrance of an increasing variety of franchises, as well as a swelling of the numbers of those already operating in the region. Major fast-food chains entered these markets aggressively. McDonalds opened its first outlet in Brazil in 1979. Since then it has marketed itself aggressively and experienced strong growth. In 2005, the firm boasted over 1,200 points of sale across the country, including 500 McDonalds restaurants plus an additional 700 smaller outlets called McCafes. Other major fast-food chains also view the Latin American market as incredibly attractive. Subway, a submarine sandwich fast-

FIGURE 2.13. A Pepsi advertisement dominates the visual landscape on the main plaza in Tucumán, Argentina. The state legislative building is on the right, 1993.

FIGURE 2.14. An office of Citibank located right off the central plaza in downtown Tegucigalpa, Honduras, 1996.

food restaurant franchise, has expanded assertively in Latin America; in 2005 it operated in 11 countries. It had almost 200 franchises in Mexico and operated franchises in all the Central American countries. Costa Rica and El Salvador each boasted over 20 outlets. Its presence in South America is less extensive. It operates in over 80 shops in four countries, but has the greatest presence in Venezuela, where there are over 70.

Nevertheless, it is not only the megachains that have entered Latin America in force. The region's tremendous growth potential has led regional and narrowly based specialty brands to enter these markets in surprising numbers. Schlotzky's, a Connecticut-based sandwich shop, has begun operating in Argentina; Au Bon Pain, a Boston-based chain specializing in sandwiches, salads, and pastry, operates 15 outlets in Chile; and Benihana, a Miami-based Japanese restaurant chain, has begun operating in Bogotá, Colombia. Franchises offering ice cream and frozen yogurt have also found promising markets in Latin America. These include ice-cream vendor Baskin Robbins and frozen yogurt sellers I Can't Believe It's Yogurt and Yogen Früz.

Typically, fast-food and specialty food shops make modest concessions to local tastes in their standard menu items. When McDonalds opened an outlet in La Paz, Bolivia, it added several local items to its menu, including *coca* tea, a spicy Bolivian tomato sauce, and a typical Bolivian snack, the *salteña* . However, the *salteña* metamorphosed into the "McSalteña." In Chile, Domino's Pizza offers a variety of seafood toppings favored by locals, including mussels and clams, as well as pizzas with no tomato sauce.

But despite such concessions to local cuisine, the rapid proliferation of North American food franchises in Latin America is having a significant impact on the region's culture. Changes in food preferences as well as in the culture of eating are ushered in by these developments. Hamburgers, sandwiches, pizza, and yogurt desserts are becoming more common in local diets. Dining patterns have been affected too, as Latinos become more accustomed to "eating on the run" instead of lingering over a long midday meal, the traditional custom in most countries. Even business practices will change. The franchise concept, which promotes economies of scale, a systematic and efficient use of resources, and a consistent product, gives franchise operators critical advantages over local independent operators. Eventually these methods will be imitated by competitive local business, dramatically changing the business culture of these economies. As in the United States, the influx of franchise chains will change the urban landscape as well, homogenizing streetscapes as advertising and retail outlets make customers from Mexico City to Buenos Aires intimately familiar with McDonalds's "golden arches" or Burger King's crown.

Mapping Latin America's Culture Regions

The identification and mapping of cultural regions can be tricky. Latin America is clearly a culture region and can be readily identified spatially on a continental scale. The dominance of Spanish and Portuguese as the principal languages provides an easy surrogate for establishing the region's geographic extent. However, attempts to identify the existence and physical limits of cultural regions at more local scales, national and subnational, can be considerably more challenging.

One approach is simply to use political boundaries. Thus, for example, one may speak of Mexican, Peruvian, Brazilian, or Argentine cultures. Although this approach has its limitations, clearly there are national cultures, and in most cases it is possible to use national boundaries to identify subcontinental culture regions—for example, Mexican, Peruvian, Brazilian, or Argentine. Within a nation's boundaries political subdivisions, like states, departments, or provinces, are often used to identify local culture regions. Thus, in Colombia, for instance, there are *Antioqueños* (people from the department of Antioquia) or in Brazil there are *Paulistas* (people from the state of São Paulo). Cardinal directions or environmental regions may also be used as identifiers of culture regions at subnational levels—in Mexico and Chile there are *norteños* (northerners) and in most Andean countries there are *costeños* (coastal folks).

No system is perfect and no approach is without its detractors. Two different efforts by geographers to identify subnational culture regions for Latin America are reproduced here.

In the first of these efforts, Louis Casagrande has divided Mexico into five "nations" distinguished by their regional cultures and economies (Figure 2.15). "Mexamerica," the largest, spans the southern United States from Texas to California and includes the vast reaches of northern Mexico. This region reflects a hybrid culture, not fully Mexican nor fully American, but something new. Another hybrid region is the discontinuous area he defines as "Club Mex," tourist enclaves along the Pacific and Caribbean coasts where foreign travelers and international investment are transforming local communities. "New Spain," centered on the Mesa Central of the Mexican

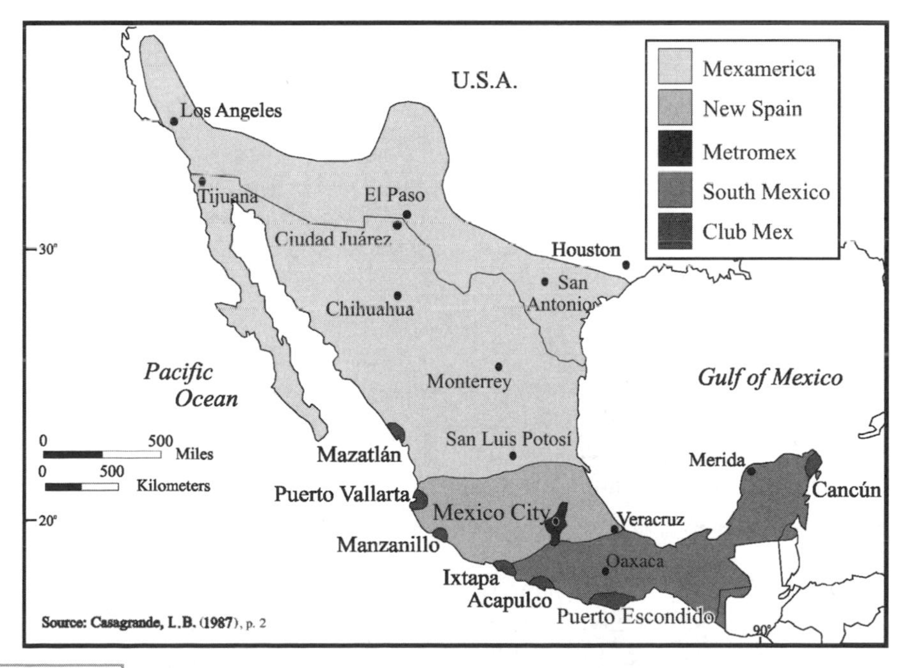

FIGURE 2.15. The five nations of Mexico. Reproduced from *Focus* with permission of the American Geographical Society.

Plateau, corresponds most closely with images of "Old Mexico" and along with "Metromex," the Mexico City metropolitan region, is the economic, cultural, and political core of Mexico. "South Mexico" includes the highlands of the south as well as the lowlands of the Yucatan Peninsula, and is united by the comparative dominance of indigenous peoples, many living traditional lifestyles. While many Mexicans would recognize the validity of Casagrande's regions, it is unlikely any would identify themselves as being "Mexamericans" or "Metromexicans"!

Caviedes and Knapp have identified a set of South American culture regions (Figure 2.16). Many of these culture regions are exclusively subnational and a few are largely urban-based—*Porteños* for those from Buenos Aires (Argentina) and *Cariocas* for the residents of Rio de Janeiro (Brazil). Some regional identities cross national boundaries, especially those that correspond with natural regions, like the *Llaneros* from the Llanos of Colombia and Venezuela and the *Chaqueños* of the Gran Chaco in Paraguay and Bolivia.

Summary

Latin America is one of the world's principal culture regions. It is distinguished from other world regions by a set of common cultural traits that include language, religion, social values, and civic institutions deriving principally from the Iberian Peninsula. Spanish and Portuguese are the predominant languages, Catholicism is practiced by the vast majority of the region's inhabitants, and social customs and civic institutions bear many similarities to those in Spain and Portugal. Nevertheless, the region is not culturally monolithic. Indigenous cultures and peoples have influenced national and subnational cultures within the region, affecting language, religion, music, food habits, social customs, and civic institutions. African immigrants originally brought as slaves also have influenced the region's culture, although their effects have been most pronounced in Brazil, the Caribbean, and coastal areas of Central America and northern South America. The cultural impact of other immigrants, including those from Italy, Asia, the Middle East, and even a few from North America, has been minor.

Improvements in transportation and communication technology since the 1950s have increased the region's contacts and interaction with the rest of the world. The widespread adoption of radio and television in the 1960s and 1970s brought new cultural perspectives to Latin America. Increasing foreign tourism to the region, as well as the immigration of millions of its citizens to the United States and other countries, has further exposed the region to diverse cultural influences. The continued reductions in costs and the ease of international travel, trade, and communications have ensured that Latin America is increasingly part of a global community.

Further Reading

Avni, H. (1991). *Argentina and the Jews: A history of Jewish immigration*. Tuscaloosa: University of Alabama Press.

Banwait, N., et al. (2000). *Television in Latin America to 2005*. London: Zenith Media.

Becker, T. H. (2004). *Doing business in Latin America: A guide to cultures, practices, and opportunities*. Westport, CT: Praeger.

Burns, E. B. (1993). *A history of Brazil*. New York: Columbia University Press.

Burns, E. B. (Ed.). (1993). *Latin America: Conflict and creation: A historical reader*. Englewood Cliffs, NJ: Prentice-Hall.

Casagrande, L. B. (1987). The five nations of Mexico. *Focus, 37*, 2–9.

Chevalier, F. (1965). The roots of *personalsimo*. In H. M. Hamill (Ed.), *Dictatorship in Latin America* (pp. 35–51). New York: Knopf.

Cole, R. R. (Ed.). (1996). *Communication in Latin America: Journalism, mass media, and society*. Wilmington, DE: Scholarly Resources.

Crist, R. E. (1968). The Latin American way of life. *Amer-*

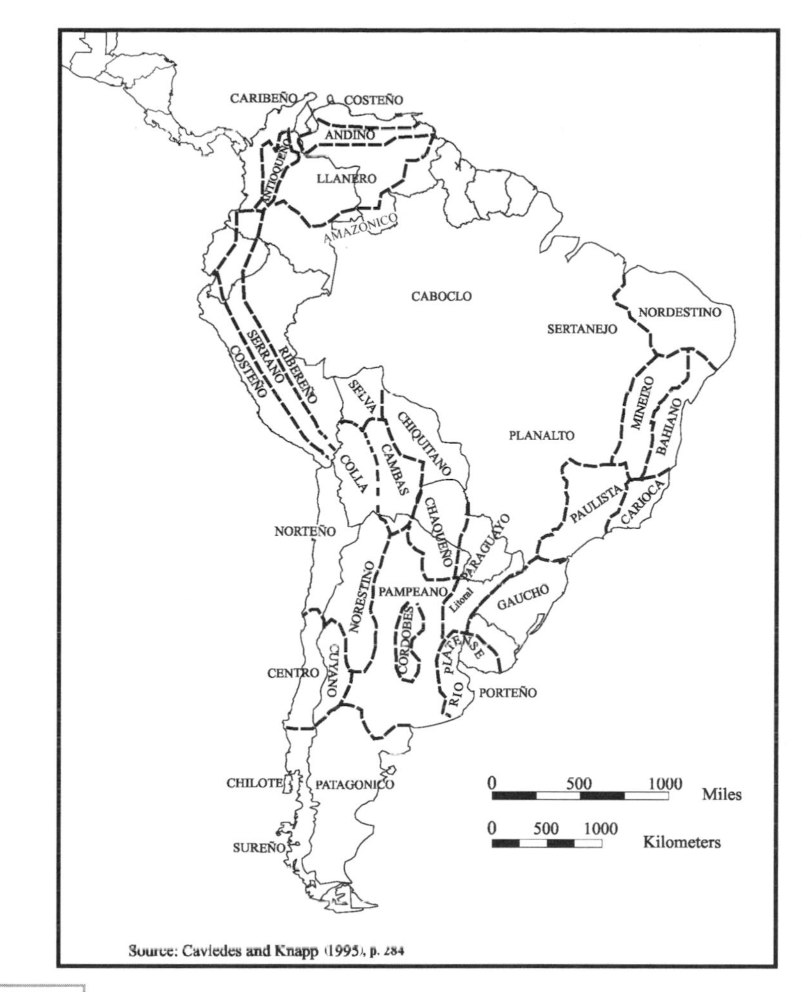

FIGURE 2.16. South American culture regions. Reproduced from *South America* with permission of Prentice-Hall, Inc.

ican Journal of Economics and Sociology, 27, 63–76, 171–183, 297–311.

Cubitt, T. (1995). *Latin American society* (2nd ed.). London: Longmans.

Davis, D. J. (Ed.). (1995). *Slavery and beyond: The African impact on Latin American and the Caribbean*. Wilmington, DE: Scholarly Resources.

Devine, E., and Braganti, N. (2000). *The travelers' guide to Latin America: Customs and manners*. New York: St. Martin's Press.

Dussel, E. (Ed.). (1992). *The church in Latin America, 1492–1992* . Mary Knoll, NY: Orbis Books.

Economist Intelligence Unit. (1999). *Consumer marketing in Latin America*. New York: Author.

Falcón, R. (1998). *Salsa: A taste of Hispanic culture*. Westport, CN: Praeger.

Foster, G. (1960). *Culture and conquest: America's Spanish heritage*. Chicago: Quadrangle Books.

Gardiner, C. (1975). *The Japanese in Peru, 1873–1973* . Albuquerque: University of New Mexico Press.

Green, D. (1997). *Faces of Latin America* (2nd ed.). London: Latin American Bureau.

Gibson, C. (Ed.). (1968). *The Spanish tradition in America*. Columbia: University of South Carolina Press.

Gonzalez-Wipper, M. (1973). *Santeria: African magic in Latin America*. New York: Julian Press.

Guttman, M. C. (1996). *The meanings of macho: Being a man in Mexico City*. Berkeley and Los Angeles: University of California Press.

Kicza, J. E. (Ed.). (1993). *The Indian in Latin American history: Resistance, resilience, and acculturation*. Wilmington, DE: Scholarly Books.

Kirchhoff, P. (1952). Mesoamerica: Its geographic limits, ethnic composition and cultural characteristics. In Sol Tax (Ed.), *Heritage of conquest* (pp. 17–30). Glencoe, IL: Free Press.

Knight, F. W. (1974). *Map of distribution of African slaves: 1500-1860* . New York: Macmillan.

Luebke, F. (1987). *Germans in Brazil: A comparative history of cultural conflict during World War I*. Baton Rouge: Louisiana State University Press.

Page, J. A. (1995). *The Brazilians*. Reading, MA: Addison-Wesley.

Stevens, D. F. (Ed.). (1997). *Based on a true story: Latin American history at the movies*. Wilmington, DE: Scholarly Resources.

Takenaka, A. (2004). The Japanese in Peru: History of immigration, settlement, and racialization. *Latin American Perspectives, 31*(3), 77–98.

Wagley, C. (1968). *The Latin American tradition: Essays on the unity and diversity of Latin American culture*. New York: Columbia University Press.

Ward, P. (2004). Globalization and the appropriation of Latin American popular music. *Latin American Research Review, 39*(1), 273–284.

Winn, P. (1992). *The Americas: The changing face of Latin America and the Caribbean*. Berkeley and Los Angeles: University of California Press.

Voeks, R. (1997). *Sacred leaves of the candomblé African magic, medicine, and religion in Brazil*. Austin: University of Texas Press.

3 The Environment

This chapter reviews several of the most important factors that influence environmental conditions in Latin America. The variation in latitude across the region contributes to the creation of distinct climatic belts ranging from tropical, to subtropical, to temperate midlatitude climates. The role of mountains in creating a wide range of environmental conditions in the region is detailed. Mountains influence precipitation patterns and have an important role in creating distinct and widely recognized altitudinal life zones within the Latin American Tropics. Finally, the chapter examines some issues related to environmental change in the region, specifically the effects of the exchange of plants and animals between the Old World and the New World, often called the "Columbian Exchange."

Latitude

In a geographical sense, Latin America covers an immense latitudinal range. The region stretches from just north of 30°N latitude along Mexico's northwestern border with the United States near San Diego, California, to approximately 55°S latitude at the southern tip of South America. This latitudinal variation itself dictates a wide range of regional climatic zones coincident with world latitudinal belts (Figure 3.1). A range of other factors create variation in these patterns, but in general terms the following regional climatic types can be identified.

The Northern Hemisphere Subtropical Belt

Subtropical climates characterize the area between 30°N and about 23.5°N, the Tropic of Cancer. This region falls entirely in northern Mexico. While much of the region is arid and warm, climatic conditions do vary considerably. Precipitation is sparse, desert conditions predominate, and temperatures are very high in the summer and comparatively mild in the winter months. Nevertheless, frosts do occur, especially in upland plateau areas and at the higher elevations of the Sierra Madre Occidental and the Sierra Madre Oriental. In the lowlands along the coast, temperatures are warmer, exhibit less variation, and rarely fall below freezing. The western coastal margins along the Pacific are generally arid. The coastal plain along the Gulf of Mexico and the east-facing mountain slopes of the Sierra Madre Oriental are well watered.

The Tropical Belt

Tropical climates predominate over much of Latin America's vast latitudinal range. The Latin American Tropics spread out over nearly 50° of latitude, running from the Tropic of Cancer at 23.5°N across the equator to the

FIGURE 3.1. Latitudinal belts.

Tropic of Capricorn at 23.5°S. Most Latin American nations fall within the Tropics. Small parts of Argentina and Chile lie inside the Tropics, but most of these nations' territory falls within the subtropical and the temperate zones.

Despite popular perceptions to the contrary, tropical climates and environments can be extremely diverse. One of the principal variations that is evident on a broad scale is precipitation. The Tropics can be divided into three major precipitation zones. The region surrounding the equator and running perhaps 10–15° north and south is characterized by a pattern of year-round rainfall. This region is coincident with a broad belt of low-pressure air evident worldwide that is known as the Inter-Tropical Convergence Zone, or ITCZ. Solar insolation along this equatorial belt creates a global convection cell in which warm, moist, low-pressure air rises, cools, precipitates, and then settles back to earth again on the margins of the equatorial belt. As a consequence, rainfall is plentiful and occurs almost all year within this portion of the Tropics.

Still within the Tropics, but falling on the margins of the ITCZ, is a second broad precipitation regime. Running roughly from around

10–15° to the Tropic of Cancer at 23.5°N and from about 10–15°S to the Tropic of Capricorn at about 23.5°S, these twin zones have a slightly different pattern of rainfall. Here, in a region that is sometimes termed "tropical wet and dry," rainfall is largely seasonal: there is a wet season and a dry season. This pattern of rainfall is a function of the annual movement of the ITCZ and other global wind and pressure belts.

The ITCZ moves northward during the Northern Hemisphere's spring and summer months. As a consequence, the rainy season is roughly coincident with the movement of the ITCZ northward and usually falls between April/May and August/September, although the precise timing varies with latitude.

A reversal of this pattern occurs in the Southern Hemisphere. Here the ITCZ moves southward from the equator beginning at the time of the equinox in September and reaches its maximum southern position at the time of the solstice in late December. Consequently, in the tropical wet-and-dry belt south of the equator, the rainy season typically begins in October/November and continues until February/March. Again, precise dates and precipitation totals vary with latitude. In the Southern Hemisphere, the rainy season is longer, rainfall is more predictable, and more precipitation falls at latitudes closer to the equator. The rainy season is shorter, rainfall patterns are more variable, and less precipitation occurs at the southern margins of the region at the Tropic of Capricorn, 23.5°S. In the Northern Hemisphere a similar pattern occurs.

Temperature, however, varies little over the region. This is true at the equator, where warm, humid conditions prevail over most of the region throughout the year. As one moves north or south away from the equator and towards 23.5°N and S, seasonal variations in temperature become more notable. At most locations in the Latin American Tropics mean temperature variation between summer and winter does not exceed 10°C. It is important to note that diurnal variations can be quite significant. An old adage summarizes this simply: "Night is the winter of the tropics." This is true in the tropical lowlands, but even more so in the tropical highlands. For example, in the tropical highlands of the Andes, diurnal temperatures can sometimes range over 30°C, from daytime highs of almost 25°C to nighttime lows below freezing.

The Southern Hemisphere Subtropical Belt

A subtropical belt, similar to that found in the Northern Hemisphere, characterizes the region that runs from the general vicinity of the Tropic of Capricorn (23.5°S) southward to approximately 35°S latitude. Uruguay, much of Paraguay, the southernmost states of Brazil, and the northern portions of Chile and Argentina all fall within this subtropical belt. Similar to the conditions in Latin America's portion of the Northern Hemisphere, precipitation in this belt varies considerably. Generally, however, precipitation is plentiful along the eastern margins, adjacent to the coast, and then decreases toward the west where truly arid conditions prevail in the interior of the continent, along the eastern piedmont of the Andes, and along the Pacific coast.

Temperatures are generally moderate throughout this subtropical belt, and freezes are rare, but there is notable regional variation. The highest temperatures and greatest variation occur in the interior of the continent, in the Chaco, and along the eastern piedmont of the Andes Mountains. Summer temperatures in these areas can be extreme and frequently exceed 40°C, whereas, during the winter months, freezing temperatures are common in the more southerly portions of the Andean piedmont. Freezes also occur occasionally in the southern portion of the Brazilian Highlands (the Paraná Plateau).

The Southern Hemisphere Temperate Midlatitude Belt

A temperate midlatitude climatic belt characterizes the far southern reaches of the Latin American region. This belt extends from around 35°S to the southernmost extreme of the South American continent at about 55°S. Temperature and precipitation conditions vary over this region, creating a range of environmental conditions. Temperatures in the north are moderate, and are similar to those in many areas of Western Europe or North America. Summers are warm and winters are cool and mild, but freezing temperatures occur frequently. Further south, summers are cooler and winter months are cold. Freezes occur regularly throughout the winter months in many locations, especially toward the interior.

The comparative narrowness of the South American continent at its southern extreme and its proximity to the Atlantic and Pacific Oceans has a major effect on temperatures in this temperate region.

Large bodies of water typically tend to mitigate the influences of latitude on regional climates, and South America is no exception. The higher specific gravity of water means it both cools and heats more slowly than adjacent land masses. Consequently, here in the southern reaches of the continent, temperatures in areas adjacent to the coastal margins are somewhat cooler in the summer months and warmer in the winter months than areas inland.

Precipitation generally declines from north to south. In the northeastern quadrant of the region, roughly coincident with the Pampa, precipitation is abundant and rain-fed agriculture is widespread. However, as one moves westward and southward precipitation diminishes. Near-arid conditions predominate along the eastern piedmont of the Andes, and much of the area to the south, Patagonia, is an arid steppe.

On the western side of the Andes precipitation patterns are almost completely reversed. Rainfall is sparse in the north and is restricted to the winter months. This area coincides with Chile's Central Valley, where rain is limited, but irrigation permits widespread agriculture. However, as one moves southward, precipitation increases dramatically as the coastal region comes under the more direct influence of the predominant winds, the Westerlies. These winds blow from west to east over the Pacific Ocean and dump large quantities of rainfall as they are pushed up over the Andes along the southern Pacific coast of the South American continent. While rainfall is seasonal in the north, in the south it occurs throughout most of the year with only a short drier season during the summer months.

An Exceptional Climatic Phenomena: **El Niño**

Although the climate patterns tend to demonstrate seasonal regularity year after year, there are some exceptions. In Latin America, one of the most well-know and significant of these is *El Niño* . The El Niño is a climatic anomaly that occurs when the waters of the east central Pacific Ocean in the equatorial belt experience greater warming than is normal and build up along South America's west coast. This buildup of warm ocean water disrupts the normal movement of air masses and ocean currents. These changes affect regional, and to some extent worldwide, weather patterns, often leading to severe disruptions in normal rainfall patterns. These disruptions can cause drought in areas that are often well watered and heavy precipitation in normally arid regions. Crop failure, flooding, and widespread human misery often follow in the wake of El Niño occurrences. This phenomena occurs irregularly, sometimes only once every 10 years, but occasionally several times in a decade. The name itself, El Niño, was given to this climatic anomaly because its onset often coincides with the Christmas season: "El Niño" is an affectionate term for the baby Jesus.

Mountains: Impacts on Regional and Subregional Environmental Patterns

Mountains, high plateaus, and rolling uplands are some of the most salient features of the physical geography of Latin America. Even a glance at a relief map of the region reinforces this observation. Mexico is perhaps the most dramatic example of this pattern: somewhere between 80 and 90 percent of the entire national territory is dominated by mountainous terrain. This includes the Mexican Plateau (the Mesa del Norte and the Mesa Central), adjoining mountain ranges (the Sierra Madre Oriental and the Sierra Madre Occidental), and a vast mountainous complex that occupies much of the southern part of the country (the Sierra Madre del Sur, the Sierra de Oaxaca, and others). The physical geography of much of Central America is also mountainous. A series of mountain ranges of volcanic origin and upland basins characterizes much of the terrain of Central American nations.

In the Latin American portion of the Caribbean region, mountains figure prominently in the physical geography of the Dominican Republic and Puerto Rico. In the Dominican Republic elevations exceed 3,000 m in the center of the island, while in Puerto Rico the highest points barely exceed 1,300 m. In both instances, these high points are not isolated physical features; much of the territory is covered by mountainous terrain. Cuba deviates from the norm. Most of the country is unusually flat. The principal exception is the Sierra Maestra mountains, which occupy the extreme eastern margin of the island nation.

Mountains, highland plateaus and basins, and rolling uplands are prominent in the physical geography of South America as well. While they do not occupy the same proportion of the continent's territory as they do in Mexico and Central America, their impact on climate and environmental conditions are nevertheless significant.

The Andes are the highest, most rugged, and most extensive mountain range on the continent. This mountain chain covers immense latitudinal distance, running from near Venezuela's Caribbean coast at about 10°N to the southern reaches of Chile and Argentina at approximately 50°S. Elevations range from 3,000 to 4,000 m over much of its extent, while its highest peaks exceed 6,000 m. Many of its highest peaks are permanently snow-covered, and glaciers are found in the Andes even within the tropical zone. The Andes encompass much of the national territories of Venezuela, Colombia, Ecuador, Peru, Bolivia, Chile, and to some extent even Argentina.

Falling entirely within the national territory of Brazil, the Brazilian Highlands are the second major upland in South America. These highlands occupy most of the southeastern portion of the country. The highlands parallel the Atlantic coast and stretch from about 5°S to 30°S, extending inland for 1,000–1,500 km. While the Andes are an orogeny of comparatively recent geological origin (100 million years), the Brazilian Highlands are much older (300 million years). Consequently, elevations are considerably lower than in the Andes and erosion has rounded and subdued the topography. The elevations of the highest peaks do not exceed 3,000 m, and most of the highlands lie below 1,500 m.

The Guiana Highlands are the third significant highland region in South America. These highlands lie in northern South America roughly between the equator and about 7°N. Much of this mountainous area falls within Venezuela, although uplands extend into the border region with Brazil and eastward into Guyana and Suriname. The highest peaks reach nearly 3,000 m, but most of the highlands reach between 500 and 1,500 m.

The abundance of mountain environments in Latin America has dramatic effects on climate and environmental conditions. While latitudinal belts of winds and pressure create broad patterns of weather, climate, and environment, the presence of mountain ranges modifies these dramatically. Indeed, it is often the case, especially within the Tropics, that environmental conditions can range from humid tropical rain forest to semiarid thorn forest or even to arid desert environments within 40 to 50 km. While such transitions are extreme, they are not unusual in much of tropical Latin America. Two factors cause these variations. One is aspect (or orientation), which refers to the direction that something faces, in this case the direction mountains face with respect to prevailing winds and weather systems. The second factor is elevation.

Aspect and Climatic Variation in Latin America

Aspect affects weather, climate, and environment through the interaction of prevailing winds and the elevation of mountain ranges. In geographical terms, *aspect* refers to the particular direction a slope or geographical feature is oriented. Aspect is an important factor in regulating the amount of sunlight that reaches a particular piece of land. In the Northern Hemisphere, for instance, a south-facing slope receives considerably more sunlight and energy than a north-facing slope. Such differences are often reflected in notable differences in climate and vegetation.

Rainfall patterns are also affected by aspect. As the prevailing winds blow an air mass across a mountain range, the air mass is forced to rise. As it rises, it cools (at the lapse rate), which decreases the capacity of the air mass to hold water vapor. If the amount of water vapor in the air is high enough and the air cools sufficiently, then the dew point is reached, condensation results, and precipitation occurs. As a consequence, the windward sides of mountain ranges typically experience considerable rainfall and thus often enjoy a luxuriant vegetation cover.

On the contrary, the leeward sides (those sides that face away from the prevailing winds) of mountain ranges receive much less rainfall. This area of scant precipitation is called a rain shadow. Two factors work to contribute to the rain shadow effect. First, most of the moisture held by the air mass is released as that air mass is pushed up over the windward side of a mountain or mountain range. By the time it reaches the other side, not much moisture is left. Second, dry descending air on the leeside of mountains heats up and absorbs more moisture. Semiarid and arid environments typified by thorn forests and xerophytic vegetation are usually found in these rain shadow areas.

The striking environmental contrasts that are created by the differing aspect of mountains and mountain ranges are commonplace over much of Latin America. In Puerto Rico and the Dominican Republic these environmental variations can be experienced over astoundingly short distances. Throughout most of the year, warm moist winds blow from the northeast to the southwest. These winds, called the "Northeast Trades," dump large quantities of rainfall along the northeastern coasts and mountain slopes that face to the north. However, on the mountains' lee, to the south and most particularly to the southwest, rain shadow conditions prevail and participation is sparse. Thus in Puerto Rico, for example, the vegetation of the island's northeastern quarter near San Juan is luxuriant and green year-round, while 60 km to the southwest in the region surrounding Ponce cacti and other xerophytic vegetation predominate (Figures 3.2, 3.3, and 3.4).

Similar patterns characterize broad swaths of Mexico and Central America, especially those regions of tropical wet-and-dry climates that fall between the Tropic of Cancer

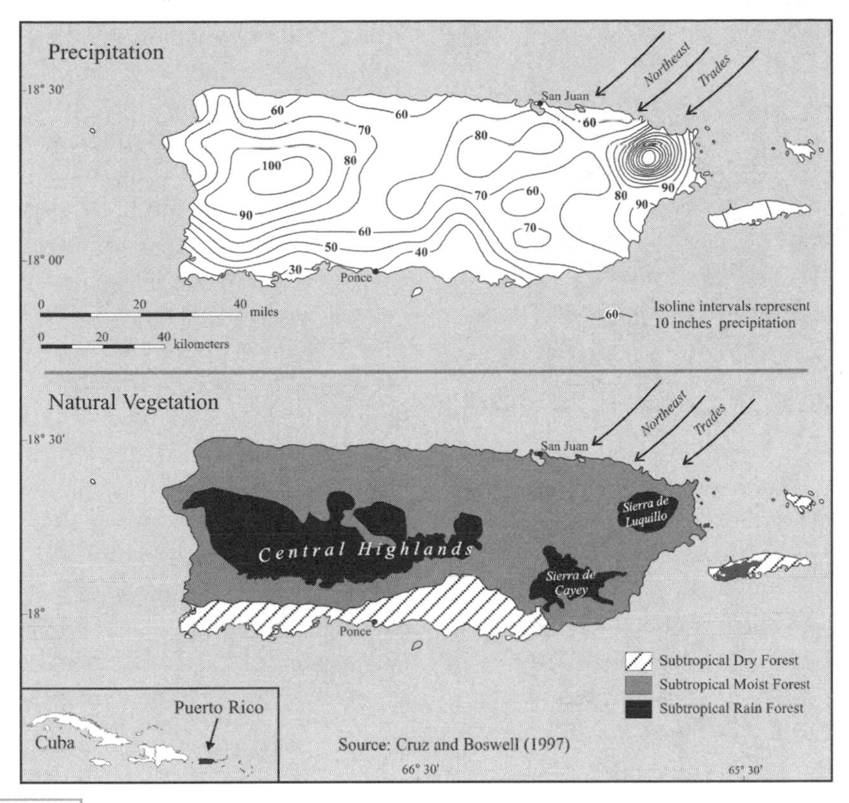

FIGURE 3.2. Precipitation, topography, and vegetation patterns in Puerto Rico.

(23.5°N) and 10–15°N. While local anomalies modify these generalizations, the east coast and east-facing cordillera receive considerable precipitation and the west-facing sides of the cordillera and the west coast are markedly drier. The highland regions are often well watered, and precipitation is adequate for rainfed agriculture.

FIGURE 3.3. Rio Prieto Falls and rain forest in the El Yunque Caribbean National Forest on the windward (northeast) side of Puerto Rico. Photo: Robert Barrett, 2001.

In South America similar patterns prevail in the tropical wet-and-dry belt that straddles the Andes Mountains between the Tropic of Capricorn (23.5°S) and 10–15°S. Here the Trade Winds blow from the southeast over the Amazon Basin toward the northwest. Again, heavy rainfall occurs on the windward eastern slopes of the Andean Cordillera, while precipitation along the Andes' leeward flank and the narrow Pacific Coastal Plain is sparse. Furthermore, the cold Peru Current runs northward along much of the South American mainland from southern Chile to central Ecuador. This current cools and stabilizes adjacent air masses, precluding the likelihood of any significant rainfall occurring along the subtropical

FIGURE 3.4. Cacti and drought-resistant vegetation near Guanica on the leeward (southwest) side of Puerto Rico. Photo: Robert Barrett, 2001.

and tropical portions of the continent's west coast. Thus, a long narrow desert region characterizes the west coast from about 30°S to 2°S, although at its northern and southern extremes semiarid conditions prevail. While much of the coastal desert is unnamed, two of the larger and more extensive desert regions are the Atacama Desert in northern Chile and the Sechura Desert in northern Peru.

Aspect, mountains, and prevailing winds also have a major effect on environmental conditions in the southern midlatitude portion of the South American continent. South of about 30°S, along the continent's Pacific coast, the Westerlies move in across the Pacific Ocean. These moisture-laden winds blow up over the Chilean Andes and dump intense precipitation on the windward slopes, giving rise to natural vegetation characterized by temperate rain forest. The rain shadow effect is pronounced to the east of the Andes in the Argentine regions of Cuyo and Patagonia. Rainfall is very limited; arid and semiarid conditions predominate in the deserts and steppes of these regions; and often intensely dry, desiccating winds, known locally as *zondas*, ravage these areas. In other parts of the world these kinds of winds are known by various names. They are called "Santa Ana winds" in southern California, "Chinook winds" in the Pacific Northwest of the United States, and "Foehn winds" in the European Alps.

In many parts of Latin America, the combination of substantial rainfall and the population's increasingly intense land-use practices contribute to widespread soil erosion, especially in fragile environments like tropical rain forests and arid and semiarid lands. Crop cultivation on steep and even just moderately sloping lands, forest clearing for agriculture and timber production, and overgrazing by cattle and sheep have led to more rapid rates of soil erosion and a worrisome depletion of this critical resource. The development of effective soil conservation measures and soil management techniques are a significant challenge in many contexts where increasing population numbers and greater demands for natural resources like timber place conflicting demands on land use.

Altitudinal Life Zones in Tropical Latin America: The Andean Example

While aspect exerts considerable influence over precipitation and consequently natural vegetation patterns throughout Latin America, elevation also plays a significant role through altitudinal zonation of natural vegetation, land use, and agricultural patterns (Figure 3.5). The principal mechanism at work is the temperature variation caused by the environmental lapse rate as altitude increases. The environmental temperature *lapse rate* refers to changes in ambient temperature caused by increasing elevation: temperature decreases by 0.5°C for every 100-m increase in elevation. When the changes in elevation are minor, such differences in temperature have comparatively little effect on environmental conditions. However, altitudinal variation in many of Latin America's mountainous regions is frequently measured in thousands of meters. Conse-

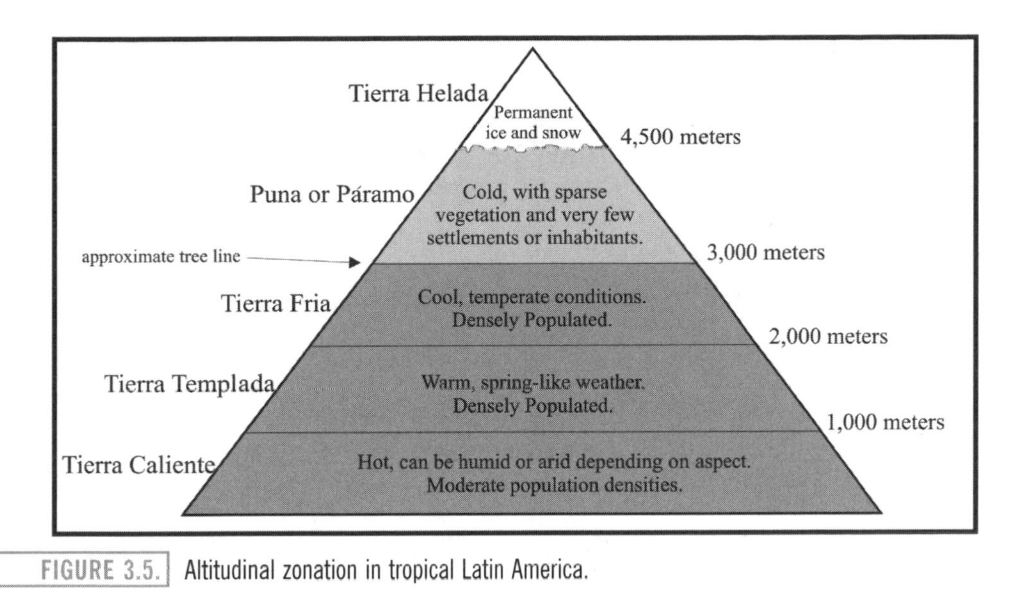

FIGURE 3.5. Altitudinal zonation in tropical Latin America.

quently, the environmental lapse rate creates steep temperature gradients in many mountain environments. While such dramatic temperature gradients are created in all mountain environments, their impact on natural vegetation, land-use, and agricultural patterns is most pronounced in tropical regions. For example, one of the most extreme contrasts can be found in central Peru at about 8°S. A series of mountain glaciers are found in the Sierra Blanca of the Andes range, some at elevations exceeding 7,000 m. Scarcely 100 km to the east in the Huallaga River Valley, rain forests and tropical crops prosper in a hot, humid, tropical environment.

While this is one of the most stunning contrasts in the Tropics of Latin America, similar examples can be found in all the major mountainous areas of the region. These include uplands from central Mexico through Central America, the Caribbean islands of Hispaniola and Puerto Rico, and all of the major mountain areas in South America, specifically the Andes, the Guiana Highlands, and the Brazilian Highlands.

The Andean Cordillera provides an excellent opportunity to examine how altitudinal zonation affects local environments and land-use patterns. The effects of the environmental lapse rate are most pronounced here because this mountain chain spans the equator and the Tropics and because some of the highest mountains are found in the region. Throughout much of the tropical Andes five distinct altitudinal zones are recognized. Nomenclature to describe these levels varies somewhat between individual countries, but nevertheless it is understood throughout the Andean Tropics and in most of Latin America as well.

Tierra Caliente, *Hot Land*

The lowest altitudinal zone, running from sea level up to about 1,000 m, is known as *tierra caliente*, or hot land. The climatic conditions here are decidedly tropical although precipation can vary depending on aspect, latitude, and other factors. Rainfall can vary from extremely heavy to almost nil. Temperatures, however, are more consistent than precipitation, with mean annual temperatures ranging between 28°C and 23°C. Many tropical subsistence crops are grown in this zone. Staple crops include yuca (manioc/cassava), rice, beans, sweet potatos, and plantains (cooking bananas). Tropical and subtropical fruits like mangos, papayas, pineapples, citrus, and bananas are cultivated in tierra caliente, as well as important commercial products such as sugarcane, cotton, cacao, rice, citrus, and, at the upper reaches of the zone, coffee. On the eastern watershed of the Andes, especially in Peru, Bolivia, and Colombia, coca, from which cocaine is derived, is also a major crop. Cattle can be raised in tierra caliente environments, although tropical breeds like Zebu and Zebu hybrids are most successful, and drier areas are more favorable to livestock.

Along the coastal areas of tierra caliente mangrove swamps are widespread. These constitute a unique tropical ecosystem: mangrove trees grow in shallow estuaries, bays, and other comparatively protected coastal areas. Mangrove trees are unusual because they grow in the intertidal zone and are constantly subjected to the effects of saltwater and ocean tides. They are found throughout the Caribbean, Mexico, Central America, and northern South America. Mangrove stands protect coasts from erosion and severe weather, but perhaps more significantly they serve as habitat and nurseries for many species of aquatic organisms and birds.

These ecosystems are increasingly threatened. In Mexico and Panama, for example, nearly 60 percent of the mangrove forests have been destroyed. They may be cut for firewood by local residents or simply cleared to make way for various forms of coastal development including rice farming and tourism. Perhaps the biggest threat has been aquaculture. As globalization has increasingly connected peoples and markets in Latin America with those in other world regions and shipping technologies have improved, mangrove swamps have been converted into farms for growing shrimp for export markets. Without concerted efforts to save these valuable and unique ecosystems, mangrove forests will disappear from many coasts in the coming decades (Figure 3.6).

Tierra Templada, *Temperate Land*

Lying between about 1,000 m and 2,000 m is the region known as *tierra templada*, or temperate land. The climate here is subtropical; some suggest that this region exhibits a climate that might be characterized as "eternal spring." Annual mean temperatures in this zone range from 24°C at the lower reaches to 18°C at the zone's upper limits. Microclimatic

FIGURE 3.6. Mangroves grow in the tidal zone on Magdalena Bay, Baja California Sur, Mexico, 1977.

variations permit a considerable range of subsistence crops to be grown by smallholders—some of which are common in the warmer tierra caliente region and others common in colder environments at higher altitudes. Staple subsistence crops include corn, beans, various tubers, and, at the upper margins of the region, potatoes, while supplemental garden vegetables like tomatoes and many squash varieties are also cultivated. Coffee is the most significant commercial crop grown in tierra templada. It is Colombia's most significant legal export commodity and is a minor export crop in Venezuela. It is not a major export crop in other Andean countries, but in Ecuador, Peru, and Bolivia, almost all national demand for coffee is supplied from farms in tierra templada.

Animal husbandry comprises a component of the agricultural economy of tierra templada, especially for subsistence farmers. Cattle, pigs, chickens, and goats frequently form part of the individual farm economies, and more prosperous smallholders may have a draft animal—an ox, horse, or donkey. Market-oriented livestock rearing is geared toward cattle raising for meat rather than for dairy production.

Tierra Fria, *Cold Land*

Tierra fria, literally "cold land," occupies the area between 2,000 and 3,000 m. The climate bears some resemblance to temperate midlatitude climates, although this description requires some clarification. Mean annual temperatures here range from 18°C to 12°C—higher, of course, at lower elevations and lower as altitude increases. Days, especially mornings, are characterized by bright intense sun and warm temperatures, while nighttime temperatures drop considerably, especially during the Southern Hemisphere's winter months when nighttime frosts occur at the upper reaches of this ecological zone. Snowfall, however, is nonexistent in this altitudinal zone. While the climate is described as temperate, a more accurate characterization might be that the days are akin to pleasant, mild, midlatitude summer days, while the nights can be likened to cool fall evenings. Seasonal temperature variations are very minor—and the seasons are marked principally by changes in precipitation.

Cropping and animal husbandry patterns in tierra fria do have strong similarities to those in temperate midlatitude climates. Traditional pre-Hispanic Andean tuberous cultivars, like potatoes, oca (*Oxalis tuberosa*), and melloco (*Ullucus tuberosus*), are widely grown as staple subsistence crops; however, potatoes are grown commercially in many areas. Other indigenous crops, like corn and quinoa, are common staples as well. Introduced grains and beans, especially wheat, barley, and fava or broad beans, grow well here and form part of the subsistence and commercial agricultural economies of tierra fria. Temperate vegetable crops like carrots, cabbage, peas, broccoli, onions, and cauliflower, which were also introduced from the Old World, also do well in this ecological zone. Temperate fruit trees, including peaches, plums, and apples, can grow here. However, the brief period of winter dormancy and the short growing season tends to produce fruit that is small and not very sweet. Commercial agriculture in this zone is practiced on larger farms that typically produce for local and regional markets the same crops produced by smallholders. In the last decade, however, nursery operations growing cut flowers, especially carnations and roses, for export to the United States and other foreign markets have established themselves as viable economic enterprises in Colombia and Ecuador. Pasture, both improved and unimproved, occupies considerable acerage in this zone. Alflafa and clover grow well and provide grazing forage for cattle in irrigated, fertile, flat basins. In constrast to the lower altitudinal zones, in tierra fria dairy cattle do well. Most production is limited to small-scale operations. While some fresh milk is produced for local sales, much of the milk is utilized for the production of homemade cheese—white, soft, and salty—known as *cuajada*. Unimproved pasture occupies much of the nonirrigated lands as well as hilly upland areas. The cattle that graze these areas—along with sheep—are raised for meat production. Tierra fria is the upper limit for most intensive crop agriculture and cattle raising (Figure 3.7).

The Puna *and* Tierra Helada, *Cold Lands*

The region lying between about 3,000 and 4,500 m is usually known as the *páramo* in Colombia and Ecuador and as the *puna* further south in Peru, Bolivia, Chile, and Argentina. The climate here can be described as cool and crisp, and frosts are commonplace. Snow does occur occasionally in this ecological zone, but it rarely lasts more than 24 hours. Mean annual temperature ranges from 12°C to 2°C. There is little precipitation and semiarid conditions prevail. Crop agriculture here is limited. At the lower reaches of this zone, wheat, barley, and potatoes can be successfully grown, although the danger of frost damage is high, and killing frosts do occur with some frequency. Traditional, pre-Hispanic crops that are frost-resistant—quinoa, cañihua, and tarwi—offer the best alternatives for cultivation here, but these are largely subsistence crops with only limited local and regional markets. The only commercial agricultural opportunity is raising livestock. Sheep are the most remunerative form of herding. Cattle do not do well and are almost entirely absent from this zone, which occurs over almost the entire span of the tropical Andean range. Although the native Andean domesticates, the llama and the alpaca, are no longer commonplace throughout the Andes, they are raised successfully in some areas—notably in southern Peru and Bolivia.

FIGURE 3.7. Small glaciers and snow cover the upper slopes of Huascarán (6,768 m) while farmers living in the *tierra fria* zone on the valley floor cultivate temperate crops like wheat and alfalfa. Huaráz, Peru, 1986.

Land above about 4,500 m is known as either *tierra helada* (frozen land) or *tierra nevada* (snowy land). These cold alpine environments are above the climatic limits of crop agriculture and livestock raising. In addition, some areas, especially the peaks of high mountains, are permanently capped with snow and ice. There are almost no permanent settlements in this zone. With the exception of a few mining operations and alpine-oriented tourism, economic activity here is almost nil.

Environmental Change

Environmental change can come in many forms. Humans may or may not play a role in the process. One of the most dramatic examples of environmental change occurred during a geological period commonly known as the Ice Ages, or, in geological terminology, the Pleistocene. During the last 2 million years periodic cooling of the Earth's atmosphere led to the formation of massive ice caps in the Northern Hemisphere and immense glaciers covered large portions of Asia, Europe, and North America. These glacial periods caused dramatic environmental changes, including the raising and lowering of ocean levels, creation of lake systems, and the molding of the physical landscape.

Human geographers, however, are concerned with the interaction of people and the environment. Environmental change occasioned by human agency is of great concern to them. Preindustrial societies are often assumed to have lived in harmony with nature and to have had little effect on their surroundings. This may have been true when sparse human populations had limited technological sophistication. However, today ecologists, anthropologists, human geographers, and other scholars know with certainty that early human societies produced significant environmental change with simple tools. For example, in North American there is ample evidence that the aboriginal use of fire to hunt game led to an expansion of prairies at the expense of forests. In other instances, subsistence-gathering habits by humans favored the survival of some plants over others.

For example, in the Yucatan Peninsula, Maya peoples prized and collected the *sapote*, the edible fruit of the rain forest tree *Acharas zapote*. Over a millennium or more the numbers of this tree increased as individuals of this species were selectively spared because of their fruit-producing potential. Meanwhile, other less desirable trees were felled for wood or for agricultural clearings.

Deforestation is a hotly debated topic in Latin America. Many ecologists and environmental scientists, including geographers, have raised continuing concerns over the rapidly diminishing forest resources in the region. They point to the critical ecological role that tropical rain forests play in the consumption of carbon dioxide and production of oxygen, as well as to the degradation of soil resources that occurs with deforestation and the loss of biodiversity and habitat for animal species in general and endangered species specifically. Although the Amazon Basin (see Figure 3.8) is frequently cited as the prime example of these problems, they are widespread through all of tropical Latin America where significant stands of rain forest remain. In addition to the Amazon, these areas would include parts of Mexico, Central America, and northern South America. It is not only tropical rain forests that are endangered. Subtropical forests in southern Brazil and temperate forests in southern Chile have also been disappearing at a rapid rate as the expansion of agricultural lands and commercial timber production put increasing pressure on forests.

Yet not everyone agrees that forest clearance is an environmental threat. These observers argue that some deforestation is an integral part of a broader pattern of economic development and a key element in improving the standard of living of Latin Americans. Some are resentful and suspicious of the motives of outsiders, especially those from industrialized countries, who suggest Latin Americans should curtail forest cutting while industrialized countries continue to pollute the environment on a vast scale.

The Columbian Exchange

The arrival of Europeans in the Americas brought two worlds, previously isolated from each other, into contact. This led to environmental changes that transformed the peoples, economies, and landscapes of both the New and the Old Worlds. Frequently referred to as

FIGURE 3.8. Forest clearing usually occurs initially along roads as has happened here in the Chapare Region of Bolivia on the western margins of the Amazon Basin near Villa Tunari, Bolivia, 1988.

the *Columbian Exchange*, these changes resulted from the introduction of "exotic biological material" from the Old World to the New World and vice versa. The exotic introductions covered a wide range of species, including plants, animals, disease germs and viruses, and human beings. Many of these introductions occurred intentionally, as the Spanish, Portuguese, and other European explorers, conquerors, and colonists brought plants and animals to the New World.

New species also arrived from the Old World unintentionally and often unbeknownst to the people who introduced them. Black and brown rats survived the voyage from Europe to the New World and jumped ship once they arrived. Other stowaways included a myriad of Mediterranean weed species, which arrived in the Americas as seeds stuck in the fur of Old World grazing animals. In many areas, these aggressive invasive weeds have almost completely replaced the native grass species.

The exchange of plant species between the New and Old Worlds is the most widespread and long-lasting impact of the Columbian Exchange, eventually affecting peoples and landscapes in almost all inhabited places on the globe. The New World produced a range of unique plant domesticates in three distinct centers of domestication: Mesoamerica, the Andes, and eastern Brazil. The rapidity with which some of these plants were integrated into the agricultural systems and lifeways of Old World peoples is nothing short of astounding.

The potato, an Andean domesticate, is a superb example. The potato was introduced into northern Europe in the 1600s. It grew well in the cool wet climates of northern Europe, and permitted an expansion of cultivated agriculture into new lands and consequently a significant growth in rural populations, notably in Ireland. In the 1840s, this key ingredient in the food supply of the Irish people essentially vanished, as a leaf virus attacked and killed potato plants, unleashing the Potato Famine. Widespread starvation and suffering caused millions of Irish to emigrate from Ireland to seek better lives in the United States, Australia, and other countries. The potato found a place in the diets and food habits of other Europeans too. German cuisine is replete with examples of potato dishes. Vodka, an alcoholic drink from eastern Europe, associated especially with Russia and Poland, is distilled almost exclusively from the potato.

Like the potato, other basic food crops also crossed the Atlantic and transformed the lives and economics of both Europeans and Africans. Manioc, a plant domesticate from the New World also known as yuca in many parts of Latin America, produces an edible tuber rich in carbohydrates that has been adopted as a food crop over wide areas of tropical Africa. In many West African societies, it is a principal food source. Corn has been widely adopted as human food in Africa, but curiously its consumption as food in European countries is limited—there it is used mainly as a feed grain for animals!

The breadth of plant introductions from the New World is staggering, as are their long-term effects. Some are simply curious, while others are profound. Consider what "Italian" cooking would be like today without the use of the tomato, a Mesoamerican plant domesticate. Or consider the peanut (*Arachis hypogaea*), a plant domesticate from tropical South America. It was introduced to the Old World Tropics and is widely grown there today. India, China, and several West African countries are major producers of this crop, much of which is crushed to produce peanut oil. While people eat raw and roasted peanuts worldwide, other uses are less common. Peanut soup is a popular dish in West Africa, while peanut butter appears only to be widely consumed in the United States and Canada.

The use of terms like "Swiss" chocolate or "Dutch" chocolate may suggest to many some kind of European origin for chocolate, but the facts are quite the contrary. Chocolate is an extract of the bean or seed of the cacao tree, *Theobroma cacao*, which was domesticated in the Americas and widely cultivated along the Caribbean coast of Central and South America at the time of contact with Europeans. Highly valued in the indigenous economy, cacao beans served as a widely accepted form of exchange for trade and commerce in Mesoamerica. Hernán Cortés and his men are reported to have learned about eating chocolate from the Aztecs when they first met them in 1520. Hence, the dark, thick, liquid chocolate consumed by Mexicans may be the closest thing anywhere to "real" chocolate, but just how close it is to being original is open to some discussion.

There is a curious irony to America's contribution of chocolate to the world. While millions of tons of cacao beans are produced annually to feed the world's voracious chocolate habit, production in the New World is modest. Two-thirds of world production comes from the West African countries of Cameroon, Ghana, Ivory Coast, and Nigeria, where cacao production is a significant factor in their national economies.

Not all plant introductions were beneficial. Consider another New World plant domesticate, tobacco. Before 1550 A.D. tobacco and the practice of smoking, inhaling, or chewing it was completely unknown in Europe, Africa, Asia, and Australia. In the Americas, indigenous peoples smoked the dried leaves of the tobacco plant as a mild stimulant and for ceremonial and ritual occasions. Tobacco cultivation and use spread quickly and widely throughout Europe during the late 16th century, a process which repeated itself more or less one century later in Asia and Africa. Today, hundreds of millions of people consume tobacco products worldwide, supporting a vast commercial network including farms, factories, transportation, and marketing, generating billions of dollars of sales annually. The human toll occasioned by tobacco consumption is, of course, staggering.

The introduction of Old World plants and the agricultural systems with which they were associated permitted a wholesale transformation of the landscapes of the New World. In Latin America, the Spanish introduced Mediterranean crops that allowed them to mimic the Old World agricultural systems with which they were familiar. Wheat, grapes, deciduous fruits, and olives—all Old World introductions—figured prominently in the colonial agricultural system wherever climatic conditions permitted. Wheat grew in the cool highlands, interior valleys, and open plains from Mexico to Chile, while grapes, deciduous fruits, and olives also prospered, but in far fewer locales. The legacy of these Mediterranean plant introductions continues to have a broad impact on the agricultural landscapes of Argentina and Chile. Since the early decades of the 20th century, the Argentine Pampa has figured as one of the principal wheat-producing regions of the world. The Central Valley of Chile, on the other hand, ships billions of dollars' worth of fresh deciduous fruits and grapes to overseas markets each year and produces many high-quality wines sold in export markets.

It is difficult to point to any animal or even insect domesticates from the New World that had any notable impact on the Old World. Llamas, alpacas, and Muscovey ducks proved of little interest or utility in the Old World except as curiosities. Turkeys and guinea pigs, on the other hand, found modest acceptance outside the New World. Turkeys are utilized as human food in some places, while guinea pigs have been adopted as caged house pets—a far cry from their use as a delicacy food in the Andes.

The transfer of exotic animals from the Old World to the New World was a major element in the dynamics of the Columbian Exchange and was nothing short of profound. The lifestyles, landscapes, and economies of the New World were completely transformed with the introduction of Old World domestic livestock. Cattle, horses, donkeys, goats, sheep, and pigs had the greatest impact. Although the absolute numbers of animals Europeans introduced were few, they reproduced rapidly. Livestock raising quickly became a key element in the economy of colonial Latin America, a pattern that has continued into the present.

The prairies, fertile grasslands, and highland meadows of the New World provided ample pasture for introduced livestock and presented few natural predators or even competitors. Cattle, and to a lesser extent horses, donkeys, and sheep, rapidly occupied these environments. On the Pampa of South America, for instance, feral cattle multiplied so prolifically that for many decades cattle raising consisted of *gauchos* (cowboys) simply harvesting free-ranging cattle.

The introduction of European livestock also transformed the lifestyles of some indigenous peoples. In south central Chile, Araucanians successfully adopted the horse, becoming excellent horsemen as well as superb strategists in the use of cavalry in combat. They employed these skills in an aggressive defense of their homeland in southern Chile against Spanish armies and in an expansion of their domain into Patagonia and the southern reaches of the Pampa during the colonial period. It was not until the 1880s that the Araucanians were successfully subjugated—they were dislodged from Patagonia and defeated in southern Chile. As introduced domestic animals occupied grazing lands, native grasses with little or no resistance to grazing ungulates quickly disappeared and were replaced by weeds from the Old World inadvertently introduced by Europeans.

Ironically, perhaps, some of the cultural icons and economic institutions that were to become synonymous with broad regions and peoples in parts of Latin America during the last 500 years represented a product of the introduction of Old World animal species into the Americas. The introduction of domestic livestock, and especially cattle and horses, coincided with the introduction of new agricultural systems and institutions. The livestock ranch, or simply the *hacienda*, in the New World traces its origins to the cattle-ranching tradition of southern Spain. Known variously as *haciendas, fazendas, estancias, fondos*, or *ranchos*, the livestock ranch played a key role in the economic system of vast rural regions of both Spanish and Portuguese America.

The hacienda also served as one of the primary social institutions in the rural milieu. Haciendas, which ranged in size from modest estates to veritable villages, often were the only appreciable human settlement on the vast reaches of the Mesa del Norte or the Pampa. The estate's owner, the *hacendado*, and family for all practical purposes represented the rule of law, and held sway over their tenants much as medieval lords ruled over their serfs. Religion too fell under the sway of the hacendado, who built the chapel, and who also more than likely paid the salary or wages of the resident or itinerant priest who represented the church in such remote outposts and settlements. The lifestyle and economy of the *gauchos* of the Pampas of South America and the *vaqueros* of northern Mexico, just like the cowboy of the American West, followed directly from the introduction of horses and cattle into the Americas.

Other exotic introductions included diseases and humans themselves. The repercussions of the introduction of disease on the aboriginal peoples of the New World were staggering; these are discussed in Chapter 5 in the section on "Disease and Conquest" and in Vignette 5.1. The impacts of the racial mixing of Native Americans with Africans and Europeans were profound and left an enduring legacy on Latin America. This topic is examined in Chapter 5 in the section on "Miscegenation and Population Change."

The Columbian Exchange is not merely an interesting historical anecdote. It is a process that continues today. In 1957 a Brazilian scientist intent upon improving the genetic characteristics of honeybees in South America brought African honeybees to his laboratories in São Paulo state. Some of the African honeybees eventually escaped from captivity and proliferated in the wild, crossbreeding with European honeybees already living in Brazil and setting off a dramatic transformation of the honeybee stocks across the continent. By the 1990s these aggressive "killer bees" had displaced prior honeybees throughout almost all of tropical and subtropical America and had found their way into the southern United States.

Summary

The physical environment of Latin America is diverse. Because it covers such a wide range of latitudes, climate varies. Examining the region from north to south, several major latitudinal zones can be identified that affect climate and environmental conditions in the region. These include the Northern Hemisphere subtropical belt, the vast tropical belt, which influences many of the region's countries, the Southern Hemisphere subtropical belt, and finally the Southern Hemisphere temperate midlatitude belt, where climatic conditions are moderated by the proximity of the oceans. Altitude also plays a major role in influencing the region's environment. Mountains, high plateaus, and rolling uplands affect the region's patterns of temperature and precipitation. The region's tropical highlands are characterized by clearly demarcated altitudinal life zones. The interaction of the mountains and prevailing winds influences both precipitation and vegetation, creating humid environments on windward slopes and arid conditions, or rain shadows, on the mountains' leeward side. The environment has also been affected by the Columbian Exchange, where Old World and New World plants, animals, and diseases have traveled between continents.

Further Reading

Coomes, O. T. (1997). Rain forest extraction and conservation in Amazonia. *Geographical Journal, 163*, 180–188.

Crosby, A. W. (1972). *The Columbian Exchange: Biological and cultural consequences of 1492*. Westport, CT: Greenwood Press.

Crosby, A. W. (1986). *Ecological imperialism: The biologi-*

cal expansion of Europe, 900-1900. Cambridge, UK: Cambridge University Press.

Enfield, G., and O'Hara, S. (1999). Degradation, drought, and dissent: An environment history of colonial Michoacán. *Annals of the Association of American Geographers, 89*, 402–419.

Faber, D. J. (1973). *Environment under fire: Imperialism and the ecological crisis in Central America*. New York: Monthly Review Press.

Hillman, R. (Ed.). (2001). *Understanding contemporary Latin America* (2nd ed.). Boulder, CO: Rienner.

Hillstrom, K., and Hillstrom, L. (2004). *Latin America and the Caribbean: A continental overview of environmental issues*. Santa Barbara, CA: ABC-CLIO.

Jenkins, R. (Ed.). (2000). *Industry and environment in Latin America*. London: Routledge.

Kellman, M., and Tackaberry, R. (1997). *Tropical environments: The functioning and management of tropical ecosystems*. London: Routledge.

Melville, G. K. (1994). *A plague of sheep: Environmental consequences of the conquest of Mexico*. New York: Cambridge University Press.

Ochoa-Gaona, S., and Gonzalez-Espinosa, M. (2000). Land use and deforestation in the highlands of Chiapas, Mexico. *Applied Geography, 20*(1), 17–42.

Place, S. E. (Ed.). (1993). *Tropical rainforests: Latin America nature and society in transition*. Wilmington, DE: Scholarly Resources.

Rudel, K. T., et al. (2000). When fields revert to forest: Development and spontaneous reforestation in postwar Puerto Rico. *The Professional Geographer, 52*(3), 386–397.

Sage, C. (1994). Coca, development and environment in Bolivia. In M. Redclift and C. Sage (Eds.), *Strategies for sustainable development* (pp. 171–186). Chichester, UK: Wiley.

Sarmiento, F. O. (2002). Anthropogenic change in the landscape of highland Ecuador. *Geographical Review, 92*(2), 213–234.

Simon, J. (1998). *Endangered Mexico: An environment on the edge*. London: Latin America Bureau.

Stadel, C. (1990). Altitudinal belts in the tropical Andes: Their ecology and human utilization. *Benchmark 1990: Conference of Latin American Geographers, 17*–18, pp. 45–60.

Timmons, R. (2003). *Trouble in paradise: Globalization and environmental crises in Latin America*. New York: Routledge.

Young, E. (2001). State intervention and abuse of the commons: Fisheries development in Baja California Sur, Mexico. *Annals of the Association of American Geographers, 91*, 283–306.

4 The Hispanic Caribbean

The modern-day countries of Cuba, Puerto Rico, and the Dominican Republic constitute the Hispanic Caribbean (Figure 4.1). The Spanish established their first colonies on the region's largest islands, Hispaniola and Cuba, and subsequently on Puerto Rico and Jamaica. The region's strategic and economic importance during the colonial period drew the attention of other European powers. They aggressively contested Spanish domination of the area and eventually weakened it. While the Spanish managed to maintain control of most of the islands of the Greater Antillies, other European colonial powers did manage to wrest the island of Jamaica and part of Hispaniola from the Spanish and to successfully occupy the Lesser Antillies.

Cuba, Puerto Rico, and the eastern half of Hispaniola, then called Santo Domingo and now called the Dominican Republic, remained in Spanish hands throughout the colonial period. Late in this period Cuba proved to be one of Spain's most profitable colonies when sugarcane cultivation transformed the island's physical and human landscapes.

The United States became the dominant power after it won the Spanish-American War in 1898. Cuba and the Dominican Republic briefly became U.S. protectorates, and thereafter experienced U.S. intervention in their internal affairs for decades. Puerto Rico became a U.S. territory and remains so to this day. A communist-inspired revolution in Cuba in 1959 brought the United States and Cuba into a pattern of confrontation that has lasted for more than four decades.

Nearly 20 million people live in the Hispanic Caribbean. Three of its metropolitan areas, one on each island, have populations that exceed 1 million. Sugar remains an important export crop, but tourism, tropical fruits, textile manufacture and assembly, pharmaceuticals, and offshore data processing are all significant earners of foreign exchange.

Immigration from the islands of the Hispanic Caribbean to the United States has been extremely important during the 20th century. Some 3.5 million people of Puerto Rican descent, about 1 million of Cuban descent, and nearly 750 thousand of Dominican descent live in the United States today.

Geography and Environment

Geologically, the Hispanic Caribbean straddles the zone of contact between two major tectonic regions, the North American Plate and the Caribbean Plate. As a consequence, two distinct landform types characterize the region. The northern portion, which corresponds with the North American Plate and almost exclusively with the island of Cuba, is comprised of a limestone plateau. This limestone plateau region extends over much of the western and northern Caribbean, with both

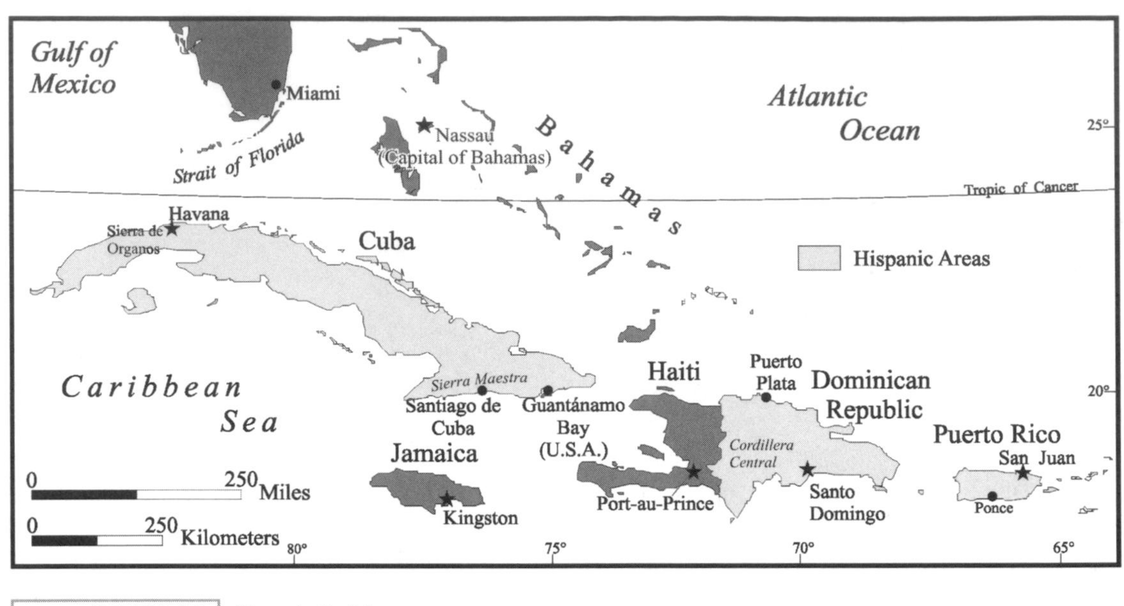

FIGURE 4.1. Hispanic Caribbean.

the Yucatan Peninsula and Florida forming part of it. A flat or gently undulating landscape is found throughout most of central and western Cuba. The area's fertile soils led early to the development of cattle raising and later to a highly productive sugarcane economy on the island. Sinkholes and caves are common here, as in other limestone regions. On Cuba's far western margin, in the Sierra de los Organos, a karst landscape predominates, with towering hills rising steeply off of flat valleys (Figure 4.2). Underground caves and subterranean rivers contribute to the dramatic effects created by this highly eroded karst landscape (it has smaller counterparts in Puerto Rico).

The extreme eastern end of the island of Cuba and the entire islands of Hispaniola and Puerto Rico lie on the northern edge of the Caribbean Plate in the contact zone with the North American Plate. These areas are comprised of uplifted fault-block mountains: the Sierra Maestra in eastern Cuba, the Cordillera Central on Hispaniola, and the Central Highlands and the Sierra de Luquillo in Puerto Rico. Generally, elevations are modest. In eastern Cuba, the most mountainous areas do not exceed 1,000 m, although the highest peak, Pico Turquino, does reach 1,975 m. Puerto Rico is hilly and mountainous, but only the highest areas exceed 1,000 m. The island's tallest peak, Cerro de Punta, is just 1,338 m. Hispaniola is distinct. Much of the island is composed of mountainous uplands and large areas reach above 1,000 m. Many peaks exceed 2,000 m and the highest, Pico Duarte, reaches 3,083 m.

Despite their tropical location, the mountainous character of the islands of Hispaniola and Puerto Rico gives rise to a range of distinct ecological zones. This environmental variation, similar to that found in the tropical regions of the Andes, is a function of temperature changes caused by the lapse rate, aspect, and direction of the predominant winds. The windward sides of the islands, facing the Trade Winds, are moist and humid and usually are characterized by abundant vegetation. A sparser vegetation pattern typifies the leeward sides of the islands. In some areas, for example, the environs of Ponce, Puerto Rico,

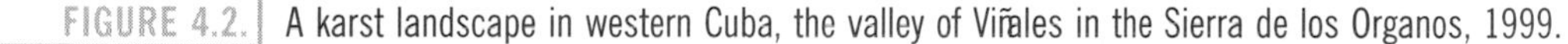

FIGURE 4.2. A karst landscape in western Cuba, the valley of Viñales in the Sierra de los Organos, 1999.

drought-resistant xerophytic vegetation, including cacti, is characteristic.

Tropical rain forest is preserved in a few protected locations, but much has been removed due to population pressure and the demand for wood and farmland. This is the case in Puerto Rico. However, in the El Yunque Rain Forest (Caribbean National Forest) impressive stands of tropical rain forest remain because the Spanish colonial authorities established a forest preserve there in the late 19th century. In the Dominican Republic significant forest cover remains, with tropical hardwoods and pines predominating. This is in stark contrast to neighboring Haiti, where the forest cover has been completely removed.

The region's climate is tropical, with little seasonal variation in temperature. Mean monthly temperature in February, the coolest month, is about 24°C in San Juan, Puerto Rico, rising to just 27°C in August, the warmest month. Temperatures for other urban centers located at sea level, like Havana and Santo Domingo, are similar (Figure 4.3). It is markedly cooler in the highlands, especially in the Dominican Republic and Puerto Rico.

Precipitation varies much more dramatically than temperature over the island geography of the Hispanic Caribbean. Seasonal variation in precipitation is marked, although some rainfall occurs throughout the year in most locations. The northward movement of the ITCZ during the summer months causes more rainfall, while its southerly migration during the winter brings the region more under the influence of the subtropical high-pressure belt and rainfall is less common. At San Juan, Puerto Rico, located on the windward side of the island, precipitation totals about 1,500 mm per year; most falls between May and December. Precipitation on the leeward side (the southwest) of the island is less plentiful due to the influence of the rain shadow effects of the interior mountains. At Ponce, for instance, annual rainfall is only about 900 mm annually.

Hurricanes mark a stunning contrast to the gentle northeasterly Trade Winds, mild temperatures, and moderate rainfall that are characteristic of the weather most of the year. Hurricane season begins in the late spring (May), usually reaches its fullest development during the late summer and early fall (August and September), and ends in November. These massive storms form over the warm waters of the western Atlantic and eastern Caribbean Sea and track in a northwesterly direction across the Lesser and Greater Antillies (Figure 4.4). Characterized by heavy rains and fierce winds, they measure some 600–800 km in diameter. Hurricanes usually take 7–10 days to move across the Caribbean and can leave considerable damage to human life and property in their wake. High winds, flooding, landslides, and storm surges are common. Low-lying areas and those open to the most direct effects of the winds are most at risk (Figure 4.5). The Spanish capital in Cuba was originally sited on the southern side of the island, but early hurricane damage convinced Spanish colonial authorities to move the city to a site on the northern side of the island with a protected harbor, Havana.

Historical Geography and Economic Development

The Caribbean islands of Hispaniola, Cuba, and Puerto Rico were the first colonial territories incorporated into Spain's New World empire at the beginning of the 1500s. The Arawak, Carib, and Taíno peoples who inhabited the islands at the time of contact with the Europeans were quickly subjugated by the Spanish. Most succumbed rapidly to the ravages of European diseases and enslavement. But miscegenation ensured the survival of some indigenous genetic stock in the resulting population. Today, many Dominicans claim Taíno heritage. A small number of Caribs have survived because their ancestors lived on remote islands with inaccessible interiors, or because their ancestors managed to escape to Central America, where they mixed with escaped African slaves—these latter are known as "Black Caribs." However, most native peoples had all but disappeared within 100 years of first con-

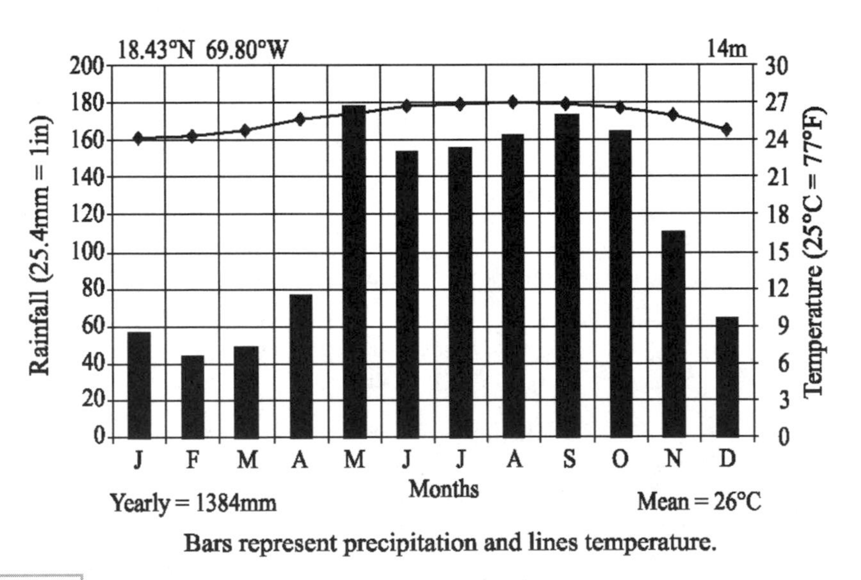

FIGURE 4.3. Climograph of Santo Domingo, Dominican Republic.

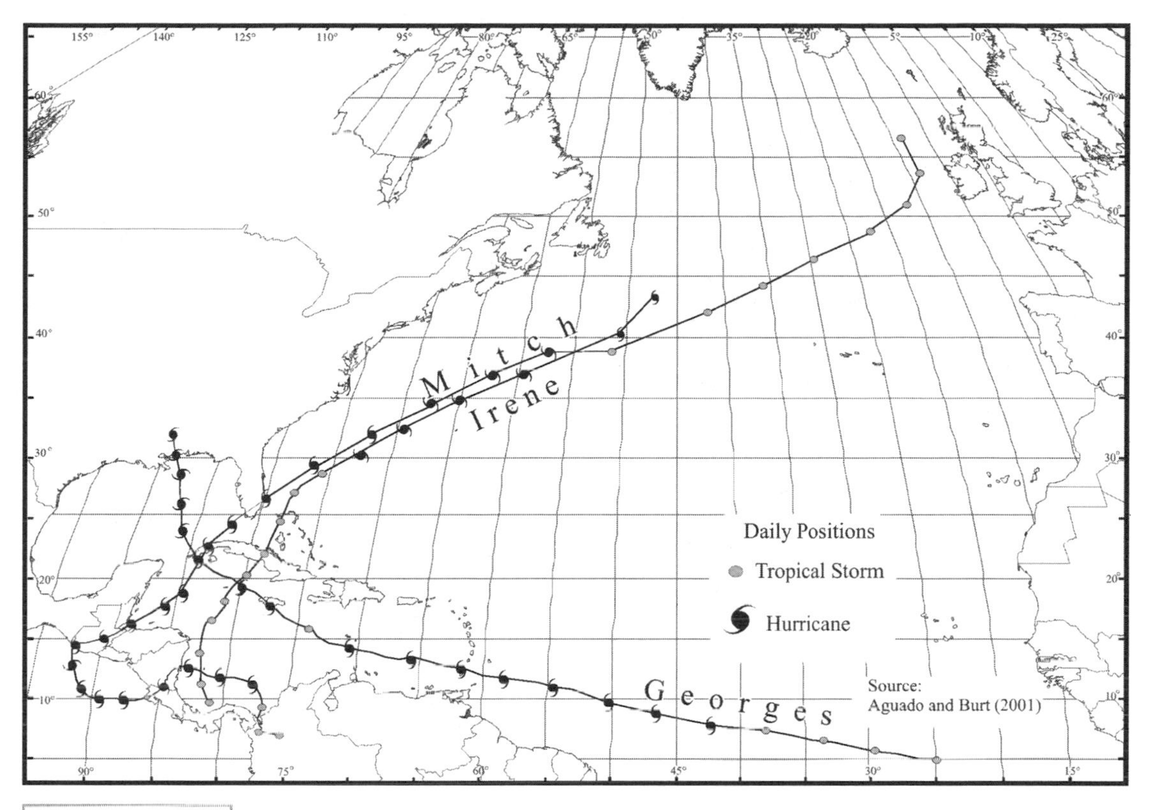

FIGURE 4.4. Paths of Hurricane Mitch and other major hurricanes in 1998.

tact with the Spanish. Language, however, includes some indigenous survivals—"hurricane" and "barbeque"—as noted previously, are indigenous terms that have survived in both Spanish and English usage to this day.

Agriculture dominated the economies of the islands. Livestock raising, especially of cattle, figured prominently in this mix, as did the labor-intensive practice of sugarcane cultivation. The annihilation of the indigenous populations led the Spanish colonist to import African slaves to provide the labor force they needed on their sugar plantations.

While the islands of the Hispanic Caribbean achieved importance in the first decades of the Spanish colonial enterprise, for much of the colonial period they were backwaters, rapidly eclipsed by the wealth of Mexico and Peru. However, Havana, Cuba's capital, did retain a certain measure of economic and political importance as the principal Spanish port during the colonial period, and it grew into one of the empire's largest urban centers. Because Spain's mercantile trading system only permitted trade between Spain and a few New World ports—Veracruz, Cartagena, Lima, Acapulco, and Colón (as a transshipment point)—no trade was permitted with foreign merchants or nations. Thus, Havana became the principal gateway to the New World for Spanish trade due to its easterly location and extremely well-protected natural harbor. The annual fleets that plied between Cádiz, Spain, and the New World would call first at Havana and then divide there for onward travel to Veracruz, Cartagena, or Colón (Lima and Acapulco). Similarly, on the return trip, the fleets originating at Veracruz, Cartagena, and Colón

FIGURE 4.5. In 1998 Hurricane Mitch caused extensive damage in the western Caribbean. These palms on the island of Utila in the Bay Islands of Honduras were toppled by the severe winds, 2002.

would sail for Havana, from whence the entire annual flotilla bound for Spain would sail. The system of controlled ports and annual fleets permitted Spanish colonial authorities to more effectively control trade and levy royal taxes on colonial export products and Spanish imports. The Crown also only authorized certain merchants and trading companies to engage in commerce between the colonies and the New World. These individuals and companies established their main offices in Havana. As a result, Havana benefited from its privileged position and colonial officials, merchants, and military garrisons located here. Massive fortresses guarded its harbor, impressive public and religious architecture graced its plazas, and the opulent homes of wealthy merchants and Crown officials lined the narrow streets of the colonial core.

San Juan, Puerto Rico, and Santo Domingo on Hispaniola both served important administrative and economic functions during the colonial period as the principal ports and governmental centers on their respective islands. But neither enjoyed the economic success of Havana and neither grew as large or as wealthy.

While all of Spain's mainland colonies were independent by 1823, the territories of the Hispanic Caribbean followed a different path. Spain ceded its colonial territories on Hispaniola to the French in 1795, and Haitian armies occupied the territory intermittently until 1844. At that time, the Dominicans sought Spanish protection from the Haitians and submitted to control by the Spanish Crown until 1865, when the country gained permanent independence. Cuba and Puerto Rico remained under Spanish control, but as the 19th century wore on independence movements brought bloody strife to both islands, and especially to Cuba.

Sugarcane cultivation boomed during the last half of the 19th century, transforming the economy, the landscape, and the human geography. This was especially true in Cuba. Although Spain was still the nominal colonial power in the islands, the increasing ascendancy of the United States as a world power after the Civil War lead to dramatic increases in U.S. investment. This investment was concentrated in the lucrative sugar industry that was rapidly being transformed by the industrialization of sugar processing and the construction of modern mills. In many regions of the islands, small family-owned plantations disappeared as land ownership was consolidated and small, inefficient, traditional mills were abandoned. Mixed farming also suffered as a crop monoculture came to dominate over broad areas. Sugar harvesting and cultivation still demanded a large labor force. The importation of African slaves continued well after 1850; slavery was not outlawed in Puerto Rico until 1878 and in Cuba until 1886. With slavery outlawed, plantation owners sought cheap labor from other sources, bringing indentured servants from several areas, including China.

Close to 200,000 Chinese immigrated to Cuba between about 1860 and 1890 to fulfill labor contracts on sugar plantations. Many stayed on after those contracts expired, giving rise to a small, but visible, Cuban Chinese community that survives to the present (Figure 4.6).

The close of the 19th century brought dramatic changes to the region's political geography. The explosion of the battleship *U.S.S. Maine* under mysterious circumstances in Havana's harbor provided the rationale for the United States to declare war on Spain in 1898 (the Spanish-American War). A U.S. Navy commission in the 1990s determined that the battleship had most likely exploded because of a malfunction in its boilers. At the time, however, the incident fed the war fever whipped up by many of the nation's newspapers, notably the "Yellow Press" led by the Hearst syndicate, and the expansionist tendencies of some politicians and commercial interests. Spain's armies and naval forces crumbled within 6 months of the start of the war. The United States forced Spain to sign a humiliating treaty in which it ceded nearly all of its overseas territories to the United States. In the Caribbean, these territories included Cuba and Puerto Rico.

The United States immediately became the region's dominant power. Its military forces occupied Cuba (1898–1902) and Puerto Rico (1898–1901) and its commercial interests held sway over the economies of both islands as well as that of the Dominican Republic. Cuba became independent in 1902 under the provisions of the Platt Amendment. The amendment, which remained in effect until 1934, reserved the right of the United States to intervene militarily in the island's internal affairs. It did so in 1906, 1912, and 1917. It also retained the right to establish military bases on Cuban soil. The legacy of this arrangement lives on over 100 years later at Guantánamo Bay where the U.S. Navy still maintains an immense military garrison despite the objections of the Cuban government. Since 2002 the U.S. military has maintained an extraterritorial prison camp at Guantánamo for alleged terrorists in an effort to avoid complying with provisions of U.S. law that protect the legal rights of prisoners.

The United States held no territorial claims to the Dominican Republic. However, it exerted immense power over the nation and U.S. Marines did occupy the country to "preserve order" and protect U.S. investments be-

FIGURE 4.6. Havana's Chinatown is a small five- or six-block neighborhood near the city's downtown. Approximately 3,000 Chinese Cubans live in the vicinity, 1999.

tween 1915 and 1924. They returned again in 1965 to restore order during a brief civil war.

Dictators, military governments, and corrupt powerful elites closely allied with the United States and U.S. business interests governed both Cuba and the Dominican Republic for much of the 20th century. Rafael Trujillo, who governed the Dominican Republic with an iron fist for some 30 years (1930–1961), and Fulgencio Batista, who ran Cuba either directly or indirectly for almost as long (1933–1944, 1952–1959) figure among the most infamous.

Puerto Rico became a U.S. territory in 1901 under the provisions of the Foraker Act. In 1917 the Jones Act granted U.S. citizenship to the island's residents. Puerto Rico's special relationship with the United States, first as a protectorate, and then after 1950 as a commonwealth with the right to elect its own governor and legislature, led to a pattern of social and economic development on the island distinct from that in Cuba and the Dominican Republic. Puerto Rican migration to the U.S. mainland had started at the beginning of the 20th century, but the number of early migrants was small. World War II fostered much increased migration, as did a population explosion on the island during the first half of the century that reduced economic opportunities. Approximately 100,000 Puerto Ricans served in the U.S. military during World War II. Their greater familiarity with the mainland and its culture contributed to increased migration to the mainland. In addition, the labor shortages created by the war on the U.S. mainland created a strong demand for laborers. Labor recruiters sought Puerto Rican laborers for unskilled agricultural and factory work. As a result, tens of thousands moved north, settling in New York (particularly in New York City), New Jersey, Florida, Illinois (especially in Chicago), and even Ohio.

The island's privileged relationship with the United States also led to significant opportunities for economic development not afforded to the other nations of the Hispanic Caribbean. Over the long term it has given the island's residents a higher standard of living than those of their neighbors. The government of the United States and the island's government embarked on an ambitious program ("Operation Bootstrap,") in 1947 to promote industrial development on the island. The U.S. government amended its tax laws to give U.S. corporations that opened factories in Puerto Rico corporate tax advantages. The Puerto Rican government also provided local tax incentives, industrial infrastructure, and worker recruitment and training services for U.S. corporations locating on the island. These led to a substantial increase in the island's industrial sector, initially with the manufacturing of consumer goods, but later with heavier industrial activity: petroleum refining, petrochemical production, and pharmaceutical manufacture. While per capita incomes in Puerto Rico still lag far behind those in the United States, they outpace those in the rest of the Caribbean by wide margins.

Contemporary Economic and Social Geography

The last half of the 20th century brought fundamental changes to the Hispanic Caribbean. The Cuban revolution in 1959, led by Fidel Castro, profoundly transformed the human and physical landscape of Cuba as few other single events have affected the landscapes of other Latin America countries during the 20th century.

One of the revolution's most immediate effects has been the emigration of hundreds of thousands of Cubans from the island to the United States, principally to south Florida. In the early 1960s, in the years immediately following the revolution, tens of thousands of upper- and middle-class Cubans, many of whom were educated professionals, left the island. Ten years later, a second wave of Cuban immigrants, numbering nearly one-quarter of a million, emigrated to the United States under the auspices of a family unification program. A third wave estimated at approximately 125,000 reached the United States in 1980. While the largest numbers of Cuban immigrants were concentrated in these three short-term flows, thousands of Cubans have risked their lives crossing the narrow Straits of Florida (150 km) in all manner of boats and rafts in the hope of reaching the United States. All told, around three-quarters of a million Cubans have immigrated to the United States since 1959.

The revolution in Cuba also transformed the economic and social landscapes of both the countryside and the city. The nation's socialist government focused intense efforts at improving the educational levels of the population and succeeded in achieving a literacy rate close to 100 percent. It also made healthcare for all citizens a national priority, lowering infant mortality rates to levels comparable with the world's wealthiest nations and extending average life expectancy to levels comparable to the United States. However, to achieve these and other ends, it imposed a police state on the island nation. It has tolerated no dissent, squashing any alternative voices and viewpoints and jailing dissidents for long periods of time. In the countryside agrarian reform reduced the size of individually owned farms to a maximum of 70 ha and resulted in the distribution of small plots of land to tens of thousands of peasant families. The revolutionary government nationalized the vast acreages of foreign-owned (usually by U.S. corporations) sugarcane plantations and converted them into state-run cooperative farms.

A more revolutionary urban landscape evolved as well. Before the revolution, Havana had been one of the favorite playgrounds of U.S. tourists and expatriates. Luxury hotels, many constructed along the Malecón, a picturesque seaside drive and walk, dotted Havana's upscale neighborhoods and served this clientele in high style. Restaurants, casinos, nightclubs, brothels, and all manner of establishments catered to these clients' demands and whims. The revolution swept away these manifestations of "capitalist excess and vice" in a heartbeat. Hotels were transformed into worker's vacation hostels or apartments, casinos were shuttered, restaurants were closed, and prostitution was outlawed. The monumental Plaza de la Revolución in Havana was constructed to immortalize the revolution and to serve as a public stage for Fidel Castro (Figure 4.7).

The dissolution of the Soviet Union at the beginning of the 1990s marked the beginning of the "Special Period" in Cuba. Simply put, the demise of Cuba's principal mentor, protector, and banker has meant a serious retrenchment for the Cuban revolution and its ideals. International tourism, especially from Europe and Canada, has become a source of urgently needed foreign exchange. This critical need for hard currency has begun to transform Havana's revolutionary landscape once again. And it is a landscape full of contradictions. Under close state direction and sponsorship, old hotels have been refurbished and new ones built, and elegant restaurants, nightclubs, casinos, and stores selling luxury products have opened to cater to the needs of foreign tourists. Ironically, almost all are off limits to ordinary Cubans. Prostitution flourishes and Cuba has become a prime destination for sex tourism.

The history of one of Havana's most elite social clubs, the Havana Biltmore Yacht and Country Club, is illustrative of the making and unmaking of Havana's revolutionary landscape. Founded in 1928 for wealthy U.S. residents and Havana's moneyed elite, it was one of a handful of such clubs operated in Havana before the revolution. The club's grounds, covering 65 ha, included a private beach, a golf course, a swimming complex, and equestrian facilities. Nationalized after the revolution, the club was utilized as a social and vacation club for workers and as a sports institute of athletes. But the demands of Cuba's Special Period have led to the rebirth of the Havana Biltmore

FIGURE 4.7. An image of Ernesto "Che" Guevara, a hero of the Cuban revolution, adorns a building fronting on the Plaza de la Revolución in Havana, 1999.

Yacht and Country Club, albeit under a new name: Club Habana. Under tight government supervision, the international Spanish hotel chain Sol-Melia operates the new club. A stiff annual membership fee allows the clubs' members access to a private beach, swimming pools, a golf course, tennis courts, restaurants, a business center, and upscale shops. Membership, however, is restricted to foreign residents. Cubans, even the very few with sufficient financial means, may not join. Indeed, Cubans may only visit the club as guests of its foreign members!

While tourism has played a critical role in sustaining the Cuban economy since the early 1990s, Cuban exiles have also played a central role in the economic survival of the island nation despite a debilitating economic embargo by the United States. Over the years, those living in the United States have sent hundreds of millions of dollars to family members in Cuba to supplement their meager incomes on the island. It is difficult to estimate the exact amount of money remitted from the United States to Cuba each year since much is sent from third countries or clandestinely in violation of U.S. laws and government regulations. A study released in early 2004 suggests that approximately $100 million is remitted annually, although other estimates have suggested that the figure is much higher.

In both Puerto Rico and the Dominican Republic migration and tourism have also been key forces in shaping the contemporary economic and social geography. Puerto Ricans have continued to migrate to the U.S. mainland, although the patterns have changed somewhat since the 1950s and 1960s. The increasing ease and decreasing cost of air travel have reduced the barriers of distance, and many Puerto Ricans move between the island and the mainland with considerable frequency. Return migration to the island is more common today, especially among educated professionals and older Puerto Ricans who chose to retire on the island after spending most of their working lives on the mainland.

Dominican immigration to the United States rose sharply after the fall of Trujillo in 1960. Often migrating first to Puerto Rico, either legally or illegally, a large number of Dominicans settled in New York City. At the end of the 1990s, some 30,000 Dominicans emigrated legally to the United States each year;

nearly 750,000 are estimated to reside in the country today. Dual-citizenship provisions in the Dominican constitution allow these immigrants to vote in presidential elections; indeed, recent candidates for the Dominican presidency have campaigned in New York City! Financial remittances from Dominicans working in the United States constitute a major contribution to the Dominican Republic's foreign exchange earnings.

Tourism brings millions of visitors annually to Puerto Rico and the Dominican Republic. High-rise hotels, exclusive resorts, and budget tourism have transformed the shorelines and skylines of remote beaches and urban areas alike. Tourists from the mainland United States are most common in Puerto Rico and the Dominican Republic.

The population of the Hispanic Caribbean exceeds 25 million. Cuba (12 million) accounts for approximately half of the regional population total, while the Dominican Republic (9 million) and Puerto Rico (4 million) account for the balance. Overall, about three-fourths of the population is urban and much of this is concentrated in large urban centers. The urban systems of Cuba, the Dominican Republic, and Puerto Rico all exhibit a high degree of urban primacy and their respective capitals dominate the urban systems. The populations of all three cities exceed one million: Havana (2.3 million), Santo Domingo (2.9 million), and San Juan (1.5 million).

Summary

The Hispanic Caribbean, made up of Cuba, Puerto Rico, and the Dominican Republic, has been important politically, socially, and economically since the colonial period. Since 1898, the United States has been the dominant power in the region. Puerto Rico is a U.S. territory. Large numbers of immigrants from all three islands have entered the United States during the 20th century. The region is tropical and local patterns of rainfall and vegetation are strongly influenced by the northeast Trade Winds and mountainous terrain, especially in the Dominican Republic and Puerto Rico. Hurricanes are common from August to November and frequently cause significant damage. Overall temperatures are tropical, with little seasonal variation.

The islands were among the first settled by the Spanish, but lagged behind Spain's mainland colonies in importance. By the mid-18th century, sugarcane transformed the economy and continued to play a pivotal role in the regional economy well into the 20th century. Since the Cuban revolution of 1959, large numbers of Cubans have immigrated to the United States to escape Castro's dictatorial regime. Tourism is important to the islands' economies. Initially, under Castro, tourism was discouraged in Cuba, but since the collapse of the Soviet Union the need for foreign exchange income has provided a strong incentive to redevelop that part of the economy.

Further Reading

Barker, D., and McGregor, D. (Eds.). (1995). *Environment and development in the Caribbean: Geographical perspectives*. Kingston, Jamaica: University of the West Indies Press.

Besson, J., and Momsen, J. (Eds.). (1987). *Land and development in the Caribbean*. London: Macmillan.

Blue, S. A. (2004). State policy, economic crisis, gender, and family ties. Determinants of family remittances to Cuba. *Economic Geography, 80*(1), 63–82.

Bourne, C. (1988). *Caribbean development to the year 2000: Challenges, prospects, and policies*. London: Commonwealth Secretariat.

Ferguson, J. (2003). *Migration in the Caribbean: Haiti, the Dominican Republic and beyond*. London: Minority Rights Group International.

James, C., and Pervilolaris, J. (Eds.). (2000). *The cultures of the Hispanic Caribbean*. Gainesville: University Press of Florida.

Knight, F. (1990). *The Caribbean*. New York: Oxford University Press.

Martínez-Fernández, L. (1994). *Torn between empires: Economy, society, and patterns of political thought in the Hispanic Caribbean, 1840–1878*. Athens: University of Georgia Press.

Pérez, L. A. (1995). *Cuba: Between reform and revolution* (2nd ed.). New York: Oxford University Press.

Richardson, B. C. (1992). *The Caribbean in the wider world, 1492–1992*. Cambridge, UK: Cambridge University Press.

Scarano, F. (1984). *Sugar and slavery in Puerto Rico: The plantation economy of Ponce, 1800–1850*. Madison: University of Wisconsin Press.

Scarpaci, J. L., et al. (2002). *Havana: Two faces of the Antillean metropolis* (rev. ed.). Chapel Hill: University of North Carolina Press.

Snodgrass, M. (2001). Assessing everyday life in post-Soviet Cuba. *Latin American Research Review, 36*(3), 204–220.

Wiarda, H., and Kryzanek, M. J. (1992). *The Dominican Republic: A Caribbean crucible* (2nd ed.). Boulder, CO: Westview Press.

Williams, S. (1994). *Cuba: The land, the history, the people, the culture*. Philadelphia: Running Press.

5 Peopling Latin America to 1820

Latin America, just like the rest of the Americas, was devoid of human settlement for much of prehistory. The Americas' first settlers migrated from Asia to North America across what is now the Bering Straits. A land bridge between Asia and North America formed repeatedly during the last 2 million years, causing sea levels to drop as much as 100 m in some places. While scholars agree that the first immigrants to the Americas initially walked across the land bridge, there is considerably less agreement about dates of arrival for the first waves of migrants and the rate with which they spread over the two continents.

Some scholars have argued that the first immigrants may have arrived in the Americas as early as 30,000 years ago, and that later major glacial periods permitted subsequent migrations. Recent research using DNA, however, suggests that the initial migrants to the New World may have arrived just 18,000 years ago. Regardless of the precise dates of these early migratory movements, archaeological evidence from both North and South America makes it clear that human beings had established themselves over the entire length and breadth of the two continents by 15,000 years ago. Their numbers were small and settlement was in no way continuous, but by that time early Americans had already explored much of the two continents.

Indigenous Cultures before Contact

On the eve of the European "discovery" of the Americas the cultural and economic sophistication of indigenous societies ranged from simple to complex. Three distinct tiers of cultural and economic development can be identified: tribal groups, chiefdoms, and advanced civilizations.

Tribal societies were (and still are) represented the simplest level of cultural and economic development. Families and clan groups were the basic units of social structure in these societies. Individual tribal units rarely surpassed a maximum size of a few hundred individuals. Their technology was rudimentary and their material culture was simple and largely impermanent. Subsistence economies characterized these societies, and surplus food production was very limited. In especially hostile environments like Patagonia, the Chilean archipelago, the Chaco, and the arid north of Mexico, hunting and gathering typified the economies of these tribal societies. Other groups, situated in more benign environments and often adjacent to more advanced peoples, practiced rudimentary forms of agriculture. Perhaps the most common of these agricultural systems was slash-and-burn agriculture, where natural vegetation is cut, cleared, dried, and burned in situ (Figure 5.1). The field is utilized for several years. When its fertility declines, it is abandoned. Much of the Amazon Basin, the rain forests of eastern Central America, and a small part of Cuba were occupied by tribal peoples who practiced these and other forms of agriculture.

Chiefdoms occupied an intermediate position in the hierarchy of cultural and economic development. These societies involved a far more complex form of social organization than did tribal societies, and their populations were vastly larger. Typically chiefdoms were comprised of many individual tribal groups that spoke the same language, shared the same culture, and were affiliated both politically and economically. They were often loosely integrated during times of peace. When threatened by outsiders, however, a more centralized form of governance, or at least closer cooperation, operated among the various tribal groups comprising the chiefdom. These forms of federation could be very effective. The Araucanian peoples, or Mapuche, of southern Chile successfully resisted the intrusion of Spanish colonial armies and subsequently those of the Chilean Republic into their territory until the 1880s.

The economies of chiefdoms depended principally upon agriculture, although hunting and gathering played a role among some groups. Agricultural systems varied from impermanent field agriculture using slash-and-burn techniques to permanent field agriculture employing simple irrigation technology. Although conditions varied, their material culture included specialized hunting and farming tools, woven cloth, and pottery utensils. The construction of permanent homes and other buildings characterized some chiefdoms, but little evidence of their former existence survives as part of the landscape or in the archaeological record. One of the few exceptions is found at San Agustín in southern Colombia where Chibcha peoples carved massive zoomorphic statues and constructed funeral chambers from hewn rock slabs (Figure 5.2).

Chiefdoms occupied diverse regions of Latin America at the time of European contact. In South America, Araucanian peoples had effectively settled the central and southern portion of what is now Chile and Chibcha peoples occupied much of the northern Andes, running

FIGURE 5.1. Slash-and-burn field on the eastern Andean piedmont of the Alto Beni, Bolivia, 1987.

FIGURE 5.2. Zoomorphic statue from a Chibcha site at San Agustín in the highlands of southern Colombia, 1975.

from what is now southern Colombia into Venezuela. In the Caribbean, Arawaks largely dominated the Greater Antilles (Cuba, Hispaniola, Jamaica, and Puerto Rico), while the Caribs, an aggressive people originally from northern South America, had successfully colonized most of the islands of the Lesser Antilles.

Advanced Civilizations

Advanced civilizations, characterized by sophisticated cultural, economic, and political development, had also evolved in Latin America by the end of the 15th century. The European "discoverers" encountered civilizations every bit as sophisticated and complex as the classic civilizations of the Old World. They also encountered urban centers, notably Tenotichítlan in Mexico and Cuzco in Peru, with populations in excess of 100,000, making them comparable to the largest contemporary cities in Europe. Advanced civilizations evolved in two regions in Latin America: in the central Andes and in the central and southern highlands of Mexico and northern Central America.

The Aztec and Inca civilizations that the Spanish conquistadores confronted in the early 16th century in the highlands of the Andes and central Mexico did not arise "whole cloth" but instead were the result of long processes. In a sense, each region served as a kind of "cradle" in which advanced civilizations arose, prospered, and eventually declined. Successive civilizations profited from the technological and intellectual achievements of their predecessors. The impressive social and material achievements of the Aztec and Inca civilizations were built on the foundations of earlier societies and represented the product of several thousand years of cultural development.

The domestication of plants and animals and the simultaneous evolution of agriculture were the earliest and most enduring achievements of these early societies. The creation of food surpluses, stability in food availability, and a more settled lifestyle all followed on the heels of these achievements. These events unfolded in Mesoamerica and the Andes region over several millennia beginning around 5000 B.C. Class stratification, occupational specialization, complex political organization, and the development of art, theology, and technology all eventually followed. Plants, and, to a lesser degree, animals were domesticated in three regions of Latin America: Mesoamerica, the Andes, and the highlands of Brazil.

In Mesoamerica the principal food domesticates comprised a triad of plants—corn (genus *Zea mays*), beans (genus *Phaseolus*), and squash (genus *Cucurbita*)—all of which were typically cultivated together in mixed-field plantings. This system exploited symbiotic relationships between the plants. For instance, corn requires great quantities of nitrogen in the soil in which it grows, while beans, as leguminous plants, fix nitrogen and replenish it in the soil. Similarly, the characteristics of each plant's growth ensure that the available sunlight is maximized. Corn grows straight up from the ground, beans grow up supporting themselves on the corn stalks, and squashes grow horizontal to the ground. Other domesticated food plants of Mesoamerican origin, well known outside the region, include chile peppers, avocados, vanilla, and cacao (Figure 5.3).

Mesoamerican plant domesticates include a wide range of lesser known plants such as several agave species from which the Mexican liquors mescal and tequila are distilled, as well as ornamental flowers like dahlias and marigolds. Animal domesticates of Mesoamerican origin are few, the most significant being turkeys and Muscovy ducks, the only birds domesticated in the New World. Domesticated insects included stingless bees kept for honey and wax, the cochineal insect harvested for a rich red dye, and the aje (*Llaveia axin*) used for natural lacquer in artisan work.

A rich mixture of plant and animal domesticates also emerged from the Andes. Of these the potato (*Solanum tuberosum*) is arguably the most important in terms of its impact on world food supply. Early agriculturalists developed a wide range of potato varieties—some of which could prosper in the dry cool environment of the Andean highlands while others prospered in the moist warm foothills of the eastern Andes. Other widely known Andean plant domesticates whose significance is widespread include the strawberry and coca, a bush from whose leaves the drug cocaine is extracted (Figure 5.4).

Other plant domesticates are less well known, but are still widely used in the Andes, especially in rural areas. Amaranthus (*quinoa*) produces an edible protein-rich seed used in soups, and several tuberous food crops like *olluco* and *papa lisa* are common dietary staples in many homes. The Andes are home to only a few animal domesticates, the largest and most significant being the llama, a beast of burden and also a producer of wool, fuel dung, and meat. Other animal domesticates include the alpaca, useful primarily for its fine wool, and the guinea pig, kept not as a pet but as a food source.

In eastern Brazil indigenous peoples also successfully domesticated a complex of plant species, although no advanced civilizations developed here. Key food plants domesticated in

FIGURE 5.3. Cacao, beans on drying trays near Limon, Costa Rica, 1977.

FIGURE 5.4. A farmer drys coca leaves along the road. Cocaine is made from its leaves. Chapare region, Bolivia, 1988.

this region include sweet potatoes (*Ipomoea batatas*) and both sweet and bitter manioc (*Manihot aipi* and *Manihot esculenta*). Other notable and widely known plant domesticates from here are peanuts, pineapples, cashew nuts, and tobacco.

Pre-Columbian Culture History

In both Mesoamerica and the Andean realm, archaeologists have identified a sequence of increasingly sophisticated cultural levels in each region culminating in an advanced civilization. This cultural sequence is most clearly understood for Mesoamerica, whose history provides a useful perspective on the timing and the characteristics of the evolution of advanced cultures throughout Latin America.

Mesoamerica

The culture history of Mesoamerica is divided into three distinct periods of cultural development: the Pre-Classic (also known as the Formative), the Classic, and the Post-Classic. It was during the Pre-Classic period, running roughly from 2500 B.C. to 300 A.D., that settled village life based on an agricultural economy became commonplace throughout Mesoamerica. Early on, slash-and-burn agriculture predominated, but this gave way to more complex agricultural systems that included the use of fallow field cycles, terracing, and irrigation. In the Valley of Oaxaca, for instance, indigenous peoples developed a system of irrigation using shallow wells and pots by about 1000 B.C. and were using irrigation canals by 500 B.C. Class distinctions were almost nonexistent at the beginning of the Pre-Classic period, but as time passed and increasingly sophisticated agricultural systems produced greater and greater food surpluses, differences in wealth, power, and social status became institutionalized. Religious life showed a similar evolution. Simple Earth Mother clay figurines measuring just 7–8 cm in height were the most common religious artifact at the beginning of the Pre-Classic period, but by about 1000 B.C. the construction of monumental ceremonial religious sites became widespread. Massive flat-topped pyramids, made of earth or stone and covered with adobe bricks, characterized these sites.

Toward the end of the Pre-Classic period, the beginnings of the establishment of advanced civilizations became clear as complex religious, social, and economic structures developed. Early civilizations arose in the mountains of southern Mexico, in the Valley of Oaxaca, and along the southern coastal plain of the Gulf of Mexico in a region now known as the Tabasco Lowlands (Figure 5.5).

The most advanced of these were the Olmec who occupied the Tabasco Lowlands. The Olmecs developed an accurate calendar, a basic system of glyph writing, and pioneered long-distance trading relationships with other societies in Mesoamerica, reaching trading partners in the highlands of Oaxaca, the Balsas Depression, and the Valley of Mexico. The Olmecs are also well known for the massive stone heads they sculpted, found at La Venta in the Tabasco Lowlands. These heads weigh several tons and are hewn from rock unavailable in the lowlands.

The civilization that blossomed at La Venta at the end of the Pre-Classic period was a harbinger of the intense cultural development that reached its zenith during the following period, the Classic. The Classic period lasted roughly 600 years, from 300 A.D. to about 900 A.D. This period was defined by theocratic societies in which an elite class of priests and a hierarchy of religious leaders ruled everyone else, either directly or through a royal class. In these societies, the state appears as a coherent political concept and an effective instrument of governance.

Refinements in agricultural technologies with important social repercussions occurred during this period, notably in the construction of irrigation systems and in terrace building. Both increased food production and led to subsequent population growth. Greater societal wealth permitted the construction of more complex and sophisticated ceremonial centers and the elaboration of monumental religious architecture, as well as a fuller development of art and the use of artistic motifs in construction. In an early example in the Valley of

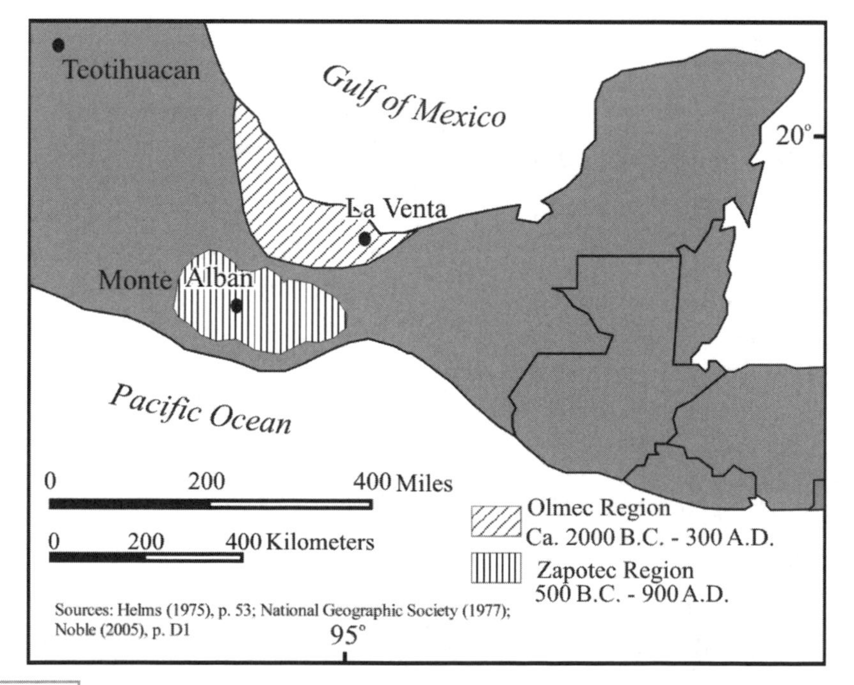

FIGURE 5.5. High culture regions, Pre-Classic period.

Oaxaca, Zapotec peoples constructed an immense ceremonial center, Monte Albán, on a strategic mountaintop overlooking the entire valley (Figure 5.6). Interregional trade, initiated in the Pre-Classic period, became commonplace and even more extensive with some goods, notably orangeware pottery and obsidian tools from the Valley of Mexico, being traded over distances of almost 1,000 km. In a few areas complex urban centers characterized by large populations, diverse economic activities, and distinct land uses evolved for the first time.

Two key centers of civilization dominated the Classic period in Mesoamerica: Teotihuacán in the Valley of Mexico and the Maya states of southern Mexico and northern Central America.

In the Valley of Mexico, the city of Teotihuacán arose as the undisputed center of a city-state that dominated the valley region and may have incorporated a population of several million people. The use of advanced irrigation techniques as well as a system of intensive gardening, *chinampa* agriculture, provided a strong agricultural base for the society's development. The city was a full-fledged urban center, probably the first to develop in Latin America, or the Americas, for that matter. At the height of its development, its population is believed to have numbered close to 100,000. The city was laid out in a rectilinear grid fashion with streets running in the cardinal directions forming regular square blocks. Clear distinctions in land-use patterns characterized the city's geography. Monumental religious structures occupied key space in the city's center, while residential, commercial, and even artisan manufacturing areas occupied distinct quarters of the city. Occupational diversity also characterized the city's population, with perhaps as much as one-third engaging in nonagricultural activities. The archaeological record shows that Teotihuacán was razed by fire near the end of the Classic period, abandoned, and never rebuilt. Several provocative theories have been advanced to explain the city's demise including a collapse of the agricultural economy and a social meltdown resulting from an overextended and unstable political system.

In the lowlands of southern Mexico and northern Central America, a distinct pattern of social and economic development evolved during the Classic period. Here, Maya peoples occupied a broad region under a much more decentralized system than that which had evolved in the Valley of Mexico (Figure 5.7). Quasi-independent theocratic states stood at the highest level of political and military authority. The Maya made considerable advances in the discovery and application of knowledge. Their agricultural systems included terracing and agroforestry farming techniques similar in many ways to those recommended today by ecologists to save tropical rain forests from deforestation. They developed two distinct celestial calendars, one for religious purposes and one for secular life. They also invented a counting system that uses base 20 and a method of writing using glyphs (Figure 5.8). Monumental architecture and a broad range of arts flourished.

The Maya constructed vast ceremonial sites for religious purposes, the most famous of which are Tikal (in today's Guatemala), Palenque (in today's Mexico), and Copán (in today's Honduras). These sites, and others like them, were essentially religious sites and not urban centers. That is to say, while some people lived in these centers permanently, their numbers were small and were limited almost exclusively to members of religious organizations and the nobility. Furthermore, while marketing or other nonreligious activities sometimes occurred, these centers were not designed to support long-distance trade, local marketing, or a diverse range of economic activities. In short, they were essentially ceremonial places for religious and secular leaders, and were not true cities.

The Maya states of the Mexican and Central American lowlands declined toward the end of the Classic period. Most of the major

FIGURE 5.6. Monte Albán is the site of long-term occupation by successive Mesoamerican cultures dating back to the Pre-Classic period. The central part of the site is shown here with the observatory in the foreground. The Valley of Oaxaca, in southern Mexico, 1977.

ceremonial centers had been abandoned by 1000 A.D. While Maya peoples still occupied these regions, the physical trappings and social structure of an advanced civilization disappeared. Excessive population pressure, depletion of the agricultural resource base, social meltdown, or a combination of all three are the reasons most scholars have suggested for the decline.

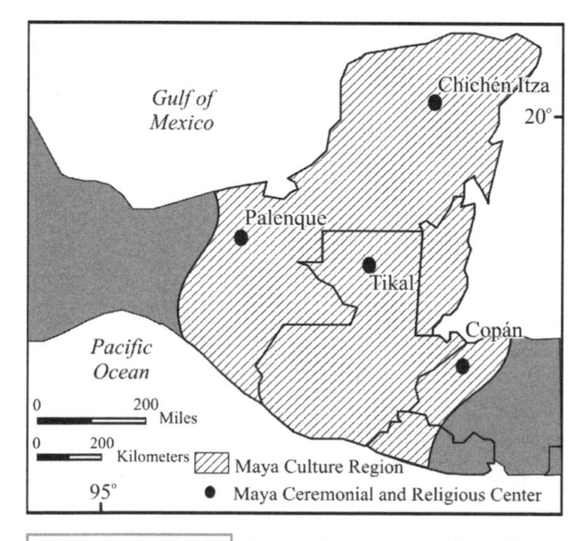

FIGURE 5.7. Maya high culture region, Classic period (300–900 A.D.) and Post-Classic period (900–1500 A.D.).

The Post-Classic period, running from about 900 A.D. to 1520 A.D., ends the prehistoric culture history of the Mesoamerican region. Few advances in technology, science, art, or architecture distinguish the Post-Classic period. Rather, this period was characterized by rising militarism, the development of a warrior class that became the ruling elite, and a consequent decline in the wealth and power of the priest class. The construction of fortifications and the use of military motifs in art became more frequent. During the Post-Classic period the Valley of Mexico continued to be the home of advanced civilizations in Mesoamerica. Here, in the vacuum created by the demise of Teotihuacán, the Toltecs consolidated an empire whose impacts were far-flung. The Toltecs colonized areas beyond the Mesa Central, extending the northern limits of the agricultural frontier into the Mesa del Norte. Their influence extended far to the south as well, where Maya temples and other religious architecture on the Yucatan Peninsula are marked with motifs of Toltec war gods, reflecting their dominance there. After several hundred years as the dominant power in the Valley of Mexico, the Toltecs were eclipsed by the Chichimecs around 1200.

Chichimecs, an aggressive and war-loving tribe from the Mesa del Norte that had been recruited initially as mercenaries by the Toltecs and others, began consolidating power as the Toltec state declined. Known subsequently as Aztecs, the Chichimecs established a vast empire, again centered in the Valley of Mexico (Figure 5.9).

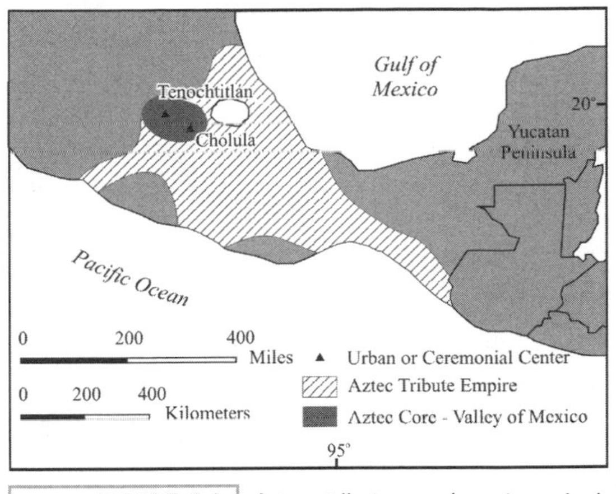

FIGURE 5.9. Aztec tribute empire at contact, circa 1500.

Its reach extended over much of the Mesa Central and the southern highlands. The Aztecs extracted tribute from conquered peoples throughout this region. The Aztecs, however, coupled their military prowess with a range of other skills to successfully consolidate their empire. They operated a broad trading network extending over southern Mexico and into Guatemala, refined the *chinampa* system of agriculture to support even greater population numbers in the Valley of Mexico, and constructed a major urban center, Tenochtítlan, with a population exceeding 100,000 in the middle of the former Lake Texcoco.

FIGURE 5.8. Maya glyphs from a stelae at the Classic period site at Copán, Honduras, 1977.

Only a few indigenous groups had the skill and tenacity to resist Aztec domination. The Tlaxcalans of eastern Mexico are a notable example. On the eve of conquest the Aztecs had amassed a long list of native enemies who were only too eager to ally themselves with Hernán Cortés and his army of Spanish soldiers.

The Andes

In the central Andean region of South America, a roughly similar pattern of cultural evolution occurred over a period of several thousand years. In the 1530s when the Spanish conquistadores confronted the Incas in the Peruvian Andes, they encountered a state and society that was geographically extensive and highly organized. As with the civilizations in Mesoamerica, the Inca civilization represented the product of a series of advanced cultures that had prospered and declined previously over a long period of time and over a wide area. This area included the warm fertile river valleys along the Pacific coastal margin, as well as the cool highland valleys in the Andes mountains.

Along Peru's Pacific coast several major civilizations developed. South of what is now Lima, the Paracas culture, best known for its intricate artistic textile designs, prospered between 1300 B.C. and 200 A.D. To the north, in the vicinity of the modern city of Trujillo, the Moche people prospered for about seven centuries before 700 A.D. Advanced irrigation technology and rich alluvial soils supported the Moche culture. Artistic endeavors also distinguished the Moche, who produced exquisite and highly decorated pottery and worked precious metals into stunning jewelry for the society's elites. Several centuries after the disappearance of the Moche culture, the Chimú arose in much the same region after 1000 A.D. The Chimú exhibited artistic skills similar to those of the Moche who preceded them. Their most impressive achievements, however, are perhaps the vast system of irrigation works they created in the coastal desert and Chan Chan, a city they constructed of sun-dried adobe bricks that housed some 50,000 inhabitants. Its ruins lie adjacent to the modern city of Trujillo. The Inca conquered the Chimú in the late 1400s, and the city was subsequently abandoned.

For nearly two millennium, a series of advanced societies preceded the Inca in the highlands of the Andes as well. The Chavin culture appeared around 1000 B.C. in the highlands of northeastern Peru and faded around 500 B.C., leaving the ruins of ceremonial sites and defensible settlements throughout the region. Later major cultural advances in the highlands occurred further south, in what is today central and southern Peru and northern Bolivia. Beginning around 600 A.D. and continuing until about 1000 A.D., at least two major cultures developed in this region. In Peru, the Huari (also spelled Wari) culture extended over much of the central Andes. Urban places, including a capital city, Huari, near modern Ayacucho, and a series of regional centers throughout the Andes, comprised a critical component in the Huari culture. These cities, located on defensible sites, were carefully planned; archeological digs reveal an orderly layout of buildings and streets. Water distribution systems, as well as infrastructure for the removal of sewage, are apparent at these sites. The Tiahuanaco culture evolved during the same period (600 A.D.–1000 A.D.) on the windswept altiplano that surrounds Lake Titicaca and includes portions of both modern Peru and Bolivia. The stark archaeological site known as Tiahuanaco is the most visible reminder of the presence of the Tiahuanco culture. Nevertheless, these people developed a religious philosophy and organizational structure upon which the Inca culture was built.

The Inca civilization and the geographic empire it spawned began to take shape between 1100 and 1200 A.D. By the time Pizarro and his soldiers confronted the emperor, Atahualpa, and his Inca armies in Cajamarca in 1532, the Incas had established a sophisticated culture and state. At its height, on the eve of the arrival of Europeans, the Inca domain stretched for nearly 2000 km from north to south, running from southern Colombia into what is today northern Argentina and Chile (Figure 5.10). The *Inca*, or god-king, ruled the empire in close cooperation with a priestly class. A series of Inca emperors and a noble elite ruled from the capital city, Cuzco, which housed as many as 100,000 residents at the time the Spanish conquered it in 1536.

The construction and organizational skills of the Inca were exceptional. An extensive system of "roads," many paved with stones, ran the length and breadth of the empire, permitting the comparatively rapid movement of people, goods, and information (Figures 5.11 and 5.12). Rope suspension bridges often joined these roads across precipitous gorges and narrow valleys. The wide geographic coverage and the high quality of this transportation system was a key element in the economic and cultural integration of the Inca Empire. Inca stonemasonry is legendary; the extant ruins of Inca architectural achievements in Cuzco and other sites in the central Andes provide clear evidence of its quality and dura-

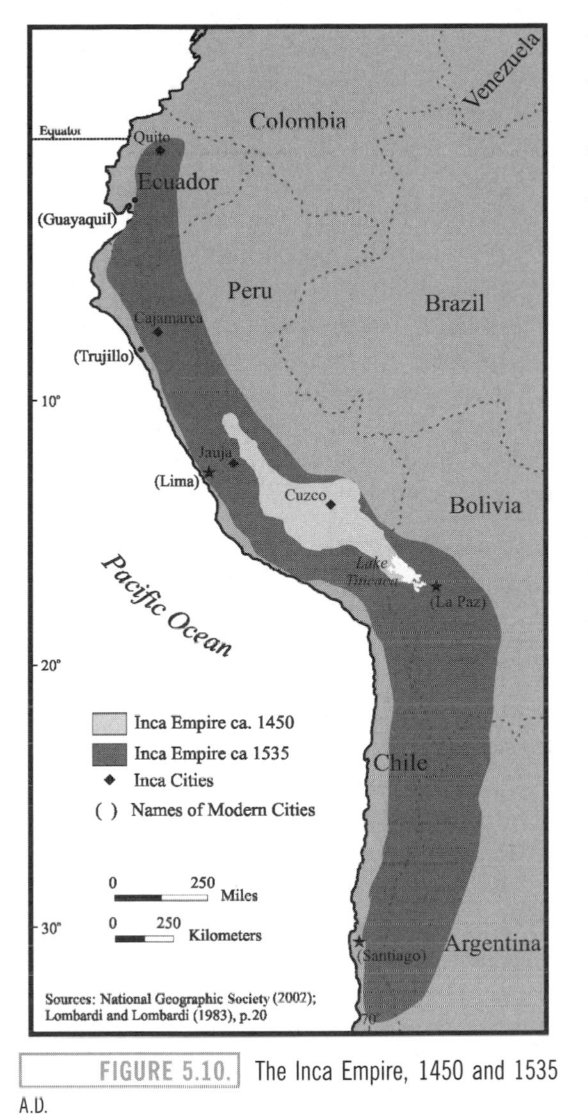

FIGURE 5.10. The Inca Empire, 1450 and 1535 A.D.

bility. The Inca developed and refined construction techniques in which they cut and then fit together individual stones with such exactitude that the use of mortar was unnecessary. Some 500 years after the fall of the Inca Empire, contemporary examples of Inca stonemasonry in the form of walls, temples, fortresses, residences, and other structures still stand in Cuzco and the surrounding region (Figure 5.13). In some areas, the engineering skills of the Inca as well as those who preceded them transformed the natural environment, often through the construction of agricultural terraces on the steep slopes of the Andes. Widespread terracing is evidenced in the canyon of the Colca River valley near Arequipa and on the slopes of the Urubamba River valley near Cuzco (Figure 5.14). In the latter case, Inca engineers also straightened and canalized the Urubamba River along some of its course to prevent flooding of prime agricultural lands.

European Conquest

At the beginning of the 16th century, both the Aztec and the Inca Empires encompassed vast territories and incorporated populations numbering in the millions. The 5–6 million inhabitants of the Aztec tributary region were settled largely on the Mesa Central of Mexico, while as many as 9–12 million people occupied the Inca realm in the Andes. Despite their vast size, large population numbers, organizational skills, and comparative technological advancement, the military prowess and formal structure of both the Aztec and Inca states crumbled in the face of only several hundred Spanish invaders. Superior weaponry was an obvious advantage for the Spanish. Protective metal armor and steel swords, lances, and knives provided the Spanish with a clear edge over indigenous warriors. It was, however, the use of firearms, including guns and cannons, that gave the Spanish an undisputed advantage. Horses and fighting dogs, both heretofore unknown in the Americas, also confounded the indigenous warriors.

Nevertheless, these military advantages cannot adequately explain the rapidity of the Spanish victories in the New World. Other circumstances and events also contributed significantly to their success. The Inca Empire was weakened and divided at the time the Spaniards arrived in the Andes because two Inca princes, Atahualpa and Huáscar, were battling with each other for the crown. The Aztecs initially thought Cortés to be the legendary god

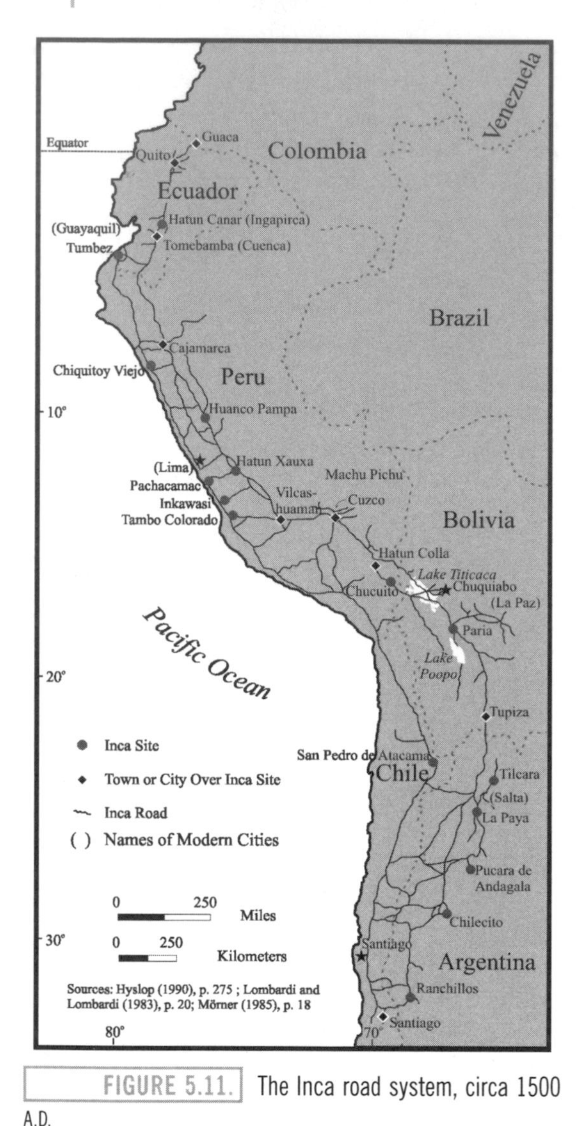

FIGURE 5.11. The Inca road system, circa 1500 A.D.

Quetzalcoatl, fulfilling an ancient religious prophecy. Uncertainty about the Spaniards' true motives led the Aztec emperor Montezuma to receive Cortés and his men in the Aztec capital, Tenochtitlán; this was an error similar to that made by the Inca leader Atahualpa, who received Pizarro in Cajamarca. The Spaniards subsequently captured, ransomed, and murdered both leaders. In central Mexico, Indian allies played a key role in the defeat of Aztec armies by the Spanish. The Tlaxcalans, longtime enemies of the Aztec, joined forces with the Spanish immediately, as did warriors from other indigenous groups whom the Aztec had conquered and forced to pay tribute. The rigid hierarchical structure of these societies also aided the conquerors because, once the native leaders had either been eliminated or co-opted by the Spanish, the vast majority of the population accepted or at least were resigned to Spanish authority. However, perhaps more than any single factor, the unintentional introduction of exotic diseases played a critical role in destabilizing, depopulating, demoralizing, and eventually subjugating the Aztec and Inca Empires to the Spanish invaders.

Disease and Conquest

Europeans, and later Africans, introduced a wide range of infectious diseases to the New World to which its native peoples had no resistance and that proved fatal for many. Influenza, typhus, measles, mumps, yellow fever, malaria, and most significantly smallpox ravaged indigenous populations, sometimes reaching them even before their first visual and physical contacts with Europeans. The impacts of the introduction of these diseases were both immediate and long term. For example, a devastating smallpox epidemic struck the Aztec defenders of Tenochtitlán as Cortés and his armies laid siege to the city in 1520 (Vignette 5.1). The city's population was reduced by nearly half, facilitating the Spaniards' eventual victory over the Aztecs. Widespread population decline began immediately after the arrival of Europeans, but continued in many areas for the next century or two. Even in the latter half of the 20th century, some remote Amazonian tribes, notably the Yanomamo in northern Brazil and southern Venezuela, are now being exposed to some exotic diseases for the first time.

Estimates of the indigenous population of Latin America at the time of European contact vary considerably. Nevertheless W. M. Denevan's (1992) calculations that over 50 million

FIGURE 5.12. An Inca road near the ruins of the famous Inca citadel Machu Picchu. Department of Cuzco, Peru, 1982.

people inhabited Latin America can be viewed as very credible (Table 5.1). He estimates that the largest population concentrations were found in Mesoamerica and in the Andes, with 17 million and 16 million, respectively. The vast tropical rain forests of South America, while not densely settled, served as home for between 8 and 9 million native peoples, far more than is often imagined. Nearly 9 million are estimated to have lived in the Caribbean Basin and Central America.

The first century of contact with the Europeans proved to be nothing short of a holocaust for native peoples. Estimates of the magnitude of population decline vary widely. However, all observers agree that the decline in population suffered by native peoples was staggering. There are cases where the population of entire villages and even regions was essentially decimated by disease, leaving vast areas of the countryside depopulated. Spanish labor demands and their physical abuse of the indigenous peoples also contributed to the depopulation. More often than not, however, a small cadre of local inhabitants did survive, providing the basis upon which population numbers grew again in subsequent centuries.

The spatial pattern of population decline over Latin America in the century after the first contacts with Europeans varied widely. In the Caribbean region, the native population, estimated to be 3 million at the beginning of the 1500s, had all but disappeared within 50 years. Despite the assertions of some residents

FIGURE 5.13. Inca stonemasonry at the fortress Sacsayhuaman. Cuzco, Peru, 1982.

FIGURE 5.14. Inca agricultural terraces in the Urubamba Valley near Pisac. Department of Cuzco, Peru, 1982.

of the Dominican Republic and Puerto Rico to be part Indian or Taíno, there are very few modern people in the Caribbean who have any credible connection with the region's original native inhabitants. A notable exception is a group of several thousand Caribs living on a reserve on the island of Dominica in the Lesser Antilles. The native population in the Caribbean declined so precipitously and so completely that native peoples never recovered as a distinct component of the population of the region.

In the face of nearly catastrophic population losses, a stubborn survival characterized the native populations of the highlands of Mesoamerica and the Andes where the majority of Latin America's indigenous population could be found. Overall, native populations probably declined about 90 percent during the first century after contact. But the pattern of population loss varied considerably in both time and space in these regions. On the Mexican plateau, for instance, the survival rate stood at 1 in 13 in the second half of the 16th century, while in the Peruvian highlands it was higher, at 1 in 4. By the mid-1600s, population decline had been checked on the Mexican plateau, and its population began a slow process of recuperation. A similar process took place in the Andes, although there it was delayed until well into the 18th century.

Population decline among the native peoples in much of the tropical lowlands fell just short of annihilation. In the late 1500s in coastal Peru only 1.5 percent of the original population is estimated to have survived, while in tropical rain forest environments the estimate is 3 percent. However, in the immensity of the Amazon Basin, some indigenous groups escaped contact with European diseases until well into the 20th century.

While the wholesale introduction of contagious diseases from the Old World decimated native populations, it is unclear if European peoples were exposed to any new diseases as a result of contact with the Americas. There is one school of thought that holds that syphilis, which is believed to have first appeared in European port cities in the middle of the 16th century, spread from early sexual contacts between European men and native women. The evidence supporting such a hypothesis is in no way irrefutable, and other observers see no clear link between the appearance of syphilis and contacts between the New and Old Worlds.

Sugarcane and Population Change

Sugarcane, a tropical grass, is an Old World crop Europeans introduced into the Americas with profound effects. The Portuguese had cultivated sugarcane on the Madeira Islands since the early 1400s and subsequently brought it to Brazil. But they did not just introduce an exotic crop plant, they introduced an entirely new economic system that left an indelible mark on the landscapes and peoples of much of Latin America. Sugarcane cultivation prospered in large part because of a voracious demand for sugarcane products—refined sugar, molasses, and rum—in western European countries. Equally critical, favorable factors of production, specifically abundant land and inexpensive labor, permitted the profitable

VIGNETTE 5.1. CONQUEST AND DISEASE CIRCA 1500

The following account, from *The Columbian Exchange: Biological and Cultural Consequences of 1492*, by Alfred Crosby, Jr., describes the smallpox epidemic that ravaged the Aztecs in 1521 as Cortés and his men laid siege to the Aztec capital, Tenochtitlán. It provides a chilling account of the impact of European disease on the native population.

"The psychological effect of epidemic disease is enormous, especially of an unknown disfiguring disease which strikes swiftly. Within a few days smallpox can transform a healthy man into a pustuled, oozing horror, whom his closest relatives can barely recognize. The impact can be sensed in the following terse, stoic account, drawn from Indian testimony, of Tenochtitlán during the epidemic.

'It was [the month of] Tepeilhuitl when it began, and it spread over the people as great destruction. Some it quite covered [with pustules] on all parts-their faces, their heads, their breasts, etc. There was a great havoc. Very many died of it. They could not walk; they only lay in their resting places and beds. They could not move; they could not stir; they could not change position, nor lie on one side; nor face down, nor on their backs. And if they stirred, much did they cry out. Great was its [smallpox] destruction. Covered, mantled with pustules, very many people died of them.'

In some places in Mexico the mortality was so great that, as Motolinía recorded, the Indians found it impossible to bury the great number of dead. 'They pulled down the houses over them in order to check the stench that rose from the dead bodies,' he wrote, 'so that their homes became their tombs.' In Tenochtitlán the dead were cast into the water, "and there was a great, foul, odor; the smell issued forth from the dead."

For those who survived, the horror was only diminished, for smallpox is a disease which marks its victims for the rest of their lives. The Spanish recalled that the Indians who survived, having scratched themselves, 'were left in such a condition that they frightened the others with the many deep pits on their faces, hand, and bodies.' "

cultivation of sugarcane in many areas of colonial Latin America. Flat tracts of fertile land along the coast, which were usually seized from indigenous groups, provided the agricultural basis for a sugarcane monoculture in northeastern Brazil, the islands of the Greater and Lesser Antilles, and coastal valleys and tropical lowlands from Mexico to Peru and into northern Argentina. An inexpensive labor supply complemented the ready availability of fertile productive land. At first, the Portuguese conscripted Indians from the coastal margin and adjacent interior areas of northeast Brazil to toil in the fields planting and harvesting cane. The arrangement proved unsatisfactory for all parties. The natives, understandably, had little inclination to cultivate sugarcane, and coercion was often necessary to keep Indian workers on these plantations. Indians often escaped, and mortality rates for those who stayed were very high.

Africa and the concept of "slavery" resolved the labor needs of the plantation economy that sugarcane cultivation spawned. A multiethnic cast of villains including Arabs, Africans, and Europeans capitalized and expanded upon the ancient African practice of slavery. Driven by demand for labor, European slave traders instituted a transoceanic slave trade between Africa and the Americas that operated for about four centuries and condemned millions of Africans to death or a life of slavery. About 10 million Africans arrived in the Americas as slaves between the mid-15th century and

TABLE 5.1. Revised Estimates of Aboriginal American Populations, ca. 1492

Region	Population
North America	3,790,000
Mexico	
Central	13,839,000
Chiapas	275,000
Yucatan-Tabasco	1,600,000
Soconusco	80,000
North	1,380,000
Central America	
Southern Guatemala	2,000,000
Honduras-Belize	850,000
El Salvador	750,000
Nicaragua	825,000
Costa Rica	400,000
Panama	800,000
Caribbean	
Hispaniola	1,000,000
Other Islands	2,000,000
Andes	
Central	11,696,000
Colombia	3,000,000
Venezuela	1,000,000
Lowland South America	
Amazonia	5,664,000
Argentina	900,000
Chile	1,000,000
Remainder	1,055,000
Total	53,904,000

Source: Denevan (Ed.). (1992).

the mid-19th century. By 1600, some 75,000 had been brought to labor in the Spanish colonies, part of a steady flow that continued until the abolition of slavery in Spain's last American colonies. Indeed, during the 19th century, over half a million African slaves reached the Spanish colonies. Untold numbers died en route between the interior of Africa and the slave markets held in major port cities like Salvador (Brazil) or Havana (Cuba). This four-century-long period ranks as one of history's most extreme and brutal forced migrations.

The colonies of Spain and Portugal received about half of these slaves, roughly 1.5 and 3.5 million people, respectively. The Caribbean colonies of the French, Dutch, Danish, and British, as well as British North America, shared this demand for inexpensive labor; between them, they received an estimated 5 million slaves. The magnitude and geographical distribution of the immigration of African slaves varied in response to the patterns of agricultural and economic development in the Americas (Table 5.2, Figure 5.15).

The Spanish utilized slaves for a wide range of tasks, so small numbers of African slaves could be found throughout the Spanish realm. However, the largest numbers and the most significant impacts on the population and cultural geography occurred on the Spanish colonies in the Greater Antilles: Cuba, the Dominican Republic (Hispaniola), Puerto Rico, and Jamaica (which was a Spanish colony until the early 1700s). Africans and their descendants also figured prominently in the population geography of the circum-Caribbean region. This was especially the case along the coastal lowlands of Colombia during the colonial period. During the late 19th and early 20th centuries, the free descendants of African slaves migrated in appreciable numbers to the coastal lowlands of Central America where they worked as laborers on banana plantations, narrow-gauge railroad lines, and the Panama Canal. At the end of the 20th century, the descendants of African slaves comprised the dominant population groups on the coast of northeast Brazil; in Cuba and the Dominican Republic; and in the circum-Caribbean lowlands in Colombia and Central America. Overall, at present, descendants of African slaves comprise nearly one-quarter of Latin America's total population of 400 million.

Miscegenation and Population Change

Miscegenation—sexual intercourse, cohabitation, or marriage between two individuals of different races—produced one of the most profound changes occasioned by the Columbian Exchange. Miscegenation became an in-

TABLE 5.2. Enslaved Africans Sent to the Americas, 1451–1870

Importing Region	1451–1600	1601–1700	1701–1810	1811–1870	Total per Region
Old World	149,900	25,100	—	—	175,000
Europe	48,800	1,200	—	—	50,000
São Tomé	76,100	23,900	—	—	100,000
Atlantic Islands	25,000	—	—	—	25,000
Spanish America	75,000	292,500	578,600	606,000	1,552,100
Brazil	50,000	560,000	1,891,400	1,145,400	3,646,800
British North America	—	—	348,000	51,000	399,000
British Carribean	—	263,700	1,401,300	—	1,665,000
Jamaica	—	85,100	662,400	—	747,500
Barbados	—	134,500	252,500	—	387,000
Leeward Is.	—	44,100	301,900	—	346,000
St. Vincent, St. Lucia, Tobago, and Dominica	—	—	70,100	—	70,100
Trinidad	—	—	22,400	—	22,400
Grenada	—	—	67,000	—	67,000
Other BWI	—	—	25,000	—	25,000
French Carribean	—	155,800	1,348,400	96,000	1,600,200
St. Domingue	—	74,600	789,700	—	864,300
Martinique	—	66,500	258,300	41,000	365,800
Guadeloupe	—	12,700	237,100	41,000	290,800
Louisiana	—	—	28,300	—	28,300
French Guiana	—	2,000	35,000	14,000	51,000
Dutch Carribean	—	40,000	460,000	—	500,000
Danish Carribean	—	4,000	24,000	—	28,000
Total	274,900	1,341,100	6,051,700	1,898,400	9,566,100

Source: Curtin (1969).

tegral part of the social and sexual fabric of colonial society in both Spanish and Portuguese colonies, although its specific characteristics varied over both time and space. Few women accompanied either the Spanish or the Portuguese settlers during the early decades of the colonial period. Even in later years, those European women who did come to the New World were generally destined to marry within the colony's social and economic elite. As a consequence, European men tended to have short-term sexual liaisons and/or marital unions with native women. The progeny of these unions, *mestizos*, or those of mixed European and indigenous American descent, were few during the early decades of the colonial period. Later in the period, the mestizo population grew to a considerable size, although the pattern was not uniform in all areas of Latin America. This process has been so complete that by the latter half of the 20th century the mestizos constituted a majority of the population in some nations. For example, this is the case in Mexico, El Salvador, Peru, and Chile.

Miscegenation also occurred between native peoples and African slaves. However, in terms of the region's overall population geography, this process was of far less significance than that of the development of the mestizo population. During the early centuries of the colonial period, miscegenation between people of African descent and native peoples was limited to areas beyond the effective control of the colonial powers. These were frontier areas where free blacks as well as runaway slaves comingled both genetically and culturally with indigenous peoples.

The terminology used to describe the product of these unions varied. In Spanish America the terms *sambo* or *zambo* enjoyed

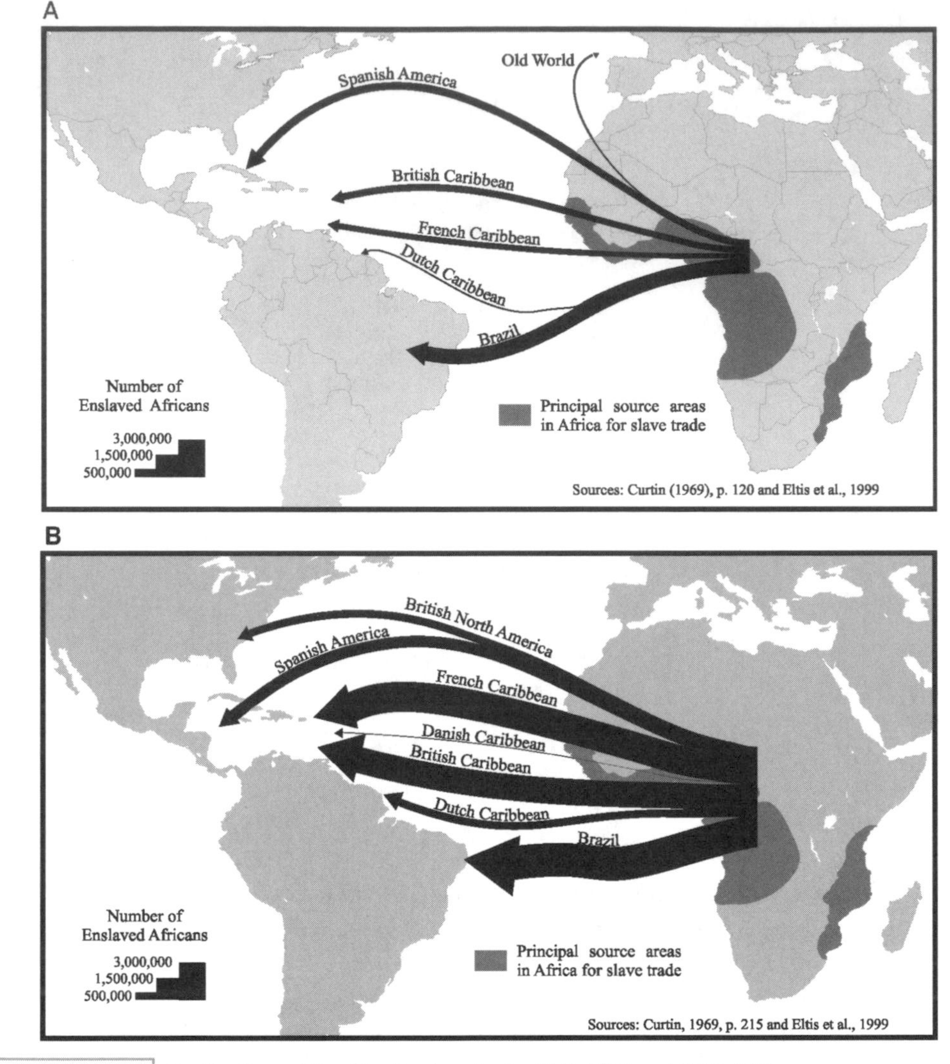

FIGURE 5.15. Destinations of the Atlantic slave trade, 1601–1700 and 1701–1810.

wide usage, while in the non-Hispanic Caribbean the word *marron* described such mixed-race peoples. In Brazil, where the mixing of native populations with those of African origin has been the most significant in Latin America, a range of terms have been used—the most common and the least precise is *pardo*, simply meaning a brown person.

The "Black Caribs" who occupy isolated sections of the eastern lowlands of Central America in Belize, Honduras, and Nicaragua represent one group of mixed African and native descent. Although the precise origins of the Black Carib settlements on Central America's Caribbean coast are murky, the British exiled nearly 5,000 Caribs from the Lesser Antilles island of Saint Vincent in 1796 to Roatán, one of the Bay Islands that lie off the coast of Honduras. The Caribs established themselves in a number of areas along the sparsely settled coast, mixing with groups of escaped slaves, and giving rise to the term Black Carib. Today descendants of these peoples are also known as the Garifuna along Honduras's Caribbean coast where they are numerous and form a distinct cultural group. Due to their geographical isolation, racial and cultural distinctiveness, and military and political alliances with the British during the 19th century, these groups remained beyond the effective control of the Central American governments within whose national territory they resided. Central America's Hispanic governments finally began to seriously integrate these regions and peoples into their respective national spheres in the middle of the 20th century.

In the Caribbean, native peoples had been almost completely wiped out by the time African slaves arrived in the early 1600s. While unions of natives and blacks did occur in the Caribbean region, their number was so small that their progeny were rapidly absorbed into the vast slave population.

Miscegenation between natives and people of African origin occurred over the broadest area and in the most significant numbers in South America. In Brazil, contact between native peoples and slaves of African origin occurred over several centuries and under a wide range of conditions. During much of the colonial period contact between blacks and Indians occurred largely in remote frontier areas beyond the control of the colonial authorities. Here, runaway slaves, sometimes operating in association or collusion with native peoples, established independent subsistence communities known as *quilombos*. Many of these, usually small and of short-term duration, grew up in the Brazilian interior, and collectively represented a real threat to the viability of the plantation system. The most famous of these independent communities was called Palmares, which prospered in the interior of Alagoas during the last half of the 17th century. Self-sufficient, self-governing, and with a population of about 20,000, Palmares successfully withstood repeated attempts by colonial authorities to overwhelm it before it finally fell to *bandierante* mercenaries at the end of the 17th century.

The New World also provided the context for another major change in population geography that was to begin in the early colonial period and continue into the present. This was the widespread miscegenation between people of European origin and those of African descent. Many of the same basic facts of demography and immigration that led to miscegenation between Europeans and native peoples also brought Europeans and Africans together. However, the institution of slavery provided the basic framework for sexual relations between most whites and blacks until the second half of the 19th century when slavery was finally abolished in Brazil and in Spain's remaining Caribbean colonies, Cuba and Puerto Rico. During the almost four centuries that slavery existed in Latin America society, the framework consisted of white men engaging in sexual relations with female black slaves under terms and conditions they dictated.

The mixed-race offspring of these unions (liaisons), *mulattos*, grew significantly in number and added to the slave population during the colonial period. For instance, in 1803 the coastal city of Salvador de Bahia, in the heart of Brazil's sugarcane-growing region, had a population of about 100,000. Blacks accounted for 40 percent of the total, while mulattos and whites each represented about 30 percent. In 1818, an estimate of Brazil's population and racial composition pegged the number of mulattos at 500,000, whites at 1 million, and blacks at about 2 million. In geographical terms mulattos figured as a visible and numerically significant component of the colonial population in those areas where slavery was practiced. These areas included the coast of northeastern Brazil, the islands of the Caribbean, the coastal lowlands of Central America, Colombia, and Venezuela, and a number of urban centers as far-flung as Lima, Montevideo, and Buenos Aires. This pattern, with some modest variations, continues to characterize the pattern of black and mulatto population distribution today in Latin America.

Early European Immigration

European immigration, principally from Spain and Portugal, occurred steadily throughout Latin America's 300 years of colonial rule. Spain adopted the most restrictive immigration policies of the two colonial powers, limiting immigration to its New World colonies to Spaniards only. Indeed, the crown's policy even restricted immigration from some regions of Spain, notably the northeast: Cataluña, Valencia, and Aragon. Such restrictions could never be fully enforced over a territory so vast and scantly populated, but overall Spain's tight immigration policy effectively limited the immigration of non-Spaniards to inconsequential numbers. Careful historical research using the archival records of the Council of the Indies (Archivo de las Indias) provides an accurate picture of Spanish immigration to the Americas during the first two centuries of the colonial period. It is estimated that between 1492 and 1600, some 300,000 Spaniards left the Iberian Peninsula to settle in the New World. The number of immigrants in the 17th century increased by 50 percent, reaching a total of about 450,000. Estimates of Spanish migration to the New World are less reliable for the 18th century, but it is generally accepted that as many as 500,000 Spaniards reached the shores of the Americas, and most remained permanently. The magnitude of this migration is impressive considering that the population of Spain at the beginning of the 17th century stood at about 8 million. While Spanish women did emigrate from Spain to its New World colonies, their numbers were small and mostly limited to upper- and middle-class women who came to marry successful Spaniards. Overall, the vast majority of immigrants were young males.

Less is known about the precise details of European immigration and population growth in Portuguese America (Brazil) during the colonial period. Absolutely no reliable data are available on the number of Indians in Brazil during the colonial period. Population estimates pertain only to that portion of the country nominally administered by the Portuguese colonial authorities. Few Portuguese immigrated to Brazil during the early colonial period. In part because of this reality, Portugal opted for a more open immigration policy than did Spain. The Portuguese required only that immigrants be Catholic. This, however, failed to set off any massive migration of Europeans into the colony. At the beginning of the 1600s, the colony's population was just 100,000. Blacks and mulattos represented 70 percent of this total and whites the balance. The population grew slowly during the next 100 years to just 300,000, with the relative proportion of blacks/mulattos to whites remaining essentially unchanged. The colony's population exploded during the 1700s and reached nearly 3 million by the end of the century. There were several reasons for this growth. First, Portuguese immigration increased dramatically due in large part to the discovery of gold and diamonds in Minas Gerais, and then of gold in Goiás. Second, there were new and lucrative opportunities for the cultivation of cotton and tobacco. Finally, these developments resulted in an increase in the demand for slave labor, and imports of African slaves rose.

Population on the Eve of Independence

At the beginning of the 19th century, Latin America was home to a vigorous and diverse population that numbered around 21 million. The colonies of Spanish America accounted for about 17 million. Indians were most numerous, at 7.5 million, while the number of mestizos reached 5.3 million. Whites, including both *Creoles*, those born in the Americas, and *Peninsulares*, those born in Spain, numbered about 3.2 million. Blacks and mulattos were least numerous, with only about 1 million. In Portuguese America, on the other hand, blacks and mulattos numbered about 2.5 million and represented the largest population group in Brazil. Whites followed with about 1 million, while Indians and mestizos together numbered some 500,000 of the colony's residents.

The region's population showed strong spatial variation. Overall, about three-quarters of the population lived in the countryside in dispersed rural dwellings or clustered into small hamlets and agricultural villages. The remaining one-quarter of the population lived in urban centers, varying in size from small towns or isolated provincial capitals numbering just a few hundred souls to large cities with populations well in excess of 100,000. The latter was the case with Spain's viceregal capitals of Lima and Mexico City, and the colonial capital of Portugal's Brazilian colony, Salvador de Bahia. The distribution of population varied considerably from region to region. In Brazil, coastal settlement accounted for a high percentage of the total population, but in New Spain (Mexico and Central America) coastal populations were sparse except for the major port cities of Veracruz and Acapulco. In Spain's Andean colonies, population concentrated in the highlands with only limited settlement on the Pacific coast.

In Brazil, the colony's population was concentrated in a long narrow band running from Santos and São Paulo in the south to near Fortaleza in the north. The most populous and economically important urban places were located on this coastal margin. Rio de Janeiro, numbering some 60,000 inhabitants, and Salvador de Bahia, with around 100,000, were the major coastal centers. The interior of the colony was only sparsely settled, although gold and diamond strikes in Minas Gerais early in the 18th century had encouraged immigration and given rise to permanent settlement based on livestock raising and crop agriculture. Subsequently a network of urban centers arose in the central highlands. The largest, Ouro Prêto, reached a population of close to 100,000 at the height of the mining boom in the middle of the 18th century, but by the end of the century its population had been reduced by as much as half. Settlement in the rest of the highlands was thin, although the raising of livestock in the Sertão in the northern highlands and the persistent exploration of the southern interior by Paulista *bandierantes* (adventurers who hunted for slaves, gold, and other valuable commodities) resulted in the beginnings of permanent settlement in these areas. The vast Amazon region was essentially devoid of any permanent European colonial settlement. This immense region, like much of the Brazilian interior, was still the almost exclusive domain of indigenous peoples.

Population distribution in Spain's colonies differed in important ways from that of Brazil. In both New Spain (Mexico) and Peru, dense rural population characterized the highlands. In New Spain, most of those highland regions had always been centers of indigenous population and continued to be so throughout the colonial period. Dense settlement characterized the Mesa Central and southern highlands of Mexico, and the highlands of Guatemala. An extensive network of mining towns dominated the northern plateau of Mexico, but overall population numbers there were small. The principal basins in the Central American highlands southward from Guatemala to Costa Rica had been settled. Population was not continuously distributed and settlements were often isolated and comparatively inaccessible. Such was the case with the settlement pattern of the central highlands of Costa Rica around Cartago and San José. Sparse settlement typified the Pacific coast of Central America at this time, and the Caribbean lowlands to the east remained essentially beyond the reach of Europeans or mestizos and firmly in the hands of indigenous peoples. A notable exception was the narrowest portion of the Isthmus of Panama, which had been occupied by Europeans since the early days of the colonial period and formed a critical link in the movement of passengers and freight between the Caribbean and the Pacific.

In Spanish South America, the population at the beginning of the 19th century was concentrated in the Andean highlands. Rural settlement was dense over most of the Andes, running from northern Venezuela into northern Argentina. Broad highland valleys and plateaus supported the largest populations, the highest population densities, and frequently the important urban centers. The Sabana de Bogotá in Colombia; the upland plateaus near Quito, Ecuador; the Mantaro and Urubamba Valleys in Peru; the Altiplano that surrounds Lake Titicaca; and the Valles de Cochabamba in Bolivia were then and have remained the most densely settled Andean regions.

The Caribbean coastal lowlands and the rain forests of the Amazon Basin were sparsely settled. A few major ports along the Caribbean, like Cartegena, and isolated mission communities in the Amazon Basin were exceptions to this pattern. Population densities along South America's Pacific coast varied strikingly. The densely forested tropical lowlands of the northwest coast remained the domain of indigenous peoples, although a few colonial outposts like Buenaventura (Colombia) had grown into cities. Further south, the Guayas lowlands in Ecuador supported a modest rural population and small urban centers, most notably Guayaquil. The arid coasts of Peru, Bolivia, and Chile effectively limited settlement to coastal valleys where perennial rivers and streams supplied sufficient water for crop agriculture. Usually, small oasis-like urban centers grew up at the mouths of these rivers, serving as agricultural marketing centers, ports for coastal traffic, and service centers for interior portions of the Andes. For example, Lima, the viceregal capital, was located in the valley of the Rimac River. The Central Valley of Chile, nestled between the Andes on the east and coastal mountains on the west, was the southernmost major concentration of population along the Pacific coast. Fertile soils and a benign Mediterranean climate supported a prosperous agrarian economy and dense rural population, while Santiago, the administrative center of the region, and the port city of Valparaiso grew into bustling urban centers.

To the east of the Andes, settlement in Spanish America was scant. With the exception of isolated religious missions and a few trading outposts along the major rivers, the Amazon Basin lacked any permanent settlements other than those of indigenous peoples.

To the south, in more temperate environments, three key areas of dense settlement can be identified. The first area ran along the eastern front of the Andes from the city of Mendoza northward to Salta and represented the heart of colonial Argentina. Rural settlement here focused on small Andean valleys, whose perennial or ephemeral streams provided critical water for irrigation and agriculture. Urban centers like Mendoza, Tucúman, and Jujuy grew up on open sites on the Andean piedmont. The well-watered and rolling terrain of eastern Paraguay was a second area of settlement that supported a vigorous rural economy and population. Asunción evolved as the region's principal urban center. Buenos Aires and the Pampa in its immediate vicinity represented the third concentration of population east of the Andes. Buenos Aires had been a nearly forgotten outpost for most of the colonial period. It did not grow significantly until the last decades of Spanish colonial rule, but by the beginning of the 19th century Buenos Aires had become a major port and commercial center, as well as the administrative center for the recently established Viceroyalty of Rio de La Plata (1777).

Summary

The indigenous population of Latin America before the arrival of the Europeans can be categorized in three ways: (1) tribal groups of small clans who had rudimentary development and subsistence-level, slash-and-burn agriculture; (2) chiefdoms of related tribes who shared a common language and culture and who affiliated both economically and politically; and (3) advanced civilizations, such as the Incas and the Aztecs, that developed over time into economically, politically, and culturally complex groups.

The cultural development of Mesoamerica demonstrated the growth of complex civilizations. During the Pre-Classic period village life based on agriculture introduced class structure, religion, writing, and trade. The theocratic societies of the Classic period that followed introduced complex urban centers and occupational diversity. The Post-Classic period saw the rise of militarism, such as that of the Aztecs. The Andes experienced similar development, with the Incas constructing a sophisticated system of roads to link together the far-flung parts of their empire.

The military strength of these indigenous civilizations could not survive the weapons and disease brought by the Spanish. The first century of exposure to European people decimated the indigenous population. The Europeans also introduced sugarcane and the slave system to work the large sugar plantations they created. The addition of Europeans and Africans altered the indigenous people even more as miscegenation resulted in the blending of races. Their descendents, mestizos and mulattos, now comprise the majority of the population in most of the region's countries. In the early 1800s, at the end of the colonial period, Brazil's population was concentrated in a narrow band along the Atlantic Coast and centered on a handful of major urban centers. Although a few coastal centers were important in Spain's colonies, most of the population lived in the highlands where rural population numbers were high.

Further Reading

Bauer, B. S. (1992). *The development of the Inca state*. Austin: University of Texas Press.

Berdan, F. F. (1982). *The Aztecs of Central Mexico: An imperial society*. New York: Holt, Rinehart, & Winston.

Borah, W., and Cook, S. F. (1963). *The aboriginal population of Central Mexico on the eve of Spanish Conquest*. Ibero-Americana, No. 45. Berkeley and Los Angeles: University of California Press.

Burkhholder, M. A. (1998). *Colonial Latin America* (3rd ed.). New York: Oxford University Press.

Cook, S. F., and Borah, W. (1971–1979). *Essays in population history: Mexico and the Caribbean*. (3 vols.). Berkeley and Los Angeles: University of California Press.

Crosby, A., Jr. (1972). *The Columbian exchange: Biological and cultural consequences of 1492*. Westport, CT: Greenwood Press.

Davis, D. (Ed.). (1995). *Slavery and beyond: The African impact on Latin America and the Caribbean*. Wilmington, DE: Scholarly Resources.

Denevan, W. M. (Ed.). (1992). *The native population of the Americas in 1492* (2nd ed.). Madison: University of Wisconsin Press.

Eltis, D., et al., (1999). The trans-Atlantic slave trade: A database on CD-Rom. Cambridge, UK: Cambridge University Press.

Hyslop, J. (1990). *Inca settlement planning*. Austin: University of Texas Press.

Klein, H. S. (1988). *African slavery in Latin America and the Caribbean*. New York: Oxford University Press.

Lockhart, J. (1992). *The Nahuas after the Conquest: A social and cultural history of the Indians of Central Mexico, sixteenth through eighteenth centuries*. Stanford, CA: Stanford University Press.

Lovell, W. G. (1992). "Heavy shadows and black night": Disease and depopulation in colonial Latin America. *Annals of the Association of American Geographers, 82*, 426–443.

Lovell, W. G., and Lutz, C. H. (1992). The historical demography of colonial Central America. *Yearbook,*

Conference of Latin Americanist Geographers, 17–18, pp. 127–138.

Mellaffe, R. (1975). *Negro slavery in Latin America*. Berkeley and Los Angeles: University of California Press.

Mörner, M. (1967). *Race mixture in the history of Latin America*. Boston: Little, Brown.

Newson, A. (1996). The population of the Amazon Basin in 1492: A view from the Ecuadorian headwaters. *Transactions of the Institute of British Geographers, 21*, 5–26.

Robinson, D. J. (1990). *Migration in colonial Spanish America*. New York: Cambridge University Press.

Smith, C. T. (1970). Depopulation of the Central Andes in the sixteenth century. *Current Anthropology, 11*, 453–464.

Whitmore, T. M. (1991). A simulation of the sixteenth-century population collapse in the basin of Mexico. *Annals of the Association of American Geographers, 81*, 464–487.

Whitmore, T. M. (1992). *Disease and death in early colonial Mexico: Simulating Amerindian depopulation*. Boulder, CO: Westview Press.

6 The Mesa Central of Mexico

The Mesa Central is the cultural, historical, and economic core of Mexico. It comprises the southern portion of the Mexican Plateau, the dominant landscape feature of the southern part of the North American continent. This plateau extends over an immense geographical area running roughly northwest to southeast some 3,000 km and at its widest point extending nearly 2,000 km from east to west (Figure 6.1). Structurally, the Mexican Plateau is divided into two principal subregions: a vast, semiarid, thinly populated, northern plateau, the Mesa del Norte, and the much smaller, more humid, densely populated, and urbanized central plateau, the Mesa Central.

Here in the high intermontane basins of the Mesa Central some of the New World's most advanced civilizations, the Toltecs and the Aztecs, arose. After the Spanish conquered the Aztecs, they selected the site of the Aztec capital, Tenochtitlán, as their capital in Mexico. Mexico City became the dominant urban center of Spain's vast colonial empire in North America, Central America, and the Caribbean. Its political, administrative, cultural, and economic dominance fueled its growth, that of its metropolitan region, and indeed even the growth of the broader region of the Mesa Central. Blessed with a benign climate, adequate rainfall and water resources for irrigation, and excellent soils in many areas, historically the Mesa Central has been one of Mexico's principal agricultural regions, a position it continues to hold in the present.

Nearly 60 million Mexicans, almost two-thirds of the nation's population, live on the Mesa Central. Approximately 70 percent of the population is urban; close to one-third of the region's total population, about 20 million, reside in Mexico City and its metropolitan region. Other major metropolitan areas, like Guadalajara and Puebla, boast populations of between 2 and 4 million. Simply put, the Mesa Central is Mexico's heartland and it dominates the rest of the nation in industrial production, transportation infrastructure, cultural and educational institutions, and government and business.

Geography and Environment

The Mexican Plateau runs from the county's northern border region southward to just south of the Valley of Mexico, the site of Mexico City. It is bounded on the east by a massive mountain range of sedimentary origin, the Sierra Madre Oriental, whose highest peaks rise close to 3,800 m. The western margins of the plateau are marked by another mountain range, the Sierra Madre Occidental, whose origin is largely volcanic and whose highest peaks exceed 3,000 m. It slopes upward gently from the north, where its elevation stands at

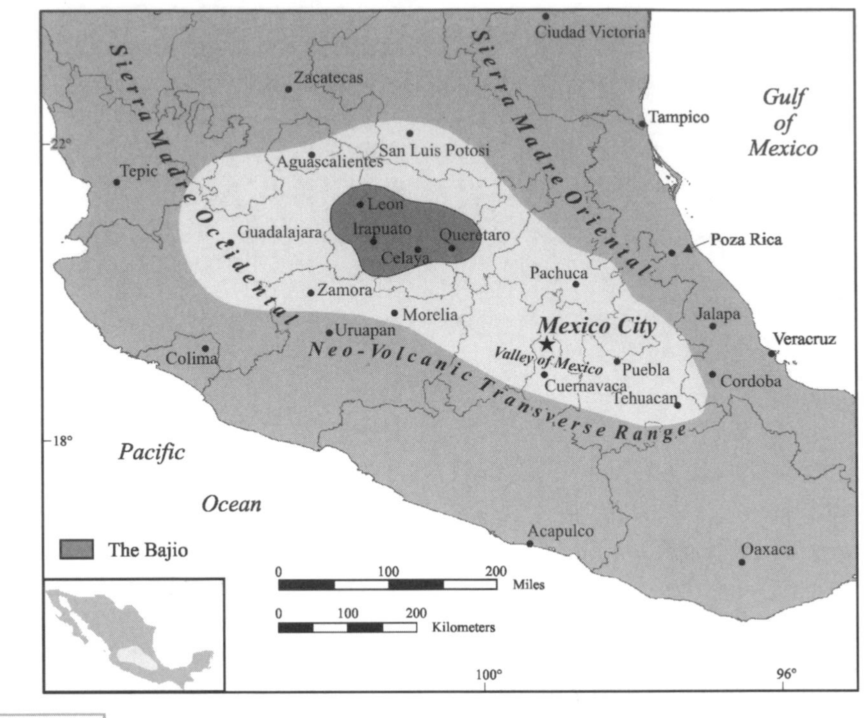

FIGURE 6.1. Mesa Central of Mexico.

about 1,000 m, to its southern border along the Neo-Volcanic Transverse Ranges, where it reaches an elevation of 2,000 m. Running just south of the Valley of Mexico, the Neo-Volcanic Transverse Ranges have some of Mexico's highest and most well-known mountains; running from east to west, these conic-volcanic peaks include Orizaba (5,610 m), Popocatepetl (5,400 m), and Iztaccihuatl (5,286 m), all of which are permanently capped with ice and snow, as well as Toluca (4,558 m), Parícutin (2,250 m), and Colima (4,240 m). This is a region of active volcanism as well as considerable seismic activity such as earthquakes. Volcanic eruptions have occurred in historic time here, the most recent in 1943, when lava that began oozing out of a farmer's cornfield eventually grew into the volcano Parícutin. Earthquakes have been more common. One of the most devastating occurred in September 1985 when Mexico City was racked by an earthquake measuring 8.1 on the Richter scale. Over 100,000 housing units were destroyed, between 5,000 and 10,000 people perished, and nearly 50,000 were injured.

The entire Mexican Plateau is an extension of an immense and complex continental orogeny created during the Mesozoic era some 60 million years ago by the collision of the Pacific and the North American Plates. This massive system of mountains ranges and associated plateaus runs from the highlands of southern Mexico, through central and northern Mexico, and into the United States and Canada. In the United States, the Rocky Mountains and the Great Basin form the central portion of this geological complex. The plateau is characteristically divided into two principal regions: the Mesa del Norte and the Mesa Central. The Mesa del Norte comprises the northern two-thirds of the plateau and runs south to the Zacatecas Mountains, while the Mesa Central occupies the southern one-third of the plateau.

The Mesa Central lies southward of the 22nd parallel. Since this region is more humid than the Mesa del Norte, rain-fed agriculture is possible here, although irrigation is widely used. The Mesa Central is comprised of a series of board-flat basins. Many of these are of lacustrine origin and as such represent the beds of ancient lakes formed during the Pleistocene, a geological period dating back about 2 million years when conditions were generally more humid. Many of these basins are endowed with rich soils derived from the ancient lake sediments as well as the ash deposits from volcanic eruptions. The largest of these basins, the Bajio, occupies much of the central and western portion of the Mesa Central and is widely acknowledged as one of the principal agricultural regions of the country—wheat, corn, beans, and temperate fruits are cultivated here. Running from the east to the west, the major basins include Puebla, Mexico, Toluca, Morelia, the Bajio, San Luis, Aguascalientes, and Guadalajara. Small mountain ranges and a few extant lakes—Lake Chapala, Lake Pátzcuaro, and Lake Cuitezo—interrupt these basins. Many more lakes occupied the Mesa Central in historic time, but most have been drained for agricultural or urban uses. For instance, a vast Pleistocene lake system comprised of five lakes (Zumpango, Xaltocan, Texcoco, Xochimilco, and Chalco) occupied the Valley of Mexico during early colonial times. These lakes have been drained to facilitate the growth of Mexico City. Only the southern portions of Lake Xochimilco and Lake Chalco remain. It is here that modern variations on the indigenous pre-Hispanic agricultural system known as *chinampa* agriculture are still practiced. A fuller discussion of the system of *chinampa* agriculture is provided in Chapter 14, on early cities and urban development.

The climate of the Mesa Central is dictated largely by its latitudinal position. Lying between about 19°N and 22°N, the region sits on the northern limits of the Tropics. Climatic conditions however, especially temperature, are modified significantly by the plateau's elevation, which is well in excess of 2,000 m over most of its extent. Temperatures at Mexico City are mild, vary only modestly throughout the year, and are characteristic of the entire southern plateau (Figure 6.2). The average annual temperature is about 15°C, with seasonal means of about 3°C above and below that figure, respectively, for summer and winter.

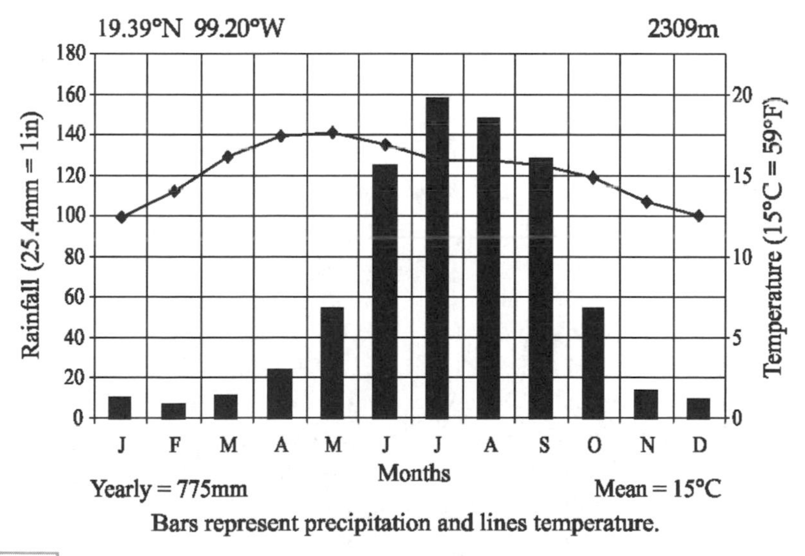

FIGURE 6.2. Climograph of Tacubaya, Federal District, Mexico City.

While seasonal temperature variation is minimal, diurnal temperature variations, characteristic of the high elevation locations in the Tropics, can be considerable. In the summer months temperatures usually reach into the mid- to upper 20°sC in the daytime and fall into the low teens at night. In winter highs reach 15°–20°C, but temperatures plummet at night into the single digits. Frost occurs occasionally on the Mesa Central during the winter months.

Seasonal variation in precipitation is marked on the Mesa Central. While total annual precipitation averages nearly 800 mm per year, most is concentrated in the summer months between June and September. During that period monthly precipitation totals range from 125 mm to 150 mm. Much of this precipitation is the result of convectional heating and cooling, which frequently gives rise to thunderstorms during summer afternoons and early evenings. Rainfall is sparse during the remainder of the year, but the driest time is the period between November and March when precipitation rarely exceeds 10 mm monthly.

Vegetation patterns across the Mexican Plateau are largely controlled by temperature, precipitation, and human land-use practices. The natural vegetation of the Mesa Central has been dramatically modified over the last several thousand years, beginning when early agricultural peoples introduced widespread crop cultivation in the fertile highland basins of the southern plateau. Prior to the establishment of dense sedentary human populations in the Mesa Central, grasses, shrubs, cacti, and other drought-resistant xeric plants dominated the basins. In the cooler and moister uplands and mountains surrounding these basins, broadleaf deciduous trees (oaks) predominated at lower elevations and conifers (pines) at higher elevations. Crop agriculture and ranching have eliminated much of the natural vegetation on the plateau, while logging for construction, firewood, and charcoal making have seriously reduced the oak and pine forests of the uplands and mountains.

Historical Geography and Economic Development

The Mesa Central and its environs have a rich culture history dating back some 5,000–10,000 years. This area is one of only a handful of centers of domestication in the world. In these regions prehistoric peoples domesticated both plants and animals and established early agricultural and herding societies. It was here that early indigenous peoples first began collecting the seeds of the wild progenitor of corn, and eventually domesticated it into the productive and adaptive plant known today as corn (*Zea maize*). Other crops followed the domestication of corn, most notably beans, squash, and chile peppers. These crops became the mainstays of the agricultural economies and cuisines of native peoples on the Mesa Central and eventually the entire Mesoamerican region.

Highly productive and nutritionally rich, these crops and the increasing intensification of agricultural practices through irrigation and terracing provided an abundant resource base for the indigenous peoples of the Mesa Central. This agricultural abundance allowed for settled village life, dramatic population growth, widespread food surpluses, occupational specialization, and eventually the rise of distinct social classes. These changes occurred during a long period of cultural development running from about 2000 B.C. to 300 A.D., commonly called the Formative or Pre-Classic period by archaeologists, anthropologists, and culture historians.

An era of great cultural and economic florescence followed, the Classic period, which ran from about 300 to 900 A.D. and which gave rise to sophisticated civilizations and urban centers on the Mesa Central and more generally throughout Mesoamerica. On the eastern margins of the Mesa Central, the earliest major civilization in this region, called Teotihuacán, evolved. Sophisticated and highly organized, these peoples built a major religious and urban center, the city of Teotihuacán, that at its height in about 600 A.D. occupied some 20 sq. km. It is believed by many to be the first large-scale planned urban center in the Americas and thought to have been home to at least 100,000 inhabitants. Lying in a small side valley of the Valley of Mexico approximately 50 km from Mexico City, the ruins of this ancient city, including the widely recognized Pyramids of the Sun and the Moon, are a major tourist attraction today. After the demise of Teotihuacán, another major civilization, the Toltecs, filled the void. The Toltecs too built a major ceremonial and urban center in the same general vicinity. The ruins of this center, Tula, lie some 65 km north of Mexico City. From Tula the Toltecs held sway over a large part of the Mesa Central and their influence extended well into southern Mexico and the Yucatan Peninsula. By 1200 A.D., the Toltec civilization had declined and disappeared as a cultural and political entity.

In the decades before the arrival of Europeans in the New World and the subsequent conquest of Mexico by the Spanish, three major indigenous civilizations flourished on the Mesa Central. The most powerful of these were the Aztecs. The ancestors of these peoples had migrated into the Mesa Central from the north around 1200 A.D. Aggressive and effective warriors, the Aztecs successfully dominated the Valley of Mexico, displacing earlier inhabitants and establishing their capital,

Tenochtitlán, on a small island in Lake Texcoco. The Aztecs practiced a highly productive agricultural system, called *chinampa* agriculture (for more on *chinampas*, see Vignette 7.1). It capitalized on the fertile sediments of the shallow lake bed, the ready availability of water for irrigation, and the moderating influences of the lake on the local microclimate and produced abundant crops. This dependable and rich resource base contributed to the Aztec's ability to consolidate their supremacy over the Valley of Mexico. Subsequently, they extended their rule over a vast tribute empire that encompassed much of the Mesa Central and also reached into the southern highlands of Mexico and even to the Pacific coast of today's Guatemala and El Salvador. All told, perhaps as many as 10 million people lived under the umbrella of the Aztec Empire by 1500. By this time Tenochtitán was a significant urban center. It housed over 100,000 inhabitants and was comparable in size and complexity to Europe's major cities at the time: London and Paris. Two other major indigenous cultures contested the power of the Aztecs on the Mesa Central during this period, the Tarascans to the west and the Tlaxcalans to the east. Neither achieved the levels of power and material wealth enjoyed by the Aztecs, though both cultures successfully resisted Aztec efforts to conquer them.

The Mesa Central remained a core region during the Spanish Conquest and the colonial period that followed. Hernán Cortés, an audacious leader and military commander, led an exploratory and expeditionary force of about 500 Spaniards from Cuba to a spot on the Mexican coast at what is now Veracruz in 1519. In a move that stunned his men, but committed them to his venture irrevocably, Cortés stripped his vessels of weaponry and ordered all of them burned. Immediately he marched his force inland up the steep escarpment onto the Mesa Central. Within 2 years, Cortés had killed one Aztec emperor and captured another; looted, laid siege to, and conquered Tenochtitlán; and achieved an unprecedented military victory over the Aztecs that brought most of the Mesa Central, much of the Southern Highlands, and Guatemala under Spanish control. Cortés established Mexico City on the ruins of Tenochtitlán, and the Spanish Crown made the city the capital of its vast viceroyalty of New Spain, a region that at its height extended northward into Texas and New Mexico, southward through Central America to Costa Rica, and also included Spain's Caribbean territories.

Disease ravaged the native populations of the Mesa Central and surrounding regions during the first 100 years of the colonial period, reducing its total populations by as much as 90 percent. The rich soils of volcanic and lacustrine origin found in the broad basins of the Mesa Central, and especially in the west in the Bajio, evolved into Mexico's agricultural heartland and breadbasket. This productive agricultural resource base helped cement the position of the Mesa Central as the dominant region in New Spain and, at least for the next 100–150 years, in all of Spain's New World colonies.

The presence of Mexico City and its role as the Spanish Crown's political, administrative, and economic capital throughout the colonial period continued to reinforce the early dominance of the Mesa Central. The colonial transportation network focused on the Mesa Central and particularly Mexico City. Roads, which often were no more than steep muddy tracks over which mules and carts traveled with difficulty, connected Mexico City with the principal colonial ports of Veracruz on the Caribbean coast and Acapulco on the Pacific coast. Similarly torturous roads connected the Mesa Central with the Southern Highlands and Guatemala to the south, while comparatively gentle gradients characterized the roads leading northward into the Mesa del Norte. By the end of the colonial period, Mexico City's population numbered about 130,000. The Mesa Central's other major urban centers at this time were Puebla (67,000), Guadalajara (30,000), and Morelia (18,000).

Spanish colonists transformed the agricultural systems of the Mesa Central, establishing haciendas, livestock raising, and the widespread cultivation of introduced European crops. Wheat, for example, grew well in the highlands and was in great demand in mining camps and Spanish settlements throughout Mexico. Native populations on the Mesa Central had been decimated by early contact with Spaniards and the Old World diseases they carried with them in the first decades of the colonial period. But by the end of the 18th century the area's population had recovered to pre-Conquest levels, reaching close to 5 million. The nature of the population, however, had been transformed. Indian peoples were still numerous, but mestizos, people of mixed European and native descent, now formed the majority of the population.

The end of Spanish colonial rule in Mexico was confused, prolonged, and bloody. Most of the key events in the war of independence played out on the Mesa Central. It was in the small town of Dolores, near Guanajuato, where the priest Miguel Hidalgo raised the call for revolution to his congregation in his famous speech "El Grito de Dolores" in 1811. While Hidalgo's effort at armed revolution failed, others followed. In the ensuing decade the countryside as well as the towns and cities of the Mesa Central were convulsed by guerrilla warfare and military conflict. Spain's colonial armies, although never decisively defeated, were worn down by the endless guerrilla warfare. They eventually capitulated in 1821, effectively ending the wars of independence in Mexico and Central America.

The years after independence were tumultuous for the young Mexican nation. Much of the drama of its growing pains took place on the Mesa Central. Governments and heads of state changed with some regularity and a pattern of political instability developed. Intellectuals and political leaders espoused the virtues of representative government and democratic participation, and elections were held, but in reality military prowess and brute force played a central role in most governments.

The Mexican-American War, which began in the summer of 1846, permanently altered the territorial extent of the nation and its political geography. While much of the war unfolded far from the Mesa Central, the region played a significant role in the war's conclusion and aftermath.

Initially, U.S. Army troops entered Mexico from the northeast, crossed the Sierra Madre Oriental, and took the city of Monterrey in late 1846. Other U.S. troops pushed southward down the Mesa del Norte, taking Chihuahua and other population centers on the northern plateau. The principal U.S. assault, however, took place on the Mesa Central. After a successful amphibious assault on Veracruz, General Winfield Scott led U.S. troops over the escarpment and onto the plateau, taking Puebla easily and then moving on immediately to Mexico City. While the Mexicans mounted a fierce defense of their capital, it fell to U.S. forces in September 1847. The next year, in a suburb of the capital, under extreme duress, the Mexicans signed the Treaty of Guadalupe Hidalgo, which among other things ceded nearly one-half of the territory claimed by the Mexican nation to the United States.

Mexico's travails with foreign powers interfering in its internal affairs did not end with the Treaty of Guadalupe Hidalgo. The nation's inability to pay its foreign debt led to an intervention by England, Spain, and France in 1861. Together they took over the port facilities and the customs house in Veracruz to collect the debts owed them. The English and the Spanish quickly abandoned the adventure, but the French pressed on, sending troops into the Mesa Central and eventually capturing Mexico City. The French installed a monarch of European lineage, Maximilian I, as emperor of Mexico. Supported by the French Army, Maximilian I held power until 1867, when the French recalled their troops. Unable to effectively maintain control of the whole nation, and barely ruling even the Mesa Central, Maximilian did little to promote social or economic development in Mexico. At the close of his reign the nation was likely worse off than at its start.

Mexico in the 1860s was still a preindustrial nation. The economy of the Mesa Central, despite its position as Mexico's core region, was largely rural, agricultural, and technologically backward. On the eve of the modern period, many agricultural and mining technologies in Mexico dated from the pre-Hispanic or colonial periods. Railroads, which had revolutionized transportation, trade, and development in the United States, Canada, and much of Europe by this time, hardly existed in Mexico. In 1860 Mexico had about 250 km of railroad lines, while the United States had 48,000 km!

The end of the French intervention and the restoration of the Mexican Republic marked the beginnings of the modernization of the Mexican nation. Benito Juárez's efforts as president of Mexico before Maximilian's rule had focused on land reform, education, and improving social conditions for Indians. But in his third term as president he began an aggressive and determined campaign to modernize the nation's economy. The centerpiece in this effort for Juárez was his determination to see the completion of the railroad line along the difficult route from the Caribbean port of Veracruz to Puebla and Mexico City on the Mesa Central. Completed in 1872, this was the first of a series of railroads that subsequently connected Mexico City and the Mesa Central with Mexico's far-flung territories in the Mesa del Norte and eventually with other regions of the country.

The last quarter of the 19th century and the first years of the 20th century were a period of profound transformation for the Mexican economy and the country's landscapes. Many of these changes were most notable on the Mesa Central, although most regions of the country were affected. Known as the "Pax Porfiriato," or the Peace of Porfirio, this period (1876–1911) corresponds to the time Porfirio Díaz ruled the country directly or indirectly. Diaz pursued an aggressive policy of economic development and technological modernization for Mexico. Immense sums of foreign investment and widespread foreign ownership played a critical role in providing the capital and the entrepreneurial skills necessary to pursue these goals. Díaz's regime emphasized the creation of a solid economic infrastructure. Early efforts focused on investments in transportation and communication.

Railroad construction represented a key component of this strategy. The completion of the Ferrocarril Mexicano from Veracruz to Mexico City in 1872 had been a remarkable engineering feat and an economic boost for the Mesa Central. But the railroad network was so limited that at the beginning of the Pax Porfiriato the nation had only 640 km of railroads, nearly all of it accounted for by this line. Later railroad construction extended the nation's network from Mexico City and the Mesa Central northward onto the Mesa del Norte and the region's principal cities, Chihuahua and Monterrey, and then on to the United States. Completed in the mid-1880s, the Mexican Central Railroad connected Mexico City with Chihuahua and Ciudad Juárez/El Paso. The Mexican National Railroad was finished several years later and linked the capital with Laredo, Texas. Investments and improvements in port facilities on both the Caribbean and the Pacific coasts also facilitated conditions for ocean-borne trade with the interior. At the same time, investments in communications infrastructure brought the telegraph and the possibility of rapid long-distance communication to hundreds of towns and villages across the Mesa Central, and to Mexico generally. The telephone appeared at the same time, but its use was largely restricted to the principal urban centers and large provincial towns.

Cities and towns changed dramatically during this intense period of economic development. The changes were most apparent in the capital and the principal cities, but many eventually reached down the urban hierarchy to provincial centers and market towns. Industrial activity and the urban landscapes associated with manufacturing increased markedly, notably in Mexico City, in Guadalajara, and to a lesser extent in Puebla, which had been a textile manufacturing center during the colonial period.

Manufacturing focused principally on consumer goods, textiles, paper, tobacco, beer, soap, and other consumables.

Modern amenities characteristic of contemporary urban centers in Europe and the United States for many decades finally began to appear in Mexico's capital and principal cities as well. Mexico City had some electrical generators as early as 1881. Electric streetlights replaced gaslights there by about 1890. Limited telephone service also began at this time. Public transportation evolved, and a train system with about 175 km of track and over 50 locomotives extended over much of the city by 1890. Similar innovations and developments found their way to other cities and towns, albeit more slowly and often on a much smaller scale. By the end of the Porfiriato, urban populations had grown steadily as their economic structures diversified with increasing manufacturing and greater trading opportunities as the transportation network improved. Mexico City's population exceeded 500,000 by 1910, while Guadalajara's stood at about 120,000 and Puebla's reached about 95,000.

The Mexican Revolution racked town and country for nearly a decade (1910–1920). Warring factions and regional caudillos, including Pancho Villa from the north and Emiliano Zapata from the south, jockeyed for power across the length and breadth of the country. But, as in so much that is Mexican, the Mesa Central figured prominently in the events that unfolded. Between 1.5 and 2 million people, including hundreds of thousands of civilians, died as a direct result of the fighting. The country's basic transportation, communication, and industrial infrastructure, built at great cost and effort during the Porfiriato, suffered serious damage during the revolution. It took at least until the 1930s before the Mexican economy fully recovered to prerevolutionary levels.

The Mexican Revolution, and the decades that followed, brought considerable changes to the social and economic geography of the Mesa Central. During the 35 years of the Porfiriato, rural elites and landowners had successfully expropriated vast acreage from Indian communities and smallholders over much of Mexico, including on the Mesa Central. As a consequence, haciendas dominated the rural landscape and held a stranglehold on the agrarian economy and the rural population. Agrarian reform and land redistribution was one of the principal aims of the revolution. The Constitution of 1917 mandated agrarian reform that led to the disappearance of the hacienda as the principal form of land ownership, as well as the dominant social and economic institution in rural areas. The Constitution of 1917 introduced a new concept and practice to landholding in rural Mexico, the *ejido*. Inspired by pre-Hispanic land-use practices and socialist ideals, the ejido was a distinct form of landholding. As part of the land reform process, privately owned hacienda lands were confiscated and these lands assigned to local communities or indigenous groups. Under the terms of the constitution, these lands could not be sold, preventing farmers from using the land as collateral for bank loans or selling them to generate ready cash. But it also protected them from losing the land to unscrupulous local elites. Local ejido councils assigned usufruct rights to agricultural lands on the ejido to community members, who in turn could cultivate them as they saw fit. Thus, the extensive low-intensity land uses that typified the hacienda—most often livestock raising on improved pastures and open range—disappeared over much of the Mesa Central. Small plots, planted with maize, beans, squash, and a wide variety of subsistence crops, became a dominant characteristic of the region's landscape. The demise of the hacienda, and a large increase in the number of peasant farmers as a result, energized and stimulated local markets and small towns and cities across the region.

The disappearance of the hacienda as a major component of the rural landscape also contributed to greater geographic mobility among rural residents. While the initial effects were modest, over time a slow but inexorable movement began from the countryside to villages, small towns, and larger urban centers. These movements, although just in their incipient stages in the first decades of the 20th century, initiated widespread population growth in urban areas, and presaged a wholesale rural-to-urban migration during the second half of the 20th century that transformed many sleepy provincial towns and state capitals into bustling urban centers by the end of the century.

Contemporary Economic and Social Geography

The social and ethnic landscape of the Mesa Central is diverse. Mestizos comprise the overwhelming majority of the region's population. Nevertheless, indigenous groups, many living traditionally in isolated communities and speaking native languages, are surprisingly common and number as many as 1 million. Whites, or peoples of European origin, are a small percentage of the region's population, perhaps numbering 1 or 2 million. They are concentrated principally in the major cities, especially Mexico City and Guadalajara. Despite their small numbers, they are disproportionately represented in senior positions in government, business, and other sectors. It is striking, for instance, that on Mexican television people with light skin and European features fill almost all the starring roles on the most popular shows and hold most of the positions as news anchors or talk show hosts. Individuals with mestizo or indigenous features are nearly invisible.

The second half of the 20th century transformed Mexico as well as the Mesa Central. Many regions of the country experienced tremendous economic and social development that energized the national economy and also served to soften the stranglehold the Mesa Central exerted over the remainder of the nation. Mining and metallurgy in the north fueled the growth of Monterrey as an industrial center on the Mesa del Norte, while the discovery and exploitation of oil reserves along the Gulf Coast brought considerable development to the states of Veracruz and Tamaulipas (Figure 6.3). In the northwest, in the 1940s, major government investments in large-scale irrigation projects in Sonora and Sinaloa produced an agricultural boom in the subsequent decades in the arid river valleys of the Pacific coast. In the southeast, in the tropical lowlands of Campeche and Veracruz, in the 1960s, massive government investments in agriculture promoted tropical land colonization, settlement, and population growth. Beginning in the 1970s, government investment and promotion of major tourist destinations like Cancún and other coastal resorts contributed to Mexico's development as a major international tourism center. Finally, the creation of the border industrialization program along the United States–Mexico frontier, which promoted Mexican manufacturing by eliminating customs duties on raw material imports and the reexport of the finished products made from those materials, fueled the proliferation of manufacturing enterprises known in Mexico as *maquiladoras*, factories that typically produce textiles and consumer electronics. Their proliferation led to the mushrooming of manufacturing centers in border cities like Tijuana, Ciudad Juárez, Nuevo Laredo, and Matamoros in the 1980s and 1990s.

These developments, while not insignificant, did little to decentralize Mexico's economic development and promote other re-

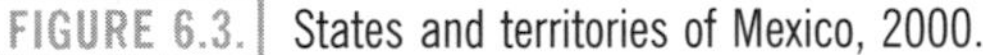
FIGURE 6.3. States and territories of Mexico, 2000.

gions of the nation. Indeed, the second half of the 20th century only enhanced the position of the Mesa Central as the core of the Mexican nation. Much of this geographical dominance is clearly linked to Mexico City and its role as the nation's political, administrative, economic, and cultural center. Agglomeration economies, stemming first from the city's role as Spain's colonial capital for its territories in New Spain, and then compounded over time as more and more of the colony's and eventually the nation's governmental, commercial, social, and cultural functions located in Mexico City, gave it and its surrounding region an insurmountable advantage over would-be rivals. The city's population grew rapidly during the 20th century, from about 500,000 in 1900 to over 20 million by 2000. As astounding as this grown in absolute numbers is, the increasing significance of the metropolitan region nationally is even more dramatic. In 1900 the city's population represented 2.5 percent of the nation's total population, but by 2000 its population accounted for fully 20 percent of the Mexican nation! If a city of similar magnitude existed in the United States, its population would exceed 60 million! This disproportionate concentration of population is mirrored in almost every sector: banking, transportation, services, cultural amenities, institutions of higher education, and a myriad of others. Manufacturing output is typical and also illustrates the general pattern of increasing concentration of many activities in the metropolitan region during the 20th century. In 1930 Mexico City accounted for slightly less than 30 percent of the nation's manufacturing output; some 70 years later, close to 50 percent of Mexican manufacturing output is produced in the metropolitan region.

While about three-fourths of the population of the Mesa Central is urban, the region's rural population is numerous. Agriculture still plays a key role in the rural economy. Close to 15 million people live in small villages and in rural areas on the Mesa Central. Rural population densities average about 100 per sq. km. However, in some particularly fertile areas, like the Toluca Basin, the Puebla Basin, and around Tlaxcala, population densities reach nearly 400 per sq. km. After the institution of the ejidos and the disappearance of most of the haciendas during the first half of the 20th century, small farms producing subsistence crops (maize, beans, wheat, and squash) have dominated much of the Mesa Central.

Historically, however, the Mesa Central has been the most productive agricultural region in Mexico. Commercial agriculture has been an economic mainstay of the region since the colonial period. Commercial agriculture continues to be a critical economic activity. Some of the most productive and intensive commercial agriculture is concentrated in the Bajio, a series of interconnected basins and lacustrine plains, situated principally in the state of Guanajuato, but including parts of other adjacent states as well: Queretaro, Michoacán, and Jalisco. A mild climate and adequate water for irrigation permit double cropping on the Bajio. During the wet summer season, farmers cultivate maize, beans, and sorghum; in the dry winter months, farmers grow wheat, alfalfa, and vegetable crops, usually with the aid of irrigation. In recent years multinational agricultural conglomerates have begun operating in the Bajio region, producing vegetables and some specialty crops for the fresh-frozen market, most of which is exported to the United States. Amendments to the Mexican Constitution in the mid-1990s eliminated the special legal status of *ejido* lands and permitted, for the first time in Mexican history, their sale. Neoliberal politicians and economists had pushed hard for these changes, arguing that these reforms would bring market forces to bear and thus rationalize landholding practices and encourage the most productive use of agricultural lands. These changes also appear to be contributing to the abandonment of farming by small-scale producers and subsistence farmers and their migration to urban centers. One notable exception has been the recent commercial success of cultivation of some species of agave, the plant from which tequila is made (Vignette 6.1).

Mexico City's immense metropolitan area covers hundreds of square kilometers, with the contiguous urbanized area extending some 35 km north–south and 25 km east–west. While the city's location at over 2,000 m altitude contributes to a pleasant climate, other site factors have made living in the city difficult at times. Ringed by high mountains, the city is particularly vulnerable to temperature inversions that trap large amounts of industrial and automotive pollutant emissions. Air pollution here is a real health hazard, with air pollution levels among the highest in the world. Some residents wear masks in an effort to reduce their exposure, and schoolchildren are forbidden to play outdoors at recess when pollution levels are extreme. The city's construction on the drained lake bed of Lake Texcoco has produced a host of problems, most notably the subsidence of the ground surface and subsequently of many of the buildings constructed on it.

By the beginning of the 21st century, the outlines of a major urban–industrial axis had taken shape across the southern end of the Mesa Central. Running east–west, the axis extends from the port city of Veracruz through Puebla and the Mexico City metropolitan area to the adjacent cities of Toluca and Cuernavaca. Manufacturing employment by state illustrates this axis-like pattern across the Mesa Central (Figure 6.6). The lion's share of Mexico's heavy industry, including steel production and automobile manufacture, are located in the east. Much of this sector of the axis is densely urbanized. This is especially evident along the backbone of the axis, coincident with

VIGNETTE 6.1. TEQUILA: FROM AN ANCIENT PLANT, A "MODERN" DRINK

On the western margins of the Mesa Central, in the state of Jalisco, one kind of specialized commercial agriculture merits note: the cultivation of various species of agave, drought-resistant plants from the botanical family *Agavaceae*, known commonly in English as the "century plant" (Figure 6.4). Several domesticated varieties of agave have been grown in Mexico since pre-Hispanic times. Some are valuable for their fibers and are used for making twine and rope~~for~~ example, henequen, which is still cultivated commercially, albeit on a small scale, on the Yucatan Peninsula.

Other agaves are used for the production of alcoholic beverages, as is the case in Jalisco and neighboring states. The region's pre-Hispanic population used the sweet sap harvested from the agave root as the basic ingredient in the production of a fermented beer-like drink called *pulque*. During the colonial period, Spaniards introduced the concept of the distillation of alcohol, and the subsequent distillation of pulque led to the production of two other drinks, tequila and its less refined cousin, mescal. Traditionally tequila has not been viewed as a high-end alcoholic beverage, but in recent years it has experienced rapidly increasing popularity and premium brands are now produced. Tequila production is big business, and large quantities are exported, primarily to the United States. A measure of the significance of tequila production is that only liquor produced from a specific agave species (called Weber blue agave) and grown in a specific geographic area centered on Jalisco state can be labeled and sold as tequila (Figure 6.5). Agave liquor produced outside of this region, or that is produced from agaves other than Weber blue agave, cannot be labeled or sold as tequila. Instead it must be marketed as mescal.

FIGURE 6.4. Agave plants in southern Oaxaca, Mexico, 1977.

the principal highways and rail lines, where the built-up areas of adjacent cities and towns have coalesced. Crude oil and natural gas pipelines from the petroleum fields on the east coast parallel this development corridor, bringing essential energy supplies to industry and consumers. The axis runs westward from Mexico City to Querétero, to León, and finally to Guadalajara, Mexico's second largest city, in the west. Overall, the western sector is less congested, less urbanized, and less industrial than the Valley of Mexico, adjacent valleys (Toluca), and those to the east (Puebla).

Mexico City is, of course, the region's preeminent urban center. With a metropolitan population of close to 20 million, it ranks as one of the world's five or 10 most populous cities, depending on whose estimates are used. The rapid growth of the city in the second half of the 20th century has created a sprawling, highly congested, and polluted urban environment. While the government has made some serious efforts to decentralize the nation's economic development and to slow in-migration to the metropolitan area, these efforts have borne little fruit and migrants continue to pour into the city at the phenomenal rate of about 2,000 per day or nearly 750,000 per year. In addition, natural population increase by the city's existing residents has compounded these numbers. The increasing population has stretched basic public services and the housing supply to beyond the breaking point.

Transportation is a nightmare. Streets and roads are clogged with traffic much of the day. Typical daily journeys to and from work can easily average 2 hours each way. Overcrowded buses, trucks, and taxi cabs compete for scarce space on the roadways with the city's 4 million passenger cars, discharging vast amounts of emissions into the atmosphere and contributing to the city's extreme levels of air pollution—perhaps the worst in the world for a major city. The first line of the city's subway system, the Metro, opened in 1969; by 2000, it consisted of a network of 10 lines covering the center of the metropolis and carrying about 4 million passengers daily. Nevertheless, the system is strained to capacity; crowding is so extreme that in order to protect riders some cars are reserved solely for women and children.

The provision of many other basic public services is incomplete or inadequate. This is especially true on the edges of the metropolitan area where new migrants often settle in vast shantytowns—*colonias*, as they are known in Mexico. Natzahualcóyotl, on the eastern

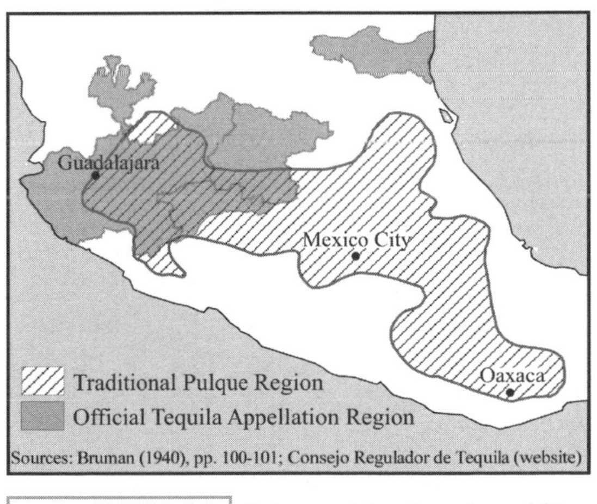

FIGURE 6.5. Pulque and tequila regions of Mexico.

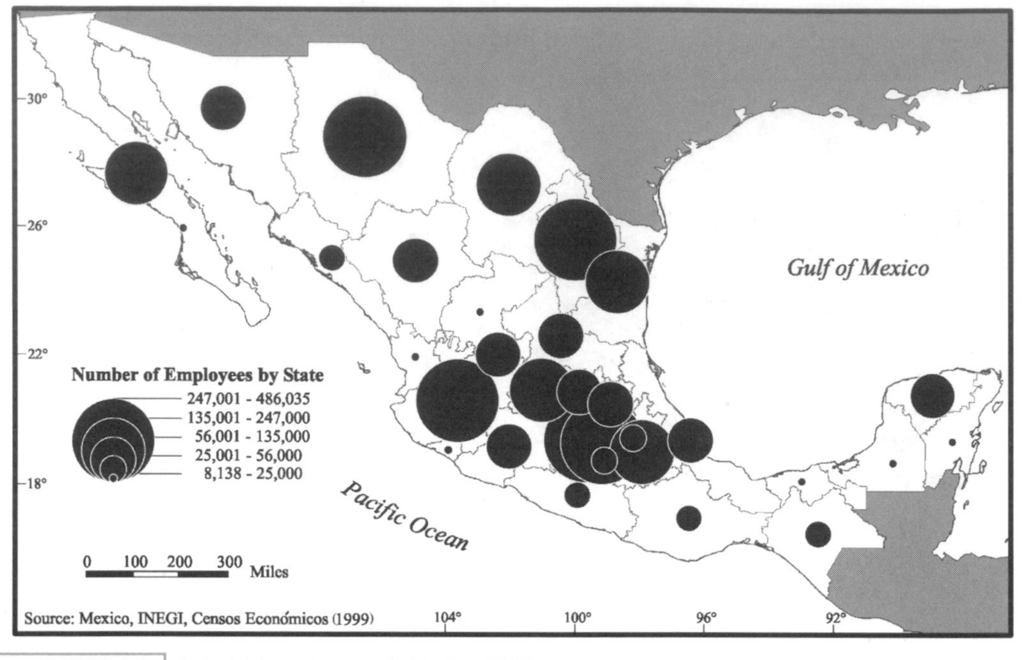

FIGURE 6.6. Industrial employment in Mexico, 2000.

side of the city, is one of the largest of these colonias. Settled initially in the 1960s, today it is home to between 1 and 2 million people. Living conditions here have improved incrementally over the four decades since its establishment, as is usually the case in shantytown settlements. However, new colonias, without basic services or adequate housing, continually grow up on the metropolitan periphery. While the city boasts lovely neighborhoods, elegant parks, and upscale social venues for its elite, many of the city's residents live in dire straits. Perhaps 20 percent of the population lives in abject poverty, with another 60 percent living on the margins of poverty. It is somewhat ironic that even with such inadequate public services and the low standard of living for so many of its inhabitants, the metropolitan area receives a disproportionate share of the nation's tax revenues from the central government.

Approximately 50 percent of all the nation's governmental social services spending occurs in the capital.

What will the future bring for one of the world's largest and most polluted cities? Although recent population data have shown a slight drop in the population of the Federal District (Distrito Federal), Mexico's capital district, and the city's downtown core, the Mexico City metropolis will surely continue to increase in population and expand geographically for the foreseeable future. Increasing pollution, transportation congestion, and difficulties securing adequate housing and services may stem population growth. Alternative employment opportunities in the north along the United States–Mexico border or even in the United States itself may also contribute to a reduction in migration to the metropolitan area. Recent election reforms implemented in the late 1990s now provide for the popular election of Mexico City's mayor, who previously was appointed directly by the nation's president. This fundamental change in how the city is governed is likely to have a powerful impact on the city's future development and growth.

The urban hierarchy of Mexico is dominated by the cities of the Mesa Central. Including Mexico City, all four of the nation's five largest metropolitan centers are on the plateau (Table 6.1). Although dwarfed by Mexico City, these cities and their adjacent suburbs and satellite towns (Guadalajara, 3.9 million; Puebla, 2.6 million; and León, 1.4 million) are major metropolitan centers.

Guadalajara, the second largest city in Mexico, sits on the western margin of the Mesa Central. With nearly 4 million residents, the city is a sophisticated cosmopolitan center with a historic core that includes churches, administrative buildings, and palaces from the colonial period juxtaposed with modern high-rise office buildings, shops, and restaurants. The city grew rapidly during the last half of the 20th century. During the 1970s and 1980s population growth and increasing transportation options permitted an enormous expansion of the city's area, as developers constructed literally hundreds of new residential subdivisions, targeted at the city's prosperous middle- and upper-class residents, on the urban periphery. Industry plays a major role in the economic structure of the city; close to 35 percent of the city's economically active population is employed in manufacturing. Besides producing a wide range of basic consumer goods like textiles, shoes, processed food, tobacco, and beverages like beer and soft drinks, Guadalajara also produces chemicals, electronic goods, photographic equipment, construction materials, and plastics. Surprisingly, some traditional handicraft industries continue to provide a notable component in the local economy. The region is renowned in Mexico for its glassware and pottery. Tourism is also an important component of the economy of the city and its immediate environs. This is due in part to the fact that the city enjoys a mild spring-like climate, where clear skies and low levels of air pollution are the norm. In addition, Lake Chapala, a large highland freshwater lake, has proved to be attractive to Mexicans as well as to North Americans, who have built retirement homes and whole communities on the lake and its surrounding slopes. In Guadalajara and its surrounding areas expatriate retirees from the United States number close to 10,000.

Puebla, Mexico's fourth most populous city, with a metropolitan population of close to 2.0 million, sits on the eastern end of the urban–industrial axis that straddles the Mesa Central. It was the nation's second most important urban center until the early 20th century, but since then it has been surpassed by both Guadalajara and Monterrey. The city retains its historical importance as an administrative and ecclesiastical center. Colonial architecture is still characteristic of the city's core and surrounding neighborhoods. Puebla was an early industrial center, even in the colo-

TABLE 6.1. Mexico's Urban Hierarchy: The Ten Most Populous Cities, 2000

City	State	Region	Population
Mexico	Distrito Federal/Mexico	Mesa Central	21,234,000
Guadalajara	Jalisco	Mesa Central	3,896,000
Monterrey	Nuevo Leon	Mesa del Norte	3,528,000
Puebla	Puebla	Mesa Central	2,586,000
Leon	Guanajuato	Mesa Central	1,434,000
Juarez	Chihuahua	Borderlands	1,294,000
Tijuana	Baja California	Borderlands	1,271,000
San Luis Potosí	San Luis Potosí	Mesa Central	954,422
Mérida	Yucatan	Yucatan	889,000
Acapulco	Guerrero	Pacific	888,000

Source: World Gazetteer (website, 2003).

nial period, when it specialized in the production of cotton and woolen textiles. At the present, like both Mexico City and Guadalajara, Puebla is an industrial center, with about one-third of its population employed in manufacturing. The production of consumer goods, including textiles, constitutes a large part of the city's industrial base today, but heavy industry, including automobile manufacture, is also significant. For example, Volkswagen's Puebla plant turned out approximately 400,000 automobiles per year in the late 1990s, employing 16,000 workers directly and outsourcing some parts manufacturing to over 20 other firms in the vicinity that employed another 4,000 workers.

The Mesa del Norte: A Vast Hinterland

The Mesa del Norte constitutes the northern two-thirds of the Mexican Plateau. This immense, thinly populated, and semiarid region has long served as a resource hinterland for Mexico's Mesa Central by raising livestock and producing immense quantities of silver and other raw materials since the colonial period.

The area is bounded in the north by the borderlands along the United States–Mexico border. The eastern margin of the plateau is marked by the Sierra Madre Oriental, a cordillera of sedimentary origin including notable amounts of shale, marl, and limestone, some of whose peaks reach close to 3,800 m (Figure 6.7). In the west, the Sierra Madre Occidental, a mountain range of largely volcanic origin with elevations exceeding 3,000 m, defines the limits of the Mesa del Norte. The southern border lies at about 22°N, and is generally defined by a series of low mountain ranges that lie north of San Luis Potosí. The elevation of the plateau surface rises gently from about 1,000 m in the north in Chihuahua to about 2,000 m in the south. The vast open plains of the plateau are often punctuated by mountain ranges trending north–south, with broad gentle intermontane valleys, known as *playas* or *bolsones*, which

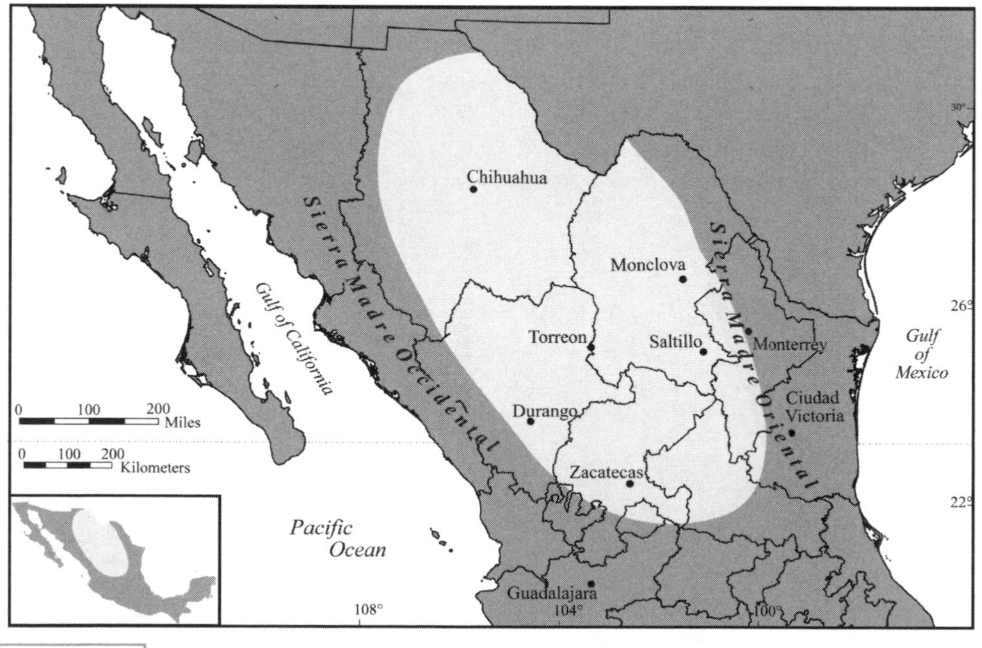

FIGURE 6.7. Mesa del Norte.

are characterized by interior drainage, as is almost the entire region.

Arid and semiarid climates prevail over much of the region. Temperature and precipitation patterns reflect this fact. Data for the city of Chihuahua in the far north are illustrative (Figure 6.8). While the mean annual temperature stands at about 18°C, this statistic masks considerable seasonal and diurnal variation on the northern plateau. Temperatures are highest in the summer months between May and September, when monthly averages are about 25°C. Daily temperature fluctuations can be large, with highs sometimes reaching and exceeding 40°C and lows falling to below 20°C. December, January, and February are the coolest months, with mean temperatures fluctuating around 10°C. Diurnal variation is also pronounced in the winter, with daytime highs reaching the upper teens and nighttime lows frequently falling to 0°C.

Rainfall is sparse on the Mesa del Norte. In Chihuahua, for example, average annual precipitation is about 350 mm. Most precipitation falls between June and September, with monthly totals ranging from 30 to 60 mm. The winter months are drier, with precipitation rarely exceeding 10 mm monthly. Rainfall is insufficient to permit the cultivation of crops without the benefit of irrigation.

The Chihuahua and Coahuila Deserts cover the northern portions of the Mesa del Norte and desert-like landscapes typify the region. Drought-resistant plants like mesquite, cacti, agave, scrubs, succulents, and low grasses figure prominently in the mix of natural vegetation. The slopes of adjacent uplands and mountains, where precipitation is greater and temperatures cooler, support oak and pine forests, although these forests have been severely reduced by logging for mining, construction, and charcoal making.

Early Settlement and Historical Geography

The pre-Hispanic culture history of the Mesa del Norte contrasts dramatically with that of central and southern Mexico. While intensive agricultural systems, high population densities, settled village life, and eventually advanced societies characterized the Mesa Central and the southern highlands, the indigenous peoples of the north were predomi-

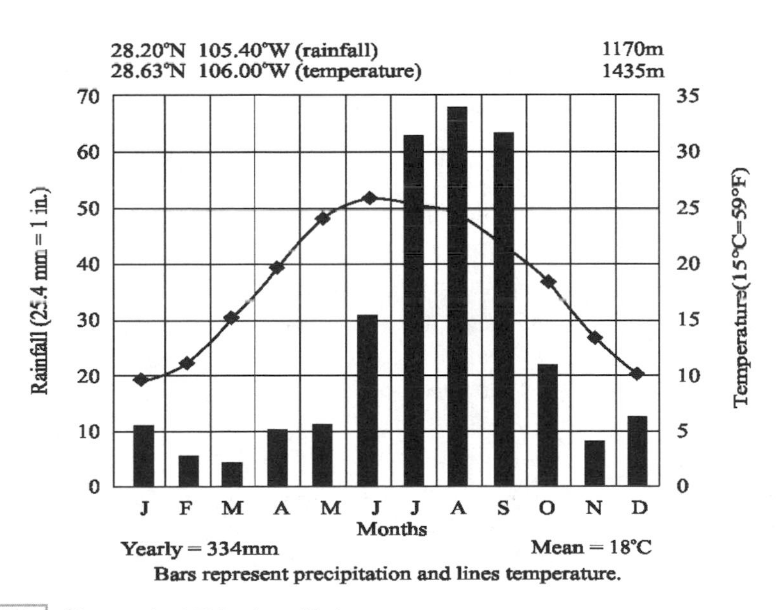

FIGURE 6.8. Climograph of Chihuahua, Mexico.

nantly seminomadic hunters and gatherers with a rudimentary material culture. For centuries these tribal peoples, often referred to generically as "Chichimecs," raided and attacked the advanced societies of the Mesa Central along the interface between the two regions, and at times successfully penetrated deep into the south. Aztecs are the most notable of these northern peoples who succeeded in breaching the northern frontier and establishing themselves permanently in the south, beginning around 1200 A.D.

The exploration and settlement of the Mesa del Norte lagged behind that of the Mesa Central in the south. Sparse indigenous populations, often hostile to the Spanish, and few readily exploitable agricultural resources made the semiarid plains and steppes less attractive for colonization. The discovery of silver in Zacatecas in 1546 marked the beginning of a century-long string of silver strikes and booms along the eastern foothills of the Sierra Madre Occidental extending north into Chihuahua. By the late 1600s, the highly productive silver mining region, or "Silver Belt," that had developed along the western edge of the Mesa del Norte eventually supplanted the mines around the Andean city of Potosí, Bolivia, as the principal source of silver for the Spanish Crown. Settlement and economic development on the Mesa del Norte depended largely on the location of productive mines. The region's principal towns—Zacatecas, Durango, Parral, and Chihuahua—were established in the vicinity of these mines.

Ranching developed initially along the piedmont region between the mountains and the plateau where rainfall, surface water, and groundwater moisture were somewhat more abundant and grasses for grazing more plentiful. Demand from mining centers and mining towns for pack animals (horses, mules, and donkeys), as well as for meat, hides, and tallow, supported the development of the haciendas on the Mesa del Norte. Small irrigated farming districts, responding to demand for foodstuffs from mining communities, also developed in the valleys of streams and rivers draining the eastern slopes of the Sierra Madre Occidental. By 1600 settlement along this western edge of the Mesa del Norte had extended into the upper Rio Grande Valley and into what is now New Mexico.

Spanish settlement along the eastern portion of the plateau lagged behind that in the west. Indian resistance was greater. While silver deposits were discovered in the east, these were less numerous and less bountiful than those of the Silver Belt in the west. The founding of the principal urban settlements in the east, Saltillo and Monterrey, occurred in the last decades of the 1500s.

Economic Development

The Mesa del Norte has remained a peripheral region within Mexico since the colonial period. Mining and cattle ranching have traditionally been the area's principal economic activities. Despite the region's vastness and marginal location within the Mexican state, a gentle relief pattern over most of the plateau did permit comparatively easy transportation. The region enjoyed adequate transportation links with the Mesa Central. In the last decades of the 19th century, railroad lines were constructed from Mexico City to the Mesa del Norte, and then northward to the United States.

Increased transportation access, substantial foreign investment, and foreign technological and entrepreneurial leadership in the moribund mining industry also played a decisive role in stimulating the regional economy. The mining industry had remained largely stagnant since the colonial period: few technological or organizational advances had occurred in mining operations. However, under the presidencies of Porfirio Díaz (1876–1880 and 1884–1911), the Mexican government granted liberal concessions and financial incentives to foreign investors to encourage economic development. This rejuvenated aging mining operations and led to the development of many new enterprises, especially for silver and lead mining in Chihuahua and Coahuila. American mining companies figured prominently in the redevelopment of mining across the Mesa del Norte. Smelters and foundries sprung up in Chihuahua, Durango, and San Luis Potosí.

Metalworking and associated industries, like ore smelting and metal foundries, concentrated in Monterrey. By 1910 the city, sometimes known as "Mexico's Pittsburgh," was the nation's center for steel production. Industry brought increased prosperity and population growth to Monterrey. The city boasted electricity by the 1890s, and was served by both telegraph and telephones by the turn of the century. By 1910 its population had topped 80,000, making it one of Mexico's largest urban centers.

Despite the development of Monterrey into one of Mexico's most populous and industrial cities, much of the regional economy of the Mesa del Norte still involves the extraction or production of raw materials for national and international markets. Agriculture plays a significant role in the regional economy, especially cattle ranching. Nearly 40 percent of the nation's cattle are found on the Mesa del Norte. Free-range cattle graze on extensive pastures in northeastern Chihuahua, in the northwest of Zacatecas, and in central Durango. Grazing on improved pastures has become more common in recent years.

Crop agriculture is also locally significant in areas where surface water from mountain streams and rivers permits irrigation. During the 1920s and 1930s government planners sought to improve agricultural productivity through the development of two large irrigation projects on the Mesa del Norte. These projects, one near Torreón in Durango and the other near Delicias in Chihuahua, brought over 200,000 ha of irrigated land into production between 1920 and 1940. Cotton, alfalfa, and wheat accounted for much of this acreage then and continue to do so today. A dropping water table from excessive groundwater pumping for irrigation and increasing soil salinization stemming from poor irrigation practices have led to reduced production in recent decades.

Forestry plays an important role in the regional economy. Major forest reserves are found in the mountains of Chihuahua and Durango. Forestry contributes to a range of economic activities including trucking, timber processing, and furniture construction (Figure 6.9).

Mining continues to play a key role in the regional economy of the Mesa del Norte just as it has done since the colonial period. Silver mining, most along the traditional Silver Belt in the Sierra Madre Occidental, accounts for

FIGURE 6.9. Logging trucks in Durango, 1989. Photo: William E. Doolittle.

two-thirds of Mexican silver production. Mexico, the world's largest silver producer, extracted 94 million oz in 2003 (Peru came in a very close second with an output of 89 million oz). Mexico's output represented approximately 15 percent of total world silver production. In 2003, Fresnillo, the world's second most productive silver mining complex, extracted 32 million oz. Modest iron ore reserves in Chihuahua, Coahuila, and Durango supply smelters and foundries in Monclova, Monterrey, and several smaller metallurgical centers (Figure 6.10). Coal is mined in the Sabinas Basin in Coahuila and supplies steelmakers on the Mesa del Norte as well as smelters and foundries in the Mesa Central.

Contemporary Conditions

At the beginning of the 21st century, the Mesa del Norte remains sparsely populated with a regional population of barely 7 million. Approximately 70 percent of the population is urban, with most of this concentrated in a few major urban centers: Monterrey (3.5 million), Chihuahua (832,000), Saltillo (668,000), Torreón (517,000), and Durango (440,000).

The metropolitan areas of Monterrey, Chihuahua, and Torreón are the principal centers of manufacturing and industry. Major iron and steel production centers are located at Monclova and Monterrey, with secondary production in Chihuahua, Durango, and Torreón. As Mexico's second most important manufacturing center, Monterrey boasts a broad range of manufactured products from durable goods and consumer durables to basic consumer goods.

The people and culture of the Mesa del Norte and much of the rest of northern Mexico are distinct from those of southern Mexico. Relatively few indigenous peoples occupied this region prior to European colonization, and few survive. Those that do are geographically isolated. Many observers describe this region as more "European" than central and southern Mexico, due in part to the limited influence of native peoples on the post-Conquest population of the region. In addition, during the late 1800s and early 1900s, the Mexican government encouraged the settlement of both Mormons and Mennonites from the United States and Canada on the Mesa del Norte in an effort to promote agricultural development and modernization. Although not numerous, their descendants are still a significant presence in some areas.

Cultural differences from the south are manifested in several ways. Wheat and wheat products, like bread, play a bigger role in local cuisine than in southern Mexico. Here for example, wheat tortillas are the staple, unlike in southern Mexico where corn tortillas are preferred. The Mesa del Norte and northern Mexico in general have always been more conservative politically. It is here that the longstanding conservative opposition party, the Partido de Acción Nacional (PAN) (the National Action Party) won its first state governorships and established its base. After several decades of incremental electoral gains in northern Mexico, the PAN elected its first president, Vicente Fox, in 2000. This was the first time since the establishment of the Partido Revolucionario Institucional (PRI) (the Institutional Revolutionary Party) in the 1920s that it had not won the presidency of the nation.

Summary

The Mesa Central, the southern part of the Mexican Plateau, is Mexico's heartland, containing the core of industry, transportation infrastructure, and cultural and educational institutions. The region, home to two-thirds of Mexico's population, contains Mexico City, the long-time dominant urban center. The arid Mesa del Norte, the northern two-thirds of the Mexican Plateau, is a vast, thinly populated,

FIGURE 6.10. Iron and steel foundries at Monclova, Coahuila, 1987. Photo: William E. Doolittle.

peripheral region that has been a mining and livestock-raising center since the colonial period.

The moist subtropical highland climate of the Mesa Central encourages agriculture, although changing agricultural practices have altered native plant life and the water supply. The area is affected by seismic and volcanic activity. The region was the birthplace and home to sophisticated early civilizations, and served as the site for the first large-scale planned urban center, Teotihuacán. The Aztecs were a dominant regional civilization until the arrival of the Spanish under Cortés. The region then became the center of New Spain until Mexico won its independence in the early 19th century.

An independent Mexico struggled against foreign influence, losing substantial territory to the United States in the Mexican-American War of 1846. By the late 19th century, the region had fallen behind economically and agriculturally. Under the rule of Porfirio Díaz, economic development progressed, including the addition of railroads and manufacturing. Agrarian reform followed the Mexican Revolution in the early 20th century.

By the late 20th century, the region had developed tourism and a more diverse manufacturing and agricultural economy. Mexico City continues to dominate the area because of its size and influence, bringing with it major urban pollution, as well as transportation and housing problems.

To the north of the Mesa Central lies the vast hinterland, the Mesa del Norte. Mining, especially for silver, and livestock raising are key economic sectors on this semiarid and sparsely populated plateau.

Further Reading

Bruman, H. J. (2000). *Alcohol in ancient Mexico*. Salt Lake City: University of Utah Press.

Chevalier, F. (1965). *Land and society in colonial Mexico: The great hacienda*. Berkeley and Los Angeles: University of California Press.

French, W. E. (1996). *A peaceful and working people: Manners, morals, and class formation in northern Mexico.* Albuquerque: University of New Mexico Press.

Hassig, R. (1985). *Trade, tribute, and transportation: The sixteenth-century political economy of the valley of Mexico.* Norman: University of Oklahoma Press.

Hedrick, B. C., et al. (Eds.). (1971). *The north Mexican frontier: Readings in archaeology, ethnohistory, and ethnography.* Carbondale: Southern Illinois University Press.

Lewis, O. (1960). *Tepoztlán: Village in Mexico* . New York: Holt, Rinehart, & Winston.

Licate, J. A. (1981). *Creation of a Mexican landscape: Territorial organization and settlement in the eastern Puebla basin, 1520–1605* . Research Paper No. 201. Chicago: Department of Geography, University of Chicago.

Malkin, V. (2001). Narcotrafficking, migration, and modernity in rural Mexico. *Latin American Research Review, 28*(4), 101–128.

Melville, G. K. (1994). *A plague of sheep: Environmental consequences of the conquest of Mexico.* New York: Cambridge University Press.

Pick, J. B., and Butler, E. W. (2000). *Mexico megacity.* Boulder, CO: Westview Press.

Ross, J. (1996). *Mexico: A guide to people, politics, and culture.* London: Latin American Bureau.

Tullis, L. M. (1987). *Mormons in Mexico.* Logan: Utah State University Press.

Ward, P. M. (1998). *Mexico City* (2nd ed.). New York: Oxford University Press.

Wilkie, J. W. (Ed.). (1990). *Society and economy in Mexico.* Berkeley and Los Angeles: UCLA Latin American Center Publications, University of California Press.

7 Early Cities and Urban Development

Pre-Columbian Urban Development

In the Americas the domestication of plants and animals and the evolution of religious and political systems set the preconditions for the development of cities long before the arrival of Europeans in the Western Hemisphere. This pattern of urban development is similar to that which has characterized the establishment and growth of the first cities in other world regions: the Middle East, Africa, and Asia. In Latin America, cities prospered in two specific subregions coincident with the evolution of advanced civilizations. Specifically, cities only developed in Mesoamerica and in the Andean realm. Until the beginning of the 16th century, these cities and urban systems in the New World developed completely independently from those in the Old World.

Human settlement takes many forms, so it is important to underscore the fact that a dense concentration of human population does not necessarily constitute a city or even an "urban" place. No definition of a city is truly satisfactory. Social scientists, demographers, and others who study cities/urban places employ a range of criteria to do so. Typically, minimum population thresholds for cities/urban places range upward from several thousand to as many as 10,000 inhabitants. However, population size is not the only critical factor.

Characteristically, cities originated as the ceremonial and administrative centers of a society's religious elite and eventually developed into centers of political and military power as the connection between religion and governance intensified. The concept of the god-king that characterized early civilizations in the Old World also applied in the Americas, where Maya kings and Aztec and Inca emperors were attributed with divine lineage and powers. The construction of monumental religious structures, royal palaces and tombs, and subsequently of civic buildings are characteristic of early urban centers in the Americas and elsewhere. Religious ceremonial centers alone, however, are not cities.

Cities evolved from these complex religious ceremonial centers, as these also became centers of economic exchange for their hinterlands. The permanent residents expanded beyond priests and those who served them and came to include a wide range of occupations, including merchants, artisans, and skilled laborers. Social stratification accompanied occupational specialization; cities in the New World, as in the Old World, were typified by a class-stratified society. An elite ruling class dominated a small commercial and artisan class and a mass of servants, laborers, and farmers. Cities evolved spatially and specialized land-use patterns appeared. Monumental religious and civic architecture usually dominated the city's central quarters, while specialized residential, production, and commercial quarters also arose within the territorial confines of the city.

Urbanism in Mesoamerica

In Mesoamerica, the earliest true cities, functionally complex centers with large populations, date from the Classic period (300–900 A.D.). Teotihuacán, in the Valley of Mexico, whose population in 450 A.D. exceeded 100,000, was the religious and administrative center for a powerful city-state. Spread over an area of about 12 sq km, the city's infrastructure included broad avenues, temples and other monumental religious buildings, a regular grid network of streets, artisan shops and production areas, and residential districts segregated by social class. Supported by a densely cultivated and highly productive agricultural hinterland, Teotihuacán prospered for several hundred years. The city and the political system that supported it collapsed around 800 A.D., when it was razed by fire and subsequently abandoned. While a range of hypotheses has been proposed to account for its demise—including an overextended political system, internal disorder, and a collapse of its agricultural resource base—none are completely satisfactory.

Other major civilization centers arose during the Classical period in Mesoamerica. In the tropical lowlands of southern Mexico, Guatemala, and northern Honduras, the Maya constructed a broad network of extensive and architecturally imposing administrative and religious ceremonial centers. Tikal, Copán, Palenque, and Piedras Negras are among the more widely known (Figure 7.1). But in many respects these Maya centers, like the Olmec center at La Venta and the Zapotec center at Monte Albán, lacked key characteristics usually associated with the definition of true cities. Permanent residences were few and populations small. These centers tended to be dominated by priests, acolytes, and other religious personnel. There is little evidence that these cities served as key points of economic exchange or centers of artisan production.

The central plateau of Mexico continued to be the preeminent center of urban development during Mesoamerica's Post-Classic period (900–1519 A.D.). The region's broad highland basins, which are dried lake beds of Pleistocene origin with rich soils, provided a sturdy resource base upon which a number of city-states of varying size and duration arose during the Post-Classic. Tula, founded about 1000 A.D., served as the capital city and central urban place for the Toltecs, who dominated much of central Mexico between the 10th and the 13th centuries. With a small system of tributary cities and towns, Tula served as an early example of a true American metropolis.

The growth of city-states and other smaller urban centers occurred in neighboring regions. Just about 75 km south of Tula, a well-integrated system of cities had been established around Lake Texcoco by the end of the 14th century. The formation of an alliance by the Aztec cities of Tenochititlán and Tlateloco, along with the city-state of Texcoco and the smaller polity of Tlacopán, provided the basic urban framework and focus for the vast empire the alliance came to rule over the next two centuries.

Tenochititlán, the Aztec capital, dominated this system and clearly reached the dimensions of a major urban center. Its population, geographical extent, and functional complexity rendered it one of the world's most imposing urban centers—comparable to Europe's largest contemporaneous urban centers—on the eve of its destruction by the Spanish conquistador Hernán Cortés. Situated on a small island and adjacent to reclaimed lands, the city occupied between 12 and 15 sq km at its zenith. About 40 percent was urbanized, and the remaining 60 percent was occupied by *chinampas*, reclaimed lake bottom used for an intensive form of agriculture (Vi-

FIGURE 7.1. Mesoamerican ball court at the Classic period Maya ruins at Copán, Honduras. Stelae stand in the background, 1977.

gnette 7.1). A regular grid of streets, canals, and causeways formed the basic outline upon which the city arose. Broad avenues converged on a monumental urban core dominated by temples and pyramids, while specialized land uses for residential, artisan, and commercial activities occupied the other parts of the built-up area. Less visible, but no less impressive, a sophisticated system of dikes and canals facilitated the control and management of the waters of Lake Texcoco, ensuring the city and its highly productive *chinampas* of an adequate supply of freshwater. Some estimates of Tenotchitlán's population place it as high as 300,000, but even the conservative estimates give the city a population greater than 150,000 at the beginning of the 16th century.

The Central Andes and the Pacific Coast of South America

The development of individual cities as well as the eventual establishment of a geographically extensive network of urban places also occurred in the central Andes and the adjacent river valleys of the arid Pacific coast of South America. By the first centuries A.D., advanced agricultural societies began to develop along the coast and in the highlands. Their development was often accompanied by the construction of ceremonial centers. Especially notable among these were the Moche culture, along Peru's northern coast, whose settlements typically included flat-topped pyramids and temples constructed of sun-dried mud bricks (adobe); the Nazca culture on Peru's southern coastal margin; and the Tiahuanaco culture centered on the shores of Lake Titicaca on the Altiplano of the Andean highlands. However, in no case did their settlements achieve the geographic extent, population size, or functional complexity required to be true urban places or cities.

The Huari (Wari) culture, which prospered between about 500 and 900 A.D. in the central Andean highlands in the vicinity of the Peruvian city of Ayacucho, gave rise to what was likely the first truly urban place in the Andean realm. Their capital, also known as Huari, housed as many as 25,000 residents. Its

VIGNETTE 7.1. *CHINAMPAS*

Many scholars identify *chinampas*, or the "floating gardens of the Aztecs," as one of the principal factors in the development of a complex urban system and a vast tributary empire based on the city of Tenochtitlán in the Valley of Mexico.

The shallow lakes that filled the highland basins of the Mesa Central provided a fertile medium for the establishment of an intensive agricultural production system that supported the Aztecs' civilization. Although sometimes known as floating gardens, the chinampas are really raised beds constructed in the shallow margins of the lakes. Using sticks, cane, and other materials, farmers enclose a small area and then proceed to fill it with sediments dredged from the bottom of the shallow lake bed. The fertile sediments and abundant water allow farmers to produce high yields and often three crops per year.

Modernization and development, however, have not been kind to the chinampas. The area cultivated in chinampas has declined since colonial times as urban expansion has consumed it. Perhaps only 15 percent of the original area in chinampas remains. The drilling of new wells that lower the water table, pollution, and continuing pressure for urban expansion threaten the remaining areas. The "Floating Gardens of Xochimilco," a small remaining area of chinampas now turned into a park, are a recreational attraction for tourists and for Mexico City's residents.

layout and operation reflected a clear understanding of urban planning; its infrastructure, for example, included systems for the distribution of water as well as the removal of sewage.

However, it was between 1200 and 1500 when the process of urbanization and urban development reached its peak. In the fertile river valleys on Peru's north coast, the Moche culture had been superseded by that of the Chimú. Expert artists who produced highly ornamented pottery and stunning gold jewelry, the Chimú also had a sophisticated understanding of irrigation and water management. Their productive agricultural system supported dense populations in the river valleys along the coast, with at least one major population and ceremonial center in each valley. The most impressive of these was the Chimú capital, Chan Chan, located on the Pacific coast near the modern city of Trujillo, Peru (Figure 7.2). The city dates from the early 1200s, when the Chimú began a process of imperial expansion along the Pacific coast, eventually consolidating an empire that stretched nearly 1,000 km along the littoral from the Rimac River in the south to the modern border with Ecuador in the north.

Chan Chan fell to the Incas in the 1460s. Today, the ruins of Chan Chan spread over about 10 sq km of desert. The aridity of the coastal zone has preserved the outlines of the ancient city. The city lacks the clear definition of an unambiguous ceremonial center, and no temples or monumental constructions are extant at the site. Ten enclosed compounds, known as citadels (*ciudadelas*), are perhaps the site's most remarkable features. Measuring 200–600 m on each side, these citadels were enclosed by adobe walls up to 10 m high and are thought to have been the palaces or royal compounds of the Chimú rulers. Its population is estimated to have reached 100,000, although population densities were low.

Cuzco and Other Inca Cities

The Inca Empire gave rise to the most integrated and geographically extensive system of

FIGURE 7.2. Ruins of compound walls in the coastal city of Chan Chan near Trujillo, Peru, 1982.

cities that developed in pre-Columbian America. The Inca, who emerged as a political and military power in the 1200s in the region around Cuzco, reached the peak of their territorial expansion just before the arrival of the Spanish in Peru. The Inca imposed an ordered world on themselves and those they conquered and sought to plan and efficiently administer an immense empire. A rigid social hierarchy fixed wealth, power, and privilege in the hands of a miniscule royal and noble elite, a cadre of priests and their acolytes, and a strong military class. Peasants farmed, contributing as much as two-thirds of their production to the state's grain silos while retaining the balance. Although Andean peoples had a long-established tradition of landscape modification, the Inca advanced the transformation of the physical environment in ways and at a rate previously unknown in the Andean realm. They terraced the hillsides, captured water and diverted it into extensive irrigation networks, constructed roads, built forts, and even planned and founded urban centers.

The very nature of the Inca Empire required the development of a system of administrative centers, indeed cities, over the length and breadth of the Inca realm. As the geographical limits of the empire expanded, especially in the last century of their power, the Inca incorporated the cities and towns of their conquered foes into their empire. Some, like Quito, Cajamarca, Tumbes, Jauja, and Pachacamac became regional capitals of the new social and economic order the Inca imposed.

Where the existing urban system proved inadequate, the Inca founded new towns. Their overarching vision of urban design gave rise to the construction of cities that included a consistent set of urban and architectural elements. A central plaza, a regular street grid, numerous storage silos and warehouses, and the use of the Inca's unique stonemasonry and trapezoidal windows were notable similarities. These planned cities included Huánuco, Tumpu, Ollantaytambo, Tambo Grande, and Incahuasi. Eventually a string of cities, each with a system of smaller towns and villages tributary to it, ran from the northern reaches of the Inca Empire southward. These centers usually were located in fertile highland basins in the Andes, although some were located on the Pacific coast.

The empire was united by an extensive road system. All roads led to Cuzco, the Inca capital and principal urban center, situated in a high mountain valley at over 3,000 m elevation in the central Andes. A monumental ceremonial and administrative city, it radiated from a large central plaza and housed grand temples, royal palaces, and nobles' residences. The city's workaday world included granaries, warehouses, and administrative buildings, as well as artisan quarters and the residential districts of all manner of commoners. An imposing stone fortress constructed of massive stone blocks, some weighing as much as 60 tons and standing as high as 5–6 m, overlooked the city. The population of the city and its surrounding suburbs reached approximately 100,000 at the height of the empire's strength in about 1500 A.D.

Colonial Urban Development

The New World, or that part of it that was to become Latin America, encompassed an immense geographical area and presented the Spanish and Portuguese with a wide range of social and environmental conditions that influenced and at times dictated the location and spatial structure of cities.

The role of the city in the process of conquest and colonization varied considerably between the Spanish and the Portuguese. For the Spaniards, the city represented one of the key elements of conquest and settlement. As the Spanish conquistadores extended the territorial limits of Spain's territory in the New World, they also established the outlines of an urban system that was to endure, largely intact, throughout the colonial period and in appreciable measure into the present as well. During the first 100 years of the colonial period, Spanish conquistadores founded cities with lightning speed, distributed land grants of the surrounding lands to the officers and soldiers who had participated in campaigns, and established the basic framework of the colonial authority and administration.

Small, remote, and often primitive, many of these early outposts of European civilization did not approach the kind of functional complexity characteristic of a true city; however, these settlements were the principal nodes of Spanish control over the surrounding regions. Spanish soldiers, royal officials, and clerics concentrated in these towns and cities, using them as the base from which the surrounding countryside and its Indian inhabitants were exploited and managed. The importance of the city as a center of power in the economic, social, and political life of the evolving colonial society was such that Spanish landowners, if they had the financial means, typically maintained a home in the city as well as one on their rural property.

In Portuguese America the colonial city evolved somewhat differently. Settlement was largely limited to a narrow band along the colony's northeastern coast. Comparatively speaking, relatively few cities were founded in Brazil. Spanish cities were often established in the midst of dense aboriginal populations, but their absence in the Portuguese colony removed one of the key functions the city served in much of the Spanish realm: defense against Indian attack. Once sugarcane cultivation was introduced along the Brazilian coast in the mid-1500s, a rural-based planter elite arose that effectively dominated the social, political, and economic life of the colony for nearly 200 years. Thus, while Portuguese colonial cities served as the administrative and mercantile centers for society, power resided principally in the hands of a rural-based planter aristocracy and not the urban elite.

On the islands in the Caribbean, it was the Spanish who established the first European settlements in the years immediately following the discovery of the West Indies in

1492. Nearly all of the native settlements on the islands were agricultural or fishing villages. No indigenous settlements achieved urban proportions anywhere in the West Indies. Most of the earliest Spanish settlements were little more than trading posts or military garrisons, many of which were abandoned or relocated later to more favorable locations.

Founded initially in 1498 and moved and reestablished on a more favorable site in 1502, Santo Domingo on the island of Hispaniola became the first real urban center in Spain's New World empire. For the first decades of the colonial period, it effectively served as the empire's capital, housing the seats of the first colonial governor and the first archbishop (Figure 7.3). Several essential elements in the city's urban design presaged the pattern of city development that was to evolve in the subsequent decades on the mainland. Notably, the colonial town was laid out with a regular street grid and included a centrally located main plaza, around which the cathedral and other public buildings were constructed. Santo Domingo's role as an economic and administrative force waned quickly after the discovery of Mexico and its conquest by Hernán Cortés in 1521.

Spanish City Founding

A long urban tradition dating back nearly a millenium characterized both Mesoamerica and the Andean realm at the time the Spanish conquistadores reached the continental mainland of the New World in the first decades of the 1500s. Tenochtitlán and Cuzco, the principal New World cities, were comparable in size and functional complexity to any contemporary European capital. A network of secondary cities, subsidiary towns, and villages formed part of a broader urban network for each of these imperial capitals. These served as the foundation for the establishment of cities and to some extent the urban network that evolved during the colonial period in these densely settled regions.

In the decades that immediately followed the conquest of the Aztecs in Mesoamerica and of the Inca in the Andean realm, Spanish conquistadores and colonists had founded, laid out, and settled cities and towns across the length and breadth of the new colonial empire. The sheer number of towns, the immense geographical area over which they were distributed, and the rapidity of their founding testify to the zeal and energy that accompanied the

FIGURE 7.3. The governor-general's home and office in Santo Domingo, Dominican Republic, constructed in the early 1500s, 1992.

early years of the colonial urban experience. For example, between 1540 and 1560, and running from the northern extreme of Spain's territory southward, the following cities were founded: Durango (northern Mexico; 1560), Mérida (Mexico; 1541), Mérida (Venezuela; 1558), Bogotá (Colombia; 1538), Cuenca (Ecuador; 1557), Lima (Peru; 1535), La Paz (Bolivia; 1548), and Santiago (Chile; 1541).

A few significant tenets guided the establishment of cities by the Spanish during the early decades of the colonial period. The subjugation of indigenous peoples played a critical role in the long-term economic development strategy of the colonial empire. Tribute from Indian towns and villages and the extraction of Indian labor were essential to the colonial economy. The Spanish first used the *encomienda*, and then later replaced it with the *repartimiento*, as legal mechanisms to ensure Indian labor for agriculture, mining, and public works construction. The history and workings of the *encomienda* and the *repartimiento* are explained in more detail in Chapter 8 in the section "Historical Geography and Economic Development." Thus, the location of existing native populations, especially those that were densely concentrated, proved an undeniable attraction for the Spanish.

Often Spanish conquistadores founded cities on the same sites upon which Indian cities existed. Occupation of these cities provided immediate access to the Indian population, but also permitted the Spanish to use the existing residential and administrative infrastructure. Indigenous temples and religious buildings were often partially razed, and Catholic churches or cathedrals built on their foundations. Three modern-day national capitals rise on the sites of indigenous cities or towns: Mexico City, Quito, and Bogotá. Similarly, a handful of secondary cities in Mesoamerica and the Andes trace their origins to pre-Columbian times. These include Tlaxcala, Oaxaca, Cholula, Quetzaltenango, and Santiago de los Caballeros de Guatemala in the Mesoamerican region and Cuenca, Cajamarca, and Tumbes in the Andean realm.

The Spaniards also founded other "new" cities in immediate proximity to dense aboriginal settlement. In Mesoamerica, Tepéc, Guadalajara, Pátzcuaro, and Puebla (all in modern-day Mexico) figure among these cities, while in the Andean realm, the same holds for Peru's three largest urban centers: Lima, Trujillo, and Arequipa.

Trade with Spain played a critical role in the operation of the colonial economic system that it imposed on its New World territories.

Consequently, good natural harbors adjacent to its most productive colonial territories proved attractive sites for city development. For example, Veracruz, Acapulco, Cartagena, Havana, Lima/Callao, and Valparaiso were favored for urban development.

In other cases, port cities grew despite a poor physical site when a booming hinterland created the potential for a sufficient volume of trade and warranted major investments in dredging and port infrastructure. In Spanish America, a number of today's largest ports, all founded in the colonial period, fall into this category: Guayaquil, Buenos Aires, and Montevideo.

In a functional sense, many of the port cities of colonial Latin America were simply satellite cities to more populous and economically and politically important interior cities. Lima's port, Callao, lies on the Pacific coast about 15 km west of the capital's center, while the port of La Guaira is sited on the Caribbean coast about 30 km north of Caracas. In other cases the distances are even greater, but, in the context of the colonial economy, the dependency of the port on the interior city was almost complete. For example, situated on the steep slopes of a coastal mountain range and adjacent to a magnificent harbor, Valparaiso, Chile, is separated by almost 100 km from Santiago, while the distances between Mexico City and her twin ports, Acapulco and Veracruz, are even greater.

After this intense period of city foundings during the first century of the colonial period comparatively few new cities were established. At the end of the colonial period, when the search for additional resources led to the colonization of peripheral regions, the establishment of new cities occurred in resource-rich regions and at key points of exchange in the tropical rain forests of Central America, the Amazon Basin, and the vast plains of the Llanos and the Pampa.

Urban Form in Spanish America

Early in the colonial period the urban morphology of Spain's colonial cities began to take on a regular pattern that eventually came to characterize most towns and cities in Spanish America. Town plans typically included a street grid laid out in a rectangular or square pattern. Streets were usually oriented in the cardinal directions, except when the city fronted a river or a major body of water. In those latter cases, the street grid was oriented with respect to the riverbank or coastline. Cities invariably included a large square sited in the center of the city and occupying an entire city block. This central square was known by various names throughout the region including the *plaza*, *plaza central*, *plaza de armas*, or *zócalo* (Figure 7.4). In coastal cities the plaza was located at the center of the urban grid, but adjacent to the waterfront.

Key public institutions located their offices and buildings in close proximity to the plaza—and the most important usually fronted on it. Typically, a Roman Catholic church or cathedral faced the plaza; in the case of the latter, the bishop's or archbishop's residence might also front on the plaza. Sometimes the monasteries and convents of Catholic religious orders occupied lots contiguous with the plaza, although this only occurred in the biggest cities. The local government, known as the *cabildo* throughout most of Spanish America during the colonial period, situated its offices on the plaza as well. If the city was of sufficient importance to warrant the presence of official representatives of the Spanish Crown, then their offices could almost invariably be found on the plaza. Finally, in the administra-

FIGURE 7.4. The principal plaza, or *plaza de armas*, in Trujillo, Peru, 1981.

tive capitals at the top of the hierarchy of governance, the residence of the viceroy or governor might also be sited on the plaza.

The method of town founding and dispensation of land grants employed initially by the conquistadores tended to establish a pattern of residential segregation in which the plaza played a role. Proximity to the plaza was viewed very positively, and consequently the highest ranking and most powerful of the conquistadores lived facing or close to the plaza. Eventually, a pattern of residential segregation based on wealth arose in most colonial Spanish American cities (Figure 7.5). The wealthy and powerful lived on or near the city's central plaza. In the urban capitals, they lived in palatial homes—really compounds—surrounding interior patios, which were often devoted to gardens.

As one's resources and social status lessened, the distance of residence from the central place increased. Early during the colonial period, peoples of mixed racial origin—mestizos, *zambos*, mulattos, and free blacks—lived on the periphery of the town. This changed, especially for mestizos, as the colonial period progressed and the number of mestizos increased, as did their opportunities for social advancement. Just beyond the formal limits of the town, often just across some kind of natural boundary like a river, creek, or ravine, a settlement often known as the "*pueblo de indios*" sprang up. This so-called pueblo, or town, was functionally really a *barrio*, or neighborhood, of the city. In subsequent decades, as colonial towns grew—and they did grow despite dramatic declines in total population during the 16th century—the Indian settlement was absorbed as part of the basic urban fabric.

The early physical plan of colonial cities, as well as their patterns of land use, remained remarkably similar throughout the immense geographical expanse of the Spanish colonies in the first decades of the colonial period, especially in the absence of any comprehensive royal directives or rules governing town planning. Such formalization of town-planning laws and regulations occurred later under a series of decrees and edicts generally known as the "New Laws of the Indies." The most definitive of these came in 1573 when the Spanish king Felipe II (Philip II) promulgated the "Discovery and Settlement Codes" governing

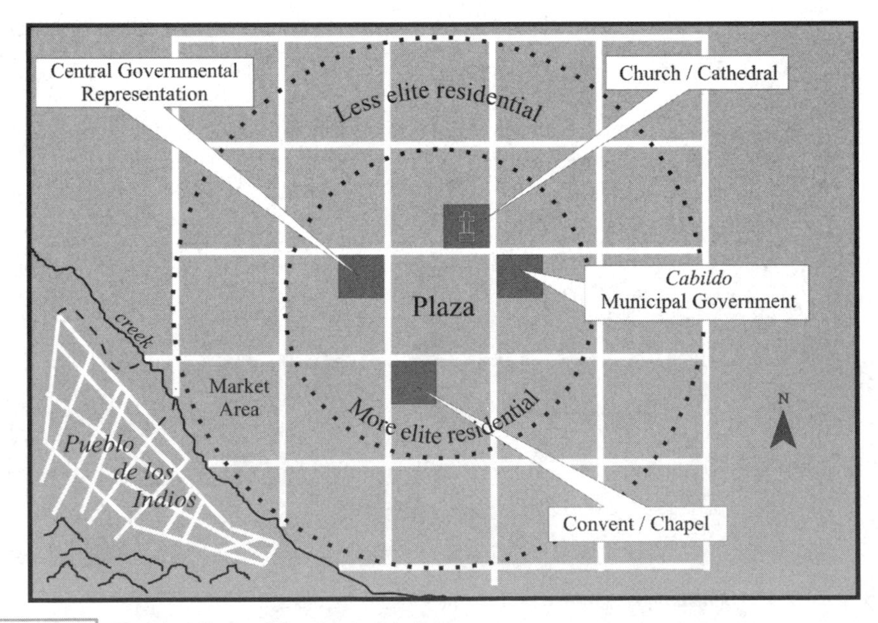

FIGURE 7.5. Model of the traditional colonial city.

the colonization and settlement of the new lands. These included the "Royal Ordinances Concerning the Laying Out of Towns," formally stating the guidelines for town planning for the colonies. These essentially confirmed and codified what had in fact been common practice for decades.

Colonial Towns and Urban Structure in Brazil

The evolution of urban places and the urban system in colonial Brazil exhibited marked differences from colonial Spanish America. One key difference was the nearly exclusive coastal orientation of almost all Portuguese cities established during the first 150 years of the colonial period. Settlement in early colonial Brazil concentrated on the Atlantic seaboard, running discontinuously from Belém in the north, situated at the mouth of the Amazon River, to Santos in the far south. Thus, coastal locations and port functions represented common denominators. Notable among these cities, beginning in the south and running northeast, are Santos (1532), Rio de Janeiro (1565), Espíritu Santo (1536), Ilheús (1536), Bahia (1549), Olinda (1537), Paraíba (1582), Fortaleza (1612), and—at the mouth of the Amazon River on its enormous estuary—Belém (1616).

In Spanish America, on the other hand, the Spanish founded literally hundreds of cities under a tremendous range of geographic conditions during the first 150 years of the colonial period (1500–1650). Settlements were established at ports, in mining districts, in fertile agricultural regions, in densely populated indigenous areas, and, for defense, at strategic locations.

During the three centuries of Portuguese colonial rule, far fewer cities were established in Brazil than in Spanish America. In fact, few towns were established in the Brazilian interior before the end of the 16th century; one notable exception is São Paulo (1558), although the city remained a secondary urban center until the end of the 19th century.

The discovery first of gold and later of diamonds in the Brazilian highlands (Minas Gerais, Goiás, and Mato Grosso) at the end of the 17th century brought an avalanche of fortune hunters into a few areas of the interior. Mining fostered the growth of a number of cities and towns across the central portion of the Brazilian Highlands. Villa Rica (Ouro Prêto; 1711), founded in torturous terrain near rich veins of gold-bearing ore, is clearly the most notable of these early interior cities, growing to a population of nearly 100,000 at the peak of the gold boom in the mid-18th century. However, it is useful to underscore the fact that urban development in Brazil's interior during the colonial period was severely circumscribed geographically, and that most of the interior remained thinly settled during this period.

While Portuguese settlers and colonists came to the New World without royal letters of instruction or royal ordinances that guided the siting and layout of towns as did the Spanish, Brazilian colonial cities do exhibit common site and structural characteristics. Defense played a dominant role in the siting of towns in the colonial period, and with good reason. The French and the Dutch vied with the Portuguese for control of parts of the Brazilian coast, and both nations occupied, colonized, and held portions of the coast for periods of up to 50 years. Thus, the Portuguese favored defensible coastal sites with good natural harbors.

Frequently, these sites gave rise to a two-tiered city: a lower city and an upper city. The lower city lay immediately adjacent to the water, focused on the port, and was usually sited on a small strip of flat land on the coastal margin. Much of the basic business of the city transpired here. Markets, warehouses, and transportation facilities could all be found in the lower city. The principal agricultural commodities of the surrounding rural hinterland were marketed and exported here. Sugar usually figured in this trade, but other agricultural products and raw materials were also exported. Slaves were imported, as were a diverse range of merchandise and commodities.

The upper city was another world sited on a bluff or promontory above the din and foul smells of the lower city. Forts occupied strategic locations, while the civilian city established itself around a series of plazas that were oriented in a linear fashion roughly parallel to the coast.

Governmental buildings, churches, and convents and monasteries were the principal landmarks in these colonial cities. Residential space occupied much of the upper city, distributed over an irregular street network imposed by the physical limitations of the terrain.

The urban system of Brazil changed little during most of the colonial period. Few new urban places were established after the initial round of city founding during the first century of colonial rule. The major cities remained fixed on the coast, oriented, on the one hand, toward Portugal and Europe, and, on the other hand, toward the Brazilian interior from which the agricultural products and natural resources that fueled the colonial mercantile system of trade originated. Indeed, as the economic significance of sugarcane production and the sugar trade declined in Brazil after 1650, the original colonial capital, Bahia (later Salvador), gradually lost power and importance, and its agricultural hinterland withered. Then when mineral strikes in the Brazilian Highlands at the beginning of the 18th century set off a boom in gold and diamond production, the nearest port city, Rio de Janeiro, grew rapidly. Its ascendancy was so complete that in 1762 the colonial capital was moved from Salvador (earlier Bahia) to Rio de Janeiro.

Cities and Urban Systems on the Eve of Independence

An extensive network of cities covered much of central Mexico, the Andes, and the northeastern coast of Brazil at the close of the colonial period in the first decades of the 19th century. Mexico City, Lima, and Salvador dominated these urban systems, although the latter's dominance was increasingly contested by the new capital, Rio de Janeiro. Havana and Buenos Aires dominated smaller urban networks, linking them through their ports to the colonial mercantile system. Cities grew during the colonial period, in both geographic extent and population numbers. Nevertheless, urban populations, even in the largest centers, rarely exceeded 100,000 inhabitants. Mexico City was likely the most populous urban center, with its population reaching approximately 130,000. Other sizeable urban centers in Mexico's urban system included Guanajuato (100,000), Puebla (60,000), and Guadalajara (30,000). In Peru, Lima's population reached about 50,000, and several of Peru's secondary cities reached appreciable size, including Arequipa (36,000), Cuzco (25,000), and Trujillo (12,000).

Along Brazil's northeastern coast a string of cities, each with a port, linked the interior agricultural hinterland with the Portuguese colonial economy. No city truly dominated this collection of urban centers, but Salvador with a population of close to 100,000, and the smaller national capital, Rio de Janeiro, with a population of about 60,000, were clearly the most important of these urban centers. Other coastal cities were appreciably less populous. For instance, Porto Alegre's population reached approximately 12,000, Recife's close to 20,000, and Belém's about 10,000.

In Spanish America, the dominance of Mexico City and Lima waned during the last years of the colonial period as administrative reforms and other changes slowly chipped away at the size of the geographic regions that comprised the Viceroyalty of New Spain and the Viceroyalty of Peru. The process was most pronounced in the Viceroyalty of Peru (originally all of Spanish South America), where the 18th century saw the carving off of the Viceroyalty of New Granada (Colombia/Ecuador), the Viceroyalty of Rio de la Plata (Argentina/Uruguay/Paraguay/Bolivia), the Captaincy General of Venezuela, and the Captaincy General of Chile. These changes stimulated urban growth in each of the capitals of these new administrative areas and contributed to the formation of city systems tributary to them. On the eve of political independence from Spain, these cities had achieved the status of key urban centers in South America. Buenos Aires, the capital of the Viceroyalty of Rio de la Plata, had about 40,000 residents; Bogotá, the capital of the Viceroyalty of New Granada, had 24,000, and Santiago and Caracas, capitals of captaincy generals, each accounted for about 30,000 inhabitants.

Although cities played a key role in colonial society and the economy, a comparatively small proportion of the population resided in urban places. But the fragmentary and tenuous nature of colonial censuses and demographic data, as well as contemporary differences in definitions and opinions as to what is considered a city or an urban place and what is not, make it difficult to say with any precision what proportion of the region's population was urban.

However, colonial census data from the last decades of Spanish rule in Peru provide a rough approximation of the relative proportion of urban population. City populations in Peru accounted for about 12 percent of the population enumerated in the "national" census. No comparable data are available for Mexico, but it is likely that the proportion of urban population in Mexico was greater, perhaps nearly 20 percent. Other major regions, including coastal Brazil, probably had urban populations ranging from 8 to 15 percent.

The basic urban hierarchy and city systems found in Latin America at the beginning of the 19th century remained, with few exceptions, unchanged until the end of that century. Then massive foreign immigration in some regions and the opening up of new settlement frontiers in others initiated changes in preexisting urban patterns.

Urban Development in the Republican Period

Social and economic processes during the postindependence period weighed heavily on the development of cities in 19th-century Latin America. A vast colonial empire disappeared and, with it, some of the political and social order it imposed. In Spanish America, nations did not arise whole cloth in the wake of the defeat and retreat of the Spanish. The formation of independent nation-states, in which a national government exerts effective political and administrative control over its territory, required decades, and in some cases continued well into the middle of the 19th century. Mexico and Argentina, two of the largest nations in Spanish America, experienced decades of political discord and intermittent armed strife, as did many other nations (Vignette 7.2).

The nation-states that did emerge generally coincided with the former administrative divisions of the Spanish colonial government, and their capitals were almost invariably the colonial capitals of civil administrative regions called viceroyalties, captaincy generals, and intendencies, or they were the capitals of judicial administrative areas called *audiencias*. Two bold experiments in the establishment of independent states uniting multiple colonial administrative units ended in failure within two decades of independence. The first, the United Provinces of Central America (Provincias Unidas de Centro América) disintegrated into five discrete political units within 15 years of its establishment. With the exception of Nicaragua, the individual countries that evolved were centered on former colonial capitals: Guatemala City, San Salvador, Comayagüela, and San José. The second, the Republic of Gran Colombia, which attempted to integrate Venezuela, Colombia, and Ecuador, evolved similarly. Within 10 years of its creation, it too disintegrated into discrete nation-states dominated by the former colonial capitals of Caracas, Bogotá, and Quito, respectively.

VIGNETTE 7.2. REGIONS, *CAUDILLOS*, AND THE POLITICAL LANDSCAPE IN THE EARLY REPUBLICAN PERIOD

In what must be described as a Herculean feat, Spain's colonial officials successfully overcame the immense distances and difficult physical geography the New World presented and imposed a reasonably orderly and productive colonial administration that lasted for three centuries. The demise of Spanish colonial rule over the Latin American mainland brought many changes. First, it left an immense economic, political, and military power vacuum. At an international scale, other major European powers, particularly Great Britain and to a lesser extent France, and somewhat later the United States, occupied the position formerly held by the Spanish. Foreign capital from these and other European nations soon played a controlling role in most of the region's principal sectors: banking, mining, agriculture, communications, and transportation. This financial and commercial power was soon translated into a regional economic and political hegemony by Great Britain, which remained the dominant power until the late 1800s. The United States supplanted Great Britain in the region as the principal mercantile and political power at the beginning of the 20th century.

In most of the new republics the early decades of political freedom were frequently accompanied by civil strife and armed conflict as competing political interests vied for control. In most Spanish American nations, the broad outlines of political debate were framed around the use of the terms "liberal" and "conservative" to describe the political parties that emerged to represent these interests. Liberals generally took their inspiration from political thought and developments originating in Western Europe and the United States. Largely a product of the Enlightenment, these beliefs emphasized the rationality of human beings, as well as individual liberty and equality. This placed liberals, who were drawn principally from among the ranks of professionals and the middle class in urban centers, squarely at odds with two of the society's most powerful social groups: the Catholic Church and rural landowners. These two groups—and other elites with a strong commitment to the maintenance of the status quo—coalesced under the banner of the conservatives. Staunch defenders of the central role of the Catholic Church in society and its privileges and property, conservatives espoused no clear political philosophy, often standing simply as the guardians of tradition and the status quo.

An early and enduring political debate, with immediate and long-term political and economic consequences, occupied Latin Americans early in the republican period. What structure should the emerging "democratic" governments take? One camp, associated with liberals, espoused a federalist solution much like the model used in the newly independent United States of America. This vision typically saw each nation's subnational political units (states, provinces, etc.) exerting regional autonomy over a range of legislative and tributary matters, including the judiciary and police. The other camp, associated with conservative political parties and thought, advocated a centralist solution to governance, where a national legislature and a strong chief executive would exercise broad authority across the entire nation. Centralists saw the subnational political units, usually dubbed provinces or departments, as no more than administrative subdivisions. Prefects or governors appointed by the nation's chief executive (president) would administer these regions, which would enjoy no local control over legislation or tax codes, while a national police force and a central judiciary would enforce the nation's laws. A decades' long struggle ensued in many of the Spanish

(cont.)

American republics, where political debates centered on the contrasting visions of federalist and centralist governance and liberal and conservative ideas and values. All of this played out at the ballot box and, as often as not, at the end of the barrel of a gun.

This volatile political environment, as well as the difficulties imposed by the physical landscape and regional isolation, contributed to the rise of the *caudillo*, who was to become a fixture on Latin America's political landscape for over a century. *Caudillos*, often described as regional strongmen, were men of action who frequently had military or quasi-military backgrounds. Many had fought in the wars of independence against Spain, in which they had gained both military experience and a certain national prominence and visibility. Typically, a caudillo drew his power from a peripheral rural region where he commanded an irregular militia whose members owed him their personal loyalty. Regional elites, usually large landowners like *hacendados*, provided successful caudillos with direct support that included money and additional armed men, permitting them to consolidate power over a vast interior region. National governments were weak and often could not effectively control much of the national territory either politically or militarily beyond the immediate environs of the national capital. Caudillos filled these spaces, offering a semblance of order and stability to the regional elites and rural landowners who supported them and whose interests they protected.

These personal fiefdoms existed over the length and breadth of Spanish America during the 19th century. Essentially apolitical, caudillos vied for personal power and wealth, allying themselves with whomever might advance their desires most effectively at the moment—whether it be liberals, conservatives, federalists, or centralists. Throughout this period, the very survival of many national governments depended upon the support of at least some prominent caudillos. Loyalties were ephemeral, however, and shifting alliances often meant frequent changes in chief executives and governments.

Many regional caudillos became nationally prominent, and the most successful of these occupied the presidency. In Argentina, Juan Manuel de Rosas achieved regional prominence and political power on the Pampa with the support of the cattlemen and *gauchos* (cowboys). He declared himself governor of Buenos Aires province in 1829, a position he held almost continuously from then until 1852, when he was driven from power by another regional caudillo, this one from the north, Justo José de Urquiza. In the absence of a national government, Rosas' hold on the province of Buenos Aires and the nation's chief port, also Buenos Aires, gave him effective control of the nation for nearly 25 years. In Mexico, General Antonio López de Santa Anna (1794–1876), occupied the presidency of Mexico 11 times between 1833 and 1855. Essentially apolitical, Santa Anna rode into Mexico City from his power base in rural Veracruz to occupy the presidency under all manner of political affiliations.

The classic regional caudillo of the 19th century served as a model for another of Latin America's stereotypical leaders, the military strongman who assumes the presidency. Porfirio Díaz, who ruled Mexico with an iron hand between 1870 and 1910, was an early example of this type. So too were Rafael Trujillo, who governed the Dominican Republic from 1930 to 1961; Juan Domingo Perón, who was president of Argentina from 1945 to 1954; and, most recently, Alfredo Stroessner, who governed Paraguay from 1954 to 1989 when his overthrow finally led to the establishment of democratic governance. In his novel of historical fiction, *Feast of the Goat*, the Peruvian novelist Mario Vargas Llosa provides a chilling and insightful account of the life and times of one such leader, the ruthless Dominican dictator Rafael Trujillo.

In these cases, and in many others, national capitals evolved into the controlling political entities in their respective nation-states, growing to wield overwhelming and uncontested power in the national context. Such primate cities (see the section "Urban Primacy and Primate Cities" in Chapter 14) often accounted for as much as one-quarter of a nation's population and exerted a pervasive influence over the nation's social and economic life.

While independence brought a certain measure of political and economic freedom to each new nation, the basic model of colonial economic development prevailed. The newly independent Latin American republics continued to produce agricultural commodities and natural resources for export in return for European and North American manufactured goods and other products. This model of "dependent development" characterized the 19th century. Although Spain's hegemony evaporated quickly, Britain, and to a lesser extent France and the United States, quickly filled the economic and political vacuum Spain's absence created.

This pattern of development reinforced the dominance of the traditional colonial ports and administrative centers. Investment tended to focus on infrastructure that would facilitate the export of each nation's principal commodities—whether it was beef, coffee, tin, rubber, or any one of a handful of other commodities. Therefore, the construction of railroads, roads, and port facilities focused on each nation's principal city/port and failed to serve the broader needs of a developing economy. In many cases, the transportation network, and especially railroads, took on the appearance of the ribs of a fan or that of a hub and spokes, with all lines converging on the nation's principal port and capital city.

The vast, open plains of the Argentine Pampa provide a classic example of this pattern. Here, at the beginning of the 20th century, an extensive network of railroad lines had penetrated the Argentine Pampa, all radiating from the capital and principal port, Buenos Aires, with few connections between the "spokes" of the network (Figure 18.5). Direct rail connections between the capital cities of neighboring provinces did not exist, and train travel between such places required a long detour through Buenos Aires, the network's hub. In Andean countries like Peru, where the physical geography did not favor the development of a national transportation system focusing on a single hub, railroad lines and roads ran from mines and agricultural production regions straight to port facilities and export markets.

The demands of an agro–export economy dependent on foreign capital and international markets directed the terms of the subsequent development of urban centers like Buenos Aires, Santiago, Lima, Montevideo, Rio de Janeiro, and Mexico City. A heavy dependence upon export duties to generate government revenues and the centralization of administrative and political control also contributed to the further concentration of population, commerce, and political activity in these major centers. This simultaneously impeded the growth of secondary cities and retarded the evolution of more balanced urban systems.

While population growth in urban centers during the first half of the 19th century was slowed by social strife and civil disorder in many countries, the second half of the century saw rapid growth in many of the region's principal urban centers. By 1900, Buenos Aires was very much a cosmopolitan center, seen by many as a kind of "Paris" in the New World, with a population of nearly 1 million inhabitants. The populations of Rio de Janeiro and Mexico City both exceeded 500,000, while the populations of Montevideo, Santiago, Havana, São Paulo, and Salvador (formerly Bahia) ranged between 200,000 and 300,000.

Summary

Before European colonization, urban development occurred in two regions in parallel with the development of advanced civilization: the Mesoamerican and the Andean realms. Cities began as religious centers and evolved into centers of political and military power. Cities had occupational diversification and class stratification. The Inca Empire gave rise to the most integrated and extensive urban system in pre-Columbian America.

During the colonial period, cities played different roles for the Spanish and the Portuguese. For the Spaniards, the city was a key element in conquest and settlement. The Spanish established cities throughout their empire populated with soldiers, clerics, and officials to manage the surrounding region. Santo Domingo was Spain's first urban center in the colonial period and included design features repeated throughout the empire. Cities were plotted with a regular street grid around a central plaza housing a cathedral and public buildings. The Spanish encountered urban centers like Tenochtitlán and Cuzco, and adapted them into their own cities, modern-day Mexico City and Cuzco, respectively. Other modern cities, such as Quito and Bogotá, were built on the sites of precolonial cities. The pattern of Spanish urban development placed wealthy residents close to the central plaza and urban core, and those of lesser social and economic standing on the city's periphery.

Early Portuguese cities, on the other hand, were developed principally on the coast. The Portuguese cities have common characteristics, including good harbor sites for defense. These locations often led to the development of upper and lower cities. Shipping, warehousing, marketing, and poor and working-class residential districts were located on the lower level by the harbor, whereas government buildings, commercial and financial activity, and upper-class residential areas developed on the bluffs of the upper city.

With the end of the Spanish Empire in the early 19th century, the former colonial capitals and administrative centers often became the capitals of the newly independent countries. The commercial functions of these cities became more diversified as the national economies of most countries became highly dependent on the export of natural resources and agricultural products to European markets.

Further Reading

Flannery, K. V. (Ed.). (1976). *The early Mesoamerican village*. New York: Academic Press.

Gade, D. W., and Escobar, M. (1982). Village settlement and the colonial legacy in southern Peru. *Geographical Review, 72*, 430–449.

Greenfield, G. M. (1994). *Latin American urbanization: Historical profiles of major cities*. Westport, CT: Greenwood Press.

Hardoy, J. E. (1975). Two thousand years of Latin American urbanization. In J. E. Hardoy (Ed.), *Urbanization in Latin America: Approaches and Issues* (pp. 3–55). Garden City, NY: Anchor Books.

Hoberman, L. S., and Socolow, S. M. (Eds.). (1986). *Cities and society in colonial Latin American*. Albuquerque: University of New Mexico Press.

Hyslop, J. (1990). *Inka settlement planning*. Austin: University of Texas Press.

Morse, R. E. (Ed.). (1971). *The urban development of Latin America, 1750–1920* . Stanford, CA: Center for Latin American Studies, Stanford University.

Nutall, Z. (1922). Royal ordinances concerning the laying out of new towns. *Hispanic American Historical Review, 5*, 249–254.

Sanders, W. T., and Webster, D. (1988). The Mesoamerican urban tradition. *American Anthropologist, 90*, 521–546.

Smith, R. C. (1955). Colonial towns of Spanish and Portuguese America. *Journal of the Society of Architectural Historians, 14*, 3–12.

Stanislawski, D. (1947). Early Spanish town planning in the New World. *Geographical Review, 37*, 94–105.

8 The Andes

The Andes is one of Latin America's most prominent environmental and cultural regions. The mountain range that defines this region extends along the continent's entire western margin from the Caribbean coast to Tierra del Fuego and includes some of the highest peaks in the Western Hemisphere. Sophisticated and highly developed civilizations arose in the Andean highlands and the region is one of only a handful of world centers of plant and animal domestication. The Inca Empire, one of the most advanced civilizations to develop in the Americas prior to the Conquest, occupied the central Andean highlands.

The region is rich in mineral resources. It was the world's principal producer of silver during the 16th and 17th century and continues to produce large quantities of a wide range of precious and industrial metals, including gold, silver, tin, lead, and copper. Significant quantities of natural gas and petroleum are found in the Andean piedmont.

The population of the Andes region, which includes the highlands of Venezuela, Colombia, Ecuador, Peru, and Bolivia, numbered between 60 and 70 million at the beginning of the 21st century. Four national capitals, Caracas (Venezuela), Bogotá (Colombia), Quito (Ecuador), and La Paz (Bolivia), all of which have populations greater than 1 million, are sited in the Andes. The region is also home for large concentrations of extant indigenous peoples. These include as many as 10 million Quechua peoples and close to 1.5 million Aymara peoples, many of who still speak their native languages.

Geography and Environment

This massive mountain chain, one of the most geographically extensive in the world, begins near Venezuela's Caribbean coast at about 10°N. The range runs westward into Colombia, and then bends south and extends south, uninterrupted, until it slowly disappears as a mountain chain at about 50°S and grades into the Patagonian steppe at the southern end of the continent (Figure 8.1).

The theory of plate tectonics, which explains the movement of massive sections of the Earth's crust, accounts for the formation of the Andes. The range rises near the margin of two of the Earth's major tectonic plates, the continental South American Plate and the oceanic Nazca Plate. The South American Plate, which comprises the area of the Earth's crust underlying the western half of the South Atlantic and the South American continent, is pushing westward. It collides with the eastward-moving Nazca Plate that underlies much of the eastern Pacific along a broad front corresponding roughly with the western coast of the continent. Along this front, called a subduction zone, the Nazca Plate is pushed under the South American Plate. The subduction process

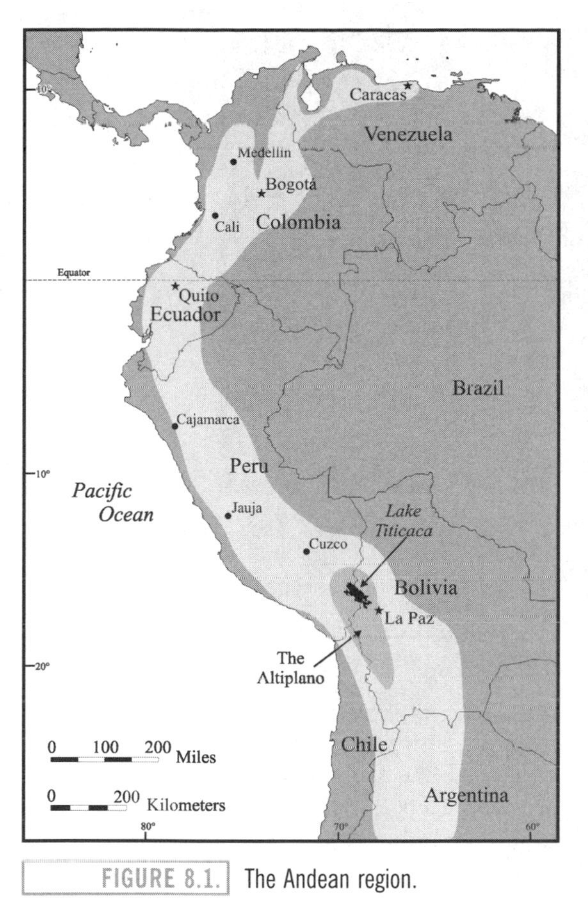

FIGURE 8.1. The Andean region.

at the zone of contact between the two plates can cause earthquakes and volcanic eruptions along the entire front. It also gives rise to the formation of mountain ranges, in this case the Andes, and a deep oceanic trench offshore, known as the Peru–Chile Trench. The Andes are overlain by sedimentary rock, eroded away in some areas to expose the igneous and metamorphic rocks that form the geological core of the range.

The active geological nature of the Andes, a relatively young mountain range, means several types of natural hazards are common. Precipitous slopes and heavy rains in some areas combine to cause frequent landslides and earth flows (Figure 8.2). In addition, the entire Andean region is classified as having high seismicity; devastating earthquakes have occurred in historic times in several highly populated areas, causing immense destruction and huge loss of life.

Volcanism occurs in three broad regions along the length of the mountain range: southern Colombia and Ecuador, southern Peru and northern Chile, and south central Chile. One of the most devastating eruptions in modern times occurred in northern Colombia when Nevado de Ruiz (5,321 m) erupted in November 1985. The initial eruption set off a series of immense landslides, or lahars. *Lahars* are rapidly moving mixtures of snow, ice, mud, and lava unleashed by volcanic eruptions. The lahars from Nevado de Ruiz, at times up to 50-m thick, moved down hill rapidly, covering nearly 100 km in only 4 hours. Downslope the town of Armero was completely destroyed, and approximately 23,000 people died.

The chemical character of the magma underlying the Andes often leads to explosive volcanic eruptions in which vast clouds of volcanic ash, rather than molten lava, are sent into the atmosphere, sometimes effecting vast territories. In 1991 a volcano in southern Chile erupted, sending volcanic ash eastward into the Argentine Patagonia, blanketing the grasslands there with fine ash, causing the deaths of hundreds of thousands of sheep, and resulting in a major blow to the region's economy.

The character of the Andes varies notably as one moves from north to south. At first it divides into several parallel ranges. These then join to form an extensive cordillera, a massive continental-scale mountain range, which in the central Andes is subsequently transformed into a vast plateau wedged between mountain chains. Finally, in the south, the cordillera coalesces into a single chain that runs to the end of the continent.

In the north, outliers of the Andes appear at about 65°W near Venezuela's Caribbean coast, and run in an east–west direction into Colombia. A low coastal range, on which Caracas sits at about 1,000 m, parallels a higher, more rugged, interior range, the Sierra del In-

FIGURE 8.2. A bulldozer works to clear a landslide from a highway in the Ecuadorian Andes as stranded travelers look on, 1975.

terior. These two ranges join in western Venezuela, forming the more formidable Sierra de Mérida, with maximum elevations reaching about 4,000 m (Pico Bolivar). Venezuela's population is concentrated here in many major towns and cities, including Valencia, Barquisimeto, Trujillo, Mérida, and San Cristóbal. Agriculture is also concentrated here, with subsistence crops as well as such commercial crops as coffee, cocoa, cotton, and sugarcane.

The Andes trend north–south subsequently. This pattern is clear in Colombia, where the Andes divide into three distinct ranges. These are the Cordillera Oriental, the Cordillera Central, and the Cordillera Occidental. The Magdalena River Valley lies between the Cordillera Oriental and the Cordillera Central, while the Cauca River Valley separates the Cordillera Central from the Cordillera Occidental. As in most Andean countries, many of Colombia's main urban centers—Bogotá, Medellín, and Cali—are located in the highlands, and nearly all of its population is concentrated there as well. The Bogotá Plateau, adjacent to the city of Bogotá, in the central portion of the Cordillera Oriental, is one area of concentrated highland settlement. To the northwest, in the Cordillera Central, in the environs of Medellín, the Antioquia Plateau is also densely settled, while in the extreme south population is densest on the Popayan Plateau. Just north of Colombia's border with Ecuador the three ranges unite in a complex topographical feature known as the Pasto Knot.

The Andes divide again in Ecuador, forming two parallel ranges, in the east the Cordillera Royal and in the west the Cordillera Occidental. Impressive snow-capped volcanic peaks punctuate both ranges, including Cayambe (5,791 m), Cotopaxi (5,897 m), and Chimborazo (6,272 m; see Vignette 8.1). More than 15 volcanic peaks in the Ecuadorian Andes exceed 4,572 m in elevation. A series of densely settled highland basins, with elevations between 2,000 m and 3,000 m, lie between the two ranges, and extend the length of the country. Many of the country's principal urban centers, including Quito, Cuenca, and Loja, are sited in these highland basins.

As the Andes reach southward into Peru, the range takes on a variety of forms. In the north, three distinct parallel cordilleras can be distinguished. The front range, which parallels the Pacific coast, is separated from the middle cordillera by the deeply incised valley of the

Marañon River; likewise, the Huallaga River separates the middle cordillera from the easternmost range. Southward a fourth range appears that is generally coincident with the upper reaches of the Marañon and Huallaga River valleys. This range, called the Cordillera Negra, and the next, known as the Cordillera Blanca, provide some of the most spectacular alpine scenery in the central Andes. Mountain glaciers cover much of the Cordillera Blanca and are the source of its name: the "White Mountains." Peru's highest peak, Nevado Huascarán (6,768 m), is also found here. Southward into central Peru the distinct cordillera that characterized the Andes in the north coalesce into a single cordillera, although distinct spurs and ranges can be distinguished in many areas. In southern Peru, however, the range divides into two distinct cordilleras, a western range and an eastern range that continue into Bolivia, where they are known as the Cordillera Occidental and the Cordillera Oriental. A vast upland plateau region, the Altiplano, lies at elevations of between 3,600 m and 4,000 m between these two Andean cordilleras. Beginning in southern Peru, the Altiplano extends southward into Bolivia for nearly 800 km, extending nearly to Bolivia's border with Argentina. Despite its high elevation, the Altiplano is densely populated, especially in its northern half. Two immense intermontane lakes with no external drainage are found here: Lake Titicaca and Lake Poopó. East of the Cordillera Oriental a series of lower ranges extend in a stair-step fashion eastward into the lowlands of the Amazon Basin. These ranges and associated valleys form a region known in Bolivia as the Yungas, an Aymara word for warm valleys, or simply *valle* in Spanish.

At about 25°S, in northern Chile and Argentina, the principal cordillera of the Andes lies immediately east of the Pacific coast, while a series of three to four smaller parallel cordillera lie further to its east. Southward, these outlying ranges become more diffuse and lower, creating a series of small ranges bordering the Pampa. One of these outliers, the Sierra de Cordoba, lies some 500 km east of the principal crest of the Andes. Some of the highest peaks in the Andean chain lie in this region. Cerro Aconcagua (6,960 m) on the Argentina–Chile border is the highest. Overall elevations descend as the Andes run southward toward the southern tip of the continent, with most peaks standing between 2,000 m and 3,000 m at the southern extreme of the range near 50°S. Vast glacial ice fields, lying largely in Chile, but also extending into Argentina, cover much of the upper elevations of the Andean range between 45° and about 50°S.

Climatic patterns in the Andes are so complex that it is difficult to offer broad patterns. Most maps of world or continental cli-

VIGNETTE 8.1. HOW HIGH IS CHIMBORAZO?

Geographical facts are often open to some dispute. Once thought to be the highest peak in South America, Chimborazo has long been the object of attention by geographers and mountaineers. Alexander von Humboldt, the famous German geographer and explorer, attempted to scale it in 1802, but only reached an elevation of 5,759 m. It was not until nearly 80 years later that the British mountaineer Edward Whymper successfully reached its summit. But how high is Chimborazo anyway? Well, it depends on whom you ask. The *Encyclopaedia Britannica* reports Chimborazo's elevation to be 6,310 m, *Webster's New Geographical Dictionary* cites a figure of 6,267 m, and the *Physiographic Diagram of South America* notes its elevation as 6,272 m. Of course, one might also ask, How much does it matter unless one is a mountaineer?

mates do not define a climatic type for the Andean highlands, instead offering vague statements like "complex highland climates" or "climate varies with elevation and latitude" to describe the region. Nevertheless, it is possible to outline some broad parameters that help one understand the complexity of climates found in this region. The Andes straddles an immense latitudinal range (10°N–50°S). As a consequence, global pressure and wind belts, specifically the ITCZ (the Inter-Tropical Convergence Zone, the equatorial low-pressure belt), the Trade Winds, the subtropical high-pressure belt, and the midlatitude low-pressure belt (the Westerlies) do affect local climates throughout the Andes. However, their impacts are mitigated and modified by elevation and aspect.

Temperature variations in the Andes are strongly impacted by the environmental lapse rate. The *environmental lapse rate* is the regular decline in temperature that occurs as elevation increases. As elevation increases, temperature declines at a rate of approximately 0.5°C per 100 m. As a result, within mountainous regions of the Tropics, like the Andes, a series of distinct climatic belts, or life zones, are created. These are widely acknowledged within the Andes, although sometimes terminology varies within different national contexts.

Three major zones, in which most of the Andean population resides and within which a wide range of agricultural and pastoral occupations can be pursued, occupy the uplands between sea level and about 3,000 m. A tropical life zone, *tierra caliente* (hot land) falls between roughly sea level and 1,000 m. A subtropical belt, *tierra templada* (temperate land), sits between 1,000 and 2,000 m. A much more temperate region, *tierra fria* (cold land), lies between 2,000 and 3,000 m.

Above 3,000 m, population is sparse and human use of the landscape considerably more limited. The *puna*, or the *páramo* as it is also known, lies between 3,000 m and 4,500 m. Here agricultural opportunities are severely restricted and grazing is usually limited to sheep and llamas (Figure 8.3). Both crop agriculture and grazing are found only along the lower portion of this life zone. Mining, however, is an important economic activity in this zone. Above 4,500 m permanent ice and snow are common, even at tropical latitudes; this altitudinal belt is called *tierra helada* (frozen land). A more detailed discussion of these altitudinal zones is provided in Chapter 6.

Rainfall patterns are complicated and defy easy description. The major global pressure and wind belts establish the basic parameters of rainfall over this region. However, aspect and elevation weave a dizzying array of microenvironmental zones where precipitation can vary immensely over very short distances. In the Andes, it is possible, in the span of some 30–40 km, to go from a well-watered temperate upland, into a dry, almost desert-like valley, and then into a upland cloud forest that is drenched in almost perpetual precipitation. Typically, however, throughout the tropical belt of the Andes, precipitation along the eastern slopes and the piedmont is heavy and occurs throughout most of the year. Precipitation in the uplands and the intermontane valleys is usually seasonal; most of these areas are well watered without being excessively humid. Precipitation patterns along the western

FIGURE 8.3. High-altitude *puna* landscape in the Andes of northern Argentina. Bunchgrass (*ichu*) is the predominant vegetation, 1993.

slopes of the Andes vary considerably over its immense latitudinal extent. In the north, roughly coincident with Colombia and northern Ecuador, precipitation is heavy. It decreases quickly through central Ecuador and near-arid conditions prevail along the entire western slope of the Andes in Peru and into northern Chile.

A snapshot of basic climatic conditions in several major urban centers across the length of the Andes provides some appreciation of the variation in conditions in this region. Caracas, Venezuela, lies in a rift valley at the extreme northern end of the Andes at about 10°N at an elevation of slightly over 900 m. It sits on the boundary between tierra caliente and tierra templada. Temperatures vary little throughout the year, with the mean temperature in the warmest month of the summer, July, standing at 20.6°C, while the mean temperature in the coolest winter month, January, is about 18°C. Precipitation is seasonal, although some rain falls in every month. Annual precipitation totals about 800 mm, with a distinct rainy season between May and November when almost 100 mm of rain falls each month. Between December and April precipitation is sparse.

Further to the south in the Andean highlands, Quito, Ecuador, lies almost squarely on the equator. It sits in a large intermontane basin at an elevation of 2,850 m, placing it in the upper margins of the tierra fria zone. Average monthly temperatures throughout the year vary little and stand at about 12.5°C. This is due to the city's equatorial location where the intensity of the sun's rays remains the same all year long. Precipitation, however, does flucuate on an annual basis, with a short comparatively dry season during June, July, August, and September, and then an extended rainy season with monthly precipitation totals over 100 mm during the rest of the year. Annual precipitation totals approximately 1,050 mm.

La Paz, Bolivia, lies at the southern end of the Andean Tropics. At 3,580 m, it is one of the world's highest elevation urban centers. The core of the city sits in a deeply incised valley, just below a high-altitude plateau, the Altiplano. It lies squarely within the puna/paramo life zone.

Although La Paz lies in the Tropics, its high-altitude location means that temperatures are comparatively cool and that frosts can occur during the nights of the coolest months of the year. Annual variations in mean monthly temperatures between the winter and summer months are significant due to the city's southerly location in the tropical belt. In the winter's coolest months, June and July, monthly means stand at about 6.5°C, while in the summer's warmest month, November, the monthly mean is about 11°C. These statistics tend to obscure the fact that diurnal temperatures vary dramatically in both summer and winter, with temperature ranges as great as 25°C between nighttime lows and daytime highs (Figure 8.4).

Average annual precipitation in La Paz is modest, about 550 mm. Its latitudinal location at the margins of the Tropics means that it is characterized by a pattern of precipitation that is tropical wet and dry. Most rainfall occurs during the summer months from December to March, when about 100 mm per month falls. Rainfall is sparse the remainder of the year, with almost none falling during the winter.

Historical Geography and Economic Development

Human settlement in the Andes dates back several millenniums. Like the southern highlands of Mexico, the central Andes region is one of a handful of world centers where sophisticated civilizations arose. The region also is one of the few centers of plant and animal domestication worldwide where agriculture developed and advanced agricultural practices

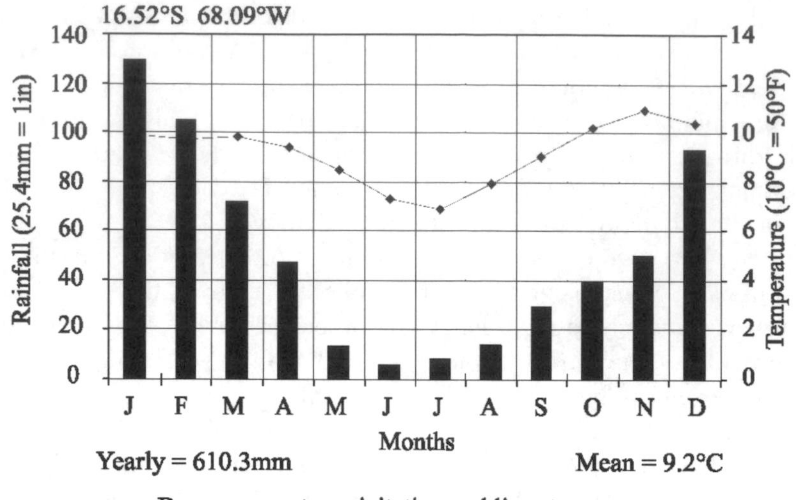

FIGURE 8.4. Climograph of La Paz/El Alto, Bolivia.

evolved. These practices, including irrigation and land terracing, created food surpluses, and consequently led to high population densities and the evolution of complex societies in which social classes and occupational specialization developed.

Agriculture and herding formed the basis for the development of early civilizations in the Andean region. Plant and animal domestication began here more than 5,000 years ago. The potato is the most well known of the Andean domesticates. Among them it has clearly had the greatest impact on other peoples and cultures throughout the world since its diffusion from the Americas to the Old World after the Conquest. Early Andean peoples also domesticated many other tubers, although most have not been widely adopted as crop plants outside the region. Nevertheless, a significant number of other plant domesticates originated in the Andes, including strawberries, quinoa (a high-protein seed crop), string beans, zinnias (a common garden flower), and coca, the plant from which cocaine is derived. Animal domesticates from the Andes are few, and generally have had little impact on other cultures beyond the region. Andean peoples domesticated two camelid species, llamas (used as beasts of burden and to some degree for wool and meat) and alpacas (used principally for wool). The guinea pig is another Andean domesticate. It is best known to North Americans and Europeans as a small docile house pet kept in cages by young children, but in Andean culture, often to the surprise of outsiders, guinea pigs, which can reproduce rapidly, are a traditional rural food raised by peasants.

A dense population of indigenous peoples occupied the Andes at the time of European contact in the first half of the 1500s. While estimates vary, somewhere between 10 and 15 million people may have inhabited the region. The most advanced of these peoples were the Inca (Quechua language speakers), who occupied a broad swath of the central Andes stretching from the southern portion of present-day Colombia through Ecuador, Peru, Bolivia, and into the extreme north of Argentina. The Inca, who had only consolidated this vast empire a few hundred years prior to the arrival of the Spanish in South America, had built upon the cultural advances of a series of earlier cultures. The Inca had established a tightly organized and highly centralized monarchy ruled by "the Inca," or emperor, and a royal elite. An extensive "road" network (paved trails and marked paths 1—2 m wide) connected the disparate points of the empire. Governance, military, and religious functions were concentrated in a series of urban centers, some of which were the precursors of such present-day cities as Quito (Ecuador), Cajamarca (Peru), Jauja (Peru), and Cuzco (Peru). Cuzco, the empire's capital, had a population of close to 100,000 at the time of contact with Europeans in the early 1500s.

Other aboriginal peoples occupied the remainder of the Andean highlands. In the north, chiefdoms, like the Chibcha, lived in the mountainous areas of Colombia and Venezuela. To the extreme south, the uplands of Chile and Argentina were largely uninhabited. On the Altiplano, surrounding the shores of Lake Titicaca, between 1 and 2 million Aymara people, a fiercely autonomous culture, sustained a measure of independence despite the constant efforts of the Inca to subjugate them.

The Spanish conquest of the Andes, and indeed of South America, began in the early 1530s when a small force of conquistadores under the command of Francisco Pizarro landed on the north coast of Peru near the modern city of Trujillo. Pizarro and his men moved inland quickly and soon reached the Inca city of Cajamarca in the Andean highlands. Here they encountered Atahualpa, one of two royal brothers feuding over the succession to the Inca throne, and his army. Despite the overwhelming numerical superiority of the Inca army, Pizarro and his men, with the aid of firearms, horses, and audacious treachery, routed the Inca army and captured and executed Atahualpa. By 1535 the Spanish had defeated the remaining Inca armies, occupied the Inca capital of Cuzco, and established effective control over the greater part of the Inca realm. While the number of conquistadores and colonists was still very small, Spanish colonial rule of most of the Andean region was effectively consolidated by the middle of the 1550s.

The siting of Spanish colonial settlements and urban development in the Andes tended to be determined by two key factors: native population concentrations and the location of precious mineral resources. Initially, the Spanish were attracted to areas with large concentrations of indigenous peoples. Dense populations were a sign of fertile land. Moreover, these dense populations were associated with population centers that often reached urban proportions.

Early on the Spaniards employed a range of techniques to extract tribute and/or labor from aboriginal peoples, including the *encomienda* and subsequently the *repartimiento.* The *encomienda* was a system in which elite colonists extracted tribute from indigenous communities. It was officially outlawed early in the colonial period, but it survived in many areas despite its official banning. The *repartimiento*, was a system of forced labor by indigenous peoples to benefit Spanish colonists and the government. The Spanish also adapted the Inca tribute system, the *mita*, to their colonial administrative system. These techniques extracted both labor services and tribute from local communities and imposed a tremendous toll on Andean native peoples. They enriched the Spanish elite and the Spanish Crown in one way or another for most of the remaining 350 years of colonial rule in the Andes region. A more in-depth discussion of the encomienda and the repartimiento is provided in Chapter 12, in the section "European Impacts on Land Tenure."

Thus, Spanish settlement and city formation followed predictable patterns in the Andes. Early population centers included the Sabana de Bogotá, a cool, moist, fertile plateau (2,600 m) in the Colombian highlands with a dense aboriginal population of Muisca peoples. Bogotá was established on the plateau in 1538, grew into a major colonial center, and later became the national capital of modern

Colombia. Other Spanish colonial centers included Quito, an Inca city, and now the national capital of Ecuador, situated in a fertile and densely populated highland valley at approximately 2,800 m. The Inca capital, Cuzco, in the central Andes of what is today Peru, was another of these colonial cities that emerged from an area of dense indigenous population. Arequipa (Peru), Cochabamba (Bolivia), La Paz (Bolivia), and Salta (Argentina) were all colonial centers that grew from similar origins as native cities or areas of dense aboriginal settlement.

Mining also played a role in the settlement of the Andean highlands and the establishment of urban centers. The exploitation of precious minerals drove the behavior of Spanish authorities for most of the colonial period. Indian labor was utilized to exploit gold and silver deposits wherever they could be found. Initially Spanish efforts were focused on the exploitation of easily reached placer deposits. Since these deposits were often limited and disbursed over broad areas, the settlements associated with them were often no more than simple camps and temporary villages. Nevertheless, in about 1545 the discovery of bountiful deposits of silver in the Bolivian highlands in a mountain thereafter named "Cerro Rico" (Rich Mountain) gave rise to the founding of the city of Potosí. Despite its location at 3,995 m and the absence of an exploitable aboriginal population or agricultural resources in the vicinity, for nearly 100 years it was one of the principal cities in the Andes, with a population that exceeded 100,000. Huancavelica, Peru, was another of the handful of city-size mining centers established during the colonial period in the Andes. Also situated at high altitude (3,800 m), the city prospered thanks to silver and mercury mining. Mercury, a key ingredient in the amalgamation process for extracting silver from silver ore, propelled the city's growth as a key export item to other mining districts in the Andes, especially Potosí.

While towns and urban centers were the primary focus of Spanish settlement, investment, and interest during the colonial period, limited rural settlement also occurred. In the Andes, the hacienda was the principal means of Spanish occupation of the countryside. Based on royal land grants to conquistadores and early colonists, these large feudal agricultural estates incorporated legions of indigenous and mestizo workers as part of their economic fabric. Remote from international markets and ports, the haciendas focused on the production of livestock on the hoof (cattle, horses, donkeys, mules, and sheep) and their nonperishable products, like wool, tallow, dried meat, and hides. Crop agriculture played a small role in their economic mix and typically was largely for subsistence needs or for sale in local markets.

Many Indian communities in the Andes remained intact throughout the colonial period. Occupying remote and/or marginal lands of little interest to the colonial authorities, some native peoples managed to maintain strong cultural traditions and vibrant communities well into the 19th and 20th centuries.

Independence from Spain in the early decades of the 19th century brought few immediate changes to the Andean landscape and settlement patterns. But, by the middle of the century, important changes in the region's economic geography initiated significant changes in land-use patterns, social relations, and settlement. The increasing importance of Andean commodities in the world economic system dominated these developments. Notable changes included the introduction of coffee in the northern Andes; the growing importance of the mining of industrial minerals, especially lead, zinc, copper, and tin; and the introduction of hybrid sheep for wool production in the southern Andean highlands of Peru.

The introduction of coffee into the Andean highlands of southern Colombia and western Venezuela in the mid-1800s energized these regional economies and transformed social relations. The labor-intensive nature of coffee production and the comparatively small areas in which coffee can be grown economically fostered the growth of a class of small farmers that eventually evolved into a rural middle class. These farmers provided the foundation for a prosperous regional economy fueled by an entrepreneurial spirit and habits of hard work and thrift. In the department of Antioquia in southern Colombia, these smallholders contributed tremendously to the creation of the conditions that encouraged the development of small manufacturing and industrial concerns, allowing Medellín to grow into one of Colombia's principal industrial centers and largest cities. However, despite the success of coffee in transforming the economic geography of the Andes of southern Colombia and western Venezuela, most of the rural Andes remained poor, inaccessible, and undeveloped well into the middle of the 20th century.

The Industrial Revolution in Europe and North America increased the demand for a wide range of industrial minerals during the 19th century. The Andes provided a ready supply of many of these commodities. In the Andean highlands, rich lodes of lead, zinc, copper, and tin supplied world markets. These mining operations promoted export-oriented growth and development and enclave economics with few linkages to local regions. National governments ceded the rights to mine to foreign investors, often British or American, allowing them wide latitude in their business affairs. Export taxes on mineral exports provided ready, easy-to-collect, and often abundant fiscal resources to the treasuries of the Andean republics. Foreign investors constructed railroads and roads, but these invariably led directly from the mining center to the port, with few or no connections to the local transportation network or linkages to the local economies. In central Peru, mining centers of this kind developed in La Oroya and at Cerro de Pasco, where lead and zinc were important commodities, but a scarred and polluted landscape were also significant by-products (Figure 8.5). In the arid Andean foothills of southern Peru and northern Chile, copper mining produced an economic boom for the national treasuries and riches for international mining companies such as Kennecott, Anaconda, and the Southern Company, but the economic benefits for local miners and local economies were

FIGURE 8.5. Smelters at the mining center La Oroya, in the central Andes in Peru, 1981.

limited. In the Bolivian highlands south of La Paz, in Oruro and other high-altitude settlements, tin mining supported thousands of miners, whose life expectancies were unmercifully short. It also supported a small mining elite, including the entrepreneurial Patiño family. After creating a tin-mining empire and amassing an immense fortune, the Patiños abandoned Bolivia for the comforts of Paris where the patriarch and his family lived in regal splendor.

The Industrial Revolution in the Northern Hemisphere spurred demand for other commodities too. Textile manufacturing transformed the physical and social landscapes of Great Britain and the northeastern United States as mill factories grew up and mass migration from rural areas to factory towns occurred. The demand for raw materials to feed these mills impacted the Andean landscape and social system as well. In the Andes of southern Peru the introduction of hybrid sheep varieties bred to produce high-grade wool proved immensely successful. Regional elites prospered and grew wealthy on the production of wool and its transport, marketing, and eventual export to overseas markets. Indian communities, however, often got the short end of the stick as hacienda owners strove to expand wool production by increasing herd size and encroaching on communal and village lands. During the latter half of the 19th century hacienda lands expanded significantly, usually as a result of illegal or questionable acquisition of native lands by local white and mestizo elites.

Contemporary Economic and Social Geography

The Andean region is one of the most traditional and lesser developed regions of Latin America. It has also remained one of the most concentrated areas of indigenous population. This has permitted the survival of native languages and cultures to a degree only matched in the highlands of Mesoamerica (southern Mexico and Guatemala) and Paraguay. Its traditional orientation and underdevelopment are also a function of an extreme mountainous environment that has made transportation and communication difficult, costly, and often slow. Few railroads penetrated or traversed the mountains even in the heyday of railroad construction and only a small number are operational at the present. Road construction and maintenance are difficult and expensive. Landslides block roads and floods wash out bridges with great frequency. Air transportation has proved effective in many areas, but it is costly and beyond the reach of the vast majority of the region's population. Until the advent of satellite communications in the last decade of the 20th century, telegraph and telephone communication was of poor quality, expensive, and frequently interrupted.

Like much of Latin America, the Andean region is one of strong social and economic contrasts. Agricultural patterns reflect these contrasts. Many of the region's residents live under economic and social conditions largely removed from modern technology, world economic markets, and contemporary cultural trends. Traditional land-use patterns, tools, crops, and animals dating back hundreds and even thousands of years characterize many of the highland's small farmers and indigenous peoples in remote inaccessible regions of the mountains. A form of the hacienda system still survives in some areas, where the owners of large estates raise livestock and crops, employing peasant laborers under conditions just marginally better than their predecessors experienced in the first half of the 20th century.

At the same time, other rural residents of the Andes produce specialty crops for export markets in North America and Europe and are thus integrally linked to the modern world economic system. Farmers in the highlands of Colombian and Venezuela produce coffee on tight margins, competing with producers throughout Latin America and indeed the world. The cut-flower market in the United States, especially that for roses and carnations, is now served almost exclusively by Ecuadorian and Colombian growers in the Andes as result of favorable trade regulations designed to promote agricultural alternatives to coca growing. The combination of these trade incentives and low-cost production in the Andes has forced most U.S. producers out of business. In highland Peru, fish farms raise high-quality trout that are frozen on-site and then shipped by air freight to consumers in Western Europe. Illegal drugs, notably coca (used to produce cocaine) and marijuana, are often grown under the most rudimentary conditions by peasant farmers in the Andes. Despite some 20 years of a much-touted "War on Drugs" by the United States and the governments of Colombia, Peru, and Bolivia, including drug enforcement, alternative development strategies, and aerial spraying of farmlands to eliminate drug production, the cultivation of coca and the production of cocaine for the illegal drug market continues unabated on a large scale. Coca finds its way into the world trading system as it is processed and transported from remote regions by drug traffickers, thus linking poor peasants from the Andes to the streetwise residents of New York and Los Angeles.

The economic and social dualism that characterizes Andean society today, as it has in the past, has given rise to great social tension. Much of this social tension plays out politically. During the last 40 years some Andean countries have been subjected to wrenching political and revolutionary violence. The efforts of Fidel Castro to export the Cuban revolution in the early 1960s brought his trusted lieutenant Che Guevara to Bolivia in 1966 in the hopes of initiating a popular revolution in that country. But Guevara and his comrades had seriously misjudged the social and political conditions in Bolivia at the time and found no popular support for their efforts. In the fall of 1967 the Bolivian army captured him and his small band of followers and executed Che, thereby creating a revered martyr of the left.

While revolutionary violence has reappeared periodically in Bolivia and in other Andean republics in recent decades, nothing compares to the bloodbaths and social disintegration have shaken the highlands of Colombia and Peru.

Colombia has been the most severely affected of all the Andean republics. Beginning in the 1960s, revolutionary guerrilla movements advocating social justice and inspired in part by Castro's successful revolution in Cuba began operating in Colombia. The Revolutionary Armed Forces of Colombia (Fuerzas Armadas Revolucionarias de Colombia, or FARC) is the most powerful of the groups currently operating in that country. Its cadres have been active since the mid-1960s. It conducts operations today throughout most of the national territory, but it maintains its greatest strength in the rural areas of southeastern portions of the Andes. In an effort to promote peace talks, in 2000 the national government ceded effective control of a region about the size of Switzerland (42,000 sq km) to the FARC. Then, in early 2002, government forces retook the area (Figure 8.6). The area lies about 100 km south of Bogotá and is centered on the town of Uribe. The FARC has an armed force of approximately 8,000 combatants. It attacks a wide range of military and civilian targets, kidnaps hundreds of persons a year for ransom, and is now involved in the illegal drug trade.

The second major guerrilla force in Colombia is the National Liberation Army (Ejercito de Liberación Nacional, or ELN), which also began fighting in the mid-1960s. Its armed fighters number about 4,000. Their activities are focused north of Bogotá in the vicinity of Bucaramanga and adjacent regions where they attack petroleum installations and extort money and supplies from local businesses and farmers. Smaller leftist guerrilla

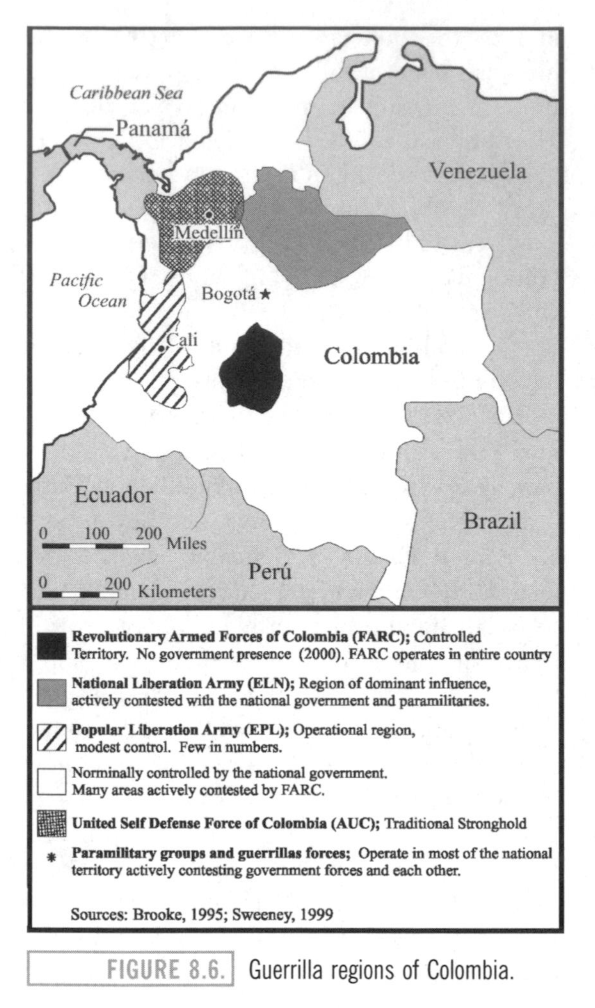

FIGURE 8.6. Guerrilla regions of Colombia.

movements also operate in Colombia, as do well-organized right-wing paramilitary organizations that support the police, military, and national government.

Forty years of civil war has left Colombia in a state of almost permanent semichaos. Large portions of the national territory remain beyond the control of the military, the police, or the national government. Other areas are constantly contested as guerrillas, police, military, and paramilitary forces vie for control. Civilians pay a terrible price. The country has 1.5–2 million internal refugees and displaced persons as a result of the war. Hundreds of thousands of Colombians, many of them the nation's best and brightest, have migrated to Venezuela, Spain, the United States, and other countries to escape the turmoil. The rampant violence cost the lives of approximately 35,000 during the 1990s. Redoubled efforts by Colombia's government to end the civil war in the early 2000s have been complicated by the appearance of powerful and well-armed right-wing paramilitaries, often linked to the nation's military command. As revolutionaries, the government, and paramilitary forces vie for control across the country, kidnappings, assassinations, and massacres continued to occur with depressing regularity. Since 2000 thousands of Colombians have lost their lives each year to the continuing political violence and civil war.

Peru has also suffered the effects of a long bloody civil war. Between 1980 and 1995 Shining Path (Sendero Luminoso) guerrillas, who modeled their ideology and military strategy on those of China's revolutionary leader Mao Tse-tung, wrought havoc on the Peruvian nation. Over 30,000 persons died and hundreds of thousands were displaced internally during the conflict. The nation suffered grievous economic damage as the guerrillas, who may have numbered as many as 5,000 at the height of their strength, and the government's military and police battled in the rural environs of the Andes and in the nation's urban centers for control of the country. Egregious human rights abuses by the revolutionaries and to a lesser extent by the government forces shook the moral fiber of the nation. The riveting movie *Boca del Lobo* (1988), directed by Francisco Lombardi, provides a moving fictionalized account of a Peruvian army unit isolated in a small Andean town and its encounters with Shining Path and the townspeople. The revolutionaries had been largely defeated by the mid-1990s. However, remnant groups survive and continue to harass government troops and police and to terrorize the citizenry in some parts of the country.

The neoliberal economic reforms of the 1980s and 1990s that swept across Latin America, including the Andean republics, have increased nations' total wealth and have brought prosperity to a small class of entrepreneurs and many in the middle and upper classes. But in the absence of a well-established social safety net, many citizens of Latin America have seen their standard of living decline. The percentage of the population living in poverty has actually increased in most countries during the implementation of these economic reforms.

A sense of hopelessness and despair overwhelms many citizens, creating conditions that promote the continuation of revolutionary violence, lawlessness, and general social unrest. Efforts to curb drug cultivation and the drug trade by the governments of Colombia, Peru, and Bolivia with the assistance and insistence of the United States may contribute to reducing the power and appeal of revolutionary armies like the FARC and the ELN. But without widespread and significant social and legal reform it seems unlikely that political and revolutionary violence will disappear any time soon.

The Andes is one of the most populous regions in Latin America. It is composed of the national territories of at least five countries. Within these countries political subunits (e.g., departments or states) are often not geographically homogenous and frequently extend from the highlands into the lowlands. These facts make it difficult to easily establish accurate population numbers for the region. Reasonable estimates, however, can be made. The five Andean republics in which most of the Andean region falls—Venezuela, Colombia, Ecuador, Peru, and Bolivia—have a total population of around 105 million. Small, sparely populated portions of both northern Chile and northern Argentina also fall in the Andes region (Figure 8.1).

Between 60 and 70 million people live in the Andean highlands. Rural population densities in the fertile intermontane valleys of the Andes are high. However, a large proportion of the population also lives in villages, small towns, and market towns with populations of 5,000–50,000.

Despite the widespread images of the Andes as a quaint alpine landscape inhabited by llamas, alpacas, and colorful natives, the Andes region is also characterized by modern urban centers many of whose residents are worldly, sophisticated, and linked culturally and economically to the world's social and political centers. These urban centers include four national capitals and at least two other cities whose populations exceed 1 million.

As the most populous of the countries in the Andean region, Colombia has three of the region's megacities, including its largest, Bogotá. With a metropolitan population of approximately 7.6 million, Bogotá sits on a fertile highland plateau at about 2,600 m, at the base of a range of towering peaks, which overlook the city. It was the colonial capital of the Viceroyalty of New Granada late in the Spanish colonial period. Its historic architecture reflects its importance during that era. Today it is the nation's capital, principal administrative and cultural center, and home to many universities and cultural institutions. It is also a major economic engine. It is an agricultural processing and marketing center for regional commodities, especially coffee, but also sugar, vegetables, and fruit. It is also a manufacturing center. Consumer goods, electronics, pharmaceuticals, and chemicals all figure prominently in the city's industrial mix. Like most major metropolitan centers in Latin America, the city is a study in stark contrasts. It has a wealthy elite upper class that lives in a world almost completely removed from the reality of most of the city's residents, many of whom are desperately poor and live in squalid conditions in shantytowns and slums. The continuing lawlessness and political violence that have wracked Colombia means that most of the upper class lives in fortress-like homes in walled and guarded neighborhoods and travels around the city with armed bodyguards in cars and SUVs that have been armored to withstand the attacks of kidnappers and revolutionaries.

Colombia, unlike most Latin American nations, is not characterized by *urban primacy*, a condition where one city dominates the urban system, and as a consequence has no primate city. Instead, it has a reasonably well-developed urban hierarchy. Two other cities in the nation, Medellín and Cali, have populations that exceed 1 million. Both are located in the Andean highlands. Medellín is located at about 1,500 m in *tierra templada* in the heart of the coffee-growing region with a population of about 3.2 million. Coffee played a key role in the economic development of Medellín and its hinterland and provided the basis for an economic boom that propelled the city's industrialization at the end of the 19th century, making it one of the nation's principal urban centers. Today coffee continues to play a crucial role in the city's economic well-being. However, the city now boasts a varied manufacturing sector and serves as major regional transportation hub.

Cali lies at a slightly lower elevation, approximately 1,000 m, in the fertile Cauca Valley on the margin of *tierra caliente* and *tierra templada*. Its population numbers near 2.6 million. Instead of coffee, sugarcane cultivation, processing, and transport provided the stimulus for Cali's growth and development. A diversified economic base characterizes Cali today. This includes a varied manufacturing sector, as well as transportation, marketing, and finance sectors. Cali, like Medellín, has also been profoundly affected by the growth in the economic importance of illegal drugs, especially cocaine, in the national economy. Drug trafficking has brought immense wealth to both cities and transformed their local economies. It has also contributed to widespread lawlessness and corruption and torn their social fabrics. Both cities are probably best known outside of Colombia for the drug cartels that have operated in them and bear their names. The Colombian director Victor Gaviria captured much of the violent street life of these cities in his film *Rodrigo D: No Futuro* (1990). Shot on location in the shantytowns above Medellín, this movie portrays the environment and life of young working-class men who are caught up in a culture of drugs, violence, and crime.

The remaining million-plus Andean cities, Caracas, Quito, and La Paz, were all regional colonial centers that evolved into national capitals at the time of independence. Caracas, Venezuela's national capital, sits on the northernmost extension of the Andes at about 1,000 m. Unlike other Andean cities, it is located close to the ocean: the port city of La Guaira that serves Caracas is only 15 km from the heart of the latter city. Its metropolitan population is approximately 3.2 million. The city spreads over the floor of a narrow rift valley and then onto the slopes of the mountains that surround it. The city is characterized by a modern city center, constructed in large measure with the country's petroleum wealth, as well as limited-access freeways reminiscent of the United States and a subway system constructed in the 1980s. As Venezuela's primate city, Caracas is the dominant urban center in the nation and the preeminent focus of culture, business, and manufacturing.

Quito, Ecuador's national capital, is one of the most attractive of Andean metropolitan centers. Much of the city's colonial architectural core has been preserved, including many impressive colonial structures (Figure 8.7). The United Nations declared the city a World Cultural Heritage site in 1978, and it is most deserving of that designation. Quito was an Inca city prior to the arrival of the Spanish in the 16th century, although evidence of its pre-Conquest history has almost completely disappeared from the landscape. The city was the capital of the Audiencia (a second-tier judicial administrative subdivision) of Quito during the colonial period and became the national capital upon the nation's independence from

FIGURE 8.7. Quito's cathedral in the city's colonial core, 1979.

Spain. The population of the Quito metropolitan area is approximately 1.8 million. While commerce and manufacturing do comprise a part of the urban fabric, the city is principally a center of government, education, and culture. Guayaquil, on the Pacific coast, is Ecuador's most populous city and its principal industrial and commercial center.

Few reminders of the colonial past are found in La Paz, Bolivia. Sited originally in a deeply incised valley on the edge of the windswept Altiplano, today La Paz and its surrounding districts have expanded onto the Altiplano (El Alto) and far down the valley at whose head it was originally located (Figure 8.8). La Paz is the national capital of Bolivia (although the high court is located in Sucre). Its current metropolitan population is approximately 1.6 million. The city's geographical site, in a narrow valley, has led to the development of an urban area that is largely lineal. A narrow urban core runs along the valley floor for nearly 10 km, with residential districts spreading up the steep slopes that surround it. Transportation in the city is difficult and the roads are almost always congested. Despite a clear need, the city has been unable to construct any kind of linear metro or light-rail system along its principal axis that would reduce congestion. Government, culture, business, and finance are the city's principal economic mainstays. Manufacturing and industrial activity are limited, and largely concentrated in the production of processed foods, beverages, and consumer goods for national consumption.

Beyond the Piedmont: Peripheral Regions on the Andes' Edge

The Andean mountain range adjoins several remote and sparsely settled regions over its extent. In addition to the Amazon Basin, which is treated separately in Chapter 19, these peripheral regions include the Llanos, the Pacific coastal lowlands in western Colombia, the Pacific coast deserts that run from northern Peru to central Chile, and the Gran Chaco. This section examines the environment and contemporary conditions in two of the most well-known and readily identifiable of these regions, the Llanos and the Gran Chaco.

The Llanos

Llano simply means an open comparatively flat place. Hence, the term Llanos describes the vast open plains that lie to the east of the Andes in northern South America (Figure 8.9). The region is characteristically delimited in the south by the course of the Guaviare River (roughly coincident with 3°N latitude). The western boundary of the Llanos extends from

FIGURE 8.8. La Paz, Bolivia, sits in a deeply incised valley on the edge of the Altiplano, 2004.

the south in a crescent-shaped arc running parallel to the Andes as it runs north and then veers eastward across the northern portion of the continent. The eastern boundary is also clearly defined, running parallel to the Orinoco River. The beginning of the Orinoco River delta marks the end of the Llanos region.

ENVIRONMENT

The region is a plain where elevations rarely exceed 300–400 m. Nevertheless, there are some notable variations in the topography. In the central portion of the plains, exposed outcrops of the Guiana Shield rise above the Andean sediments and produce significant relief. The isolated Sierra de Macarena also arises in this central region. Elevations exceed 2,500 m and a unique alpine ecology has developed in these mountains.

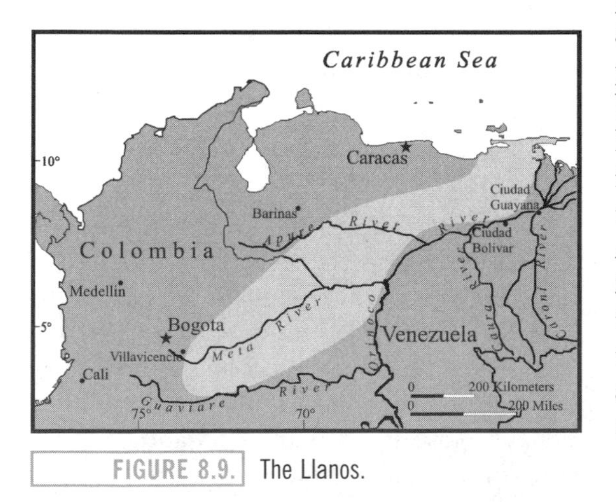

FIGURE 8.9. The Llanos.

The Llanos are often described as being divided into two sections, the Llanos Altos and the Llanos Bajos, the High and Low Llanos, respectively. The Llanos Altos refers to that section of the plains just beyond the Andean piedmont. Here the watercourses are incised 50–75 m below the general surface of the plains. To the east, on the Llanos Bajos, braided streams and rivers characterize the hydrography, and seasonally large areas are flooded. Especially notable is the region between the Meta River in the south and the Apure River in the north. Measuring about 200 km north–south and 300 km east–west, this vast region has an ecology that is totally transformed each year as the seasonal rains and runoff turn an arid savanna into an inland sea teeming with fish and birds.

The climate of the Llanos is tropical.

Monthly mean temperatures vary little over the course of the year. Seasonal variation ranges from 4°C to 5°C, but can be less. At Ciudad Bolivar, the seasonal range is 2°C (from 26°C to 28°C). Diurnal ranges, however, often exceed 10°C. Annual mean monthly temperatures average 23°C, but vary locally by several degrees. In the warmest months (November–April), mean daily highs can reach 35°C, while nocturnal lows average between 18°C and 24°C. Two clearly defined seasons divide the year, a wet season running from May to October and a dry season running from November to April (Figure 8.10).

The rainy season begins earlier and ends earlier in the south, and begins later and ends later in the north, reflecting the northward migration of the ITCZ. Precipitation totals vary considerably. Along its western margin, at the foot of the Andes, annual rainfall in the Llanos can be very high, averaging 4,500 mm at Villavicencio. To the east, precipitation declines quickly in the open savanna. At Ciudad Nutrias, in the central region of the Llanos, annual precipitation reaches only 1,100 mm.

Grasses, sedges, low shrubs, and isolated trees and palms are the natural vegetation of the Llanos. However, dense gallery forests often occupy the riparian habitats along stream and river courses, particularly in the Llanos Altos. Elsewhere, where seasonal flooding occurs, only plants adapted to tolerating high levels of moisture can survive.

Early Spanish explorers found little in the Llanos region to attract their attention. Indigenous peoples, living at a subsistence level, with no resources or material wealth that could be readily exploited, occupied the plains. The founding dates of the region's principal cities suggest just how peripheral the Llanos were in the colonial, republican, and even modern period: Ciudad Bolivar (1764), Villavicencio (1840), and Ciudad Guayana (1961).

CONTEMPORARY CONDITIONS

Rural settlement in the Llanos, led largely by mestizos, began sparingly during the late colonial period when a cattle-ranching economy evolved on the plains. Isolated ranches, called *hatos*, formed the basic economic and settlement unit. Cattle grazed the savannas of the Llanos Bajos during the dry season, moving to pastures on higher ground when seasonal flooding occurred. This basic pattern of settle

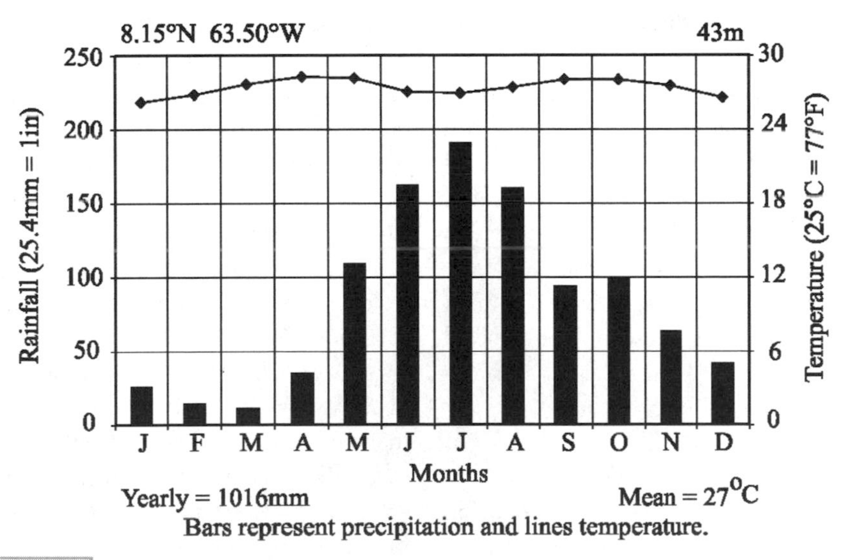

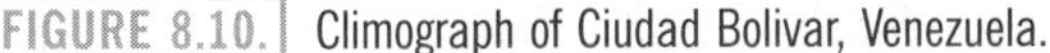
FIGURE 8.10. Climograph of Ciudad Bolivar, Venezuela.

ment, economy, and resource exploitation continues to characterize much of the Llanos at the end of the 20th century.

Extractive industries began to diversify the economic base of the Llanos by the mid-20th century. Extensive petroleum fields discovered in the eastern quarter of the Llanos around Tigre in the 1930s continue to produce, and important fields have also been opened in the west on the Andean piedmont near Barinas. Rich iron ore deposits along the eastern banks of the Orinoco, mined since the 1940s, have contributed to a regional industrialization process.

The population of the Llanos numbers roughly 2 million. About one-third of this number occupy the Colombian Llanos, while the balance are found in the Venezuelan portion of the plains. Most of the population is concentrated along two distinct axes, both of which lie along the periphery of the Llanos. The first axis, which began to develop in the late colonial period, corresponds roughly with the Llanos Altos and lies east of the Andes, running roughly parallel to the mountains in both Colombia and Venezuela. The lack of annual flooding on these elevated plains allowed permanent rural settlement. The small towns established here capitalized on their strategic position between two distinct regions, the Llanos and the Andes, becoming important points of commercial exchange. A second axis, also dating from the late colonial period, but receiving its principal impetus in the last half of the 20th century, runs along the lower reaches of the Orinoco, roughly between Ciudad Guayana and Ciudad Bolivar. With few exceptions, all urban places of any consequence are located along these two peripheral axes. Over 50 percent of the region's population is urban. The interior reaches of the Llanos are thinly populated.

Ciudad Guayana is the most important urban center in the region. It is a key exception to the dominant pattern of natural resource exploitation in the Llanos. Established in 1961 by the Venezuelan government, Ciudad Guayana is one of the best examples of state-directed growth-pole industrialization in Latin America. The city is located on the banks of the Orinoco River and is accessible to ocean-going vessels. Economic development planners in Venezuela sought to capitalize on the site's access to natural resources, hydroelectric power, and a small set of existing industries. Comprehensive urban planning for the city included residential, commercial, recreational, and industrial quarters. Extensive deposits of iron ore and bauxite mined from the adjacent Guiana Highlands provide the raw material for the production of iron, steel, and aluminum. Inexpensive power generated at the Guri hydroelectric facility about 80 km south on the Caroni River produces the energy required by the urban–industrial complex. Ciudad Guayana grew rapidly; population growth rates reached 20 percent in some years. By 1981 the city's population had reached 200,000 and by the mid-1990s it had surpassed 500,000.

Ciudad Bolivar and Villavicencio are the only other cities of consequence in the Llanos region. Both cities are located on the region's periphery. Ciudad Bolivar sits on the Orinoco River about 100 km upstream from Ciudad Guayana. It is a key transportation hub situated on the eastern bank of the Orinoco at the river's narrowest point (0.8 km), the Angostura Narrows, the head of navigation for oceangoing vessels. The construction of the Angostura Bridge in 1967, the first span to cross the river, further contributed to its role as a transportation hub. Its exports include cattle, horses, hides, gold, and timber. Its population stands at about 250,000.

Over 1,000 km to the southwest, the city of Villavicencio lies on the extreme western margins of the Llanos. It is sited at the base of the Andes just 100 km southeast of Bogotá. Its location, at the margin of the Llanos and the rain forest of the eastern Andes, has made it the principal commercial center for the entire

Colombian portion of the Llanos and adjacent rain forest areas. Agricultural processing industries located here include brewing and distilling, soap manufacturing, rice milling, coffee roasting, and tanning and leather goods production. Despite its regional importance, the city is comparatively small with a population of about 100,000.

The Gran Chaco

The Gran Chaco, also known as simply "the Chaco," lies in the central portion of South America's Southern Cone. Geographically extensive, it sits to the east of the Andean piedmont in southern Bolivia and northern Argentina. The region includes portions of three countries: Argentina, Paraguay, and Bolivia (Figure 8.11). In a territorial sense, about 50 percent of its area lies in Argentina, about 40 percent in Paraguay, and only about 10 percent in Bolivia. In the south the Chaco grades into the Argentine Pampa at about 30°S latitude. It extends northward through Paraguay and into Bolivia to approximately 17°S. The eastern margin of the Chaco is defined by the Paraná River in the south, and then by the Paraguay River northward from its confluence with the Paraná. Along its western boundary, the Chaco grades into the Argentine Northwest in the south and into the margins of Amazonia further northward in Bolivia. Its east–west dimension varies from 500 to 700 km.

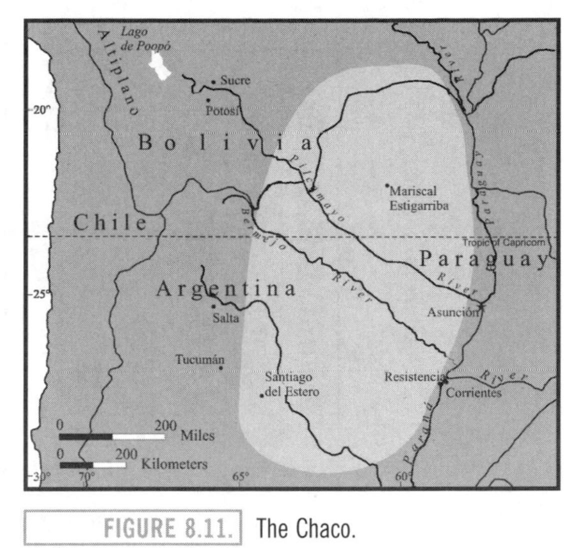

FIGURE 8.11. The Chaco.

ENVIRONMENT

Geologically, the Chaco is the descendant of an ancient oceanic embayment. This shallow, comparatively protected sea was joined to the ocean, but over millions of years it was filled by alluvium eroded off the Andes on the west and the Brazilian Highlands on the east. These unconsolidated sediments, composed largely of sand and silt, extend to depths as great as 3,000 m. The Chaco is aptly described as an interior lowland. Elevations along its western margin stand around 500 m and descend gradually to only about 50 m in the east along the Paraná River. The region is drained by the western tributaries of the Paraná–Paraguay river system, which run roughly parallel in a northwest–southeast direction. Only four of the rivers that drain the region are permanent: the Río Dulce, the Río Salado, the Río Bermejo, and the Río Pilcomayo; the remainder are seasonal. All watercourses in the Chaco, permanent or seasonal, are characterized by shifting and braided courses, seasonal flooding, narrow reaches, and sandbanks. All of these features are typical of rivers in arid and semiarid environments, with shallow gradients, and running over unconsolidated deposits.

The climate of the Chaco is subtropical (Figure 8.12). Semiarid conditions prevail in the east and the region becomes progressively arid moving westward. Average annual precipitation is greatest in the east, reaching between 1,300 and 1,400 mm and declining progressively toward the west to less than 500 mm. Rainfall is seasonal and is concentrated in the summer months (October–April). Seasonal flooding occurs during these periods. In the southern portion of the Chaco, along the lower

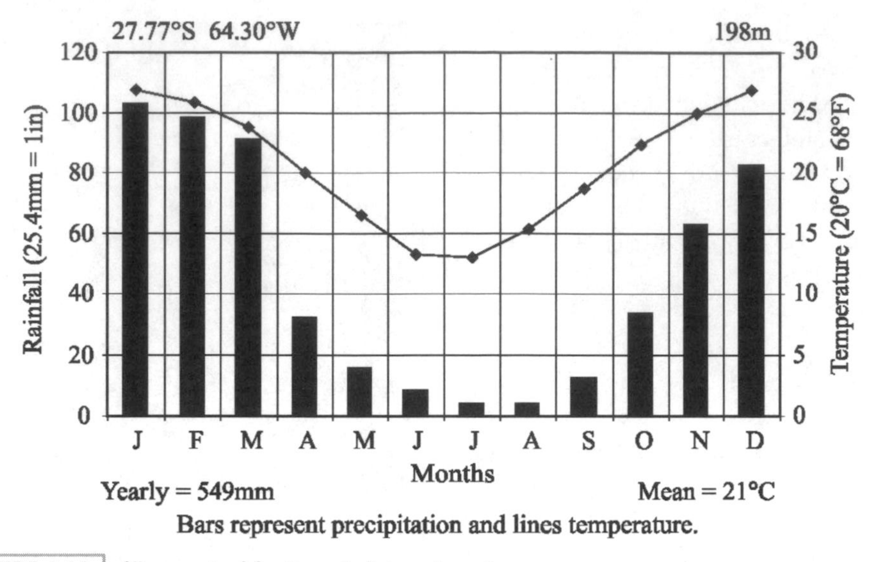

FIGURE 8.12. Climograph of Santiago de Estero, Argentina.

river courses, this flooding maintains extensive marshes, known as *esteros* in Argentina. Further upstream, the flooding fills shallow ephemeral basins called *bañados* that evaporate in the dry season and are then often cultivated during that period.

Like precipitation, temperatures also vary considerably over the region. Seasonal variation is not extreme. Temperature maxima in the summer months frequently reach into the high 30°sC, while winter temperature maxima stand in the low 20°sC. Two distinct temperature gradients affect environmental conditions in the Chaco. The first runs from the south to the north and the second from the east to the west. In the first instance, occasional frost can occur during one or two of the winter months (July and August) in the southern half of the Chaco, but almost never further north. Similarly, summer temperatures are slightly lower in the south than in the north. Second, a comparable pattern pertains from the east to the west, with more moderate temperatures in the east that increase to the west as the effects of climatic continentality increase. *Climatic continentality* refers to the fact that interior continental locations, distant from the moderating effects of large bodies of water, demonstrate greater seasonal fluctuations in temperature. Typically, such interior locations are hotter in the summer and cooler in the winter than locations adjacent to oceans and seas.

The variations in the natural vegetation of the Chaco follow the patterning of temperature and precipitation, although local changes in soil conditions modify them. Along the region's eastern margin, extending perhaps 100 km westward from the Paraná and Paraguay Rivers, the natural vegetation is characterized by dense forests on the uplands and interfluves, with marshes and swamps in the low-lying areas. Here, where precipitation is highest and temperatures most moderate, species diversity is great. This area, which lies mostly within Argentina, is known as the Chaco Bajo. Agriculture is practiced extensively here; consequently, the loss of natural vegetation is widespread.

To the west, the vegetation cover is sparser. Semideciduous trees cover the watercourses, while savannas and open grasslands, interspersed with extensive stands of *quebracho*, cloak the interfluves. This area is often referred to as the Chaco Boreal (Northern Chaco) or the Chaco Alto (High Chaco). The vegetation becomes more xerophytic to the west. While *quebracho* is still found, deciduous thorn forest tree species like the *algorrobo* (*Prosopis* spp.) and drought-resistant brush and shrub species dominate.

EARLY SETTLEMENT

Sparse settlement and low population densities marked the Chaco in the pre-Columbian period. Native peoples subsisted on hunting, fishing, and gathering. The sweet protein-rich pods of the *algorrobo* were one of their principal food sources. Agriculture also figured in the aboriginal economy, although it was limited to plots on small floodplains along stream courses where the inhabitants cultivated basic food crops like corn, manioc, squash, and beans. Seminomadic tribal groups with a simple social structure typified the region. Their material culture was unsophisticated—simple thatched-hut dwellings, rudimentary pottery, and basic hunting and tillage tools. Metallurgy was unknown.

At the present, few aboriginal peoples remain in the Chaco. Those that do belong to small tribal groups, each totaling a few thousand members. Most of these are found in the Paraguayan portion of the region where about 30,000 are distributed among some 13 distinct communities.

Europeans traversed the Chaco early in the colonial period. Santiago de Estero (1553), one of the earliest towns founded in the southern portion of South America, was in the southern Chaco. Other early colonial centers founded on the southern and eastern margins of the Chaco included Asunción (1537), Santa Fe (1573), and Corrientes (1588). Indian resistance and an inhospitable environment limited Spanish exploitation and settlement of the Chaco throughout the colonial period and well into the 19th century. Logging of white quebracho (*Aspidosperma quebracho-blanco*) for tannin and other species (including red quebracho, *Schinopsis quebracho-colorado*) for timbers to be used in Andean mines and extensive cattle raising were the most profitable economic activities and formed the principal basis for settlement of the region well into the 20th century.

The Chaco was the unlikely scene of international conflict in the 1920s and 1930s. Long frustrated in its efforts to find an outlet to the sea, Bolivia coveted access to the Paraguay River. After a series of border skirmishes in the late 1920s, Bolivia undertook a full-scale invasion of Paraguay in 1932. The resulting war, the War of the Chaco (1932–1935), was costly to both sides, resulting in approximately 100,000 casualties. Paraguay eventually prevailed and not only defeated the Bolivian invaders, but secured most of the Chaco Boreal as Paraguayan territory in what had previously been Bolivia.

CONTEMPORARY CONDITIONS

The economy of the present-day Chaco is dominated by extractive industries, cattle raising, and agriculture. *Quebracho* continues to be a mainstay of the regional economy in the Argentine and Paraguayan portions of the area, but long-term exploitation and the absence of any efforts to reforest *quebracho* stands threatens the continued exploitation of this resource. In the more accessible parts of the Chaco, principally in the east, industrial processing plants extract tannin, an essential element in tanning leather. Cattle raising is widely practiced in the Chaco. Hardy *criollo* cattle with some resistance to endemic diseases and ticks constitute almost all herds, rather than the purebred herds that characterize the Pampa.

Agriculture is concentrated chiefly along the eastern margin of the Chaco, in a 100-km-wide belt running along the western banks of the Paraná and Paraguay Rivers, where precipitation is greater and temperatures more moderate than to the west. The lower courses of the floodplains of the major rivers that traverse the Chaco, principally the Bermejo and the Pilcomayo, also are cultivated. Cotton is widely grown in the Argentine and the Paraguayan Chaco; traditionally, it has been the dominant crop. The Chaco produces nearly all of Argentina's domestic cotton, a low-quality grade with no prospects for competing in the international market. In recent years a more diverse mix of crops have been grown, including corn, sorghum, sunflowers, and soybeans. In Paraguay, Mennonite settlers raise dairy cattle and farm a diverse mix of crops, including cotton and peanuts, deep in the Chaco Boreal. Agriculture is absent from the Bolivian Chaco.

The Chaco is sparsely populated. Total population stands at about 2.2 million inhabitants, with the vast majority located in the Argentine Chaco, few in Paraguay, and fewer in Bolivia. Nearly 2 million people inhabit the Argentine Chaco; they are found in the provinces of Formosa, in Chaco, in Santiago del Estero, and in the northern portion of Santa Fe. Most are located in urban and agricultural areas on the Chaco's eastern periphery. Resistencia in the east and Santiago del Estero in the southwest both number over 250,000 inhabitants and are the most populous urban centers in the Chaco proper. The vast Paraguayan Chaco, accounting for over half of Paraguay's land area, is peopled by only about 60,000 inhabitants, about 1 percent of the nation's population. The largest urban center in the Paraguayan Chaco, Mariscal Estigarribia, numbers about 6,500 people, while the remaining towns have populations of only a few thousand each. The total population of the Bolivian Chaco numbers less than 50,000.

Summary

The Andes are one of Latin America's most prominent environmental and cultural features. Although the mountains lie almost exclusively within the tropical belt, climatic conditions are complex and range from alpine to tropical. A series of altitudinal life zones can be identified and are widely recognized across the region: tierra caliente, tierra templada, tierra fria, puna/páramo, and tierra helada. The region's population is subject to a wide range of natural hazards ranging from landslides and earthquakes to volcanic eruptions.

Agriculture and herding formed the basis for the development of early civilization in the Andes. Andean peoples domesticated the potato and other plants and animals nearly 5,000 years ago. A series of progressively more sophisticated civilizations arose in the Andes, culminating in the Inca who had developed an imperial system stretching from Ecuador to northern Argentina by the time of the Conquest. The dense population, agricultural productivity, and mineral wealth of the Andes made it a focus of Spanish colonial rule. Although vast numbers of people continued to live at a subsistence level after independence, the region developed export economies tied to international markets. Mineral products and agricultural commodities led production.

Some 60 million people inhabit the Andes. Their lives and the region's economy and society are characterized by strong socioeconomic dualism. Modern city dwellers in the region's million-plus cities like Bogotá or La Paz use cellphones, e-mail, and the Internet, while their brethren in small towns, villages, and rural areas eke out near-subsistence existences frequently without access to clean water, electricity, or medical care. Violent revolutionary movements have racked Colombia and Peru during the last 40 years and the continuing violence in Colombia shows no sign of ending soon.

The Andes are flanked by several remote and sparsely populated peripheral regions over the course of its extent. The most extensive of these is the Amazon Basin (discussed in

Chapter 19), but other notable areas include the Llanos and the Gran Chaco.

Further Reading

Bebbington, A. (2000). Reencountering development: Livelihood transitions and place transformations in the Andes. *Annals of the Association of American Geographers, 90*, 495–520.

Bowman, I. (1916). *The Andes of southern Peru*. New York: Holt.

Brush, S. (1977). *Mountain, field, and family: The economy and human ecology of an Andean valley*. Philadelphia: University of Pennsylvania Press.

Denevan, W. M. (Ed.). (2001). *Cultivated landscapes of native Amazonia and the Andes*. New York: Oxford University Press.

Gade, D. W. (1992). Landscape, system, and identity in the post-Conquest Andes. *Annals of the Association of American Geographers, 82*(3), 461–477.

Gade, D. W. (1999). *Nature and culture in the Andes*. Madison: University of Wisconsin Press.

Galvis, C. A. (2000). *The heart of war in Colombia*. Broomfield, CT: Kumarian Press.

Holmes, J. S. (2002). Terrorism, drugs, and violence in Latin America. *Latin American Research Review, 37*(3), 217–230.

Jokish, B. D., and Lair, B. M. (2002). One last stand?: Forests and change on Ecuador's eastern cordillera. *Geographical Review*, 92, 213–234.

Kent, R. B. (1993). Geographical dimensions of the Shining Path insurgency in Peru. *Geographical Review, 83*(4), 441–452.

Klaren, P. F. (2000). *Peru: Society and nationhood in the Andes*. New York: Oxford University Press.

Klein, H. S. (1992). *Bolivia: The evolution of a multi-ethnic society* (2nd ed.). New York: Oxford University Press.

Knapp, G. (1991). *Andean ecology: Adaptive dynamics in Ecuador*. Boulder, CO: Westview Press.

Korovkin, T. (2003). Cut-flower exports, female labor, and community participation in highland Ecuador. *Latin American Research Review, 30*(4), 18–42.

Miller, E. S. (Ed.). (1999). *Peoples of the Gran Chaco*. Westport, CT: Bergin & Garvey.

Palmer, D. S. (Ed.). (1992). *The Shining Path of Peru*. New York: St. Martin's Press.

Parsons, J. J. (1982). The northern Andean environment. *Mountain Research and Development, 2*, 253–262.

Rausch, J. M. (1999). *Colombia: Territorial rule and the Llanos frontier*. Gainesville: University Press of Florida.

Saeger, J. S. (2000). *The Chaco mission frontier: The Guaycuruan experience*. Tucson: University of Arizona Press.

Stadel, C. (1992). Altitudinal belts in the tropical Andes: Their ecology and human utilization. *Conference of Latin Americanist Geographers: Benchmark 1990, 17–18*, pp. 45–60.

Young, K. R. (2004). Environmental and social consequences of coca/cocaine in Peru: Policy alternatives and a research agenda. In *Dangerous harvest: Drug plants and the transformation of indigenous landscapes*, Steinberg, M. K., et al. eds., pp. 249–273. Oxford, UK: Oxford University Press.

9 Migration, Population Change, and Race

The century between the wars for independence in Spain's American colonies and the onset of World War I brought little noticeable change in the basic patterns of population composition and distribution that characterized some countries but altered that of other nations dramatically. Mexico, the nations of Central America, the Andean republics, and the isolated and landlocked nation of Paraguay experienced only modest alterations in their population composition and geography during this century. But in the Southern Cone, especially in Brazil, Uruguay, and Argentina, and in Spain's remaining Caribbean colonies of Cuba and Puerto Rico, substantial modifications in the size, composition, and geographical distribution of their populations occurred.

Slavery and African Immigration

The importation of African slaves continued at a surprisingly rapid rate during the first half of the 19th century (Figure 9.1). Around 2 million reached the Americas between 1810 and 1870 (Table 9.1). This represented nearly 20 percent of all Africans brought to the New World as slaves. Nearly all of these individuals were destined for Brazil, Cuba, and Puerto Rico. Brazil, where slavery remained legal until 1888, received almost 1.2 million. Most remained along Brazil's northeastern coast, the traditional focus of the slave population, but the central and southern highlands, where a booming coffee economy was developing, also received appreciable numbers of African slaves.

Slavery disappeared slowly in Spanish America. Mexico, Chile, and the Central American republics abolished slavery in the 1820s shortly after their independence from Spain. But the institution's economic importance and the political power of slave owners delayed abolition in Venezuela, Colombia, Ecuador, and Peru until the middle decades of the century. Consequently, most of the 600,000 Africans slaves brought to Spanish America during the 1800s were taken to Puerto Rico and Cuba, where slavery was finally abolished in 1878 and 1876, respectively. During this period the number of slaves brought to Cuba, and to a lesser extent Puerto Rico, was so great that Spanish authorities sought to reduce their proportion of the population by enthusiastically promoting white emigration from Spain.

While the forced importation of African slaves impacted the population geography of Cuba and Puerto Rico during the 19th century, the migration of free blacks from a number of Caribbean islands produced important

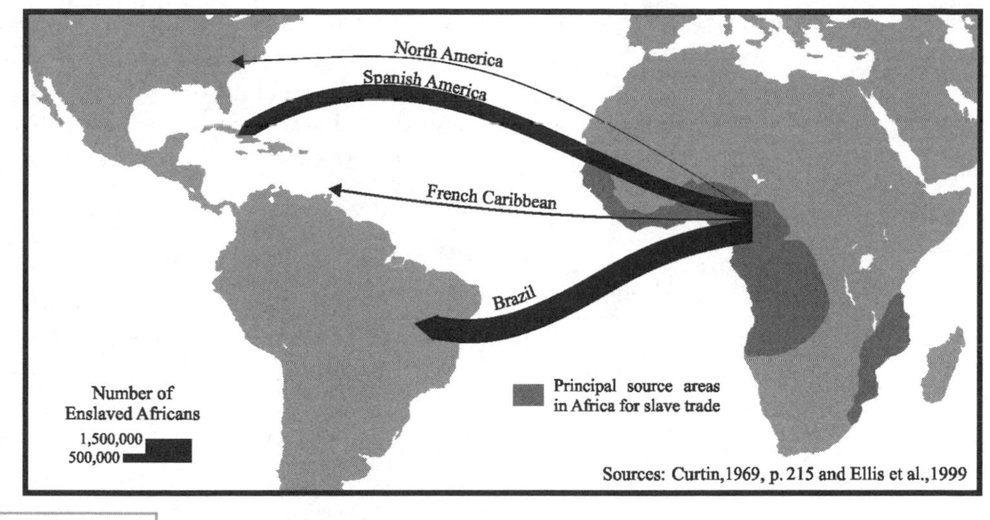

FIGURE 9.1. Destinations of the Atlantic slave trade, 1811–1870.

changes in the population geography of Central America. The establishment of vast banana plantations, the building of railroads, and perhaps most significantly the construction of the Panama Canal (1904–1914) created a strong demand for labor in the isolated tropical lowlands and encouraged immigration of blacks from many English-speaking islands, especially Jamaica and Barbados. Many of these laborers settled permanently. Thus blacks came to account for a large percentage of the population in a few areas along the Caribbean coasts of Nicaragua (Bluefields), Costa Rica (Limón), and Panama (Colón).

European Immigration

Latin America did not experience the repetitive waves, large numbers, or wide variety of national origins of European immigrants that characterized the immigration history of the United States or Canada during the 19th century. However, European immigration to some areas of Latin America had a permanent impact on the landscape, culture, and population characteristics. During the 19th century, Brazil, Argentina, and Chile, among other Latin American countries, actively encouraged European immigration for several reasons. These included the colonization of unsettled frontier regions, increasing demands for inexpensive agricultural labor on the Pampa and in the booming coffee plantations of southern Brazil, and often a desire to "whiten" a country's population.

Italians

Italians were the largest group of European immigrants who settled in Latin America during this period. Between the early 1880s and 1920, millions of Italians left their homeland in search of a better life and many settled permanently in the Southern Cone of South America. Their numbers totaled close to 3.5 million by the time this tidal wave of immigrants had subsided. Brazil and Argentina were the principal destinations. Brazil alone absorbed close to 1.5 million Italian immigrants. Like the German immigrants who preceded them in Brazil, the Italians settled in the south, principally in the states of Paraná and São Paulo. Many toiled initially on the coffee plantations that had spread rapidly over the southern highlands as world demand for coffee boomed

TABLE 9.1. Africans Sent to the Americas as Slaves, 1811–1870

Importing region	1811–1870	Total 1451–1870
Old World	—	175,000
Europe	—	50,000
São Thome	—	100,000
Atlantic Islands	—	25,000
Spanish America	606,000	1,552,100
Brazil	1,145,400	3,646,800
British North America	51,000	399,000
British Caribbean	—	1,665,000
Jamaica	—	747,500
Barbados	—	387,000
Leeward Is.	—	346,000
St. Vincent, St. Lucia, Tobago, & Dominica	—	70,100
Trinidad	—	22,400
Grenada	—	67,000
Other BWI	—	25,000
French Caribbean	96,000	1,600,200
Saint Dominque	—	864,300
Martinique	41,000	365,800
Guadeloupe	41,000	290,800
Louisiana	—	28,300
French Guiana	14,000	51,000
Dutch Caribbean	—	500,000
Danish Caribbean	—	28,000
Total	1,898,400	9,566,100

Source: Curtin (1969), p. 268.

at the end of the 19th century. Eventually, most of the Italians migrated to urban centers, especially the city of São Paulo, leaving an indelible mark on the city's population composition and culture.

It was in Argentina, however, that Italian immigration was to have its greatest impact. It is no exaggeration to say that Italians arrived in Buenos Aires by the boatload during these years. Many came only temporarily as seasonal workers for the wheat harvest. They were known as *golondrinas*, or "swallows," because they migrated southward during the Northern Hemisphere's winter to work the wheat harvest during the Southern Hemisphere's summer and then returned to Italy for the next season's harvest, thus mimicking the annual migration of swallows. But in the end, nearly 2 million Italians settled permanently in Argentina. This process profoundly transformed Argentina because of its small overall population, and also because almost all the Italian immigrants disembarked in the port and capital city of Buenos Aires and most remained there.

Spaniards

Spanish immigration figured significantly in the evolution of the population of Argentina during the late 19th century and the early 20th century. Cultural affinity, language, and economic opportunity all contributed to continuing Spanish migration to Argentina, which totaled 200,000–300,000 during these decades. While Andalusia and Extremadura in southern and western Spain had provided most of the immigrants to Latin America during the colonial period, Galicia, Asturias, and Santander in northern Spain sent the vast majority of immigrants to Argentina at this time. So many Spaniards emigrated from northern Spain, especially Galicia, that to this day Spaniards are known pejoratively in Argentina as "*Gallegos*." During much of the 20th century, regional clubs and associations, which frequently were the focus of social life for Spanish immigrants, maintained a strong presence in Argentina's cities and towns.

Spanish immigration was also significant in the population geography of Spain's two remaining Caribbean colonies, Cuba and Puerto Rico, during the last half of the 19th century. In Cuba the sugarcane business had boomed during much of the century, and African slaves had continued to arrive on the island in large numbers. As blacks came to represent an increasing proportion of the population—reaching about 45 percent in the 1840s—colonial officials responded by encouraging additional immigration from the Iberian Peninsula. The flow of Spaniards to Cuba during this period totaled between 200,000 and 300,000. It continued at a vigorous pace even after the United States defeated Spain in the Spanish-American War in 1898 and after Cuba's independence in

1903. By 1930, fully 15 percent of the population of Cuba held Spanish citizenship.

Portuguese

In Brazil, even after independence from Portugal, Portuguese immigration continued to be a major element in the country's overall immigrant population. Close to 1 million entered Brazil between 1820 and 1920. This longstanding pattern of Portuguese immigration long after the colony's political independence served to deepen and enrich the cultural affinity established between Portugal and Brazil during the colonial period.

Germans

German immigrants began arriving in South America in small numbers during the 1840s and 1850s. The cool forested valleys and mountains of Chile's "Lake District" between Valdivia and Puerto Montt in the far south of the country attracted agricultural colonists from a number of German regions, including Moravia, Bohemia, and Westfalia, between the late 1840s and the early 1890s. Less than 10,000 Germans settled in this frontier region during the last quarter of the 19th century, yet in the absence of other significant populations this region became something of a German *ecumene* reflecting Teutonic agricultural and building practices.

It was in southern Brazil, however, especially in the states of Santa Catarina and Rio Grande du Sul, that German immigration was both much larger in numbers and endured over a longer period. German immigrants began to arrive in southern Brazil in the middle of the 19th century. They settled heavily forested frontier regions, clearing the forests and establishing subsistence farms based on livestock raising and crop cultivation. German immigrants continued to arrive in southern Brazil in appreciable numbers until the early 1920s. The cumulative impact of this migration flow, which reached a total of perhaps 150,000, resulted in the effective agricultural colonization of broad areas of Brazil's southern forests, the region's integration into the national economy, and a locally distinctive German Brazilian culture that is still extant today (it is most typified by the city of Blumenau and its hinterland).

Other Europeans

While Italians, Spaniards, Portuguese, and Germans were the most significant immigrant groups to settle in Latin America during this period, they were not alone. A surprisingly wide range of nationalities and ethnic groups immigrated to the region. The total number of immigrants from any one of these secondary groups rarely exceeded 100,000 and was often less, sometimes only a few thousand. Nevertheless, the impact of these groups was often regionally or locally important, and in some cases left an indelible mark on the present-day population geography of Latin America.

For example, Welsh settlers colonized the lower course of the Chubut River valley in the Argentine Patagonia in the 1870s and 1880s. They prospered; established a series of towns, including Rawson, Trelew, and Gaiman; and created a viable regional economy by keeping sheep, growing row crops, and raising deciduous fruit.

Jewish immigrants arrived in Latin America in modest numbers. Typically these immigrants settled in large urban centers, mostly national capitals, although in both Argentina and Brazil Jews settled some provincial and state capitals as well. Their numbers in most cities were small, but usually sufficient to support a synagogue. Of those who immigrated to Latin America, many came to Argentina. Although some settled in agricultural colonies in Santiago del Estero and provincial capitals, eventually most Jews located in the capital, Buenos Aires. At the beginning of the 20th century, Buenos Aires claimed a Jewish population of between 50,000 and 100,000. Many lived in Once, a neighborhood where kosher butchers and restaurants are still found. The Jewish population in Argentina reached a high of 300,000 in the early 1960s. However, emigration to Israel and other countries had reduced the country's Jewish population to about 200,000 in 2002. There are close to 100,000 Brazilian Jews, concentrated in São Paulo and Rio. Of other Latin American countries, only Mexico, Chile, and Venezuela have significant Jewish populations, of which none exceed 50,000.

While British entrepreneurs, capital, and technology had a major impact on the economy of much of Latin America during the 19th century, few Britons immigrated to the region. Argentina is an exception to the rule. Close to 100,000 Britons eventually settled in the country. These British Argentines became part of the nation's economic elite, amassing fortunes in cattle ranching, railroads, finance, and business. But they distanced themselves from Argentine society at large by forming private social clubs, where they played polo and indulged in other "English" pastimes, and established English-language schools to educate their children. The St. Andrew's Scots School in Buenos Aires, founded in the 1840s, continues to educate many young British Argentines, although today in both outlook and culture these children are becoming more and more Argentine.

Although few in number, Middle Eastern immigrants from the eastern margin of the Mediterranean Basin played a key role in transforming commerce and retail business in several Latin American regions. These immigrants—typically Christians—hailed from Syria, Lebanon, and Palestine and immigrated primarily to Argentina and Colombia. In Argentina, they energized retail commerce by becoming itinerant traders, moving from railway station to railway station across the vast Pampa and then traveling by foot to serve isolated farmers and ranchers. This initial toehold allowed many to establish permanent commercial enterprises in towns and cities, especially in the far west—in cities like Rioja, San Juan, and Catamarca. In Colombia, many immigrated to the Cauca River valley and the surrounding Andean highlands. They concentrated particularly in the region around the city of Medellin, contributing business and commercial know-how as well as great energy to the economic transformation of the region from primarily coffee production into one of Latin America's principal diversified urban/industrial centers.

Asian Immigration

Over 500,000 Asians immigrated to Latin American destinations during the last half of the 19th century and the beginning of the 20th century. Most settled in Peru, Mexico, Brazil, and Cuba.

Chinese

Chinese laborers, known then as "coolies," were first brought to Peru beginning in the 1840s on long-term labor contracts to work mining guano deposits along Peru's coast. Later, in the 1860s and 1870s, as the Peruvian economy expanded and capitalist business entrepreneurs invested in mining and agricultural enterprises in the sierra, railroad construction demanded additional labor. Chinese coolies again filled this demand, with perhaps as many as 100,000 Chinese laboring in Peru. Many eventually returned to China, but large numbers remained. They located in Lima or other coastal cities, gradually merged integrated into Peruvian society, intermarried with Peruvians, adopted Spanish, and largely disappeared as a distinct cultural group.

Labor contracts also brought Chinese coolie labor to Cuba in the 1870s and 1880s.

Sugar planters, anticipating an end to slavery and consequent labor shortages, sought to fill the need for laborers. Between 100,000 and 150,000 Chinese reached Havana under these conditions, many leaving again when their labor contracts had been fulfilled. Others stayed, but their numbers were further reduced after the Cuban revolution in 1959, when many migrated to the United States. Nevertheless, Cuba still boasts a small Chinese Cuban population concentrated almost exclusively in Havana.

Chinese immigration also figured as regionally important in northwestern Mexico at the end of the 19th century as labor demand in the lower Colorado River valley in the vicinity of Mexicali brought Chinese coolie labor to Latin America. Although their numbers were not large, and many eventually migrated to the United States, the descendants of those Chinese who stayed form a visible, albeit minor, part of the local population.

At the end of the 20th century, some 100 years after most Chinese immigrated to Latin America, most of their descendants have been largely assimilated into their host societies. Nevertheless, a Chinese presence can often be noted in some parts of Latin America (Figure 9.2). Small "Chinatowns" have survived in Lima, Havana, and Mexicali in northern Mexico. In both Lima and Havana local governments have attempted to revitalize these neighborhoods in an effort to capitalize on their uniqueness and promote them as tourist attractions. In Central America the Chinese presence is notable, often because of the frequency of Chinese restaurants, not only in capital cities, but in regional centers as well (Figure 9.3). The increased opening of China has fueled a tiny stream of Chinese immigrants to Latin America since the 1990s, although most of these immigrants are seeking to emigrate to the United States and Canada and a stop in Latin American is only a way station on that journey.

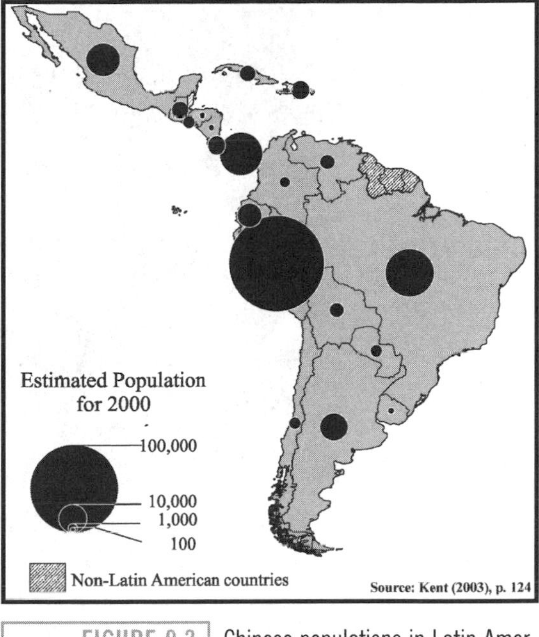

FIGURE 9.2. Chinese populations in Latin America, circa 2000.

Japanese

The Japanese are the only other Asian nationality to migrate to Latin America on a large scale during the 19th and early 20th centuries. Japanese immigrants totaled between 150,000 and 250,000 during this period. They arrived in both Peru and Brazil in the final decades of the 19th century and continued to come into the first decades of the 20th century. Agricultural labor contracts brought Japanese settlers to Peru in the 1870s and 1880s; this pattern continued by fits and starts for several decades. Their numbers rarely exceeded 1,000 per year, but many stayed after concluding their labor contracts, settled, and brought wives from Japan. Many remained in rural areas, establishing themselves as small-scale producers of fruit and vegetables for urban markets, particularly Lima. Although numbering less than 100,000 by the early 20th century, Japanese Peruvians formed a distinct element

FIGURE 9.3. A Chinese restaurant in Tegucigalpa, Honduras, 2002.

of the population of Lima and the coastal region surrounding the metropolitan area.

A similar pattern of Japanese immigration occurred in Brazil. The rapid transformation of the Paraná Highlands from Araucanian pine forests to coffee plantations that began in the 1850s and continued for nearly 50 years fueled a demand for agricultural labor that Japanese immigrants helped fill. Once established, many industrious and thrifty immigrants purchased their own farms and began producing fruits and vegetables for urban markets, particularly São Paulo. Over 200,000 Japanese eventually settled in the Paraná Highlands, mostly in the state of São Paulo, establishing a visible and cohesive Japanese Brazilian community that endures to the present. Indeed, the Brazilian population of Japanese descent totaled over 1 million in 2000 and is the largest concentration of ethnic Japanese in the world outside of Japan.

Population Developments in the 20th Century

By the middle of the 20th century, Latin America's population stood at about 160 million (Table 9.2). The countries of South America accounted for almost two-thirds of this total, slightly over 100 million. Mexico and Central America accounted for most of the remainder, with their combined populations standing at about 36 million. In the Caribbean Basin, the combined populations of the three Hispanic nations, Cuba, the Dominican Republic, and Puerto Rico, stood at almost 11 million. In the early 1950s only a handful of Latin American countries had populations greater than 10 million—and most did not exceed 5 million. In national terms, Brazil, with almost 55 million inhabitants, had fully one-third of the entire region's population. Mexico followed a distant second with some 28 million, and Argentina was third, with 17 million. With a population of 11 million, Colombia was the only other Latin American nation whose population exceeded 10 million.

Immigration Patterns

The character and patterns of ethnicity, race, and national origin changed only modestly in Latin America during the 20th century. The pattern of the mass migration of tens or even hundreds of thousands of immigrants to Latin

TABLE 9.2. Urban Population in Latin America in 1950

Country	Urban (in thousands)	Percent urban	Total population (in thousands)
Argentina	11,206	65	17,150
Bolivia	1,025	38	2,714
Brazil	19,407	36	53,975
Chile	3,553	58	6082
Colombia	4,431	37	11,946
Costa Rica	289	34	862
Cuba	2,889	49	5,850
Dominican Rep.	559	24	2,353
Ecuador	957	28	3,387
El Salvador	712	36	1,951
Guatemala	876	30	2,969
Honduras	243	18	1,380
Mexico	11,832	43	27,737
Nicaragua	384	35	1,098
Panama	308	36	860
Paraguay	514	35	1,488
Peru	2,711	36	7,632
Puerto Rico	900	41	2,218
Uruguay	1,746	78	2,239
Venezuela	2,385	47	5,094
Total	66,927	42	158,985

Source: United Nations (1998), pp. 116–118

America that characterized the last decades of the 19th century slowed and then generally ceased to be of consequence after the 1930s.

Venezuela is the only case where tens of thousands of immigrants have arrived in Latin America from overseas in the second half of the 20th century. Here, a booming petroleum industry in the 1940s and 1950s prompted an increasing demand for labor. Government policies encouraged European immigration. By the 1970s nearly 1.2 million Europeans, mostly Portuguese, Italian, and Spanish, had come to Venezuela. Most eventually settled in the capital city, Caracas, transforming it into a cosmopolitan center with a European air.

The comparatively minor changes that have occurred in the ethnicity and national origins of the population of Latin American countries during the 20th century have hinged on a variety of types of migration. Often, although the relative number of immigrants was small, they have had significant impacts on the host nation. These immigrants have ranged from agricultural colonists to political refugees.

Mennonite colonists, seeking both economic opportunity and religious freedom, established agricultural colonies in a handful of Latin American countries. National governments, seeking to effectively occupy their territory and stimulate economic development, encouraged Mennonite immigration to frontier regions on the national periphery. Mennonites from Anglo-America, increasingly concerned by the intrusion of the governments of Canada and the United States into their lives, established colonies in a number of countries, but the largest were in Chihuahua in northern Mexico, in the Chaco of Paraguay, and in the tropical lowlands of Santa Cruz in Bolivia. These communities prospered, supplying new agricultural products for local consumption and sometimes for export, and energizing and transforming the regional economy. Their populations expanded too, growing into the tens of thousands by the end of the 20th century. The Mennonite colony in Chihuahua, Mexico, numbers approximately 50,000. The population of the Chaco colonies in Paraguay is slightly less, about 30,000. The geographic isolation of their settlements from other population groups makes them stand out, as do their uniqueness and Anglo-American origins (Vignette 9.1).

Spanish political refugees fleeing the Spanish Civil War—and subsequently the fascist government of the Franco dictatorship—figured prominently in the flows of Latin American immigrants between the 1930s and 1950s. Although their numbers were small, these immigrants were usually highly educated and talented. Among them were some of Spain's best artists, writers, and scholars. Mexico and Argentina received most of them; they gravitated to cosmopolitan Mexico City and Buenos Aires, where their contributions to regional culture proved especially significant.

VIGNETTE 9.1. MENNONITE COLONISTS IN LATIN AMERICA

One of the more anomalous features of the human geography of Latin America are the Mennonite settlers. Bearded men dressed in blue-denim overalls and wide-brimmed hats and women wearing plain dresses and bonnets who speak German (actually Plattdeutsch, or Low German) stand out from a crowd anywhere in Latin America. Although not numerous, the Mennonites, thanks to their distinctive culture and lifestyle and their economic contributions, are notable elements in the region's evolving cultural landscape.

Mennonites are Anabaptists who follow a Christian theology that calls for adult baptism rather than baptism as a child. The denomination originated in Holland and Switzerland in the 1500s. Mennonites' unorthodox religious views brought them under repeated scrutiny and harassment by ecclesiastical and state authorities in Europe. Seeking to avoid persecution and pursue their way of life unhindered by others, Mennonites sought isolated geographical locales first in Europe, then later in the United States and Canada, and most recently in Latin America.

Mennonites settled in Paraguay's central Chaco in 1927, establishing the first of what was to become a total of three independent colonies. These first settlers arrived from Canada, where the government had refused to allow the Mennonites to continue to live in isolation from mainstream Canadian society. The Canadian government had banned German-language education in Mennonite schools and had demanded that Mennonite men provide compulsory military service. The government of Paraguay, in the midst of a long-term territorial dispute with Bolivia over the Chaco, welcomed the immigrants, seeing them as a means of establishing a firm Paraguayan hold on the Chaco. In return, the Paraguayans granted the Mennonites broad rights of independence within their colony under an agreement called a "privilegium"—a legal concession granting them religious liberty, total control over their schools, and exemption from military service in recognition of their beliefs in pacifism and nonviolence. Subsequently, two other Mennonite colonies were founded, one by refugees from the Soviet Union in the early 1930s and the other by German Mennonites from the Ukraine in 1947.

The Mennonites succeeded in taming the formidable "Green Hell," a local term used to describe the Chaco's dense, thorny, and at times impenetrable vegetation. Their agricultural practices have proven sound as well as productive, and their colonies have prospered. The colonies supply about half of all the nation's dairy products—milk, cheese, yogurt, and butter—as well as 10 percent of the meat consumed nationally. Commercial crops include cotton, long a staple cash crop for the colonies, and more recently peanuts.

While the recently completed Trans-Chaco Highway has contributed significantly to the Mennonites' recent economic successes, it also exposes the communities to potentially disruptive "outside influences." Alcohol and tobacco are now sold openly in Mennonite towns, although ostensibly only non-Mennonites consume them. In recent years a diverse range of groups has begun to put increasing pressure on Mennonite communities. Some indigenous communities are demanding a return of lands from the Mennonites. These lands were granted to the Mennonites by central government concession, although the government is likely to have originally usurped them from indigenous communities. Others grumble that the Mennonites have received special treatment long enough and that it is time to end their privileged status. The Mennonites have responded by electing a provincial governor and a representative to the national congress. This is an unusual move because the Mennonites have traditionally shunned political involvement.

(cont.)

During the first half of the 20th century Mennonites also immigrated to Mexico, Bolivia, Argentina, and Costa Rica in their search for religious and social autonomy as well as economic opportunity. Latin American governments usually actively encouraged Mennonite colonists, seeking settlers for peripheral regions to establish a strong national presence and at the same time being mindful of the Mennonites' reputation for diligence and economic productivity. Although not to the same extent as in Paraguay, Mennonites in Mexico's far northwest and Bolivia's 'Oriente" near Santa Cruz have had strong regional economic impacts and have generated multipliers. Like their brethren in Paraguay, all of Latin America's Mennonites are feeling the increasing crush of modern life. In Cauchatuame, a 50,000-strong Mennonite colony on Mexico's Mesa del Norte in Chihuahua, small numbers of young Mennonite men were implicated in trafficking drugs for export to the United States in the late 1990s.

Migration Patterns within Latin America

During much of the 20th century significant migration flows have occurred between Latin American countries. This process has been most significant in areas where dramatic differences in standard of living exist between Latin American nations. Two types of migrants characterize these migration flows.

Small numbers of well-educated individuals, who immigrated legally by their own choice or who were actively recruited to migrate, typify one group. Often these individuals are professionals, managers, highly skilled technicians, or business entrepreneurs who are drawn to another country by economic and career opportunities. Characteristically, their abilities and achievements bring them respect in the host society, and they easily merge into the mainstream.

The situation for most immigrants is completely different. Most immigrants are of peasant or working-class origin, are undocumented, and toil as agricultural workers or manual laborers. Agricultural labor, construction work, and domestic service have all figured prominently as employment opportunities for this type of immigrant. Some return repeatedly to a country as seasonal migrants working in the agricultural sector, while others see themselves as temporary migrants who plan only to stay for a relatively short period, perhaps a few months or a few years. Many eventually stay on permanently. While the magnitude of the migration of well-educated professionals is measured in terms of thousands, the movement of unskilled and semiskilled workers between Latin American countries in the 20th century is measured in hundreds of thousands.

The three main destinations for Latin American migrants within the region have been Argentina, Venezuela, and Puerto Rico. All three areas have been attractive to immigrants because of their higher standard of living and the comparative availability of well-paid jobs.

The Great Depression and resultant changes in Argentine immigration policies marked the end of the waves of southern European immigrants who had come to Argentina for almost 50 years. But the country's wealth and robust economy, especially during the war years of the 1940s, served as a magnet for skilled and unskilled workers from neighboring countries. This flow was strongest in the 1940s and 1950s and has since subsided considerably, although Argentina continues to attract immigrants from neighboring countries. In northwest Argentina, the vast sugarcane plantations of the Tucúman region have drawn thousands of seasonal (mostly male) migrants from Bolivia during the harvest season for decades. The lure of jobs as laborers in construction and other industries have brought many more thousands of permanent (again, mostly male) Bolivian immigrants to Buenos Aires. Buenos Aires has also been a key migration destination for Paraguayans, with males seeking jobs as unskilled laborers and women seeking work as domestics and hotel employees. The vibrant and diverse economy of Buenos Aires has also acted as a magnet for a range of professionals and entrepreneurs from Bolivia, Paraguay, and other neighboring countries—although their total numbers are small.

Venezuela has proved to be a favored destination for legal and illegal immigrants from other Latin American nations since the 1940s. The comparative wealth of the nation's petroleum-based economy has been the foundation upon which this immigration has been based. A dramatic surge in the nation's prosperity in the 1960s and 1970s fueled a range of labor demands. Physicians, engineers, and other professionals arrived in small numbers. Unskilled laborers who found employment doing construction work in Caracas, harvesting coffee beans in the Andes, or doing the messiest and most dangerous jobs in the oil fields came by the tens of thousands. All told, nearly 1.3 million Latin American immigrants have settled in Venezuela; Colombians account for nearly 1 million of this, while Peruvians, Ecuadorians, and Dominicans account for most of the balance. At the start of the 1980s, nearly 20 percent of Venezuela's total population was foreign-born.

In the Caribbean Basin neighboring countries and islands produce migration patterns that mirror sharp variations in economic conditions. Haiti, which has the lowest level of economic development in the Americas, shares the island of Hispaniola with the Dominican Republic. Destitute Haitians cross the border into the Dominican Republic to work on sugarcane plantations in demanding low-paid jobs that the Dominicans shun. Similarly, the Dominicans migrate illegally to Puerto Rico to seek employment and take the jobs that the Puerto Ricans avoid. Meanwhile, labor demands on the U.S. mainland attract Puerto Ricans, who as U.S. citizens may move freely to the United State, which they do in large numbers.

During the 1980s and 1990s, Puerto Rico had become an increasingly attractive destination for illegal immigrants from the Dominican Republic. Puerto Rico's comparatively vigorous economy, its status as a commonwealth territory of the United States, and its geographic proximity to the Dominican Republic have all worked to stimulate the illegal immigration of Dominicans. Some 200,000–300,000 Dominicans have immigrated to Puerto Rico since the 1970s. Many have used the island as a stepping-stone for subsequent migration to the U.S. mainland, but many others have stayed—in the island's capital, San Juan, some neighborhoods, like Santurce, are now largely Dominican.

Demographic Change

In geographical terms, the distribution of population in Latin America changed little in the 19th century. The highlands of Mexico and Central America and the Andean highlands all remained densely populated. The Brazilian Highlands, especially the southern reaches on the Paraná Plateau, experienced intense population growth as a consequence of coffee cultivation and the dramatic growth of São Paulo as an industrial center. Dense settlement continued to characterize the coast of northeastern Brazil, but many coastal margins in the region remained thinly populated and unattractive to permanent settlement. Nevertheless, major urban centers, like Lima, Buenos Aires, Montevideo, and Guayaquil were coastal cities.

During the late 19th century and the first half of the 20th century, the percentage of urban population increased steadily, so that by the beginning of the 1950s more than 40 percent of the region's population could be classified as urban (Table 9.2). Many countries remained decidedly rural, especially in Central America. For example, in Honduras and Guatemala, the percentage of urban population was only 18 and 30 percent, respectively. The countries of the Southern Cone presented a completely different panorama. Most of the population lived in urban centers, with fully three-fourths of Uruguayans and two-thirds of Argentines living in cities. At the time, cities like Buenos Aires and Montevideo were among the world's largest metropolitan centers.

The 20th Century

One of the most telling characteristics of Latin America's population history during the 20th century was the tremendous growth in the region's population (Table 9.3). In the 1930s the regional population total barely reached 100 million, and only three countries had populations that exceeded 10 million: Brazil, Mexico, and Argentina. By the end of the century, Latin America's population had grown by a factor of about 5 to surpass 500 million by the year 2000.

This rapid regional population growth has placed tremendous pressure on the region's natural resources and urban systems. In rural areas, this growth has led to high rates of forest clearance, the expansion of the agricultural frontier, a reduction in long-term soil fertility, increased erosion, and the degradation of fragile ecosystems. Urban places have often grown at even more rapid rates than the population at large since they serve as catch basins for most of the rural population growth. This population growth has strained urban systems beyond the breaking point, particularly in the areas of housing, transportation, and providing basic public services.

TABLE 9.3. Population Growth in Latin America, 1930–2000

	Population (in thousands)					Average annual grown (percent)			
Country	1930	1950	1965	1985	2000	1930–50	1950–65	1965–85	1995–2000
Argentina	11,896	17,150	22,283	30,564	37,074	1.83	1.75	1.58	1.28
Bolivia	2,153	2,766	3,841	6,371	8,317	1.25	2.19	2.53	2.12
Brazil	33,568	53,444	84,292	135,564	171,796	2.33	3.04	2.38	1.36
Chile	4,424	6,091	8,571	12,038	15,224	1.6	2.28	1.7	1.38
Colombia	7,350	11,597	18,114	28,714	42,120	2.28	2.97	2.3	1.78
Costa Rica	499	858	1,482	2,600	3,929	2.71	3.64	2.81	2.46
Cuba	3,837	5,858	7,808	10,038	11,202	2.12	1.92	1.26	0.43
Dominican Rep.	1,400	2,409	3,729	6,243	8,353	2.71	2.91	2.58	1.67
Ecuador	2,160	3,310	5,162	9,378	12,420	2.13	2.96	2.99	1.71
El Salvador	1,443	1,940	3,005	5,552	6,209	1.48	2.92	3.07	1.82
Guatemala	1,771	2,969	4,568	7,963	11,423	2.58	2.87	2.78	2.71
Honduras	948	1,401	2,304	4,372	6,457	1.95	3.32	3.2	2.72
Mexico	16,589	27,376	43,500	78,996	98,933	2.5	3.09	2.98	1.64
Nicaragua	742	1,098	1,750	3,272	5,073	1.96	3.11	3.13	2.73
Panama	502	893	1,326	2,180	2,950	2.88	2.64	2.49	1.99
Paraguay	880	1,371	2,019	3,681	5,470	2.22	2.58	3	2.5
Peru	5,651	7,632	11,467	19,698	25,952	1.5	2.71	2.71	1.7
Puerto Rico	1,552	2,219	2,594	3,451	3,816	1.79	1.43	1	0.71
Uruguay	1,704	2,239	2,693	3,012	3,342	1.37	1.23	0.56	0.75
Venezuela	2,950	5,009	8,970	17,317	24,277	2.65	3.88	3.29	2.07
Total	102,019	157,630	239,478	391,004	504,337				

Sources: Centro Latinamericano de Demográfia (CELADE; 1981 and 1985); United Nations (1968); United Nations (1986); United Nations (2003); University of California, Los Angeles. UCLA Latin American Center Publications (2000).

The rate of population change among Latin American countries has been quite uneven. Some countries have experienced steady, comparatively measured and manageable population growth. For others, the rate of increase has been nothing short of breathtaking. For instance, over the 70-year period between 1930 and 2000, Argentina's population increased by a factor of about 2.7, Puerto Rico's by a factor of about 2.5, and Uruguay's by a factor of about 1.9. But population totals in Brazil and Mexico, the region's most populous nations, increased by factors of 5 and 6, respectively. Similar increases have characterized most other nations in the region. In Central America population growth has been especially high, with population numbers increasing by a factor of about 7 in Guatemala, Costa Rica, and Nicaragua. More recently, the increasing availability of a range of birth control methods, as well as significant changes in social customs, has had a major effect on population growth in many countries (Vignette 9.2).

Latin American's Population and the Demographic Transition Model

One of the most useful ways of putting Latin America's past, present, and future population growth into context is by examining it in the context of the demographic transition model. This model is a conceptual tool long used by demographers and others concerned with understanding the characteristics of population change in societies with different socioeconomic characteristics. Originally developed to explain population change in Western Europe during the last 400 years, the model has since proved useful in explaining population growth and change in other regional contexts.

The model examines three key demographic factors: the birthrate, the death rate, and the natural rate of population increase (note that migration is not considered in establishing the natural rate of population increase). The model suggests that as societies evolve and are transformed from what can be characterized as traditional and rural, with primarily agricultural economies, to modern and urban, with primarily industrial and service-based economies, dramatic change takes place in the patterns of birthrates and death rates and consequently in the rate of population growth. The model suggests that with respect to population growth rates, societies (which can be conceptualized in a variety of ways—e.g., countries, regions, or other geographic areas) move through four stages or phases (Figure 9.4). In Stage 1, *traditional economies*, birthrates and death rates are both high, but are also in rough equilibrium, and consequently the population growth rate is very small or stagnant. In Stage 2, *developing economies*, social and economic factors are transforming traditional, rural, agriculturally based societies into modern, urban, industrial societies. Death rates fall precipitously as the benefits of modern medical science (e.g., vaccines) and a vastly improved understanding of public health and the measures that can be taken to improve it affect the society at large. But longstanding cultural norms and the perceived economic benefits of having many children tend to keep birthrates high. Consequently, net population growth rates are high, typically over 2.0 percent annually but often close to 3.0 percent.

Stage 3, *developed or industrialized economies*, refers to societies that are increasingly urban and in which a smaller and smaller percentage of the economy is based on agriculture. These societies are typified by steadily declining death rates and a significant dropoff in birthrates as changing cultural values, urban living, and the effects of consumerism make large families with many children less desirable. Population growth rates in Stage 3 typically range from 1.0% to 2.0% per annum.

The model's final stage, Stage 4, describes societies that might be called *mature industrial (or postindustrial) economies* in which the service sector increasingly domi-

VIGNETTE 9.2. POPULATION MYTHS: LATIN AMERICANS AND BIRTH CONTROL

Latin Americans are predominately Roman Catholic, or at least nominally so. Stereotypical perceptions suggest that the behavior of Latin American women with respect to birth control and abortion will be tightly proscribed by the dogma of the Roman Catholic Church. But social science research since the mid-1990s demonstrates that the reproductive practices of Latin American women are far more diverse than commonly believed.

In practice, many Latin American women ignore the directives of the hierarchy of the Roman Catholic Church and employ various methods of birth control, and, in many cases, also have abortions. Since the 1970s, Latin American women have chosen to adopt "modern" methods of birth control in increasing numbers (Table 9.4). These include birth control pills, diaphragms, intrauterine devices, and several other techniques as opposed to traditional methods (the rhythm method or abstinence). The pattern of adoption of modern methods of birth control has, of course, varied from country to country. The rate of adoption has been higher in countries with higher standards of living and large urban populations, and lower in nations with low per capita incomes and high percentages of rural population. By the early 2000s, over 70 percent of the married women in Brazil, Costa Rica, and Cuba, and over 50 percent of married women in nine other Latin American countries including Mexico, used modern methods of contraception. While the data only reflect the birth control practices of married women and data are not available for all countries, it is clear that when it comes to family planning, personal and family considerations strongly outweigh Catholic religious doctrine for many Latin American women.

Abortion, a highly charged social and political issue in the United States, does not ignite the same intense controversy in Latin America. And it need not, because, with the exception of Cuba, abortion is illegal in all Latin American countries. This is hardly surprising considering that Latin Americans are overwhelmingly Roman Catholic, and that Catholicism is the official religion of several Latin American nations. In 1994, researchers from the Alan Guttmacher Institute in New York published some unexpected results from a study of abortion patterns in Latin America. They analyzed government hospitalization statistics and conducted fertility surveys of women in several Latin American countries. Their research showed that despite its illegal status, the rate of abortion among women between 15 and 49 in many Latin American countries far exceeds that in the United States and Western European countries where abortion is legal. The annual abortion rate for women in this age group is 2.7 percent in the United States and less in Canada and most of Western Europe. However, it stands at 5.2 percent in Peru, 4.5 percent in Chile, 4.4 percent in the Dominican Republic, 3.7 percent in Brazil, and 3.4 percent in Colombia. Of those Latin American countries analyzed, only Mexico, at 2.2 percent, had a lower rate of abortion than the United States.

nates the economy. In these societies population growth rates drop to below 1.0% annually as birthrates and death rates both decline and again come close to equilibrium. Eventually, population growth rates fall below the level necessary for even a replacement of the population.

Examining the population growth rates of Latin American nations over the last century (Table 9.3) in light of the demographic transition model helps put the data in a more meaningful context.

Countries like Bolivia, El Salvador, and Peru that have been transformed in the last century from rural agricultural societies (Stage 1) to developing countries (Stage 2) reflect this transformation in significant change in their population growth rates: these rates were below 2.0 percent annually before 1950, but increased to between 2.5 percent and 3.0 percent annually in the years up until 1985. Other nations, like Brazil and Colombia, which began the process of industrialization and massive urbanization earlier, seem to have moved from Stage 2 (growth rate over 2.0 percent) in the years before 1950, to Stage 3 (growth rates below 2.0 percent annually) in the period after 1985. Population growth rates in Argentina, one of the most industrialized and urbanized nations in the region, have remained between 1.0 percent and 2.0 percent since the 1930s.

Puerto Rico, Cuba, and Uruguay are the only nations in Latin America with population growth rates below 1.0 percent per annum. All three are interesting anomalies. Puerto Rico's low population growth rate is surely related to its political and economic status as an integral part of the United States and the steady stream of migrants from the island to the U.S. mainland. Cuba's status as the only socialist nation in Latin America, where the influence of the Catholic Church is almost nil and birth control is widely practiced, explains low population growth rates on that island nation. The situation in Uruguay is more difficult to explain, although its stagnant economy for many decades may be a key factor.

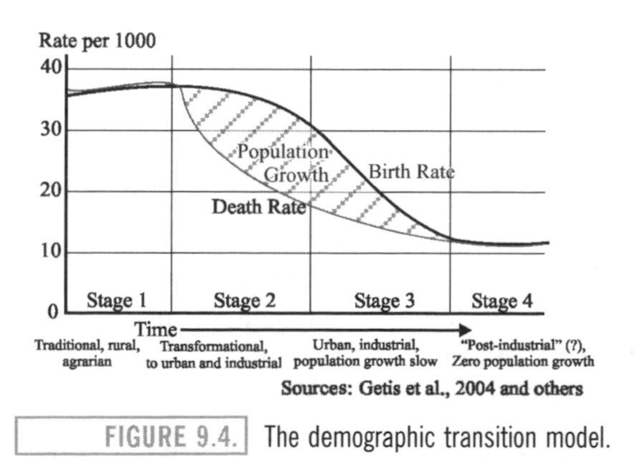

FIGURE 9.4. The demographic transition model.

TABLE 9.4. Contraceptive Methods among Married Women of Reproductive Age, ca. 2003

Country	Total percent using all methods (2003)	Percent using modern methods (2003)
Argentina	—	—
Bolivia	48	25
Brazil	76	70
Chile	—	—
Colombia	76	64
Costa Rica	80	72
Cuba	73	72
Dominican Republic	70	66
Ecuador	66	50
El Salvador	60	54
Guatemala	43	34
Honduras	62	51
Mexico	68	59
Nicaragua	69	66
Panama	—	—
Paraguay	57	48
Peru	68	50
Puerto Rico	78	68
Uruguay	—	—
Venezuela	—	—

Source: Population Reference Bureau (2004).

Venezuela is another anomaly. It had one of the highest rates of population growth among all Latin American countries during the last half of the 20th century. Growth rates exceeded 3.0 percent annually between 1950 and 1985. While natural increase played some role, much of this growth can be attributed to the immigration of hundreds of thousands of Spaniards and Portuguese, as well as other South Americans, who were seeking work during the boom years of the petroleum industry in Venezuela.

Latin America's Population: 2000 and Beyond

Population growth rates in many Latin American countries appear to be slowing at the beginning of the 21st century (Table 9.5). Nevertheless, population growth rates continue to be very high in some countries. This is especially the case in Central America, where rates range between 2.5 percent and 3.0 percent in several nations. Barring dramatic change, those countries will likely double in population in less than 25 years, while populations will likely double in many other Latin American nations in 35 years or less. Most of these nations are poor and have limited resources with which to confront such precipitous population growth. Increasing social discord and unrest may become widespread in these nations in the coming decades if rapid population growth is not reduced. Declining living standards, as well as increasing environmental degradation and pollution, are likely to ensue if the rate of economic development does not increase in tandem with population increase and its benefits are not shared more equitably by the region's population.

Race and Race Relations in Latin America

Latin American countries are marked by distinct and indeed often sharp differences in the racial characteristics of their population. While these differences are evident to even the most casual traveler as he or she moves from one Latin American nation to another, efforts to quantify these characteristics are fraught with difficulty. Simply defining racial categories is difficult since these definitions vary from nation to nation. Furthermore, in many Latin American countries, census data do not distinguish among racial categories. Hence, the best "data" that are available are simply the informed estimates of demographers and country specialists rather than concrete statistical information. The data presented in Table 9.6 provide one such approximation for most Latin American countries; the Caribbean countries of Cuba, the

TABLE 9.5. Population Totals and Rates of Increase in Latin America, 2000–2050

Country	Population mid-2000 (millions)	Natural increase (annual,%)	"Doubling time" in years at current rate	Projected population (millions) 2025	Projected population (millions) 2050
Argentina	37.0	1.11	62	47	55
Bolivia	8.3	2.04	34	12	16
Brazil	170.1	1.54	45	221	244
Chile	15.2	1.29	54	20	22
Colombia	40.0	2.01	34	58	73
Costa Rica	3.6	1.76	39	6	7
Cuba	11.1	0.67	103	12	11
Dominican Republic	8.4	2.18	32	12	15
Ecuador	12.6	2.09	33	18	21
El Salvador	6.3	2.36	29	10	14
Guatemala	12.7	2.94	24	22	32
Honduras	6.1	2.76	25	9	11
Mexico	99.6	1.95	36	133	152
Nicaragua	5.1	3.02	23	9	12
Panama	2.9	1.68	41	4	4
Paraguay	5.5	2.67	26	9	13
Peru	27.1	2.14	32	39	48
Puerto Rico	3.9	0.92	75	4	4
Uruguay	3.3	0.65	107	4	4
Venezuela	24.2	2.02	34	35	42
Total	503.1			683	799

Source: Population Reference Bureau (2005).

TABLE 9.6. Ethnic Profiles of Selected Countries in Latin America: Percentage of Total Population

Country	Amerindians	Mestizos	Europeans	Africans/Mulattoes
Argentina	>1	2	97	—
Bolivia	54	32	14	—
Brazil	<1	—	50	50
Chile	5	70	25	—
Colombia	1	50	20	29
Costa Rica	0.5	7.5	92	—
Ecuador	40	40	15	5
El Salvador	10	89	1	—
Guatemala	54	42	4	—
Honduras	5	92	1	2
Mexico	10	75	15	—
Nicaragua	3	76	10	11
Panama	5	67	14	14
Paraguay	3	76	20	1
Peru	46	42	12	—
Uruguay	—	7.5	90	2.5
Venezuela	2	7	20	8

Source: Collier, Skidmore, and Blakemore (1992), p. 161.

Dominican Republic, and Puerto Rico are absent from this list.

These data suggest that *mestizos*, people of mixed European and indigenous ancestry, are dominant in the population geography of Latin America, accounting for 70 percent of the region's total. Mestizos usually account for between two-thirds and three-fourths of the total population of most Latin American countries. Notable exceptions are Argentina, Uruguay, Costa Rica, and Brazil, where they represent less than 10 percent of the population.

People of European descent—whites, almost exclusively from southern and western Europe—are the second largest racial group in the Americas. They account for some 20 percent of the region's total population, and thus are a distant second in terms of their numerical significance. Whites are present in all Latin American countries, but their relative numbers vary sharply—even between neighboring countries. Over 90 percent of the populations of Argentina, Uruguay, and Costa Rica are of European descent, while less than 5 percent of the populations of Honduras, El Salvador, and Guatemala are of European ancestry.

The distribution of peoples of African origin—blacks, as well as mulattos and *pardos*—is more restricted than that of mestizos or whites. In terms of the region's overall population geography, peoples of African heritage are most important in Brazil, where they account for about 50 percent of the total national population (some 170 million). In Cuba, the Dominican Republic, and Puerto Rico, the proportion of peoples of African or mixed African and European descent is large, ranging from around 40 percent in Puerto Rico to close to 90 percent in the Dominican Republic. The coastal margins of several circum-Caribbean nations also have dense populations of blacks and mulattos. Notable among these are Venezuela, Colombia, Panama, and Nicaragua, where the percentage of African Americans ranges from 8 percent in Venezuela to 29 percent in Colombia. In a handful of countries blacks and mulattos are small minority populations accounting for less than 5 percent of the total population. There is no notable or demographically significant presence of blacks or mulattos in almost one-third of the region's nations.

Amerindians, or indigenous peoples, are

only a small percentage of the region's population, just about 2 percent. Yet, despite their small numbers, Amerindian populations are found in almost all Latin American countries with the exception of the Caribbean states. In most countries their relative proportion of the national population is miniscule. However, in some countries, like Guatemala and Bolivia, native peoples form a majority of the total population; in other countries, they form an appreciable proportion, particularly in Ecuador and Peru (Figure 9.5). Furthermore, while Amerindians only account for 10 percent of the population of Mexico, this figure represents approximately 10 million people. In absolute terms this is the largest population of indigenous people in any Latin American country.

Racial prejudice and racial discrimination are among the unavoidable facts of life in almost all Latin American nations. The characteristics and dynamic of the racial hierarchy, who is favored and who is not, and the nature of racial discrimination vary somewhat from country to country. Nevertheless, the overall pattern is roughly similar throughout Latin America. Whites sit at the top of the hierarchy and enjoy the most power and prestige, while indigenous peoples and the darkest, most African-looking blacks sit at the bottom. Both groups are systematically disadvantaged and experience widespread discrimination. Mestizos and mulattos occupy intermediate positions in this racial hierarchy, although differing national contexts produce distinct patterns. At the same time, peoples' perceptions about who they are and their personal identification often do not reflect their racial origins (Vignette 9.3).

Race in Brazil and the United States: Some Comparisons

The racial dynamics of Brazil, Latin America's most populous nation, deserve closer examination. Here peoples of African or mixed descent account for almost 50 percent of the population. Brazil is sometimes touted as a "racial democracy" where whites, blacks, and everyone in between live in relative social harmony; racial prejudice and discrimination are largely nonexistent. This "ideal" is often compared to the situation in the United States where the plight of African Americans is perceived to be

FIGURE 9.5. Women of indigenous descent in El Alto, Bolivia, 2004.

VIGNETTE 9.3. RACE AND PERSONAL IDENTIFICATION

Latin America is characterized by significant racial diversity. Race does matter in most Latin American societies, influencing one's social position and chances for economic advancement. However, very few Latin Americans identify themselves or even consider themselves primarily in racial terms. Beyond one's family and economic class, most Latin Americans will identify themselves based primarily on their nationality-Brazilian, Mexican, Bolivian, and so on-and secondarily on their region of origin in their country (sometimes known as one's 'little country,"or *patria chica*). This secondary identity may be focused on the town or city of origin, on a subnational political division, or on a broader regional basis. Thus, a Brazilian from Rio de Janeiro will be a *Carioca* (a native of Rio de Janeiro), a Mexican from Monterrey or Chihuahua will be a *norteño* (a person from the north), and a native of Mexico City will be a *Defeño* (a person from the Federal Distirict, *Distrito Federal*)—or a *Chilango*, used pejoratively by those from outside the capital to describe its residents.

Geographical regions often provide the basis for personal identification. In most Andean countries, for instance, people make a clear distinction between *serranos*, mountain people, and *costeños* , people from the coast. In Bolivia, which has no coast, a person from the eastern lowlands is a *camba*.

Popular writing, textbooks, and scholarly treatments of Latin America by North American and European authors emphasize the significance of mestizos and mulattos as major components of the region's population. It may come as a surprise to first-time travelers to the region that Latin Americans simply do not identify themselves as mestizos, mulattos, or even blacks, regardless of their ancestry. A Cuban historian recounts the story of a visiting North American colleague who persisted in referring to her as an Afro-Cuban, despite her insistence that she was 'Cuban."The same historian, who subsequently moved to Spain, noted that she had never considered herself as black or mulatto until she left Cuba and came to Europe where she was immediately categorized as such.

There are few exceptions to the tendency of Latin Americans to identify themselves by their country of origin, geographic region, or patria chica. On a regional scale, indigenous peoples, or Indians, are the most likely to identify themselves as members of a particular tribe or indigenous or aboriginal community. Nevertheless, something of a continuum exists among native peoples. Those who live in isolated regions peripheral to the national core and continue to keep their language and culture intact will usually identify themselves first as members of a particular tribal or linguistic group, then more generally as Indian or native (*nativo*), and lastly by nationality. Typically, as the geographical isolation of indigenous groups decreases and their level of integration/acculturation into the national society grows, identification with the nation-state increases correspondingly. However, in the last quarter of the 20th century, there has been an increasing consciousness by almost all native peoples in Latin America of their distinct cultures and histories, and a growing self-identification as Indians or native peoples. Thus, even indigenous peoples who are essentially acculturated into the mainstream of national society will now increasingly identify themselves as Indian or native. A rising racial, ethnic, and cultural consciousness among native peoples has given rise to the establishment of new institutions and organizations. These include regional social movements that span national borders. This is the case in the Amazon Basin. Other examples include indigenous political parties in Bolivia and the Zapatistas in southern Mexico.

The last decades of the 20th century have also seen an increased consciousness of African origins and a

(cont.)

growing political awareness among blacks and mulattos in some parts of Latin America. This has been most pronounced in countries like Brazil, Colombia, Ecuador, and Venezuela where blacks and mulattos form a sizable, but not dominant, segment of the total population. Generally, however, these groups are not as well organized, nor do they carry the kind of political clout, as similar organizations wield in the United States. On the other hand, in Cuba and the Dominican Republic, political and social consciousness based on blackness and African origins is of little consequence.

far worse than in Brazil. However, rosy words and platitudes cannot disguise the fact that prejudice and discrimination based on race and racial characteristics are deeply ingrained in the Brazilian culture and psyche. The heritage of discrimination caused by nearly 300 years of slavery has proven hard to shake and will undoubtedly remain part of the country's national culture for decades to come.

Racial consciousness is manifested in many ways. The terminology used in Brazil to describe race and "shades" of race is instructive. Brazil is a vast nation, and regional variations in the mix of racial groups and categories can give rise to a wide range of terms. In a detailed study of a small town in the Brazilian Highlands, the American anthropologist Marvin Harris notes that, in addition to whites and blacks, local folk recognized five other racial categories: *moreno*, *chulo*, *mulatto*, *creolo*, and *cabo-verde*. As Harris notes, variations in skin tone and hair provide the key elements in this system of classification. A Ghanaian anthropologist who has studied race relations in Brazil claims that Brazilians recognize up to 20 different racial categories ranging between the extremes of white and black. Officially, however, the national census apparatus (the Instituto Brasilero de Estatística y Geografia) recognizes only four racial categories: *branco* (white), *pardo* (brown/mulatto), *prêto* (black), and *amarelo* (yellow/Asian and indigenous) (Figure 9.6).

As the multitude of racial terms utilized in Brazil suggests, the categorization of individuals into racial groups is no simple matter. Perhaps one of the most important distinctions that must be made is that, in Brazil and generally throughout Latin America, racial categorization exists along a continuum, with black and white at the extremes, and many shades and gradations in between. Thus, racial categorization is not simply a matter of "black and white," as it is in the United States, where the dividing line between black and white is clearly defined.

A useful way to understand this distinction is as follows. In Brazil, racial prejudice and discrimination are based on "prejudice of appearance" (also referred to as "prejudice of mark"). Hence, skin color (its degree of darkness), hair type and color, and facial features, especially noses and lips, are critical variables in assessing and categorizing a person's race. An overt practice up until the 1950s and 1960s, but since discontinued, was for job advertisements for female employees to clearly indicate if only white or "whiter" job candidates were acceptable by inserting code phrases like "of good appearance" into the ad.

In the United States, in contrast, racial categorization between blacks and whites is based on "prejudice of origin." The critical issue is not so much one's appearance, but rather one's ancestry. There is no racial spectrum between black and white in the United States—you are one or the other. Thus, in the United States, a person of mixed European and African heritage, even an individual with light skin and European facial characteristics, is typically classified as black or African American. However, in the population census of 2000, in response to pressure from an increasing proportion of the U.S. population of mixed parentage who do not

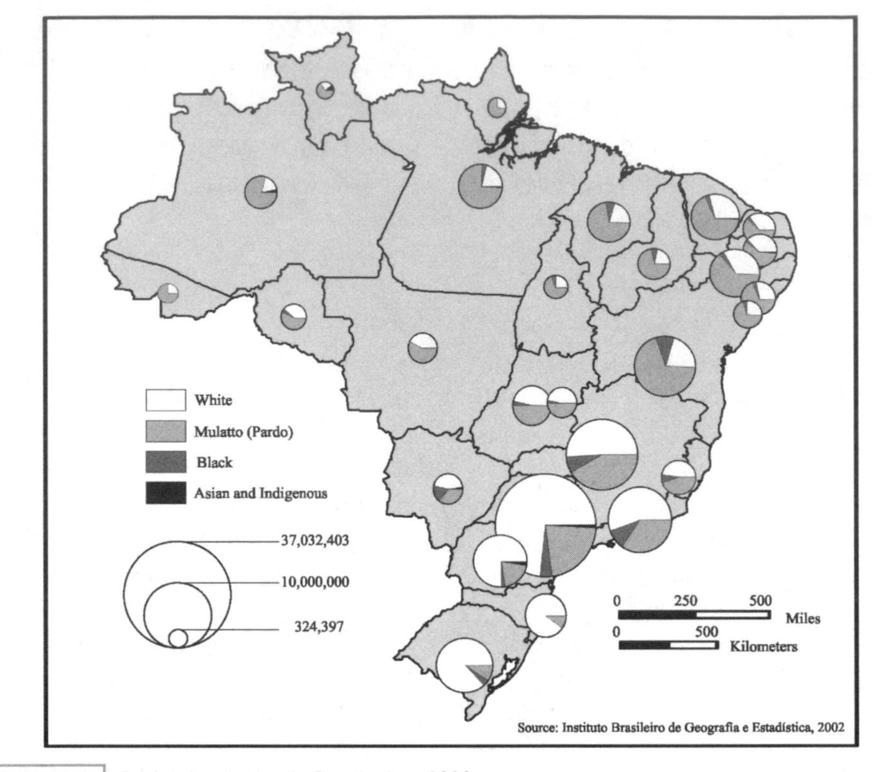

FIGURE 9.6. Racial distribution in Brazil, circa 2000.

view themselves as falling within these tidy racial categories, the U.S. Census Bureau began to allow census respondents to classify themselves in terms of a wide range of biracial and multiracial categories. The extent to which these categorizations might displace traditional practice is uncertain, but is seems unlikely that a dramatic change in perceptions and customs will occur soon.

Latin American immigrants of mixed European and African heritage in the United States often find the system of racial classification they encounter bewildering and profoundly disturbing. On the one hand, individuals who never considered themselves black nor were considered to be black in their country of origin are unequivocally classified as blacks in the United States. American blacks, on the other hand, frequently expect blacks of Latin American origin to identify with them and their causes, and are disturbed and put off by them when they do not, as is usually the case.

In Brazil, however, racial classification is often more than simply an assessment of one's physical characteristics. A whole range of additional factors can weigh heavily in how an individual is perceived and the treatment to which he or she is accorded. "Money whitens" is a well-known phrase based on a real phenomena. Simply put, economic success and its trappings of fine clothes, luxury cars, and real property bring one immediate respect and elevate one's racial status. Professional status, intellectual achievements, and artistic, musical, and athletic talent also whiten. Thus, a wealthy educated black (a *prêto*) will be elevated in status to *moreno* (a light brown). Similarly, an individual with the physical characteristics of a light mulatto who is a successful professional is likely to be referred to as a *blanco* (a white) and to be socially accepted as such. While whites absolutely dominate the general staff of all military branches and the senior officer corps, some mulattos and blacks do reach the ranks of the later, especially in the army. "Rank whitens" in this context and gave rise to the adage that in the Brazilian army "There are no black colonels." While economic or professional success can elevate one's racial standing, it is also true that low-status employment, scant educational achievement, or a marginal lifestyle all tend to reduce one's racial status.

Brazil is sometimes touted as a "racial democracy" where people of all races mix freely and live in harmony. Most thoughtful observers, however, find that these kinds of statements fall far short of reality. Blacks and mulattos do rise in Brazilian society, but the number of those who do is truly minuscule given that blacks and mulattos represent about 50 percent of the nation's people. The ironic Brazilian aphorism, "In Brazil, there is no racism: The Negro knows his place," summarizes much about the nature of race relations in Brazil. Overt discrimination is unnecessary if the social norms and conventions of racism are accepted by the affected population.

João Jorge Santos Rodrigues, a black leader in Salvador, Bahia, in the mid-1990s, recounted his experiences with racism there during the 1960s to a *New York Times* reporter (Brooke 1993, p. 4/7). "Our parents and grandparents warned us: Don't go to that neighborhood, don't go to that beach, don't go to that apartment building—you will have to go up the service elevator," Rodrigues said. Then continuing, he noted: "In South Africa, it was clearly written: 'No Blacks.' In Brazil it is the social fabric that proscribes, that says: Listen, you are prohibited from coming here." Sometimes however, discrimination can be more overt. Municipal governments in the mostly all-white and prosperous south (Santa Catarina) have at times taken to canvassing bus stations and establishing municipal checkpoints to discourage or turn back "dark" migrants.

Apologists for the Brazilian racial system sometimes suggest that it is economic class rather than race that truly divides Brazilians. In many ways this is true, as literally millions of Brazilians of all races, perhaps as many as 50 million, live in poverty. Blacks and mulattos (*prêtos* and *prados*), however, represent a disproportionate share of these numbers, and census data further illuminates the economic disparities. In the mid-1990s, the average monthly income of a black male (equivalent to $163) was just 41 percent of the average monthly income of a white male.

Saying this, it is probably also fair to say that somehow racism in Brazil is more muted and softer than it is in the United States. Although significant, race is but one factor taken into account in assessing an individual in Brazilian society—education, employment, and personal appearance all matter as well. Brazilians of different races do mix socially, at least on a superficial level, and with greater ease than do blacks and whites in the United States. It is also true that while marriage between blacks and whites is exceptional and frowned upon in Brazil, marriage between white people and those of mixed race is acceptable and commonplace.

The dynamics of the relationship between people who are black, white, and all the hues and shades in between is not the full story of race relations in Brazil, but it is clearly the most important one. People of indigenous origins, especially those who live on the margins of Brazilian society and whose culture and economy remain aboriginal, are at the bottom of the racial pecking order and suffer the greatest prejudice and discrimination. Acculturated Indians, *caboclos*, *pardos*, and *prêtos* , sit in a somewhat intermediate position. As a racial group, whites occupy the most privileged position—all other things being equal, whites are likely to receive more deferential treatment than that accorded individuals of other racial groups.

Race in Hispanic America

In Hispanic America, the racial hierarchy and dynamics of racial prejudice and discrimination have much in common with the pattern found in Brazil. Blacks and indigenous peoples are at the bottom of the social hierarchy. Mulattos and mestizos occupy intermediate positions, while whites sit at the top.

However, while the basic pattern of discrimination is similar, there are notable differences. First, in the majority of Hispanic American countries, the fundamental racial divides are between Indians (native peoples), mestizos, and whites. Second, the relative proportions of these three groups vary considerably among these countries (Table 9.3). Consequently, the dynamics of racial prejudice and discrimination can be quite distinct in different countries. In Guatemala, for instance, where indigenous peoples comprise the majority in many highland regions, mestizos, known locally as *ladinos*, dominate the regional power structure—occupying key social, economic, and political positions. In contrast, in Argentina, where whites account for about 95 percent of the total population, mestizos are found in large numbers only in the northwestern provinces, forming a high proportion of that region's urban and rural working class. In Mexico, where mestizos are the vast majority of the population, racial prejudice and discrimination occur primarily on two fronts. The nation's small "white" population tends to receive deferential and preferential treatment in almost all contexts, while the country's native peoples, representing some 10 percent of the nation's population, are subject to both racial and cultural prejudice and discrimination.

Throughout Hispanic America the definition and usage of the word *mestizo* has evolved considerably since the colonial period, when it was first employed. Originally, the term *mestizo* defined a person who was the offspring of the union between a European and an Indian, usually a European male and an Indian female. As the colonial period progressed, and the myriad of racial permutations multiplied, so did the terminology. For example, the offspring of a *mestiza*, a female *mestizo*, and a Spaniard was known as a *castizo*. This nomenclature varied among the various colonies, and as the population of mestizos grew and the number of unions between whites, mestizos, mulattos, blacks, and Indians increased, a dizzying array of racial mixtures and permutations occurred. Indeed, by the middle of the 17th century, mixed-race peoples accounted for the majority of the population in most towns and mining camps, and their presence in rural areas was increasingly visible. The cumbersome task of racial categorization in Hispanic America eventually collapsed under its own weight. A more generic term, *casta* or *castas*, came into widespread usage during the later part of the colonial period to describe all peoples of mixed ancestry.

The term *mestizo* continued to be used to identify people of mixed European and Indian parentage, but increasingly its usage reflected the cultural affinity of individuals rather than their genetic makeup or racial appearance. Consequently, individuals who adopted the trappings of Spanish culture and society in the New World, including speaking Spanish, worshiping in Catholic churches, wearing European clothes, engaging in urban occupations and trades, and following a range of other "Western" behaviors would be classified as mestizos, even if their racial origins were clearly indigenous.

At the present, in countries like Peru, Mexico, and Guatemala, where Indian peoples still comprise an appreciable segment of the population, this dualism in the use and understanding of the term *mestizo* continues to be significant. Eventually throughout much of Hispanic America, the terms *gente decente* (decent people) or *gente de razón* (rational people) came into wide usage to distinguish individuals or families who had acculturated and had adopted Spanish customs and values from those who had not. The terms still enjoy broad usage in many parts of Hispanic American today, although they are used principally to distinguish the middle and upper classes from the working classes, the peasants, and the poor.

The dynamics of race and racial prejudice and discrimination varied considerably in the Hispanic colonies on the islands of the Caribbean. Within less than 50 years of contact with Europeans, the indigenous population had been almost completely eliminated—they were the victims of introduced diseases, enslavement, and multiple abuses at the hands of the colonial authorities. The importation of African slaves filled the void created by the disappearance of native peoples and led to the development of a racial hierarchy very similar to that which developed in Brazil.

Nevertheless, despite their common colonial heritage, divergent social histories during the 19th and the 20th centuries have given rise to distinct racial demographics as well as distinct patterns of prejudice and discrimination in Cuba and Puerto Rico. Both continued as Spanish colonies until the end of the 19th century, when slavery, a key economic and social institution, was abolished. A socialist revolution in Cuba in 1959, which included racial justice and nondiscrimination as two of its primary social goals, has succeeded in dramatically reducing racial prejudice and discrimination in that country. Many Cubans claim that racism is effectively nonexistent in Cuba today. The emigration of nearly 1 million Cubans, mostly whites, to the United States after the revolution in 1959 undoubtedly facilitated this transformation. In Puerto Rico, despite its status as a U.S. commonwealth subject to U.S. civil rights laws, change has come more slowly. The vestiges of the racial hierarchy left by three centuries of slavery are still evident in Puerto Rican society, with whites and near-whites occupying the top of the racial hierarchy and mulattos and dark-skinned Puerto Ricans at the bottom.

A completely different dynamic characterizes race relations in the Dominican Republic. The overwhelming majority of Dominicans, perhaps as much as 90 percent, are fair-skinned mulattos; the rest of the population consists of a small number of blacks and an equally small number of whites. Characteristically, whites enjoy the greatest privileges, but what is so distinctive about the Dominican Republic is the extreme prejudice and discrimination that blacks experience in a country where almost all citizens have some African ancestry. Much of this can be attributed to the history of the Dominican Republic's relations with its neighbor Haiti, a former French colony, whose population is almost exclusively black. Dominicans feel tremendous antipathy toward Haitians, who ruled the Dominican Republic from 1822 to 1844, and with whom they fought a bloody war subsequently to preserve their independence. Thus, in the Dominican Republic, to be black is to be Haitian. The military strongman, Rafael Trujillo, who ruled the Dominican Republic for more than 30 years (1930–1961), engaged in a systematic policy of denying any vestiges of the nation's African roots and closing the country's frontier with Haiti. There is a greater consciousness and tolerance of the nation's African connections today, but blacks still experience intense discrimination in the Dominican Republic. If a racial identifier is required, typically Dominicans will not identify themselves as mulatto but rather will identify themselves as *trigueño* (light brown, the color of wheat) or as "Taíno," the name of Hispaniola's indigenous inhabitants who became extinct during the initial decades of contact with Europeans in the 16th century.

Summary

Population growth and change in the 19th and 20th centuries was influenced by multiple factors. African slavery continued throughout most of the 19th century, especially in Brazil, Cuba, and Puerto Rico, finally disappearing in the late 1880s. European immigration was limited, although millions of Italians and Spaniards settled in Argentina, and Italians and Germans created influential communities in Brazil. Asian laborers also came to work in Latin America, with Chinese populations settling in Peru and Cuba, and Japanese populations settling in Peru and Brazil. During the latter part of the 20th century, petroleum-rich Venezuela attracted immigration from Europe, especially Portugal and Spain, and migration from other Latin American countries. When population growth in Latin American countries is examined using the demographic transition model, Bolivia, El Salvador, and Peru are Stage 2, developing economies, and Brazil and Colombia are Stage 3, developed economies. Although 70 percent of the population can be characterized as mestizo, Europeans, Africans, mulattos, and Amerindians also contribute to the racial mix of Latin America. Racial discrimination is a problem. Those of European descent tend to be at the top of the social order, with Africans and indigenous people at the bottom.

Further Reading

Burkholder, M., and Johnson, L. (1998). *Colonial Latin America* (3rd ed.). Oxford, UK: Oxford University Press.

Davila, J. (2000). Expanding perspectives on race in Brazil. *Latin American Research Review, 35*(3), 188–198.

Dzidzienyo, A. (1995). An obsession with whiteness and blackness. In G. H. Summ (Ed.), *Brazilian Mosaic* (pp. 162–166). Wilmington, DE: SR Books.

Field, L. W. (1994). Who are the Indians?: Reconceptualizing indigenous identity, resistance, and the role of social science in Latin America. *Latin American Research Review, 29*(3), 237–248.

Green, D. (1991). *Faces of Latin America*. London: Latin America Bureau.

Hanke, L., and Rausch, J. (1993). *People and issues in Latin American history: The colonial experience—Sources and interpretations*. New York: Wiener.

Hunter, J., Thomas, R., and Whiteford, S. (Eds.). (1983). *Population growth and urbanization in Latin America: The rural-urban interface* . Cambridge, UK: Schenkman.

Martine, G., Das Gupta, M., and Chen, L. (Eds.). (1998). *Reproductive change in India and Brazil*. Delhi, India: Oxford University Press.

Merrick, T. (1985). *The demographic history of Brazil*. Albuquerque, NM: Latin American Institute.

Otero, G. (2003). The "Indian question" in Latin America: Class, state, and ethnic identity construction. *Latin American Research Review, 38*(1), 248–266.

Page, J. A. (1995). *The Brazilians*. Reading, PA: Addison-Wesley.

Perreault, T. (2003). Changing places: Transnational networks, ethnic politics, and community development in the Ecuadorian Amazon. *Political Geography, 22*(1), 61–88.

Price, R. (1996). *Maroon societies: Rebel slave communities in the Americas* (3rd ed.). Baltimore: Johns Hopkins University Press.

Rahier, J. M. (2004). The study of Latin American "racial formations": Different approaches and different contexts. *Latin American Research Review, 39*(3), 282–293.

Ravuri, E. D. (2002). Life-time and recent migration to Bolivar State, Venezuela, 1990: The effect of the Guayana program on migration. *Journal of Latin American Geography, 1*, 69–81.

Salzano, F. (2002). *The evolution and genetics of Latin American populations*. Cambridge, UK: Cambridge University Press.

Sansone, L. (2003). *Blackness without ethnicity: Constructing race in Brazil*. New York: Palgrave Macmillan.

Sawatzky, H. L. (1971). *They sought a country: Mennonite colonization in Mexico*. Berkeley and Los Angeles: University of California Press.

Skelton, R. (1990). *Population mobility in developing countries: A reinterpretation*. London: Delhaven Press.

Skidmore, T. (1974). *Black into white: Race and national identity in Brazilian thought*. Oxford, UK: Oxford University Press.

Sundberg, J. (2003). Conservation and democratization: Constituting citizenship in the Maya Biosphere Reserve, Guatemala. *Political Geography, 22*(7), 715–740.

Tapinos, G., Mason, A., and Bravo, J. (Eds.). (1997). *Demographic responses to economic adjustment in Latin America*. Oxford, UK: Clarendon Press.

Wright, W. R. (1990). *Café con leche: Race, class, and national image in Venezuela*. Austin: University of Texas Press.

10 The Central Valley of Chile

The Central Valley of Chile is the heartland of the Chilean nation and one of the principal population concentrations along South America's Pacific coast. Approximately 80 percent of the nation's inhabitants, about 12 million people, live in the Central Valley. Santiago, the region's primate city and the national capital, accounts for half of the regional population total. It is the country's major industrial, commercial, and cultural center. The Central Valley enjoys a subtropical Mediterranean climate and is characterized by a productive agricultural sector dominated by high-value crops, many of which—especially temperate fruits—are raised for export markets. Agroindustrial processing also contributes to the regional economy, and the region is well known for its high-quality wines.

Geography and Environment

The Central Valley and its environs encompasses an area that runs from approximately 30°S latitude, near the present-day city of La Serena on the Pacific coast, south to about 37°S latitude at the city of Concepción, where the Bío Bío River meets the Pacific Ocean (Figure 10.1). As its name implies, the principal geographic feature of this region is a long structural depression hemmed in between the Andes on the east and low coastal ranges on the west. Known in Chile as the Valle Central, or Central Valley, it is actually not a single valley, but rather a series of valleys, or basins, lying within the same structural depression. This sequence of valleys runs southward from the Aconcagua River, at about 33°S latitude, to the Bío Bío River. While the valley itself runs roughly north–south, it is drained by rivers that run from east–west, rising in the Andes and dissecting the coastal ranges before discharging into the Pacific Ocean. North of the Aconcagua River, the foothills of the Andes extend almost uninterrupted to the ocean and there is almost no coastal plain. Several small rivers running east–west drain this region and provide some areas suitable for agriculture and raising livestock.

The climate of the Central Valley region is aptly characterized as Mediterranean. The summers are warm, even hot at times, and dry (Figure 10.2). Cloudless skies are common throughout the summer. Mean temperatures in the warmest month, January, average about 20°C throughout the region, but temperatures are slightly cooler near the coast and warmer in the interior valleys that are protected from cooling marine influences. Temperatures in the spring and fall are mild, and winters are cool. In July, the coolest month, mean temperatures stand at around 7°–8°C in the southern and central parts of the region, but are warmer, between 10° and 12°C, in the north and near the coast.

Rainfall is concentrated in the winter

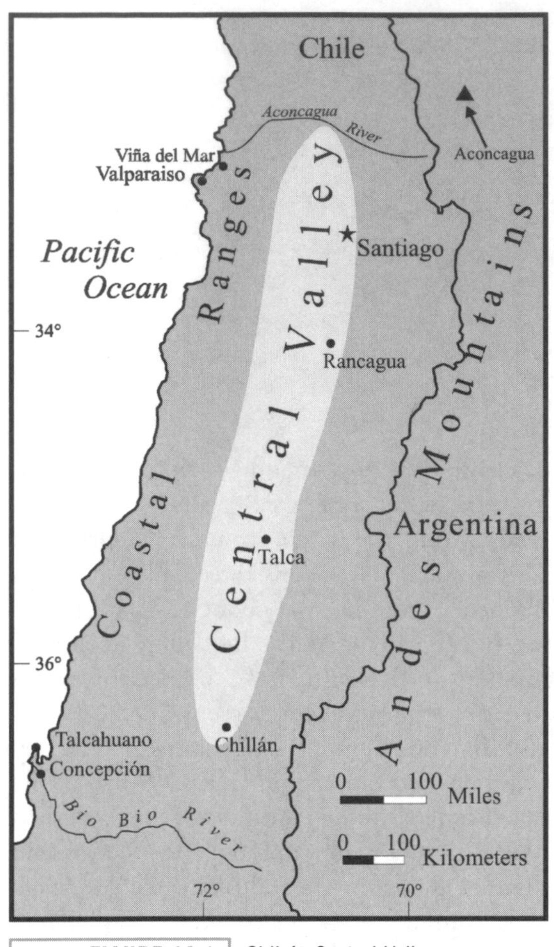

FIGURE 10.1. Chile's Central Valley.

months, and most falls between April and August. Nevertheless, in the south, some rainfall occurs even in the summer months, whereas in the north the summer months are rainless. Annual rainfall exhibits a clear north–south pattern. Nearly 1,300 mm of precipitation falls at Concepción in the far south. This declines to about 360 mm at Santiago, and diminishes to only 135 mm at La Serena on the region's northern margin. In the south, the coastal ranges act as a barrier to the westerly storms that provide most of the moisture. Consequently, rainfall totals on the coast and in the transverse valleys that open to the Pacific are higher than in the interior.

The vegetation cover of this region has been severely impacted by long-term human occupation. Indeed, much of the region is cultivated or dedicated to pasture. In the far south, around the Bío Bío River, where rainfall is most abundant, the forest cover consists of both deciduous and evergreen broadleaf species as well as conifers. Northward, however, as precipitation declines, these forests quickly yield to a vegetation cover characterized by drought-resistant trees, shrubs, and grasses that grow during the wet winter months and go dormant in the long dry summers. This vegetation cover, known as *matorral*, is similar in both form and characteristics to that of the chaparral of the southwestern United States. North of the Aconcagua River, annual precipitation diminishes and the vegetation cover becomes sparse, taking on a noticeably more arid appearance. Various species of cacti, mesquite, and other drought-resistant plants comprise the vegetation cover. The vegetation is more luxuriant and a wider range of species are found in river valleys and streams that traverse the area as well as along the coastal margin where higher humidity occurs.

Historical Geography and Economic Development

Spanish conquistadores and colonists arrived in the Central Valley region early in the colonial period. In 1535, Diego de Almagro mounted a large expedition that originated in Cuzco, Peru, and numbered several hundred Spaniards and several thousand Indians. This group reached as far south as the Bío Bío River, but food shortages and hostile Indian attacks exhausted the force. Almagro and the survivors of his expedition returned to Cuzco in 1537. Even though the region lacked vast Indian populations and precious metals, Pedro de Valdivia mounted another, more successful, expedition in 1540.

Santiago, founded in 1541 by Valdivia, be-

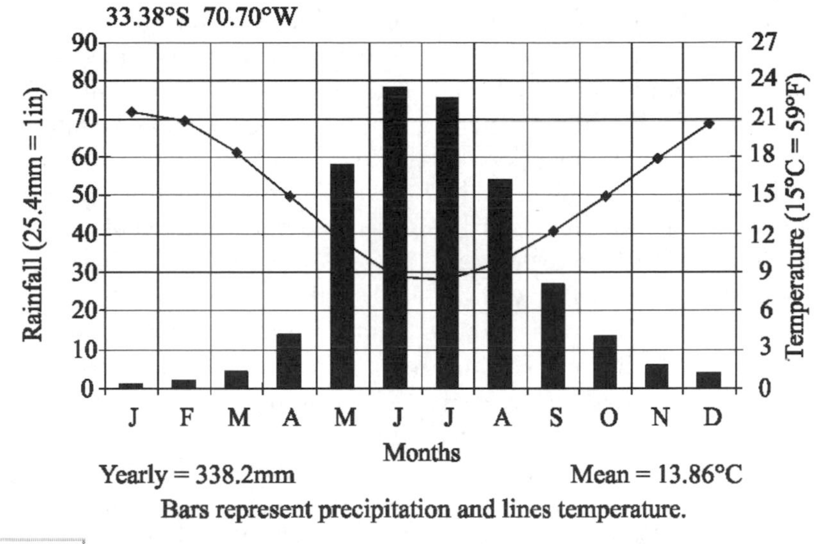

FIGURE 10.2. Climograph of Pudahuel, Chile.

came the political center and the principal administrative and economic center of the new colony. Hostile Aconcagua Indians attacked and razed the settlement during its first year, but it was reconstructed immediately under Valdivia's direction. The Spaniards consolidated their tentative hold over the region around Santiago, and at the same time Valdivia's forces pushed further south. They founded Concepción on the Bío Bío River in 1550 and a series of towns and forts to the south of it by 1553.

The indigenous inhabitants of the Central Valley cultivated a range of domesticated crops, including corn, potatoes, and beans, and employed rudimentary irrigation techniques. In stark contrast to the Araucanian peoples in the southern forests, the natives of the Central Valley were rapidly subjugated and incorporated into the emerging colonial economy.

The settlement of the Central Valley region proved attractive for the Spanish colonists who found many similarities between its climate and landscape and that of their homeland. A mild climate, fertile soils, water for irrigation, and a readily exploitable Indian labor supply permitted the rapid establishment of a successful agricultural economy. The Spanish Crown made large land grants, *mercedes*, of the best agricultural lands with access to water for irrigation to a small group of the privileged elite. Indians, and later mestizos, worked these large estates, known as *fundos* in Chile, under a variety of exploitative labor systems. This pattern of land tenure dominated the Central Valley well into the latter half of the 20th century, and it still characterizes some areas at the present. The economy of the Central Valley focused on subsistence production as well as the production of meat, hides, tallow, and wheat for export to the mining areas of Peru and Bolivia.

The Central Valley was the undisputed core of the Chilean colony throughout the colonial period. After the colony's independence from Spain in 1818, this dominance continued. Until the 1880s, the territorial limits of the Chilean nation largely coincided with those of the Central Valley. Chileans, however, eventually moved aggressively into Bolivian territory in the Atacama Desert to exploit its rich nitrate deposits for booming export markets. Disputes with Bolivian authorities over resources and taxation escalated to armed hostilities in 1879.

Peru joined Bolivia in defense of its territory in the Atacama Desert, but Chile, with the strong support of its British allies, defeated Bolivia and Peru in the War of the Pacific (1879–1883). Chile annexed sizeable portions of both Peru and Bolivia as a result, areas that later proved to hold immense and valuable copper deposits. Bolivia lost its coastline and access to the sea. Shortly thereafter, Chilean armies finally defeated the Araucanian Indians, extending the effective national territory southward as well. Despite this dramatic territorial expansion at the end of the 19th century, the Central Valley continued, and remains today, the heartland of the Chilean nation.

Agriculture has played a dominant role in the economy of the Central Valley since the colonial period. Grain, grapes, and livestock were the principal agricultural products throughout most of the 19th century. Wheat and other cereal exports brought in valuable foreign exchange earnings. The establishment of a flour-milling industry in the middle of the century marked the beginnings of industrialization and a slightly more diversified economy. Farmers also produced basic food commodities and subsistence crops for local and regional consumption. The Mediterranean climate of the Central Valley is ideal for the cultivation of deciduous fruits including grapes, peaches, apricots, and plums (Figure 10.3). However, the region's geographical isolation limited fruit production to meeting local demand and to selling dried fruit or preserves in the neighboring region of Cuyo across the Andes in Argentina.

Other economic activities, principally mining, transportation, commerce, and manufacturing played secondary roles during the 19th century. Foreign investment, mostly from British sources, fueled much of the development of the region. The mining, processing, and shipping of copper ore from the adjacent Andes, as well as that of silver and other metals mined in the coastal ranges and foothills, contributed to Chile's economic growth. Valparaíso, located on the coast some 140 km west of Santiago, consolidated its position as the nation's principal port and one of the three major ports on South America's Pacific coast. As the national capital and principal city, Santiago became the focus of much of the nation's commercial and manufacturing activities.

Manufacturing, however, was limited, and much of it was rudimentary. The tremendous foreign exchange earnings generated by the

FIGURE 10.3. Grape vineyards alongside the Panamerican Highway with the Andes in the background, 1993.

development of the mining industry in the northern coastal deserts supported the purchase of imported manufactures by the country's small elite, and demand for locally produced products was small. This situation prevailed until the early 1930s, when the worldwide economic depression reduced the supply of manufactured products and eliminated much of the foreign exchange earnings used to purchase them. This spurred the development of manufacturing, often aided by government subsidies and protective tariffs.

Contemporary Economic and Social Geography

The economic geography of the Central Valley has changed dramatically during the last quarter of the 20th century. Many of these changes resulted directly from the model of free trade and export-oriented economic development pursued by the military government that seized power from the democratically elected government of Salvador Allende in 1973. The military government overturned the land reform programs implemented in the 1960s and early 1970s. However, it did not allow the reestablishment of the traditional hacienda system that had dominated the Central Valley since colonial times. Rather, its policies promoted the development of an active land market that encouraged more intensive land use and production of crops for export markets.

Agricultural land-use practices changed quickly in the Central Valley. The cultivation of traditional crops like wheat, potatoes, corn, oats, barley, and sugar beets decreased or only grew slowly. Meanwhile, acreage dedicated to the production of deciduous fruits, especially peaches, nectarines, plums, apples, and pears, and table grapes expanded rapidly. So did the cultivation of cut flowers and some specialty vegetables like asparagus.

Chile's location in the Southern Hemisphere favored this kind of production for export, as many of these fruits and vegetables mature between December and February, when the supply in its primary Northern Hemisphere market, the United States, is limited. Improvements in postharvest handling and processing as well as transportation contributed most to this agricultural transformation.

Wine production also increased as Chileans worked assiduously and successfully to improve the quality of their wines as well as to promote their visibility on the world market. They also produced increasing quantities of high-quality varietal wines like cabernet sauvignon and chardonnay at bargain prices and found ready markets for these, especially in the United States (Figure 10.4).

These transformations also had profound social impacts. Despite the many egregious abuses of the hacienda system, the haciendas did provide a permanent home for many of the Central Valley's rural population. The demise of the hacienda system dislocated these people, pushing some into urban areas and others into the migratory and seasonal labor markets required to harvest and process fruits and vegetables for export.

The effects of the export-oriented development model followed by the military government had severe repercussions for Chile's

FIGURE 10.4. Wine makers like Santa Rita Vineyards in Chile's Central Valley produce high-quality wines for export to North America and Europe, 1996.

manufacturing sector, which was concentrated in the Central Valley's three major urban centers: Santiago, Valparaíso, and Concepción. Manufacturing had grown between the 1930s and the early 1970s; initially it prospered because the Great Depression and World War II had restricted the supply of manufactured imports and later it continued to expand because subsequent governments promoted import substitution. Free trade, through the elimination of tariff barriers, exposed Chilean industry to the world market. Inefficient firms folded and many workers lost their jobs. Shantytowns on the periphery of urban centers grew, especially in Santiago, as unemployed or underemployed urban workers slid down the economic ladder and displaced rural workers moved to urban areas in search of work.

The Central Valley is the most heavily settled region of South America's Pacific coast. It is home to approximately 12 million people, nearly 80 percent of the population of Chile. The historic dominance of this region shows no signs of waning.

Over 80 percent of the region's population are urban residents. Most live in just three large metropolitan centers: Santiago, Valparaíso/Viña del Mar, and Concepción/Talcahuano. These three urban centers also dominate the manufacturing sector and produce roughly 80 percent of the nation's industrial output, respectively accounting for 50, 20, and 10 percent.

Santiago, the nation's capital and the focus of almost all government activity, is preeminent among these three metropolitan areas, with a population near 6 million. It is the country's principal commercial, industrial, and cultural center, and it dominates all aspects of Chile's national life. Sited on the Andean piedmont, the city enjoys a spectacular visual backdrop of mountains. However, increasing atmospheric pollution, largely from automobiles and buses, and frequent temperature inversions combine during the summer months, often leaving the city under a pall of haze.

Santiago is marked by strong contrasts. On the one hand, it is a modern cosmopolitan center boasting a vibrant stock exchange, a new subway, and exclusive residential and business zones like Los Condes on the eastern margin of the city. Although the historic city center has lost much of its colonial architecture, some colonial structures, especially churches, survive alongside elegant examples of late-19th-century architecture patterned on French and other European models. The shantytowns, known as *callampas* (mushrooms), that occupy many peripheral sectors of the metropolitan area provide a striking contrast to the prosperity exhibited in the city's upscale residential districts (Figure 10.5).

The other key metropolitan center areas, Valparaíso/Viña del Mar and Concepción/Talcahuano, share several characteristics. Both are comprised of two cities that have now coalesced. The populations of both metropolitan areas are nearly the same, about 600,000, and both are located on the Pacific coast. As might be expected, these two metropolitan areas boast major ports, and indeed Valparaíso and Talcahuano are the nation's most important port cities.

Valparaíso/Viña del Mar is sited on the Pacific Coast about 140 km west of Santiago. Although the bay at Valparaíso is small and not especially well protected, it has been Chile's principal port since the colonial period. The city's urban core, port facilities, and industry occupy a narrow coastal plain beside the bay, with residential neighborhoods, many reached by funicular trams, perched on the steep hillsides that surround the bay. Viña del Mar, located 5 km to the north, is distinguished by attractive beaches that make it one of the preeminent vacation resorts in South America. Tourism dominates the economy here, and the city boasts attractive 19th-century villas, modern high-rise apartments and hotels, fine restaurants, and pleasant parks, plazas, and pedestrian promenades. Industrial and manufacturing activities in the metropolitan area in-

FIGURE 10.5. Modern architectural design, like this Hyatt hotel, characterizes many of the buildings in Santiago's exclusive suburb of Los Condes, 1996.

clude food processing, petrochemical production, and textiles.

The metropolitan area comprised of the cities of Concepción and Talcahuano is located at the southern extreme of the Central Valley region near the mouth of the Bío Bío River. Founded early in the colonial period (1550), Concepción evolved as the principal city in southern Chile. An attractive city with a modern university and the seat of regional government, Concepción serves as the principal commercial and financial center of this portion of the Central Valley. Talcahuano is located scarcely 10 km north of Concepción on the massive and well-protected Talcahuano Bay. The harbor at Talcahuano is the best harbor on the South American coast: indeed, there is no comparable harbor along the Pacific coast until one reaches San Diego in the United States. The port is home to a large fishing fleet, an active merchant marine, and the formidable Chilean navy. Industrial activities concentrate around Talcahuano; especially noteworthy are the iron and steel works at Hauchipato developed under government sponsorship in the late 1940s. Capitalizing on favorable site factors, including access to port facilities, inexpensive waterborne transportation, an abundant freshwater supply, hydroelectric power, and local coal reserves, iron and steel production here has encouraged a wide range of industries in the metropolitan region. These include petrochemical production, food processing, and wood products manufacture.

Summary

Chile's Central Valley forms the heartland of the country and is its principal population center. The Central Valley is not one valley, but rather a series of valleys trending north–south flanked by the Andes Mountains on the east. Much of the region is dedicated to pasture and agriculture. The Mediterranean climate, with hot dry summers and cool winters, makes it a prime agricultural area. The Central Valley was colonized by Spain during the mid-16th century, with Santiago as the capital. Under colonial rule, large agricultural estates, called *latifundia*, or simply *fundos* in Chile, dominated the rural regions of the Central Valley. The land was owned by a wealthy elite comprised largely of whites and farmed by Indians and later mestizos. In the late 20th century, the military government introduced neoliberal economic reforms including widespread privatization of state enterprises and open market reforms, and also transformed agriculture by promoting active land markets and crops for export, such as fruits, flowers, and wine. Major urban manufacturing centers include Santiago, as well as the port cities of Valparaíso/Viña del Mar and Concepción/Talcahuano.

Further Reading

Bauer, A. (1975). *Chilean rural society from the Spanish Conquest to 1930*. New York: Cambridge University Press.

Bethell, L. (1993). *Chile since independence*. Cambridge, UK: Cambridge University Press.

Caviedes, C. (1979). *The politics of Chile: A sociogeographical assessment*. Boulder, CO: Westview Press.

Crowley, W. K. (2000). Chile's wine industry: Historical character and changing geography. *Yearbook, Conference of Latin Americanist Geographers, 26*, 87–101.

Drake, P. W., and Jaksic, I. (1991). *The struggle for democracy in Chile, 1982–1990*. Lincoln: University of Nebraska Press.

Hojman, D. (1993). *Chile: The political economy of development and democracy in the 1990s*. London: Macmillan.

Hudson, R. A. (Ed.). (1994). *Chile, a country study* (3rd ed.). Washington, DC: Federal Research Division, Library of Congress.

Kay, C., and Silva, P. (Eds.). (1992). *Development and social change in the Chilean countryside: From the pre-land reform period to the demographic transition*. Amsterdam, The Netherlands: Centre for Latin American Research and Documentation.

Kelly, P., and Child, J. (1988). *Geopolitics of the Southern Cone and Antarctica*. London: Lynne Rienner.

Loveman, B. (2001). *Chile: The legacy of Hispanic capitalism* (3rd ed.). New York: Oxford University Press.

McBride, G. (1936). *Chile: Land and society*. New York: American Geographical Society.

McKenna, M. K. L., and Murray, W. E. (2002). Jungle law in the orchard: Comparing globalization in the New Zealand and Chilean apple industries. *Economic Geography, 78*(4), 495–514.

Monteón, M. (1982). *Chile in the nitrate era: The evolution of economic dependence, 1880–1930*. Madison: University of Wisconsin Press.

Pendle, G. (1960). *The land and people of Chile*. New York: Macmillan.

Schurman, R. (2001). Uncertain gains: Labor in Chile's new export sectors. *Latin American Research Review, 36*(2), 3–30.

11 The Central American Highlands

The Central American Highlands are the population and economic core of Central America. Over 30 million people reside in the highlands, constituting between 80 and 90 percent of the Central American nations' total population. In contrast to most other regions of Latin America with dense populations, most of this region's population is still rural. Agriculture plays a key role in the economy. Historically, coffee, which is cultivated in the highlands, either dominated or contributed a significant proportion of export earnings in all Central American nations. This pattern continues into the present. Most of the region's large urban centers as well as its national capitals are located in the highlands. The region's limited industrial base, largely food processing and the production of consumer goods, is concentrated around these centers.

The northern portion of the Central American Highlands falls within the Mesoamerican culture region. It was densely populated by Maya and other indigenous groups during the pre-Columbian period. Only vestiges of these groups are found in El Salvador and Nicaragua. In Honduras indigenous populations are more numerous. The Guatemalan Highlands are still densely settled with Maya peoples, whose proportion of the nation's population is about 50 percent.

Poverty is endemic throughout the region; the rate exceeds 50 percent of the population in all countries except Costa Rica, where it is 20 percent. Rural poverty is even more extreme, reaching 60–80 percent in most countries. Population growth rates are among the highest in Latin America, ranging from 2.0 to over 3.0 percent per year. Rapid population growth has contributed to difficult economic and social conditions. Poverty as well as an extremely unequal distribution of wealth fueled violent social revolutions in the 1970s and 1980s in Guatemala, El Salvador, and Nicaragua. While these armed conflicts have now ended and democratic governments hold sway in these nations, the underlying socioeconomic conditions that helped produce these revolutions have not been ameliorated. Migration to the United States from these countries has totaled more than 1.5 million since 1970. About 20 percent of the total population from El Salvador emigrated to the United States during that period!

Physical Environment

The Central American Highlands lie fully within the Tropics between approximately 16°N and 8°N (Figure 11.1). The physiography of the region varies over its extent. In Guatemala, the northern highlands are characterized by three clearly defined mountain ranges all trending roughly east–west and with elevations reaching about 3,500 m. The northernmost of these ranges is the Sierra de las

FIGURE 11.1. Central American Highlands.

Cuchumatanes. The Sierra de las Minas lies to its south and runs roughly parallel to it. The third range, the Cordillera Central, lies even further to the south and is separated from the northern ranges by the valley of the Motagua River, Guatemala's principal river. These ranges grade into broad, sometimes rugged, uplands, between 1,000 and 2,000 m over much of Honduras, El Salvador, and Nicaragua (Figure 11.2). In Nicaragua the uplands are concentrated across a wide swath that occupies the northern third of the national territory. Southward of these highlands, a wide geological structural depression, called a rift valley, divides the Central American highlands into northern and southern sections. Large freshwater lakes, Managua and Nicaragua, lie in this structural depression. Lake Nicaragua holds the unusual distinction of being one of the few places in the world that freshwater sharks inhabit. The San Juan River, one of the longest in Central America, crosses the southern end of the rift valley and is roughly coincident with the border of Costa Rica. The river, which nearly transects the isthmus, was a popular route across the isthmus in the mid-1800s as thousands of 49ers streamed across it on their way to the gold fields in California. It remained a notable route across the isthmus until the Panama Canal was finished; indeed, it was considered a viable option for a canal prior to the construction of the Panama Canal (Vignette 11.1).

In Costa Rica the highlands are more massive and rise to higher elevations than in Nicaragua, Honduras, and El Salvador. Rising out of the Cordillera de Guanacaste in the

FIGURE 11.2. Rural settlement in the highlands of western Honduras, near Guajquiro, 1996.

VIGNETTE 11.1. THE PANAMA CANAL

Nearly 100 years after its completion in 1914, the Panama Canal remains one of the most impressive engineering feats on the planet. Its construction in just 10 years is equally astounding. No less amazing is the tale of international intrigue, geopolitical maneuvering, and military intervention that surrounded the independence of Panama from Colombia in 1903. In a 1904 treaty Panama granted the United States absolute power in perpetuity over a 16-km wide swath of land extending 80 km from the Atlantic to the Pacific, the right to build and operate a canal, and the right to garrison troops there at will.

There had been interest in developing a canal across the Central American isthmus since the 1850s. The San Juan River in Nicaragua had been considered as a possible site for a canal, as had the narrow land bridge in Panama used by the Spanish since colonial times. French entrepreneurs and engineers began work on a route through Panama in the early 1880s, but immense engineering and financial challenges as well as the ravages of tropical disease doomed their efforts to failure. In the aftermath of the Spanish-American War (1898), military concerns about the difficulties of fighting another two-ocean war created strong U.S. interest in the construction of a canal. In 1902 the U.S. Congress authorized the purchase of the French company's assets and the negotiation of a treaty with Colombia to build the canal. When the Colombians balked, the United States supported insurgents in Panama, granted diplomatic recognition to the new country days after it declared its independence, and prevented Colombia from crushing the independence movement.

The construction of the Panama Canal was a phenomenal feat. Tens of thousand of workers, most of them blacks from the English-speaking islands of the Caribbean, labored on the project. As many as 20,000 are estimated to have died from accidents and the ravages of tropical diseases, especially yellow fever and malaria. The building of the 80-km-long canal involved the damming of the Chagres River and the creation of an immense artificial lake, Lake Gatún, in the middle of the isthmus. It also required excavating an 8-mile trench, called the Culebra (or Gaillard) Cut, across the continental divide. A series of massive locks raise ships entering the canal from the Caribbean Sea to Lake Gatún and a second series of locks lowers vessels to Lake Miraflores and subsequently to sea level on the Pacific Ocean (Figure 11.3). A superb documentary film produced by Carl Charlson *A Man, a Plan, a Canal, Panama* (1974), chronicles the canal's construction with thoughtful narration and original archival film.

Since its completion in 1914, the Panama Canal has been vital to world trade. It was also a symbolic flash point for protests against U.S. activities in Latin America and especially Panama. In 1977, the United States signed the Panama Canal Treaties, agreeing to return the canal and the Canal Zone to Panamanian control at the end of 1999. At the time many Americans feared that Panama could not effectively operate the canal. These fears have proved completely unfounded. Under Panamanian control the canal has continued to be operated effectively and efficiently. In 2003 nearly 12,000 oceangoing vessels made the 24-hour canal crossing successfully. Panamanian authorities are now exploring options to enlarge the canal to accommodate the larger ships and oil supertankers that cannot pass though the locks currently in service, and to ensure the canal's continuing relevance in the 21st century.

FIGURE 11.3. An oceangoing vessel, the *Kanuka Forest*, its decks loaded with wood, passes through the locks of the Panama Canal, Panama Canal Zone, 1979.

northwest, the central highlands of Costa Rica center on a plateau, the Meseta Central, which is surrounded on three sides by volcanic peaks reaching about 3,500 m. These highlands extend southward in Costa Rica into the Cordillera de Talamanca and even into western Panama where the highest elevations still reach close to 3,500 m. Elevations decline as the cordillera continues eastward in Panama, where in the Serrania Tabascará elevations drop to 1,000–1,500 m before the mountains disappear completely just to the west of the Panama Canal.

The highlands of Central America are roughly coincident with a volcanic axis that extends along the western edge of the isthmus. Skirting the Pacific coast, the axis crosses the rift valley that divides the highlands into northern and southern sections in Nicaragua and extends southeastward into Costa Rica and western Panama. Some 80 volcanoes appear along this axis, many of them active in historic time. Active volcanos are found in Guatemala (Agua, Fuego, Pacaya, and Santa Maria), in El Salvador (Izalco), in Nicaragua (Concepción and Negro), and in Costa Rica (Arenal, Irazú, and Poas). Irazú (3,432 m) erupted in 1723, destroying the city of Cartago and also showering fertile volcanic ash over a vast region of the Meseta Central. The volcano erupted again between 1963 and 1965, causing a number of deaths and once more spewing volcanic ash over much of the Meseta Central. Ashfalls of several inches occurred in many areas, and total ash deposits in the most affected areas exceeded a foot.

Climatic conditions in the Central American Highlands are controlled by the region's latitudinal position in the Tropics and by altitude. As with other tropical highlands, a clear patterning of climatic regions occurs (see Figure 3.5). Hot tropical conditions typify *tierra caliente*, the first of these ecological regions, lying between sea level and about 1,000 m. Only areas at the upper margin of this ecological region could be considered part of the highlands. A tropical wet-and-dry climate predominates on the narrow coastal plain and foothills of the Pacific Lowlands to the west, while in the east the climate of the Caribbean Lowlands is humid tropical.

Tierra templada lies in the highlands, falling between 1,000 and 2,000 m. Its climate is characterized by mild temperatures and moderate rainfall. The region's climate is sometimes describe as "eternal spring"; this in no small measure has contributed to this zone being the most densely settled in the Central American Highlands. The region's most important agricultural export commodity, coffee, is grown in tierra templada and occupies much of the agricultural land. In addition, four of the region's six national capitals are located in this ecological zone. Most of the highlands in El Salvador, Honduras, Nicaragua, and Panama fall within the upper reaches of tierra caliente and in tierra templada. Only a few areas exceed 2,000 m. However, in Guatemala and Costa Rica, the highlands are more extensive and at least two other ecological zones can be identified, *tierra fria* and the *páramo* . *Tierra fria* occupies the elevations between about 2,000 and 3,000 m and is typified by a cool temperate climate. Daytime temperatures are moderate and usually pleasant, while nights are cool and frost and freezes occasionally occur. Temperate crops like cabbage, carrots, wheat, and even deciduous fruit trees can be successfully cultivated in these areas in Guatemala and Costa Rica. Successful dairy farming is also possible here, while it is not at lower elevations in the Tropics. The *páramo* , lying above 3,000 m, is cool and humid. Some trees are found at the lower elevations, but low shrubs, bushes, and grasses predominate. *Tierra helada*, or frozen land, where permanent ice and snow can be found, lies above 4,000 m. In Central America this area is limited to a handful of high volcanic mountain summits in Guatemala and Costa Rica.

In the Central American Highlands diurnal temperature changes are often greater than those that occur over the course of a year. At Guatemala City, for instance, the average daily temperature during the winter months (December, January, February) is about 17°C, while temperatures at Tegucigalpa, Honduras, and San José, Costa Rica, both located at lower elevations, average about 19°C. Temperatures climb ever so slightly in summer, with Guatemala City averaging about 19°C and Tegucigalpa and San José averaging between 20° and 22°C (Figures 11.4 and 11.5).

Most of the Central American Highlands are well watered. Precipitation is generally adequate for farming and grazing. Nevertheless, precipitation totals do vary considerably over the region, with less falling in the north

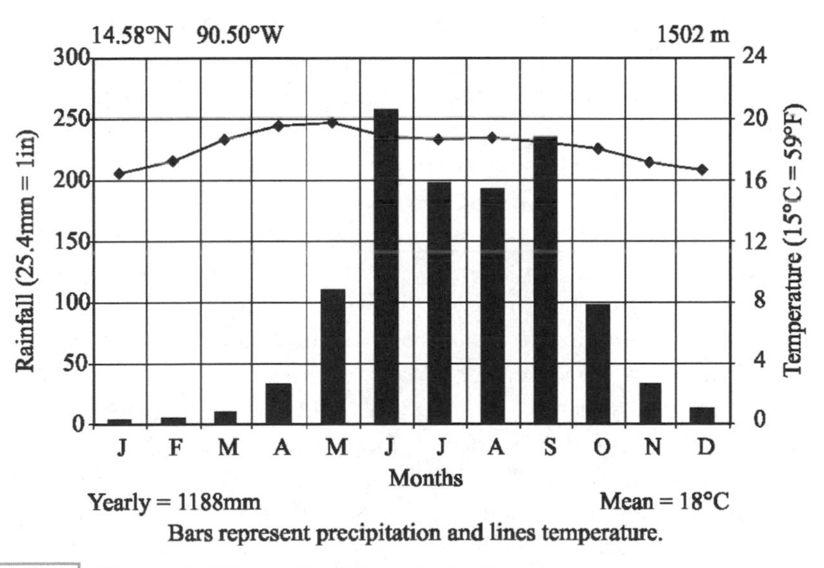

FIGURE 11.4. Climograph of San José/La Sabana, Costa Rica.

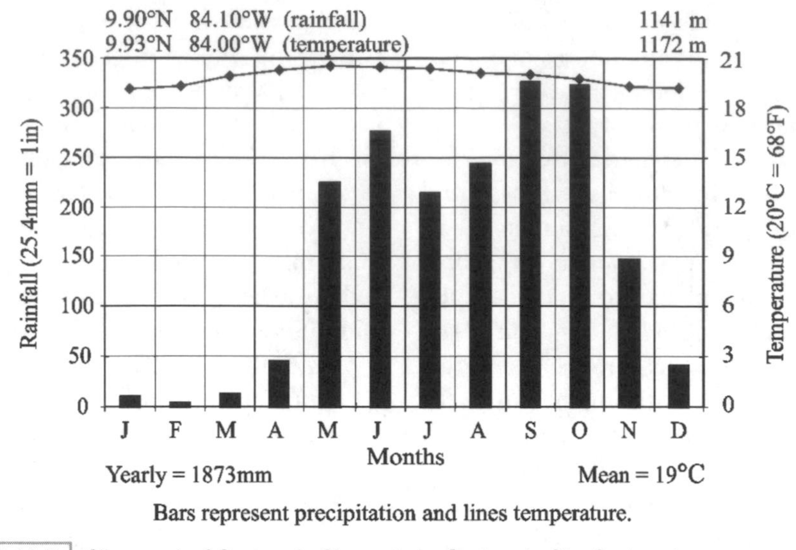

FIGURE 11.5. Climograph of Guatemala Observatorio, Guatemala City, Guatemala.

and more in the south. In Guatemala and Honduras, in the northern highlands, annual precipitation is less than 1,000 mm, but over most of the remainder of the highlands precipitation totals stand between 1,000 and 2,000 mm. Precipitation is greatest in the highlands of southern Costa Rica, where it exceeds 3,000 mm. Precipitation tends to be heavier on the eastern margins of the highlands because of the moisture-laden northeast Trade Winds that blow in off the Caribbean Sea and slightly less on the western margins that lie on the leeward side of the highland's highest peaks.

Well-defined wet-and-dry seasons, caused by the seasonal movement of the ITCZ (Inter-Tropical Convergence Zone) over the isthmus, characterize the Central American Highlands. Almost all precipitation falls during the summer, with little or none coming in the winter. In the southern highlands on the Meseta Central of Costa Rica, the rainy season begins in April or May and continues until November, with totals of almost 2,000 mm. Some 200–300 mm of precipitation falls monthly during that period (Figure 11.4). Further northward, in Guatemala, the rainy season is slightly shorter, usually beginning a month after it has begun in the south and total precipitation is also less, reaching about 1,200 mm (Figure 11.5).

Much of the natural vegetation of the Central America Highlands has disappeared as a result of forest clearing, livestock grazing, and agriculture. Deciduous tropical forests cloak the western margins of the highlands. These are often called tropical wet-and-dry forests where trees lose their leaves during the dry season. Further to the east, and often at higher elevations, the remaining natural forest vegetation includes extensive stands of pine in Honduras and Nicaragua, as well as evergreen and deciduous oaks. In the southern highlands, at the highest elevations, cloud forests occur on the eastern slopes of the highlands, grading into humid tropical rain forests at lower altitudes.

The region lies just east of the contact zone (the subduction zone) between two massive tectonic plates, the Cocos Plate and the Caribbean Plate. Earthquakes and volcanic eruptions are common along this subduction zone and create significant and recurring natural hazards. Devastating earthquakes and volcanic eruptions that have shaken the Guatemalan highlands and razed its various capitals are illustrative. Ciudad Vieja, the capital from 1527 to 1541, was destroyed by earthquakes and flooding caused by the eruptions of the nearby volcano Agua. The new capital founded subsequently at Antigua was shaken frequently by earthquakes and finally destroyed by a series of earthquakes in 1773. Although much of Antigua was subsequently rebuilt, the capital was moved to what was perceived to be a safer location on a broad mountain plain. Founded in 1776, Guatemala City, however, has not escaped the ravages of earthquakes. It was destroyed during a 6-week period in 1917–1918 when repeated earthquakes rocked the capital. The city was rebuilt on the same site, but another severe quake in 1976 caused widespread destruction and close to 23,000 deaths. Active since the 1990s, the nearby volcano Pacaya erupted in 2000, causing evacuations in surrounding villages, but no deaths.

Similarly, most of the region has been affected by earthquakes at one time or another during the last century, and small tremors are frequent. At the same time, many quakes have taken a heavy toll in life and property. Costa Rica suffered severe earthquakes in 1910 and again in 1989. A earthquake in Nicaragua killed nearly 2,500 in 1931 and about 5,000 died when the capital, Managua, was leveled by another in 1976. Most recently over 1,000 died in 1986 in El Salvador when an earthquake rocked the country.

Historical Geography and Economic Development

Pre-Columbian settlement in the Central American Highlands divided the region into two dominant culture regions. The highlands that fall within the present-day borders of Guatemala and El Salvador, as well as the western one-quarter of Honduras, western Nicaragua, and northwestern Costa Rica, fell squarely within the Mesoamerican culture region (see Figure 2.5). Dense populations, sophisticated material cultures, ceremonial religious sites, permanently settled villages, and intensive agriculture practices typified these peoples. In Guatemala, the Highland Maya terraced steep mountain slopes to bring more land under cultivation and to increase productivity. Corn, beans, squash, and chile peppers were the principal agricultural crop complex of these peoples, as they were throughout the rest of Mesoamerica. Corn-based foods like tortillas, tamales, and *pupusas* (thick corn cakes filled with beans or other foods) were the dietary staples. The Mesoamerican culture region in the Central America Highlands extends into the low-lying Sierra de Guanacaste on the Pacific watershed and on the Nicoya Peninsula. Here one finds the southernmost extension in Latin America of the indigenous use of corn tortillas as the principal foodstuff.

South and eastward of this area, the pre-Columbian indigenous peoples were strongly linked culturally to peoples in the Caribbean and South America. Few in number and organized at the tribal and band level, they lived in the highlands of central Costa Rica and southward into the Cordillera de Talamanca and the Serrania de Tabasará. Their material culture was sparse. Hunting-and-gathering economies characterized some tribes, while others practiced subsistence slash-and-burn agriculture. They did not employ advanced agricultural techniques like irrigation or terracing. Villages and settlements followed resource availability. Agriculture here depended on tuber crops dependent on vegetative or asexual reproduction (i.e., via plant cuttings), such as sweet manioc (*yuca*), arrowroot, and sweet potato.

The Central American Highlands lay on the periphery of the Spanish colonial empire. Geographical isolation from Mexico, a comparative lack of easily exploited mineral and human resources, and the difficulties of communication in the mountainous terrain between its disparate regional capitals—Guatemala, San Salvador, Tegucigalpa, and San José—all combined to keep the region on the margins of the colonial system. However, the Indian villages and settlements of the densely populated highlands of Guatemala and to a lesser extent El Salvador were an attractive resource to the Spanish conquistadores and colonists who received the *encomiendas* that incorporated these highland communities and the right to tribute payments. The colonial export economy of Central America depended largely on crops grown along the coastal lowlands of the Pacific coast, first cacao in the 16th and 17th centuries, and then indigo in the 18th century.

Nevertheless, besides agricultural production for subsistence and regional demands, the highlands did contribute to the colonial economy. Silver mining in the highlands of Honduras near Tegucigalpa began in the mid-16th century and continued at a steady, although not spectacular, pace throughout the colonial period, and indeed continues to be locally significant in the 21st century. In the 16th and 17th centuries, the production of cochineal, a small domesticated insect cultivated on the *Opuntia* cactus (prickly pear) that produced a rich red dye that was in great demand in Europe, was introduced into the Central American Highlands from Mexico. Although the development of chemical dyes in the 19th century eventually reduced the use of the natural dye from the cochineal, it continues to be in demand as an ingredient in cosmetics, notably lipstick. In the highlands of western Honduras tobacco was a commercial crop, while on the Meseta Central of Costa Rica, especially in the area around Cartago, farmers specialized in wheat production and exported it throughout the region.

Independence from Spain came in the early 1820s. After fighting off Mexico's efforts to incorporate them into its territory, the five former colonies joined together to form the Provincias Unidas de Centro América. This loose federation endured from 1823 until 1840, when it broke up into its constituent parts: Guatemala, El Salvador, Honduras, Nicaragua, and Costa Rica. Regional differences, political conflicts, and economic rivalries pulled the countries apart. Distinct regional cultures, affected in large part by the strength of the pre-Columbian indigenous cultures, evolved.

In Guatemala, dense settlements of indigenous Maya who retained their language and cultural identity accounted for the vast majority of the highland's population. A Hispano (*criollo*) elite in the capital and *Ladinos* (mestizos) in the countryside held the reins of power and used it to concentrate wealth, often to the detriment of the Maya. In Costa Rica, on the other hand, indigenous populations had been sparse on the Meseta Central and in the surrounding highlands. Few survived the Conquest and the ravages of previously unknown diseases like smallpox, influenza, and measles. Colonists of Spanish descent peopled the towns and settled the rural areas. A comparatively equitable distribution of land resources occurred in Costa Rica, with small- and medium-sized farms accounting for a large proportion of the rural economy. This contributed to the creation of a more democratic and egalitarian society than might otherwise have developed. In El Salvador, Honduras, and Nicaragua, Spanish-speaking peoples of mixed European and indigenous roots, mestizos, accounted for majority of the population, while a small class of *criollo* elite controlled most of the agricultural lands, often through traditional landholding forms like haciendas.

Coffee trees, introduced to the Meseta Central of Costa Rica at the beginning of the 19th century from Cuba, eventually caused an agricultural transformation of the highlands. Although coffee was not exported from Costa Rica until the 1830s, by midcentury coffee farms (*fincas*) covered the Meseta Central and coffee had become the country's most significant export. Coffee cultivation spread into El Salvador by the 1840s, into Nicaragua by the 1850s, and onto the Pacific slopes of the Guatemalan highlands by the 1860s. Coffee cultivation began in Honduras much later and did not become common there until the middle of the 20th century (Figure 11.6). Central America produced a fine coffee (*Coffea arabica*) that enjoyed strong demand in Europe and the United States throughout the 19th and 20th centuries. By 1910, for instance, coffee accounted for over two-thirds of all exports in Guatemala, El Salvador, and Nicaragua, and about one-third in Costa Rica (see Table 18.1).

Coffee cultivation took on several forms. Production was most sophisticated in Guatemala, El Salvador, and Costa Rica (Figure 11.7). In Guatemala it was cultivated on large plantations, often owned by foreigners, particularly Germans, that depended on seasonal migrant Indian labor from the highlands for the harvest. In Costa Rica, on the other hand, most coffee was cultivated on small family-owned *fincas* that depended entirely on family labor for its cultivation and harvesting. Coffee cultivation did not become widespread in Honduras until the 1940s; its production there still suffers from a more rudimentary character than in other Central American countries (Figure 11.8).

The Monroe Doctrine and U.S. foreign policy during the 19th century ensured that the Central American countries fell clearly within the sphere of influence of the United States. By the end of the 19th century, the United States exerted considerable political and economic clout in the region, and this continues to the present.

Entrepreneurs from the United States succeeded in securing concessions for vast tracts of land to develop banana plantations on the Caribbean Lowlands of Central America beginning in the 1880s. As banana exports increased and tax revenues from these concessions became more significant, these governments became increasingly beholden to U.S. interests. If pressure and persuasion failed to get governments to comply with their wishes, or when the U.S. government feared the political orientation of one regime or another, then military options could be pursued. The United States sent Marines to intervene in Honduras

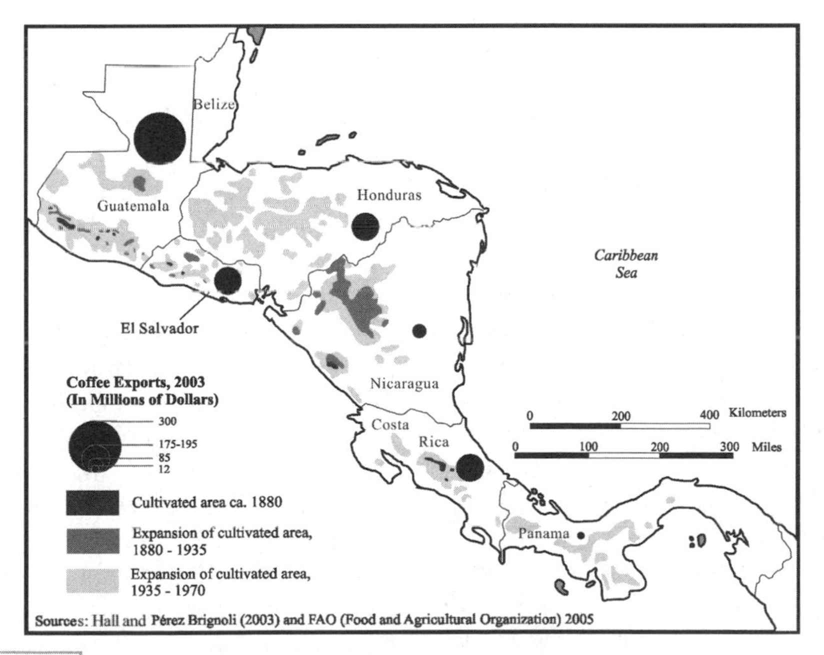

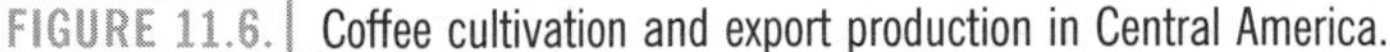
FIGURE 11.6. Coffee cultivation and export production in Central America.

FIGURE 11.7. Coffee trees in Central America are grown under an overstory of shade trees, as seen here in El Salvador, 1975.

in 1911 and 1912 to protect the banana plantations of U.S. companies when it seemed political unrest might threaten those investments. When Nicaragua failed to comply with the terms for the repayment of U.S. government loans, the Marines occupied the country almost continuously for over 20 years (1912–1933) and the United States ran the customs office and the national bank.

Panama became an independent country in 1903, after declaring independence from Colombia. The United States promoted and supported Panamanian independence because the Colombians had refused to agree to the terms of a treaty the United States proposed to build the Panama Canal. The Panamanian revolutionaries promised to sign the treaty once Panama became independent from Colombia. When Colombian military forces attempted to suppress the revolution, U.S. troops intervened and prevented Colombia from asserting its national sovereignty over the territory. The canal treaty was subsequently signed and a large tract of land ceded in perpetuity to the United States for construction of the Panama Canal. It was completed in 1914 and remained in U.S. hands until sovereignty was returned to Panama in 2000.

The United States has continued to intervene in Central American affairs throughout the second half of the 20th century. In Guatemala the United States supported the military overthrow of the duly elected government of Jacobo Arbenz in 1954 because the CIA viewed Arbenz as a leftist and a threat to the United States. In the 1980s the United States intervened directly in Nicaragua in an effort to topple its government. In El Salvador during the same period it provided billions of dollars of military equipment, training, and advisors to the right-wing government to fight leftist insurgents in a civil war. In an amazing demonstration of arrogance and an absurd assertion of extraterritorial legal rights, the United States invaded Panama in 1989 to arrest the head of the national guard and alleged drug trafficker General Manuel Noriega.

Contemporary Economic and Social Geography

The Central American Highlands have experienced rapid population growth since the 1950s. Between 1950 and 1985, annual growth rates ranged between 2.8 and 3.5 percent for all Central American nations. During this period population grew rapidly, doubling in some countries and tripling in others. For example, in Guatemala, the region's most populous country, the population rose from about 3 million in 1950 to almost 8 million in 1985 (see Table 9.5). Annual population growth rates moderated at the end of the 20th century, but nevertheless remained high in most countries—exceeding 2.5 percent in all countries except Costa Rica, where it stood at 1.4 percent (see Table 16.6). Rapid population growth has placed severe strains on the region's natural resource base and the respective national economics.

FIGURE 11.8. A makeshift coffee-drying yard in the Honduran highlands near Guajquiro, 1996.

In contrast with most other countries in Latin America, the countries of Central America have high proportions of their populations still living in rural agricultural areas. For instance, while the percentage of rural population in Argentina, Chile, and Venezuela is less than 20 percent, in Guatemala it is about 60 percent. It is greater than 40 percent in all the remaining countries on the isthmus (see Table 12.6)! Agriculture is a key economic sector and accounts for about 50 percent of all foreign exchange earnings.

The characteristics of rural life and living standards vary considerably over the extent of the highlands. In the Guatemalan highlands, land ownership is concentrated in the hands of a few. About 65 percent of the land in the country is owned by only 2 percent of the population. Most land held by wealthy landowners is dedicated to a variety of agricultural pursuits. Most significant are coffee plantations producing excellent yields of high-grade coffee for export to Europe and the United States. These modern plantations employ advanced technology, benefit from significant capital inputs, and tap a seasonal labor force for harvesting. Other commercial farms produce a range of nontraditional specialty crops for export.

Dense populations of Maya peoples also inhabit the highlands and practice intense agriculture on small plots where they produce subsistence foods for home consumption, including traditional staples like corn and beans, as well as products for local and regional consumption like potatoes, cabbages, and strawberries. *Minifundia*, peasant farms, are common here—farm size rarely exceeds a couple hectares, and many are much smaller. Some produce coffee on these small farms. As nontraditional export opportunities opened up with the United States during the 1990s, increasing numbers of farmers have taken to producing specialty crops like snow peas, cauliflower, broccoli, French beans, and berries.

Rural conditions are distinct in the highlands of Honduras. Population densities are lower, agricultural practices are less intense, and rural living conditions are somewhat better than in Guatemala, although rural poverty is endemic. Land ownership is concentrated in the hands of a small elite, although not to the extent that characterizes the Guatemalan highlands. Although some significant pockets of indigenous Lenca peoples are found in western Honduras, most of the residents of the Honduran highlands are mestizos who cultivate a range of subsistence and commercial crops (Figure 11.7). Livestock raising and forestry also play important roles in the rural livelihoods of highland Hondurans.

Rural living conditions on the Meseta Central in Costa Rica differ sharply from those in Guatemala and Honduras. The distribution of agricultural land is much more equitable here and commercially viable medium-sized family farms are the rule. Coffee cultivation predominates. The yields are extraordinary, six times greater than in Honduras or Nicaragua, and the quality is excellent. Some nontraditional specialty crops are also commercially viable. At higher elevations, cool-weather crops like carrots, cabbage, and even some deciduous fruit trees are grown. Rural poverty rates are lower, only about 20 percent. Nearly all residents of the small towns and villages enjoy the full complement of public services, including schools, electricity, potable water, and sewers. Agricultural extension services and community health programs are also widely available to rural residents.

In part due to rapidly growing populations in urban as well as rural areas, economic and social conditions for many residents of the Central American Highlands are challenging. When the countries in Central America are compared to those in other regions in Latin America, they rank as some of the most impoverished. Nationwide poverty rates exceed 50 percent in El Salvador, Guatemala, and Nicaragua and reach almost 80 percent in Honduras. Rural poverty rates are astronomical and range from 65 to 85 percent in these four countries (see Table 18.3)! Costa Rica is the only exception to this rule, where poverty rates stand at about 20 percent in both rural and urban areas.

Poverty, rapid population growth, limited opportunities, and an extremely inequitable distribution of rural lands and wealth fueled frustration and tremendous pressure for social change on the isthmus. For much of the 1970s and 1980s, Nicaragua, Guatemala, and El Salvador were racked by political violence and civil war. Both Honduras and Costa Rica were used as safe havens and staging areas during the 1980s by counterrevolutionaries fighting in Nicaragua, but neither country was drawn into the vortex of revolution and widespread political violence that affected the other Central American countries.

In Nicaragua, revolutionaries under the leadership of César Sandino launched a guerrilla war against the U.S. Marines that occupied the country almost continuously from 1912 to 1933. Subsequently, General Anastasio Somoza, the head of the national guard and later president and dictator-for-life, successfully conspired to have Sandino assassinated. While Sandino's assassination eliminated any direct challenges to the Somoza dictatorship and the elites that supported it, social conditions and dissatisfaction with the regime grew more intense over the following decades. Then, in the 1960s, a new generation of revolutionaries organized the Frente Sandinista de Liberación Nacional (FSLN) (the Sandinista Front for National Liberation). These insurgents began guerrilla operations against the Somoza government in the highlands along the Honduras border in the early 1960s and then later became active in the highlands of the coffee-growing region around Matagalpa and Jinotega (Figure 11.9).

Although not initially successful, these early operations by the Sandinistas provided organizational and combat experience upon

FIGURE 11.9. Guerrilla wars in Nicaragua, from the 1960s to the 1980s.

which the revolutionaries built. A decade later, in 1977, the Sandinistas initiated a broadly based insurgency. Within 2 years, the Nicaraguan National Guard disintegrated and Somoza's dictatorial dynasty fell, leaving the reins of power to the Sandinistas.

The Sandinistas faced a Herculean task in rebuilding Nicaragua after so many years of civil war. They embarked on a daunting socialist revolution that emphasized literacy and social and economic justice. Within 2 years of taking power, this task was further complicated by the rise of a counterrevolutionary guerrilla movement, the Contras, which incorporated elements of the former national guard as well as die-hard supporters of the former regime. The Contras enjoyed critical support from the United States under President Ronald Reagan. Beginning in 1981, citing fears of a Cuban-style communist government on the Central American isthmus, Reagan broke off diplomatic relations with Nicaragua and initiated clandestine funding of the Contras and support for their operations. The United States continued to support the counterrevolution throughout the 1980s by providing financial and military assistance to an armed force that numbered 15,000 at its height. However, when the U.S. Congress refused to continue military funding of the Contras in 1987, the movement withered and eventually disappeared, leaving tremendous destruction in its wake.

Violence convulsed both Guatemala and El Salvador for years as revolutionary guerrilla movements, government troops, police, and counterinsurgency forces confronted each other. Urban and rural insurgents initiated armed resistance in the early 1960s in Guatemala City and in the Motagua Valley. By the mid-1970s a tenacious insurgency had gripped much of the western highlands of the country. Government forces and the revolutionaries confronted each other there for years. Brutal counterinsurgency tactics led to the disappearances of thousands of Indians and included massacres of the residents of at least seven Indian communities between 1979 and 1981. Several different revolutionary groups, united in the late 1980s to form a united command and operations structure under the auspices of the Unidad Revolucionario Nacional Guatemalteco (URNG) (the Guatemalan National Revolutionary Unity).

Violence escalated during the 1990s as the URNG confronted police and army forces across the highlands. In 1991 the United States, which had heretofore provided considerable military assistance to the government, suspended all military aid because of severe human rights abuses by the government. In 1996 a cease-fire agreement led to negotiations which led to the end of the 35-year civil war and the signing of a peace agreement, the Accord for a Firm and Lasting Peace. Although estimates vary, most observers believe that between 150,000 and 200,000 people died during the civil war. Civilian noncombatants accounted for the vast majority of these deaths. Many died as a result of the Guatemalan military's genocidal campaign against the country's native peoples, the Maya. While the accord brought relative peace to the country and the URNG began to transform itself into a political movement and party, political violence, assassinations, and kidnappings continue to plague the country during the first decade of the 2000s.

In El Salvador centuries of oppression and social injustice, and 50 years of brutal military rule, reached the flash point in the late 1970s. Continued social and economic repression and human rights abuses by the government led to increasing political violence by opposition groups, which in turn prompted more violence by right-wing paramilitary groups that supported the government. These paramilitaries assassinated the country's archbishop, Oscar Romero, in 1980 because they perceived him to be too sympathetic to calls for social and economic reform. Romero's life, his assassination, and conditions in El Salvador at the time are portrayed with surprising accuracy in the Hollywood movie *Romero* (1989), directed by John Duigan.

By 1979 several revolutionary groups had initiated armed insurrections against the government. In 1981 they joined forces to form a united command structure under the name Frente Farabundo Martí de Liberación Nacional (FMLN) (the Farabundo Martí National Liberation Front), which fought the government for over a decade. During the 1980s the United States provided the right-wing Salvadoran government with billions of dollars of military assistance, training for elite Salvadoran counterinsurgency troops in the United States, and U.S. military advisors in El Salvador. The war dragged on for nearly a decade without either side achieving military victory. A peace accord signed in 1992 ended full-scale armed conflict and brought the former combatants into the political process. The United Nations oversaw a 3-year transition period as the country struggled to develop democratic institutions and began to heal the wounds of a civil war in which over 70,000 Salvadorians, mostly civilians, had lost their lives.

Poverty, social injustice, and above all civil war and political violence have led hundreds of thousands of Central Americans to abandon the region. Most have emigrated to the United States, although some have gone to Mexico, Costa Rica, and other Latin American countries. Prior to 1900, immigration from Central America to the United States was insignificant. It was not until the 1940s, when slightly more than 20,000 immigrated, that significant numbers began to reach the United States.

In the 1960s over 100,000 Central Americans immigrated to the United States. This flow skyrocketed in the following decades as violence and poverty pushed even more to leave their homelands. Approximately 1 million Central Americans migrated to the United States between 1980 and 2000! California,

Florida, and the Washington, D.C., metropolitan area have been the principal migration destinations. Nearly one-half of these migrants have come from El Salvador; Guatemala, Nicaragua, and Honduras, in that order, contributed most of the other half. Emigration from Costa Rica and Panama has been modest: only about 50,000 people from these two countries migrated to the United States during the 1990s. The positive correlation between those countries that sent large numbers of migrants to the United States and those with high rates of poverty and prolonged civil wars is striking.

Although rural populations are large in the Central American countries, most of the region's population does live in villages, towns, and metropolitan centers. Guatemala is the sole exception. The urban hierarchy of the Central American Highlands follows a pattern common in much of Latin America. One primate city, with a population many times larger than other cities in the country, dominates political, social, and economic life. The primate city is almost always the national capital.

All of the four capital cities situated in the Central American Highlands—Guatemala City, San Salvador, Tegucigalpa, and San José—are primate cities. Over one-quarter of Costa Rica's national population lives in San José; in El Salvador, over 20 percent of the population lives in the capital San Salvador; and in Honduras, Tegucigalpa is home to about 15 percent of the national population. While Guatemala City does not concentrate such a large proportion of the nation's total population—only about 8 percent of the population lives there—it is 10 times larger than the next largest urban center, Quezaltenango. Population estimates for these metropolitan areas suggest that by 2005 all will have achieved populations of at least 1 million. Guatemala City is the most populous, with almost 1.9 million; it is followed by San Salvador, with 1.5 million inhabitants. San José and Tegucigalpa both have approximately 1 million residents.

Services, manufacturing, and to a lesser extent tourism are the mainstays of the region's urban economies. Formal- and informal-sector services dominate the urban economies. This includes public-sector services of many types ranging from public education and police services to a variety of government ministries and agencies. These government bureaucracies are often large and account for a substantial proportion of total employment. Private-sector service provision ranges from high-end modern financial services in banking and insurance, wholesaling, and retailing in the formal sector, to informal-sector money changers and food vendors on street corners. Manufacturing accounts for less than 20 percent of the region's gross domestic product (GDP). Ten urban centers, including the national capitals, account for almost all manufacturing activity. Among them Guatemala City dominates, accounting for one-fifth of the region's total manufacturing establishments. Tourism, long deterred by the region's civil unrest, still plays a notable role in the regional economy, generating about 10 percent of regional foreign exchange earnings. Tourism in the Central American Highlands is most significant in Guatemala and Costa Rica.

The Caribbean Lowlands of Central America on the Periphery

The Caribbean Lowlands of Central America are an isolated and sparsely populated peripheral region lying between the Central American Highlands and the Caribbean Sea. Traditionally, this hinterland region has been only loosely connected with the Central American Highlands. Indigenous peoples have maintained a distinct identity in many areas. The long-term influence of the British and English-speaking African Caribbean migrants is notable.

Geographically, the region extends from

the Gulf of Honduras along the entire eastern shore of the Central American isthmus until reaching the northern coast of South America at the Gulf of Darien (Figure 11.10). In northern Honduras, the lowlands are narrow and are squeezed between a series of low mountain spurs that extend from the Central American uplands to the Caribbean Sea. Near the Laguna de Caratasca in eastern Honduras the lowlands widen dramatically into a broad coastal plain extending inland for as much as 100 km. This lowland is known as the Mosquito or Miskito Coast (or even just Mosquitia). It includes the extreme eastern section of Honduras and the eastern lowlands of Nicaragua. The Mosquito Coast is characterized by three broad ecological regions: a swampy coastal margin interrupted by lagoons and vegetated with extensive mangroves; savannas and scrub vegetation inland in Honduras and northern Nicaragua; and tropical lowland rain forest further inland running from Honduras through Nicaragua and into Costa Rica. The coastal lowland continues southward into Costa Rica, where it again narrows to no more than 10 km in width near Puerto Limón. South of Puerto Limón and continuing along Panama's coast to the Gulf of Darien, the lowlands are restricted to a narrow coastal plain hemmed in between the mountains on the west and the Caribbean Sea to the east. Here the rain forest covers much of the coastal margin.

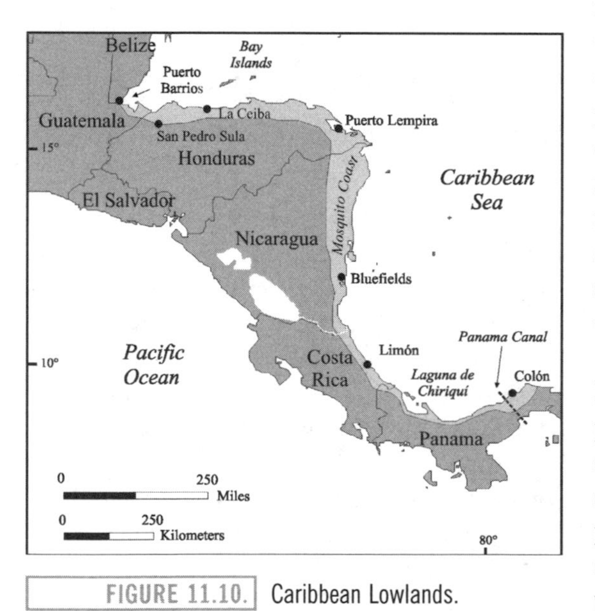

FIGURE 11.10. Caribbean Lowlands.

The climate is humid and tropical on the Caribbean Lowlands. Mean temperatures vary little over the course of the year. Bluefields, with an annual variation of about 2 degrees, is typical (Figure 11.11). Precipitation is plentiful along the entire coastal region. The annual precipitation total rarely falls below 2,000 mm or exceeds 4,000 mm. Rainfall totals tend to be higher in the south than in the north. Precipitation is common in most months, but the most intense rains occur during the summer months between May and November. The rainy season's onset is sooner in the south as the ITCZ moves northward in the summer and later in the north.

The Mosquito Coast is probably the most well-known area in the Caribbean Lowlands, although not necessarily the most significant economically. The region is named for the Miskito Indians who formed alliances with British loggers and the British Crown during the colonial period and who, with their assistance, fended off attempts by the Spanish to incorporate the region effectively into Spain's Central American colonies. Fiercely independent, the Miskito Indians still inhabit large portions of the lowlands, especially in northern Nicaragua. In the wake of the Sandinista revolution in Nicaragua, they doggedly refused to succumb to the dictates of central control and still maintain a significant degree of local autonomy within the Nicaraguan nation.

Historically, the Caribbean Lowlands have been sparsely populated. Indigenous peoples, typically hunters and gatherers and subsistence farmers, have inhabited the region since the pre-Columbian period, but their population numbers have always remained small. British loggers maintained a tentative presence throughout the Mosquito Coast re-

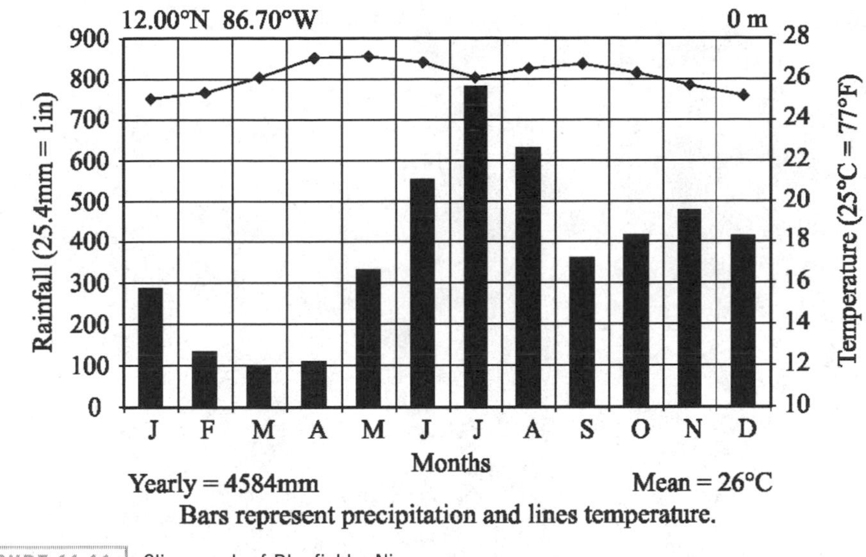

FIGURE 11.11. Climograph of Bluefields, Nicaragua.

gion during much of the colonial period. The British never effectively occupied this region, as they did in Belize, but they did exploit its rich mahogany woods for export and established alliances with Indian groups, especially the Miskito Indians. With the exception of a few key ports, Puerto Cortes (Honduras), La Ceiba (Honduras), Puerto Limón (Costa Rica), and Colón (Panama), Spanish authorities exerted almost no control over the region during the entire colonial period. Escaped black slaves from British colonies in the Carribean settled along the coasts of British Honduras, mixing with the indigenous peoples. Their descendants, often called the Black Caribs, or Carifuna, are still an important population group in these areas today.

It was not until well after independence from Spain, at the end of the 19th century, that economic and social change began to occur in the Caribbean Lowlands. Geographical proximity to the United States, a tropical climate suited to banana cultivation, and national governments eager to profit from the development of the Caribbean Lowlands initiated widespread change here (see Figure 18.4). U.S. companies, including the United Fruit Company and its precursor, the Standard Fruit Company, secured vast territorial concessions in the lowlands of Guatemala, Honduras, Nicaragua, Costa Rica, and Panama where they established banana plantations. These plantations were enclave economies with few linkages to the host nation's economy. However, they did provide revenues to the national treasuries, some railroad and port infrastructure, and limited employment opportunities for nationals. Particularly important, especially in Honduras and Costa Rica, they sponsored the immigration of significant numbers of English-speaking blacks from Jamaica and other West Indies islands to work on the plantations, which contributed to a notable change in the population in these areas (Figure 11.12). Almost simultaneously in Panama, the U.S. government began the construction of the Panama Canal, which also depended on English-speaking blacks from the West Indies for much of its unskilled labor. As in Honduras and Costa Rica, many of these immigrants remained permanently in Panama, contributing to further ethnic and racial diversity in these nations.

The banana plantation economies that

FIGURE 11.12. A rural home in Costa Rica's lowlands near Limón. The architecture reflects West Indian and Anglo-American influences in this region, 1975.

characterized the Caribbean Lowlands during the first half of the 20th century succumbed to crop diseases and changing production economics by midcentury. The multinational corporations that ran these plantations largely abandoned them or converted to a system where independent growers produced the crop and the corporations handled its packing, transportation, and marketing. In Costa Rica, considerable acreage was converted to the cultivation of the cocoa tree for chocolate production, but this has not occurred in other lowland areas. A range of other economic activities filled the void left by the lowland plantation economies: logging, fishing, livestock raising, and both small-scale subsistence farming and commercial agriculture. Tourism has begun to become important since the 1980s as the Caribbean Lowlands have become more accessible, with beaches, rain-forest ecology and wildlife, and "picturesque" indigenous groups providing the principal attractions.

Summary

The Central American Highlands are home to over 30 million people. The region is distinguished by high population growth rates, a high proportion of rural population, and widespread poverty. The ethnic and cultural composition of the region varies. In Guatemala, indigenous Maya peoples account for about half of the nation's population, while mestizos occupy most of the highlands of El Salvador, Honduras, and Nicaragua. A population predominately of European descent occupies the Meseta Central of Costa Rica. Agriculture remains the principal source of foreign exchange, as it has for the last 150 years, accounting for about 50 percent of the total. Coffee, the traditional export crop, continues as the single most valuable agricultural pursuit in the highlands. Political violence and civil war have racked Guatemala, El Salvador, and Nicaragua during the last three decades and emigration from these countries to the United States has been heavy.

The Caribbean Lowlands is a peripheral region to the east of the Central American Highlands. Its culture and traditions provide strong links to the Caribbean, specifically with Anglo- and African-American influences. Indigenous influences continue to be important in some areas. It was the focus of commercial banana production during most of the 20th century and now has a more diversified economy including tourism.

Further Reading

Arden, P. (1997). *Post-war reconstruction in Central America: Lessons from El Salvador, Guatemala, and Nicaragua*, D. Eade (Trans.). Oxford, UK: Oxfam.

Byrne, H. (1996). *El Salvador's civil war: A study of revolution*. Boulder, CO: Lynne Rienner.

Coates, A. G. (Ed.). (1997). *Central America: A natural and cultural history*. New Haven, CT: Yale University Press.

Davidson, W. V. (1974). *Historical geography of the Bay Islands, Honduras: Anglo-Hispanic conflict in the western Caribbean*. Birmingham, AL: Southern University Press.

Faber, D. J. (1993). *Environment under fire: Imperialism and the ecological crisis in Central America*. New York: Monthly Review Press.

Hall, C. (1985). *Costa Rica: A geographical interpretation in historical perspective*. Boulder, CO: Westview Press.

Hall, C., and Pérez Brignoli, H. (2003). *Historical atlas of Central America*. Norman: University of Oklahoma Press.

Hamilton, N., and Stoltz Chinchilla, N. (1991). Central American migration: A framework for analysis. *Latin American Research Review, 26*(1), 75–110.

Hamilton, S., and Fischer, E. F. (2003). Non-traditional agricultural exports in Guatemala: Understandings of risk and perceptions of change. *Latin American Research Review, 38*(3), 82–102.

Hinshaw, R. E. (1975). *Panajachel: A Guatemalan town in thirty-year perspective*. Pittsburgh, PA: University of Pittsburgh Press.

Lovell, G. (2000). *A beauty that hurts: Life and death in Guatemala*. Austin: University of Texas Press.

Mathewson, K. (1984). *Irrigation horticulture in highland Guatemala: The tablón system of Panajachel*. Boulder, CO: Westview Press.

McCreery, D. (1984). *Rural Guatemala, 1760–1940*. Stanford, CA: Stanford University Press.

McLeod, M. J. (1973). *Spanish Central America: A socioeconomic history, 1520–1720*. Berkeley and Los Angeles: University of California Press.

McSweeney, K. (2004). The dugout canoe trade in Central America's Mosquitia: Approaching rural livelihoods through systems of exchange. *Annals of the Association of American Geographers, 94*, 638–661.

Nietschmann, B. (1989). *The unknown war: The Miskito nation, Nicaragua, and the United States*. New York: Freedom House.

Paige, J. M. (1989). *Coffee and power: Revolution and the rise of democracy in Central America*. Cambridge, MA: Freedom House.

Perez-Brignoli, H. (1989). *A brief history of Central America*. Berkeley and Los Angeles: University of California Press.

Revels, C. S. (2000). Coffee in Nicaragua: Introduction and expansion in the nineteenth century. *Yearbook, Conference of Latin Americanist Geographers, 26*, 15–32.

Seligson, M. A. (1995). Thirty years of transformation in the agrarian structure of El Salvador. *Latin American Research Review, 30*, 43–47.

Walker, T. (2003). *Nicaragua: Living in the shadow of the eagle* (4th ed.). Boulder, CO: Westview Press.

Wall, D. L. (1993). Spatial inequities in Sandinista Nicaragua. *Geographical Review, 85*, 1–13.

West, R. C., and Augelli, J. P. (1989). *Middle America: Its lands and peoples* (3rd ed.). Englewood Cliffs, NJ: Prentice-Hall.

Woodward, R. L., Jr. (1999). *Central America: A nation divided* (3rd ed.). New York: Oxford University Press.

12 Land and People since the Conquest

At the time of contact, European settlers encountered native peoples in Middle America, South America, and the Caribbean whose levels of social, cultural, and economic development varied widely. Advanced civilizations, most notably the Aztec, Maya, and Inca, dominated central Mexico, northern Central America, and the central Andean highlands of South America, respectively. Advanced agricultural techniques, including irrigation and terracing, formed the cornerstone of these societies, providing abundant food surpluses and promoting the development of urban centers, the building of monumental architecture, and the flowering of art, science, and religion.

A second tier of social, cultural, and economic development was represented by chiefdoms. More rudimentary agricultural systems, often practiced in permanent fields but without the benefit of irrigation and terracing, provided the principal mainstay of these societies. Villages, not urban centers, typified the settlements of these peoples, who left almost no permanent structures in their wake. They occupied the largest islands of the Caribbean, the northern highlands of Andes (Colombia), and the southern margins of South America's Pacific coast (Chile).

Those who practiced shifting field cultivation (known variously as slash-and-burn, swidden, or *milpa* agriculture), as well as hunters and gatherers, occupied the balance of the region, including the vast arid lands of northern Mexico, the Caribbean tropical lowlands in Central America, the Amazon Basin, the Pampa, and Patagonia. Elementary forms of social and political organization typified these peoples, who lived in small tribal groups with only tenuous links to other groups. Rudimentary tools and technologies, little social or occupational specialization, and a simple material culture were also characteristic of these societies.

Indigenous Land Tenure

Land tenure and the understanding of land as a resource varied considerably among the wide range of indigenous groups and societies the European conquerors encountered in Latin America. Generally, those indigenous peoples living in the most elementary tribal societies, practicing shifting cultivation, and hunting and gathering did not view land as a commodity or as a resource that could be owned or held. Nevertheless, a strong sense of territoriality existed in most groups, and they would vigorously defend their territories against intrusion by others. Land might be used temporarily by individuals, as was the case with the crop fields of shifting cultivators, but was not owned by individuals. Indeed, the economies of these peoples provided little need to view land as individual property, as fields were cultivated for a few years and then abandoned. Villages were impermanent and once all the arable land in the vicinity had been cropped, village sites were abandoned and reestablished at a new location.

Land and the concept of land ownership took on a slightly different dimension within chiefdoms. Agriculture formed the backbone of these societies and permanent settlements were the rule. Individual land ownership did not exist; rather, individuals and families typically enjoyed use rights (usufruct) of land parcels for agricultural purposes, with the land understood as a resource controlled by a chief, council of elders, or lineage groups or clans. The use rights for land parcels used for crop cultivation typically could be passed from one generation to the next, but land could not be alienated (i.e., bought or sold). Surrounding lands, held in common and not cropped, provided a range of other products, including foodstuffs, fuel, and wood.

In advanced societies, such as those found in Mesoamerica and the central Andes, the concept of land ownership had more formal dimensions. In most cases, the king or the extended royal family owned all land but granted lands to nobles, warriors, and their families. Communities also held land, and again, at least theoretically, the land comprised part of the empire and possessions of the monarch.

The Aztecs in central Mexico at the time of the Conquest provide a good example of how landholding practices worked in advanced societies. Landholdings fell into three key classifications. In the first instance, local communities, known as *calpulli*, held land in common. These included croplands as well as peripheral common lands, which were often forested. Usufruct rights remained within a family and could be passed from generation to generation; however, each *calpulli* chose a village elder or head, who was responsible for overseeing land use and was empowered to make decisions about its allocation among community members. A second category of land, sometimes called "public lands," occupied some portion of almost all Aztec communities. These lands belonged to the Aztec state (the monarch/ruler), but members of every community were required to farm them for the benefit of the military and the Aztec elite. Finally, the Aztec nobility also held land as personal property. These lands were worked for the nobility, either by vassals bound to the land or by slaves captured in wars. At the time of conquest in the early 1500s, similar systems of land tenure characterized other advanced states in Mesoamerica and the central Andean highlands.

European Impacts on New World Land Tenure

The arrival of Europeans in the New World transformed the patterns of land tenure and introduced the concept of individual land ownership. The effects of these changes occurred first in those regions inhabited by advanced civilizations: Mesoamerica and the central Andean highlands. The dense populations and wealth of these regions attracted the Spanish conquistadores initially and Spanish colonists subsequently. Their impact on land tenure patterns and land ownership transformed the landscape within a few decades. Peripheral regions were also affected as colonial powers extended their rule beyond the colonial cores. This process occurred slowly, continuing in some areas until the end of the 20th century.

Central Mexico again provides a useful example of how Spanish colonial rule transformed landholding patterns and the concept of land ownership. At first the Spanish did not focus on land as a source of wealth or view land ownership as especially desirable. Initially the Spanish recognized the payment of tribute as the most important source of wealth, and they co-opted and adapted the traditional Aztec system in which villages and communities paid annual or biannual tribute to their Aztec lords.

In short order, however, the Spaniards realized that Indian labor would be an essential requirement in effectively exploiting the region's resources and so imposed the same system of labor exploitation they had used earlier in the Caribbean, the *encomienda*. The *encomienda* (a term that derives from the Spanish word *encomendar*, which means to entrust something to the care of others) was a system that had been utilized in the reconquest of Spain from the Moors to reward successful knights and to re-Christianize conquered territories. In the New World, the Spanish Crown made grants to individual Spaniards to protect and Christianize Indian communities. In return, the Spaniard or *encomendero* had the right to collect tribute and labor services from the communities under his protection.

The abuses of the encomienda system, so pernicious in the West Indies, led some vocal colonial and religious authorities to advocate its elimination. The crusading works of the former colonial administrator and priest Bartholome de las Casas contributed to the demise of the system. Its gradual elimination by the Crown began in the mid-1500s when this was mandated by the New Laws of the Indies (1542). The institution was slow to disappear, however. Elite Spanish colonists who had been granted *encomiendas* held on to them tenaciously, sometimes even managing to successfully pass them on from one generation to the next contrary to royal policy.

The Crown replaced the encomienda with the *repartimiento*. The *repartimiento*, sometimes known as the *mita* in the Andes, transferred control of Indian communities from the hands of the Spanish colonial elite into the hands of colonial administrators. Under this system, colonial administrators extracted labor services from indigenous communities. It lasted in one form or another from the late 1500s until the end of the colonial period. Work requirements typically stipulated that all able-bodied men between ages 16 and 60 work 2 weeks per year and that no more than about 5 percent of a community's male population be absent at one time. Colonial administrators had wide latitude in determining the kind of work that might be performed; it could include public works as well as work on agricultural estates or in the mines of the Spanish elite. Abuses of this system were widespread. For example, many Indian laborers in the Andes of Peru and Bolivia were marched off to the mines in Huancavelica and Potosí, and never returned.

In Central Mexico and the central Andes, Spanish conquistadores and subsequent colonists and settlers quickly realized that with the demise of the encomienda and its replacement with the repartimiento, land ownership was an essential strategy in the creation and protection of personal wealth. However, Crown policy, at least in theory, respected the property rights of indigenous peoples. Thus, in the Aztec region of Central Mexico, the land rights of the aristocracy as well as those of individual communities, the *capulli*, were recognized. Initially, then, only those lands that had been "public lands," the lands belonging to the Aztec state, were available for distribution to the Spanish. Within a few years of the conquest of the Aztec state in 1521, the Crown had begun the process of making land grants, or *mercedes*, often of immense tracts, to conquistadores who had rendered important service to the Crown. In the first decades of settlement, however, official Crown policy was ignored, and Spanish elites with the support of colonial administrators simply usurped Indian lands by force or used "legal" means to subvert Indian land ownership. In the Valley of Mexico and its immediate environs—one of the most desirable areas for settlement in the eyes of the Spanish—almost all land had been alienated from the Indians by the end of the 16th century. A similar pattern typified most other highland areas the Spanish found attractive for early settlement: the highlands of Guatemala, the Sábana of Bogotá (Colombia), the Quito Basin (Ecuador), and other fertile highland basins in Peru and Bolivia.

Not all Indian land loss in the early colonial period can be traced to usurpation or legal manipulation. Throughout Spanish America three factors contributed to the loss of Indian lands. First, the introduction of exotic diseases by Europeans and later African slaves during the early colonial period led to steep population declines among indigenous peoples. In Central Mexico the population decline among indigenous peoples may have been close to 90 percent, with the population dropping from around 25 million to perhaps 2–3 million. This immense population decline left many communities nearly depopulated and facilitated the alienation of vast acreage of Indian lands.

Second, in many areas the Spanish colonial authorities pursued a policy of Indian resettlement that sought to move Indians into concentrated population centers—be they villages, missions, or *reducciones* (settlements of Indians converted to Christianity)—where they might be controlled and taxed with greater ease. Catholic missionaries pursued a roughly similar policy toward their Indian charges in some areas. For example, the Jesuits established missions along the Paraná River in Paraguay and Argentina and the Franciscans and Dominicans founded many missions through northern and western Mexico. These resettlement policies effectively depopulated many rural areas and again simplified the alienation of indigenous lands (Figure 12.1).

Third, the repartimiento also contributed indirectly to the abandonment of land claims by Indians in rural areas and eased Spanish appropriation of the indigenous peoples' land. While most who labored under the repartimiento returned to their communities after the termination of their labor requirement, some Indians who were introduced to the system of wage labor eventually migrated permanently to Spanish towns, further reducing rural Indian populations.

The alienation of Indian lands proceeded

FIGURE 12.1. 18th-century Jesuit mission (*reducción*) at San Ignacio in eastern Paraguay, 1994.

at different rates and with different effects throughout Latin America during the colonial period. In core regions, where indigenous populations had been dense, the vast majority of Indian lands were transferred to Spaniards, and later, but to a lesser extent, to mestizos. On the periphery of these regions, where Spanish interests were less pronounced and fewer readily exploited resources were available, Indians retained significant land resources. In the Valley of Oaxaca, for example, only one-third of the land held by formal title belonged to Spaniards.

During the colonial period, despite commonly held perceptions, Indians were not entirely passive participants in the process of land alienation. Indians frequently sold land to Spaniards. Often these were communal lands that Indian peasant families had used for generations. This history of usufruct provided the basis for establishing ownership of the land in the eyes of the Spanish legal system. In these cases Indians may not have been fully aware of the ramifications of the process. In contrast, Indian nobles who sold their lands likely understood the nature of these transactions. Generally, many Indians came to understand the nature of the Spanish legal system, and both individuals as well as Indian communities often petitioned Spanish courts to obtain legal title for their properties.

While the process of Indian land alienation occurred frequently during the colonial period, in some ways colonial administrations, especially in Spanish America, provided some protections for Indian rights and property. An often-held view that contributed to these practices was that Indians were unsophisticated and simple, akin to children, and hence required protection. After independence in the early decades of the 19th century, this legal and moral perspective changed in almost all Latin American countries, and Indians no longer enjoyed any special protections or considerations. Especially during the latter half of the 1800s, rural landowners often moved aggressively to alienate Indian lands, especially those held by indigenous communities. In the Andes of southern Peru, for example, the introduction of productive breeds of sheep for wool and the development of a booming export market for this product brought incessant and often successful efforts by large landowners to incorporate Indian lands into their holdings in order to increase the size of their flocks.

Changing Production Systems

The colonial experience in Latin America radically transformed indigenous production systems in those regions that came under the most direct control of the Spanish and the Portuguese. The nature of the colonial enterprise, especially in Spanish America and to a lesser extent in Portuguese America, led to the concentration of large tracts of land in the hands of a comparatively small number of individuals and families. Throughout Spanish America, these large tracts of land are known generally as *latifundia*. A range of other terms, specific to particular regions and nations, sometimes denote differences in land use and economic orientation that are found throughout Latin America. These include use of the terms *hacienda* in Mexico, Central America, and throughout much of the Andes; *fazenda* in Brazil; *fundo* in Chile; and *estancia* in Argentina.

In stark contrast to the settlers who arrived in North America seeking religious and political freedom, many of the Iberian colonists and settlers who immigrated to the New World came in search of material wealth with the hope of eventually returning to Spain or Portugal. Few were inclined toward crop farming. They viewed farming as a low-status occupation unsuitable for individuals of their stature. Instead, Iberian settlers viewed the raising of cattle and horses as occupations more appropriate to their station in life. Indeed, the term commonly used for a gentleman throughout Spanish America is *caballero*, literally a "horseman" in Spanish. The widespread consolidation of these large latifundia in the New World during the first decades of colonization and settlement left a permanent mark on the agricultural landscape and permitted the establishment of agricultural production systems that reflected the values of the colonists and their economic needs.

Latifundia

In Latin America the latifundia evolved into manorial estates that bore much in common with the feudal estates that dominated Europe during the Middle Ages. The social and economic relationships between estate owners and workers closely resembled those between "masters and serfs" in the European context. Estate owners, *patrones*, held almost complete power over their Indian and later mestizo laborers, *peones*. Workers received usufruct rights to a small plot of cultivable land for their personal use and sometimes the right to graze one or two head of livestock on the estate's land. In return, the *peon* was obligated to work a set number of days on the estate and to turn over a portion of his own crops to the owner. Laborers, who had few options, were held to the land through a variety of mechanisms including debt peonage, coercion, and the inability to find viable alternatives within the local region, as other landowners were unwilling to accept *peones* who had fled another estate. On occasion, peasants who fled an estate would be pursued and, if caught, returned by force. The geographical isolation of these estates—they were typically distant from towns, villages, and other estates—contributed to the inability of peasant laborers to leave. Laborers tended to remain on these estates for generations. It was not until late in the 19th century, when urban centers began to grow, that landless peasants could abandon the countryside on a large scale and find opportunities elsewhere.

THE HACIENDA

The *hacienda*, perhaps the most common type of latifundia, dominated the rural landscape of many Latin American countries from the middle of the 1500s until the first half of the 20th century. Geographically extensive, often encompassing thousands of hectares, haciendas were typically located in interior regions and were often isolated from settlement centers and even other haciendas. This geographical isolation contributed to the development of the hacienda as a self-sufficient agricultural enterprise. Larger haciendas were really small settlements with populations sometimes reaching over 100, which in some cases included a resident priest. The principal focus of production was usually horses and cattle, although a diversified range of foodstuffs and other crops were usually raised to provide for the subsistence needs of the hacienda's population. The hacienda's principal products—livestock on the hoof, meat, and hides—were usually sold in local and/or regional markets. Haciendas usually employed traditional tools and techniques, and little effort was made to modernize or to reinvest profits in the productive enterprise. A second house in the provincial or regional capital, where the hacienda owner and his family often lived for extended periods, and a tendency toward conspicuous consumption of imported products often consumed most of the profits.

As an economic production unit, the hacienda had abundant land and labor, both of which were usually underutilized, and scarce capital resources. Initially, haciendas remained almost exclusively in the hands of individual families: they were passed from one generation to the next. However, the ownership of haciendas by the Catholic Church became more common as the colonial period wore on; wealthy patrons bequeathed their estates to the church, or the church assumed ownership of haciendas that had served as collateral for loans the church had made to their owners.

A personal quality characterized relationships between the owners of the hacienda and their laborers. Payment for labor services came "in kind," with laborers receiving the use of land in return for their services. Indeed, these relationships were typically paternalistic. The *hacendado*, or the *patron*, provided a measure of protection and care for the *peon*. At births, baptisms, weddings, and deaths or at times of personal crisis, the *peon* looked to the *hacendado* for help, whether by providing candles for a baptismal mass or liquor for a wedding or a wake. The *peon* and his family, on the other hand, were expected to give their complete loyalty to the *patron* and the hacienda and fulfill their labor and production obligations faithfully.

One enduring legacy of the hacienda has been the stagnation of many rural regions of Latin America. This occurred for at least three reasons. First, the profits generated from haciendas were rarely reinvested in the productive enterprise, so few new technologies were incorporated, hampering any significant increases in productivity. Second, because the simple fact of land ownership was viewed in such a positive light, land was hoarded and consequently underutilized. The focus on livestock grazing also led to the underutilization of land resources. Flat, fertile, and well-watered lands were often used for grazing instead of vastly more productive crop agriculture. Finally, the hacienda's agricultural labor force—the *peones*—were chronically underemployed, often with little productive work to do for extended periods.

THE PLANTATION

The plantation was another agricultural production system that European colonists introduced into the New World. Its introduction and development came as a response to intense demand for sugar in Europe and the immense profits to be made satisfying this demand. Its consequences were both immediate and long term. Immediately, it led to the introduction of sugarcane and the transformation of the natural landscapes of northeastern Brazil and many Caribbean islands into agricultural monocultures focused on its production. It also led to the enslavement and forced migration of millions of Africans to the New World and dramatic long-term changes in the population characteristics and culture of those areas where they were settled.

The plantation characterized a completely distinct type of agricultural production system from that of the hacienda. Almost invariably, plantations were located in coastal regions with easy access to oceangoing shipping. Coastal locations, or proximity to coastal locations, were essential for plantations since they had to ship to European markets to sell their sugar. Plantations also practiced crop monoculture and used land resources intensely. Production focused on sugarcane to the near total exclusion of other commercial crops. In contrast to the hacienda, plantations employed abundant capital, practiced effective business methods, and utilized advanced technology. Thus, during the colonial period, the plantation was a highly specialized modern agricultural business venture that was integrated into world markets.

The plantation and the hacienda did share some notable characteristics. Both required extensive land resources (latifundia). In addition, a similar type of physical infrastructure characterized the main compounds of both. The owner's home, typically a large expansive structure, dominated both haciendas and plantations. A few smaller intermediate-sized residential structures provided shelter for the overseers, managers, and possibly a resident priest. The housing for peasant laborers or plantation slaves was simple, small, and often consisted of little more than crude hovels or crowded barracks. Both plantation and hacienda compounds included buildings used for workshops, the processing of agricultural products, and storage for supplies and commodities to be marketed. Finally, during the first couple centuries of the colonial period, the ownership of both types of latifundia remained almost exclusively in the hands of individual families that had held them from generation to generation.

The market orientation of plantation production and their dependence on world markets forced plantation owners to view their enterprise in stark financial terms. As the colonial period passed, plantation owners were increasingly forced to respond to competitive pressures created by producers in other regions. Thus, plantations were more likely to employ new technologies as they became available in order to maximize profits, and their owners eventually mechanized as many aspects of their production systems as they could.

However, the institution of slavery, introduced into the New World when efforts to compel and recruit indigenous laborers failed, provided the labor force necessary to sustain the plantation system from the late 1500s until about the middle of the 19th century, when most Latin American countries abolished slavery. The abolition of slavery brought efforts to import indentured labor to replace black slaves, but also further encouraged the use of capital to improve productivity through increased mechanization. Indentured laborers failed to adequately replace slaves, and wage labor became increasingly dominant on plantations throughout the region. Other mechanisms, including sharecropping and debt peonage also functioned to hold the labor force on many plantations. In these cases, the workers, while legally "free," endured living conditions not unlike those of the slaves who had preceded them.

The patterns of land ownership also began to change during the 19th century, as the pressures of global competition and the need to employ more sophisticated technologies in the processing of sugarcane led to the consolidation of family-owned plantations and their replacement with corporate ownership models.

LATIFUNDIA INTO THE 21ST CENTURY

The hacienda remained an enduring economic and social institution in Latin America from the colonial period well into the 20th century. Indeed, in some ways the durability of this institution and the social and economic relationships that characterized it are quite remarkable. For example, in the early 1950s, a team of anthropologists studied a hacienda in the Peruvian Andes in great detail. At the time of the anthropologists' research, the hacienda, called Vicos, was home to nearly 400 peasant families. Each household head was required to provide labor 156 days a year (3 days a week) to the *hacendado*, and in return received a small cash payment and use of a garden plot and a house. The peasants' labor obligation might be met on the hacienda or anywhere else the hacienda's owner specified. Peasant living conditions were difficult. The farm plots they were assigned were small, highly fractionalized, and often a distance from one another. The 400 families cultivated nearly 10,000 separate plots. Individual peasants were often assigned to work multiple plots several kilometers from one another. The peasant diet was often inadequate—many peons sought additional wage labor off of the hacienda because they could not produce sufficient food for their families on the plots assigned to them.

Land reform movements in some Latin American countries broke up the haciendas and removed the vestiges of this almost feudal institution in many countries and regions. However, in other countries land reform programs faltered or were never aggressively applied, and haciendas, albeit with some modern characteristics, continue to exist in some areas.

The plantation also proved to be an enduring economic and social institution in Latin America, but its survival reflected a continual evolution rather than the stagnation that typified the hacienda. Insatiable demand for sugar provided the initial impetus for the development of the plantation, and slave labor ensured the initial profitability of the enterprise. But as the colonial period progressed, other crops, including indigo, a natural plant dye, and cotton, proved suitable to the industrial type of agriculture the plantation system represented.

The demise of slavery in the early decades of the 19th century forced plantation owners to seek alternative labor sources and models. The recruitment and immigration of indentured laborers proved an intermediate solution to the labor crisis. Indentured laborers were recruited from a wide range of countries, including European nations, but the greatest numbers reaching Latin America came from China. Cuba and Peru were the principal destinations; approximately 150,000 Chinese reached each country in the last half of the 19th century.

Imported labor could not, however, resolve the demand, and plantation owners also employed sharecropping schemes and debt peonage as mechanisms to keep emancipated black slaves as part of the labor supply. Labor shortages also encouraged increasing mechanization on sugar plantations as owners sought to replace labor, which had become increasingly expensive, with machinery. The demand for more capital set off additional changes in the structure of the plantation, especially sugar plantations. Traditionally, family ownership of plantations had been the rule, but mechanization and greater capital requirements forced the consolidation of landholdings and the construction of large sugar mills, known sometimes as *centrales*. By the beginning of the 20th century, corporate or business ownership of sugar plantations had largely replaced family ownership in Brazil, Peru, Cuba, and other sugar-growing regions. In Cuba foreign ownership, especially U.S. interests, dominated the industry.

The plantation production system proved suitable for a wide range of tropical crops. During the late 19th century, as Latin American economies became entwined more directly with the world economy, plantations specializing in a range of other products prospered. Most notable among these new plantation crops were coffee and bananas, but other crops like cacao (chocolate) and henequen (for rope and twine) were prominent too (Figures 12.2 and 12.3).

Coffee cultivation exploded during the last half of the 19th century. Distinct from most

FIGURE 12.2. A henequen plant that has been harvested many times. Henequen leaves contain fibers that can be woven into rope and twine. The Yucatan Peninsula was home to a significant henequen plantation economy at the beginning of the 20th century, 1987.

FIGURE 12.3. The processing mill and transport center for henequen plantations near Mérida, Mexico, on the Yucatan Peninsula, 1987.

other crops grown on plantations, coffee could also be profitably cultivated by farmers with small landholdings using only family labor. Thus, on the Meseta Central of Costa Rica and in the Antioquia region of Colombia, coffee production boomed on small farms. In other regions, notably Guatemala and Brazil, however, coffee cultivation prospered on large estates. In Guatemala, foreign owners, especially Germans, played a major role in the development of coffee plantations, while seasonal migration by Indian laborers from the highlands provided the necessary labor supply. In Brazil, immense coffee estates prospered in the south, notably in the state of São Paulo. The size of some of these operations is truly mind-boggling. The Dumont *fazenda* occupied over 12,000 ha. It included a private railroad line, produced as much as 4.5 million kilos of coffee annually, and employed over 5,000 people. Brazil aggressively encouraged foreign immigration during the second half of the 19th century to meet the burgeoning demand for labor on the coffee plantations. Italian immigrants were most common, but Japanese immigrants also figured prominently in this process.

Banana plantations transformed the economic geography of the Caribbean lowlands of Central America during the early decades of the 20th century (Figure 12.4). Large-scale banana production for export began in Costa Rica initially when railroad builders sought a means of defraying construction costs and utilizing the lands adjacent to the rail lines prior to the completion of the line from Limón to San José. Steam-powered ships made the journey from the Caribbean coast to ports like New Orleans and even Boston possible in a week or less, and the newly available fruit found a ready market in North American urban markets. When scientific advances reduced the danger from endemic tropical diseases like yellow fever and malaria, North American entrepreneurs obtained lucrative long-term land leases from Central American governments for immense acreages. Later, when disease crippled banana production on the Caribbean coast and when the Panama Canal (1914) was completed, the U.S. corporations that controlled banana production established new plantations in disease-free areas along the Pacific coast of Central America and later in Ecuador.

The plantation system remains as a viable

FIGURE 12.4. A banana plantation on the Caribbean lowlands of Costa Rica near the Panama border, 1979.

agricultural production unit in Latin America today. Traditional plantation crops like sugarcane, cacao, and bananas are still grown on plantations. However, since the 1980s, improvements in transportation infrastructure, refrigeration, and shipping have also permitted the cultivation of more highly perishable tropical crops, like pineapples, mangos, and melons, on corporate plantations for markets in North America and Europe.

The Minifundia: Subsistence Farming

The *minifundia*, a small landholding unit, is and has been a common component of the Latin American agricultural landscape since colonial times. Its agricultural, social, and economic characteristics stand in strong contrast to those of latifundia. In the case of latifundia, a small, privileged class of colonial elites, usually whites but also some mestizos, acquired vast estates during the colonial period through royal land grants, land purchases, and outright usurpation. During the republican period, the rural oligarchy that evolved in almost all Latin American countries continued to consolidate large land holdings, often incorporating Indian lands, again through land purchase, usurpation, and legal maneuvering. Many Indian and mestizo peasants were incorporated into the labor force on these latifundia. Others, however, remained independent and occupied small farm plots (minifundia) on what was often marginal land.

In most regions, the number of peasants exceeded the land resources available. As a consequence, small farm units predominated. Over time, land parcels tended to become smaller and smaller as inheritance patterns led to the repeated subdivision of farm plots upon the death of household heads. Increases in rural populations only tended to exacerbate this pattern and placed increasing pressure on the limited resource base. Invariably the minifundia evolved as subsistence farms, often supporting large peasant families. The intensive agriculture practiced by *minifundistas* in order to secure even modest yields from these farms strains soil resources and degrades the resource base, typically resulting in smaller yields as time progresses. Historically, a scarcity of resources limited farmers' abilities to utilize anything but the most rudimentary tools and techniques (Figures 12.5, 12.6, 12.7, and 12.8). During the last half of the 20th century continuing poverty has restricted their access to fertilizers, pesticides, improved seed varieties, and technical know-how.

Labor is chronically underutilized on minifundia. As a practical matter many *minifundistas* are forced to seek employment elsewhere, frequently as laborers on nearby latifundia or as seasonal migrant laborers on plantations. The migration of peasants from the Andean valleys of southern Bolivia to the sugarcane plantations of Tucumán in northwest Argentina is one example, as is the movement of Indians from the Guatemala highlands to lower elevations to work on coffee estates and banana plantations. Off-farm employment is usually not simply a matter of generating additional income. It is often absolutely essential for the survival of farm families since many minifundia are so small and their production so limited that the subsistence needs of the families who live on them cannot be met.

Minifundia dominate much of the high-

FIGURE 12.5. A peasant's home in the Andes of northern Peru near Cutervo, Cajamarca. Home-made adobe bricks, a wheelbarrow, and a ladder all point to a rural subsistence economy, 1985.

lands of Guatemala. A careful study of minifundia was completed in the late 1960s by researchers from the Land Tenure Center at the University of Wisconsin. Their study provides an insightful look at conditions typical of minifundia in most parts of Latin America in the closing decades of the 20th century. In the Guatemala study, much of the population was marginalized from the mainstream national culture. Only about 25 percent of household heads were literate, and about 15 percent did not speak Spanish. Among women, the percentage of non-Spanish speakers rose dramatically, to about 70 percent.

FIGURE 12.6. A simple peasant plow that breaks the soil but does not turn it. This kind of plow is used throughout Latin American by many peasant farmers. Near Cutervo, Cajamarca, Peru, 1985.

Small farm size, low levels of capital investment, and rudimentary technology typified these peasant minifundia in the Guatemalan highlands. Landholdings, often on hilly land, averaged about 3 ha per household, with crops occupying about half of the acreage and the remainder devoted to livestock grazing. Most farmers operated with only the most basic tools, a hoe and a machete, although some farmers owned handheld sprayers and crosscut saws. Total capital investment in the farm unit, including land, buildings, tools, and livestock, averaged less than $1,400. Access to credit for fertilizer, seed, or other input was severely limited, with farmers receiving an average of less than $20 per annum. Almost without exception, family members provided all the labor required on farm units. Considerable underemployment was common among most families. Basic subsistence crops—beans and corn—occupied much of the available cropland on minifundia, while wheat predominated as the principal cash crop.

FIGURE 12.7. A farmer plows with a traditional plow and an oxen near Misque, Bolivia, 1987.

Abysmal living standards characterized the families living on minifundias. Annual family incomes totaled just $268, with about a quarter of that total coming from off-farm income. Although subsistence crops formed a significant component of farm production, nearly two-thirds of family income was spent on supplemental foods purchased from local markets and stores. Tortillas, beans, tamales, raw cane sugar (*panela*), and coffee formed the basic components of rural diets. Nearly two-thirds of farm families reported eating the same foods at all meals. Rarely did fresh vegetables, meat, eggs, cheese, or milk comprise part of the typical diet of *minifundistas*. Most farm families lived in simple adobe houses, many of only one or two rooms. Running water and electricity were unimaginable luxuries for most families, and less than 10 percent even had simple toilet facilities such as latrines.

FIGURE 12.8. Peasant farmers thresh wheat with horses and pitchforks. Cumbe Mayo, Cajamarca, Peru (photo, Allen G. Noble, 1985).

At the beginning of the 21st century, minifundia are still common in the rural areas of most Latin American countries, and the socioeconomic conditions of their residents are still abysmal.

MINIFUNDIA AND LAND REFORM

Minifundia are a widespread phenomenon in Latin America. The extreme concentration of land in the hands of a few has been a problem throughout the 20th century. At the beginning of the century, Mexico had one of the most extreme patterns of land concentration of any nation in the region. In 1910, approximately 1 percent of the population owned 97 percent of the nation's land, while 96 percent of the population owned just 1 percent of the nation's land. The characteristics of the rural labor force underscore the highly skewed nature of wealth and employment in the agricultural sector at that time. Landless peasant laborers comprised 88.4 percent of the agricultural labor force; small and medium landholders, sharecroppers, leaseholders, communal owners, and free workers accounted for 11.6 percent; and estate owners represented just 0.02 percent.

An extremely unequal distribution of agricultural land characterized Peru in the early 1960s, just prior to the implementation of two distinct land reform programs (Table 12.1). Over one-third of all farm units consisted of less than 1 ha; this third accounted for just 0.6 percent of all farmland. Farms of between 1 and 5 ha constituted almost 50 percent of all farm units and accounted for approximately 5 percent of all farmland. At the other end of the spectrum, over 60 percent of all farmland was held on farms of over 2,500 ha representing just 0.1 percent of all farm units. Including farms of smaller dimensions does not change the panorama appreciably. Farm holdings of over 100 ha accounted for almost 85 percent of all agricultural land, but represented just 1.2 percent of the total farm units.

Access to land resources remains a critical problem to peasants throughout much of Latin America today, despite continuing efforts at land reform and rural development since the 1950s. For example, in 1988, in Brazil, land ownership, especially of arable land, remained highly concentrated. Some 5 percent of the landowners owned two-thirds of the arable land, and just 1 percent held fully one-half of the arable land. At the other end of the scale, a miniscule amount of arable land, 2.4 percent, was owned by 50 percent of the rural landowners.

In El Salvador, despite a protracted civil

TABLE 12.1. Farm Size and Land Holding Patterns in Peru, 1961

Farm size (ha)	Number of holdings	Percent	Total area of farmland	Percent
Less than 1	290,900	34.2	127,869	0.6
1–5	417,357	48.9	926,851	4.9
5–10	76,829	9.1	481,631	2.6
10–20	30,370	3.6	397,754	2.1
20–50	17,414	2.1	506,745	2.7
50–100	7,214	0.8	474,313	2.5
100–200	4,606	0.5	589,567	3.2
200–500	3,475	0.4	1,035,076	5.5
500–1,000	1,585	0.2	1,065,157	5.7
1,000–2,500	1,116	0.1	1,658,639	9.3
More than 2,500	1,091	0.1	11,341,301	60.3
Total	851,957	100.0	18,604,500	100.0

Source: Delavaud (1980), p. 39.

war (1979–1990), fought in large part in the name of landless peasants and the rural poor, the condition of these people appeared little changed, and indeed was perhaps worse, at the war's conclusion. While agricultural census data are not available on landholding patterns for the 1990s, an extensive survey of the occupational characteristics of the agricultural population over 16 years of age conducted in 1991 and 1992 provides telling insights. The unemployed, unpaid family laborers, temporary day laborers, and permanent day laborers represented about 62 percent of the agricultural population. Land-poor peasants, defined as those owning less than 0.7 ha of land, accounted for 14 percent of the population, while small farmers and members of agricultural cooperatives totaled 13 percent. Farmers who employed labor accounted for just 10 percent of the total. While these data do not provide direct information on agricultural land tenure in El Salvador, they do demonstrate that, at the end of the 20th century, the vast majority of the rural population either owns no land whatsoever or owns only subsistence farm plots, minifundia.

Traditional thinking about landholding patterns in Latin America has repeatedly held that both the latifundia, with the possible exception of the plantation, and the minifundia are inefficient farming units. Each in its own way, it is argued, has contributed to the stagnation and underdevelopment that characterizes most rural environments in the region. The cultural values as well as the economic behavior of many latifundia owners have limited the agricultural production of these large estates. For example, land resources on latifundia have typically been underutilized. Often perceived by rural elites as a noble pursuit, livestock raising, especially the raising of cattle and horses, has caused significant proportions of these large estates to be dedicated to grazing, with comparatively little acreage devoted to the more productive pursuit of raising agricultural crops (Figure 12.9). This pattern is often even more acute on the largest estates that tend to account for the largest proportion of agricultural lands. The percentage area devoted to crops on the haciendas in the highland province of Chimborazo, Ecuador in the early 1960s illustrates this pattern quite clearly (Table 12.2). Haciendas of less than 200 ha devoted almost one-third of all land to crops, but

FIGURE 12.9. Cattle graze on fertile flatlands in the valley of Cajamarca in northern Peru. Cultural preferences and land-use practices of the wealthy have often meant that the land most appropriate for high-intensity crop agriculture is dedicated to cattle raising instead, 1985.

TABLE 12.2. Proportion of Land Cropped for Different-Sized Haciendas, Chimborazo Province, Ecuador, 1961

Size of hacienda (ha)	Percentage of area under crops
100–199	32.91
200–499	24.31
500–999	18.91
1,000–2,499	10.14
Over 2,500	8.3

Source: Sylva (1986), cited in Bebbington (1996), p. 121.

on those with between 200 and 500 ha, this proportion dropped to just one-quarter. When the acreage exceeded 1,000 ha, the proportion of cropland dropped to one-tenth.

The tendency of *latifundistas* to hold land without exploiting it to its fullest extent dates from the colonial period. Owning land confers social prestige, and land ownership continues to be one of the safest means of holding wealth. Land ownership tends to insulate rural oligarchies from financial loses due to bank failures, inflation, unstable governments, and changing national economic priorities and policies. Perhaps surprisingly, at the same time, *latifundistas* have tended to not reinvest profits from these estates in their agricultural operations, preferring instead to invest in urban homes and conspicuous consumption. Consequently, investment in agricultural technology and infrastructure improvements has languished, and a broad pattern of rural disinvestment can be discerned in many areas.

Minifundia too have been characterized as inefficient, but these inefficiencies stem from distinct causes. Rural population growth has tended to force the size of minifundia to become smaller and smaller over time. This, in turn, has encouraged property owners or renters to attempt to squeeze as much production from each farm as possible. Thus a very high percentage of all available land is cultivated, including marginal lands on steep slopes subject to erosion and fragile lands in arid regions often prone to desertification (Figure 12.10). The degree of land-use intensity on minifundia is illustrated again with data from Chimborazo province in the sierra of Ecuador (Table 12.3). During the mid-1950s, for instance, over 95 percent of all land on farm units of less than 5 ha was devoted to crops, declining to about 66 percent of farm units of less than 50 ha, and dropping even further on large properties. The need to maximize yields also encourages the near-continuous cultivation of minifundia lands, leading to declining yields and soil exhaustion over time. This problem is compounded by the inability of most *minifundistas* to purchase fertilizers or invest in other inputs to preserve or enhance soil fertility.

FIGURE 12.10. Potatoes are cultivated on small plots on steep hillsides in the Central Andes, Department of Junin, Peru, 1982.

TABLE 12.3. Proportion of Land Cropped for Different Farm Sizes, Province of Chimborazo, Ecuador, 1954

Farm size (ha)	Total area (ha)	Area cropped (ha)	Area cropped (percentage of farm)
0–4.9	52,300	49,900	95.4
5–49.9	48,600	31,900	65.6
Over 50	214,700	37,500	17

Source: Haney and Haney (1989), cited in Bebbington (1996), p. 121.

Land Reform

Land reform has been a recurring theme throughout much of Latin America during the 20th century. It is ironic that, in the 19th century, after almost all Latin American nations had achieved political independence, the consolidation of some of the largest latifundia occurred.

The Mexican Example

The situation in Mexico is illustrative of what happened in many countries, especially those with significant indigenous populations. During the early decades of the 19th century the Catholic Church was the most important landowner in Mexico, dominating much of the rural sector and monopolizing vast acreages. Liberal reforms promulgated by the government of Benito Juárez in 1857 provided for the breakup and expropriation of these vast estates if the lands could be demonstrated to be vacant and unused. During the nearly 40 years of the government of Porfirio Díaz, commonly known as the "Porfiriato," these laws were expanded to include lands held by Indian communities. Enticed by generous provisions that permitted survey companies to claim up to one-third of all "vacant" land they surveyed, Catholic Church lands and lands belonging to indigenous communities were usurped by the government and subsequently sold to wealthy landowners. The amount of acreage incorporated into haciendas increased dramatically, as did the number of Indians who became landless laborers on these haciendas through sharecropping or debt peonage. By the beginning of the 20th century, vast expanses of rural land were concentrated in the hands of very few. While the specific mechanisms varied, a similar pattern of increasing land concentration occurred in the last decades of the 19th century in Peru, Guatemala, Ecuador, and many other Latin American nations.

The first significant effort to redress the concentration of land in the hands of very few and improve the living conditions of the rural poor through a widespread program of land reform occurred in Mexico.

In 1910 demands for land and land reform led tens of thousands of Indians and poor mestizo peasants to take up arms in southern Mexico under the leadership of Emiliano Zapata. Zapata and his Indian armies, armies from the north under Pancho Villa, and other groups fought a bloody civil war with the central government. Later they fought wars with each other that dragged on for nearly 10 years and cost the lives of nearly 1 million Mexicans. Land reform remained one of the key rallying points for the revolution.

The Mexican Constitution of 1917, still in force today, institutionalized land reform as a cornerstone of the nation's political landscape. The key tenet of the constitution, Article 27, held that all land belonged to the nation, but that the nation had the right to transfer land to individuals as private property. Thus, while the constitution provided the legal justification for land reform through redistribution, it also protected the concept of private property. As a practical matter, the expropriation of haciendas did not mean that the owner lost all his property. The owner typically was permitted to maintain the core of the hacienda—house, buildings, and other infrastructure—as well as a small amount of land. While the exact amounts varied somewhat, an owner could retain ownership of between 100 and 150 ha of irrigated land or up to as much as 300 ha of unirrigated property. A variety of mechanisms, including deeding small properties to family members, often permitted *hacendados* to preserve comparatively large properties.

The agrarian reform program the constitution set in motion continued in earnest for nearly 20 years, reaching its peak in the late 1930s during the government of Lázaro Cárdenas. The reform program distributed nearly 30 million ha of land during this period, much of it to peasants and landless workers in small parcels of less than 5 ha.

THE EJIDO

Much of the land that was distributed went to communities, usually Indian communities, in the form of *ejidos*. The *ejido* was a form of communal ownership of land resources. Under this system of land tenure, the state ceded the right to use the land to individual communities. Communities enjoyed usufruct rights to the land in perpetuity, but legally the land could not be rented, used as collateral for loans, or be bought and sold. This mechanism was intended to prevent the widespread alienation of community and Indian lands that had been one of the hallmarks of the last half of the 19th century. It also imitated indigenous land tenure practices. Typically, community leaders or councils granted use rights for individual parcels to household heads and families. Providing such parcels were effectively utilized within the parameters of community norms, the land-use rights were essentially permanent. Although no specific hereditary rights pertained to these usufruct users, the usufruct rights to these parcels usually remained in the hands of family members on the death of household heads.

Two distinct types of ejidos evolved out of the agrarian reform experience in Mexico, individual ejidos and the communal ejidos. In terms of total land area, as well as the number of ejido properties and beneficiaries, the individual ejido predominated, representing over 95 percent of the total. Concentrated in central and southern Mexico, in areas with a high proportion of Indian peoples, individuals or families farmed these properties. Characteristically, these farm units are typical of the minifundia found throughout most of Latin America: they are small, undercapitalized, subsistence operations.

In contrast, the communal ejido evolved largely along Mexico's arid northwest Pacific coast in the river valleys in the states of Sonora and Sinaloa. Here ejidos benefited from the Mexican government's push to improve rural infrastructure and increase agricultural productivity through the construction of roads and, more significantly, an extensive network of reservoirs and irrigation works. Communal ejidos functioned as large agricultural cooperatives. The entire property operated as a single agricultural unit, with the *ejidatarios* working under labor conditions similar to hired workers but sharing in the annual profits should they occur. Consequently these ejidos benefited from scale economics, capital investment, and more efficient use of land resources. During the last 50 years, they have proved so highly productive that they grow a large proportion of Mexico's agricultural exports.

Mexico's emphasis on agrarian reform diminished markedly after the 1940s. Then the country's policymakers increasingly turned to economic development strategies that emphasized industrialization as the most effective route to economic development. Nevertheless, Mexico's agrarian reform did successfully distribute agricultural lands to an appreciable proportion of the rural population and effectively checked the rural unrest that the land-hungry peasant revolutionaries had incited. It did not, however, transform the basic structure of economic and social relationships or the power of the rural oligarchy. With the exception of the communal ejidos of the northwest Pacific coast, the government made few effective efforts to provide technical assistance, extension services, or credit to *ejiditarios*. In addition, the ability of many hacienda owners to hold on to the hacienda's core property, productive infrastructure, and at least 150 ha of irrigated land, and often more, allowed them to maintain a dominant position in rural economies. Many diversified their economic activities to include agricultural marketing and transportation, and thus maintained control over key sectors in rural economies.

THE EJIDO AND NAFTA

The process of agrarian reform and the role of the ejido in Mexico's economy continued to evolve throughout the last decades of the 20th century. Steady growth of the rural population, the frequent subdivision of ejido lands, and a perpetual lack of capital and agricultural extension services all contributed to the declining productivity and the increasing marginalization of the ejido sector. Economists trained in the United States increasingly dominated Mexico's national governments, and even the presidency beginning in the late 1970s. The policies they pursued for the nation's development emphasized neoliberal economic reforms, including a decreased role of the state in the economy, privatizing state enterprises, and free trade. They moved to change the government's agrarian policies from its traditional emphasis on the distribution of agricultural land to increasing the productivity of the agricultural sector.

The signing of the North American Free Trade Association (NAFTA) in 1992 provided the impetus for a series of major reforms in Article 27 of the Mexican Constitution. These reforms transformed the ejido with the explicit intent of stimulating agricultural production by increasing productivity and decreasing costs. The reforms stipulated that the government's constitutional responsibility to distribute agricultural land had ended. They essentially privatized ejido grants, making them the property of the grant holders. The reforms permitted *ejidatarios* "to sell, rent, sharecrop, or mortgage their land parcels as collateral for loans." They allowed ejido members to buy the properties of other members, although no individual could own more than 5 percent of the ejido's lands. They also permitted land sales to outsiders, but such sales required the permission of two-thirds of the ejido's general assembly.

These changes will lead to increases in farm sizes in the rural sector as well as to some productivity gains. The opening of Mexico's markets to less expensive U.S. grain imports, especially corn and wheat, as a result of NAFTA will likely accelerate the transformation of the ejido and rural communities. It is clear that many farmers will be forced out of agriculture and will migrate to Mexico's urban centers in search of employment in the service and industrial sectors. Migration to the United States is another option many will continue to pursue.

Land Reform in the 1950s: Bolivia and Guatemala

With the exception of Mexico, land reform was almost nonexistent in the rest of Latin America until the 1950s. In Bolivia, years of peasant agitation, land invasions, and confrontation with rural landowners preceded the revolution of 1952 and subsequently brought widespread and enduring land reform beginning in 1954. The Bolivian reforms sought to eliminate the latifundia that dominated much of the rural landscape, end the feudal labor arrangements under which many *campesinos* toiled, and contribute to the diversification of agriculture. Foreign ownership of agricultural land was minimal and the reforms faced no significant opposition outside the country—indeed, the U.S. government provided significant financial and technical support for the process of agrarian reform. The reforms transformed the rural landscape and economy and distributed land to nearly 250,000 peasant families.

In Guatemala, on the other hand, the agrarian reforms implemented at almost the same time threatened the property interests of local as well as foreign landowners. Especially prominent were the corporate interests of the United States-based corporation United Fruit Company. In the early 1950s United Fruit was the nation's largest property holder, yet only worked about 15 percent of its lands. Under the reforms, nearly two-thirds of the company's landholdings were nationalized, setting off a torrent of protest to U.S. government authorities. The reforms were short-lived. The United States embarked on a campaign to destabilize the government of Jacobo Arbenz, portraying it as communist. Shortly thereafter a small force of 200 men with the support of the U.S. Central Intelligence Agency (CIA) invaded Guatemala from Honduras. Aircraft believed to have been owned and operated by the CIA attacked the nation's major cities in support of the invaders, and the Arbenz government fell (1954). The agrarian reforms were dismantled almost immediately, and almost all lands distributed during the reform process were returned to their previous owners.

The Cuban Revolution and Land Reform

The successful Cuban revolution in 1959 and the land reform program that followed caused major repercussions far beyond the island's shores. The government of Fidel Castro moved quickly in the aftermath of the revolution's success. Land reform legislation was implemented immediately in 1959. The goals varied somewhat from the objectives in other parts of Latin America. Large latifundia, usually sugarcane plantations and cattle ranches, dominated the rural landscape. Foreign corporate owners, mostly from the United States, predominated. Small private farms were not widespread in Cuba. Some land redistribution did take place: approximately 100,000 tenant farmers who had rented land on large estates were the beneficiaries of this land reform, each receiving title to 27 ha.

However, the redistribution of land to landless peasants or *minifundistas* was not a significant goal of the Cuban reforms. The vast majority of large estates that were expropriated were not subdivided, but were incorporated into the state economy as state farms or cooperatives. Contrary to popular perception, the reforms did not negate the concept of private property. The first reforms promulgated in 1959 permitted private farm ownership of especially efficient units over 1,300 ha and of properties up to almost 500 ha if the units met efficiency and productivity standards. However, subsequent revisions of the law in 1963 reduced the limits on farm size to 67 ha per individual owner and in addition required producers to sell a specific proportion of their output to the state at fixed prices. Overall, the agricultural reforms left over 70 percent of the crop and pastureland in the country in the hands of the state.

While the agrarian reforms in Cuba did little to redistribute land to poor peasants and landless laborers, it completely restructured the rural economy and society as no other reforms in Latin America had done before or since. The marginal living conditions of most rural inhabitants improved significantly as investments in schools, healthcare, and sanitation focused on their plight.

The Cuban experience stunned the United States. The success of a communist revolution so close to its shores unsettled the Americans. Cuba made no efforts to compensate owners of agricultural lands or businesses whose properties had been expropriated. This infuriated U.S. business interests. As a consequence, decades later the United States still presses these claims and uses them as justification for a wide range of punitive legislation aimed at Cuba, including an economic embargo.

The success of the Cuban revolution led the United States to initiate the Alliance for

Progress in Latin America in the early 1960s. This program sought to promote economic development on a wide front in the remainder of Latin America and to stem the threat of successful communist revolutions in other nations.

Agricultural development and agrarian reform programs represented a major component in the early strategy of the United States's Alliance for Progress. Most beneficiaries of U.S. aid in Latin America responded by implementing some kind of land reform in the 1960s. The Alliance for Progress promoted legislation for land reform programs and agrarian reform agencies to implement them in some of the most populous and geographically extensive nations, including Venezuela (1960), Colombia (1961), Chile (1962), Brazil (1963), and Peru (1964). Smaller Latin American nations including Ecuador, Costa Rica, the Dominican Republic, Panama, and Guatemala pursued similar agendas and strategies.

The Results of Land Reform

The results of agrarian reform programs in Latin America have been mixed. In Mexico and Bolivia the reforms reached a significant proportion of each nation's farm families, almost one-half and three-quarters, respectively. However, in the remaining nations, under the best circumstances the reforms benefited a maximum of just one-third of farm families. In several countries the percentage did not exceed even 10 percent (Table 12.4). Often the beneficiaries were not the poorest or most marginal farm families, but those that were better off than average. At the same time, the lands distributed to reform beneficiaries were of low quality. Tragically, the poorest farm families, often landless laborers who lived on the periphery of the haciendas and latifundias, were frequently excluded from the reform process and often ended up in an even more marginal economic position after the reforms than before. In most countries, even where a large percentage of farm families benefited, the reform programs made little effort to provide other inputs essential for a major transformation of the agricultural sector. Agricultural credit, extension services, fertilizers, improved seed varieties, education, and a rural road network connected to market centers did not figure prominently, if at all, in the reform programs. In most countries the pattern of land ownership still remains highly skewed, with hundreds of thousands of families farming minifundia and a handful of families owning immense estates that occupy hundreds of thousands of hectares.

The limited success of land reform programs in most countries can be traced to the

TABLE 12.4. Selected Countries of Latin America and the Caribbean: Areas Affected by the Agrarian Reform and Number of Peasant Families Benefited ca. 1985

	Forest and agricultural surface hectares (in thousands)			Number of farming families		
Country	Total	Affected	Percent	Total	Benefited	Percent
Chile	28,759.00	2,940.00	10.2	412,000	38,000	9.2
Costa Rica	3,122.40	221.6	7.1	155,200	8,349	5.4
Dominican Republic	2,676.70	374.6	14.0	697,800	59,411	8.5
Ecuador	7,949.00	718.1	9.0	749,000	78,088	10.4
Mexico	139,868.00	60,724.00	43.4	4,629,400	1,986,000	42.9
Panama	2,253.90	493.2	21.9	132,800	17,703	13.3
Peru	23,545.00	9,255.60	39.3	1,419,400	431,982	30.4
Venezuela	26,470.00	5,118.70	19.3	561,800	171,861	30.6

Source: Inter-American Development Bank (1986), p. 131, as cited in W. C. Thiesenhusen (1989), p. 10.

fact that beyond the peasantry support for reforms was often weak, while opposition or at least limited opposition came from several powerful quarters. The situation in 1970 showed little change in most countries (Table 12.5). Wealthy landowners, conservative politicians, and foreign business interests, often based in the United States, opposed them stridently. Some believe that even the structure of the programs themselves was designed to obstruct their rapid and effective implementation. Cumbersome enabling legislation, difficult to interpret and equally difficult to implement, impeded the successful execution of the programs. Operational obstacles were often placed in the path of land reform agencies that limited their effectiveness and the speed with which reforms could be carried out. Limited funding and frequent changes in program directors and senior personnel contributed to ineffective administration as well. Even the transfer of property titles to beneficiaries produced interminable delays. This process normally lasted many years and often dragged on indefinitely. The position of the U.S. government also made the process difficult and ineffective. On the one hand, the United States had supported the principle of land reform through the Alliance for Progress and provided funding aimed specifically at its implementation through foreign assistance programs managed by the U.S. Agency for International Development. On the other hand, U.S. business interests often stood to lose property, and any attack on the concept of the sanctity of private property did not find a sympathetic audience in the United States or among U.S. policymakers.

EL SALVADOR: REVOLUTION, LAND REFORM, AND CONTRADICTIONS IN U.S. POLICY

The equivocal position of the United States with respect to land reform played out again two decades later. During the 1980s Central American rural population growth, grinding poverty, and social injustice fueled civil wars and revolutions in Guatemala, El Salvador, and Nicaragua and created political instability in Honduras. The contrast between the U.S. policy followed in El Salvador and that pursued in Honduras illustrates clearly how United States-supported land reform policies had little to do with a concern for social justice or rural development, and much more to do with global and regional geopolitics. At the end of the 1970s a bloody and protracted civil war began in El Salvador that lasted for over a decade. The opponents of the government were based largely in peripheral rural regions. Agrarian reform and a demand for a more equitable pattern of land ownership figured heavily in their demands. While the U.S. government had rarely supported land reform programs that included the expropriation of existing landholdings and their redistribution to landless peasants, it did during the 1980s in El Salvador. This change in traditional policy

TABLE 12.5. Minifundias and Latifundias: Agrarian Structure of Selected Latin American Countries, 1970

Country	Minifundias percentage of farms	Percentage occupied land	Latifundias percentage of farms	Percentage occupied land
Argentina	43.2	3.4	0.8	36.9
Brazil	22.5	0.5	4.7	59.5
Colombia	64	4.9	1.3	49.5
Chile	36.9	0.2	6.9	81.3
Ecuador	89.9	16.6	0.4	45.1
Guatemala	88.4	14.3	0.1	40.8
Peru	88	7.4	1.1	82.4

Source: Todaro (1985), p. 295.

occurred largely because of the revolutionaries' strength on the battlefield and the inability of the El Salvadoran army and its U.S. military advisors to quash the insurgency. Both President Carter's Democratic administration and President Reagan's Republican administration that followed encouraged and supported land expropriation and redistribution as a critical measure to blunt one of the most egregious inequities of rural life in El Salvador. "National security" interests, specifically the fear of a successful insurgency and the installation of a socialist or communist government, drove U.S. policymakers to support a land reform program that, at its conclusion, had distributed agricultural lands to nearly one-quarter of the rural population.

In the neighboring country of Honduras, the foreign assistance policy the U.S. government pursued differed dramatically. Here, despite an extremely inequitable pattern of land tenure, social unrest was minimal, no serious armed opposition to the government existed, and the national security interests of the United States were in no way threatened. Under these conditions, U.S. support for land reform was tepid and was limited to a program to improve the mechanisms for land titling.

LAND REFORM INTO THE 21ST CENTURY

Land reform continues to be a contentious and politically charged issue in many places in Latin America. Demands for an equitable distribution of land resources contributed to civil wars that ravaged not only El Salvador, but Guatemala and Nicaragua. Calls for reform figured prominently in the accords that brought these armed conflicts to an end. But after nearly a decade little progress has been made on seriously addressing these issues in Guatemala and opposition to any form of land redistribution continues to be intense.

In Nicaragua a complicated scenario played out in the wake of the Sandinista victory in 1979. Sandinistas seized and redistributed huge tracts of coffee plantations, cropland, and pasture from private owners in the wake of the civil war. Much was distributed to peasants. However, when the Sandinistas lost power in subsequent elections, some unscrupulous party and government officials appropriated properties for their personal gain. Subsequently, as a result of the original property seizures and the later land grabs, the countryside has been racked by endless claims and counterclaims for these lands among the bickering and sometimes violent claimants.

Hugo Chavez, elected president of Venezuela in 1998, has made land reform a touchstone of his presidency and reform policies. Chavez has faced tough opposition from ranchers and farmers in his efforts to force the redistribution of cultivated lands and to seize land that is deemed idle for redistribution. At the same time, he enjoys immense support from peasant farmers and landless workers in this country where previous land reform legislation and programs from the early 1960s brought no meaningful reforms and were largely forgotten. While Chavez's rhetoric has been strong and some lands have been redistributed, 7 years into his presidency the basic structure of the agrarian economy and landholding patterns remains largely unchanged.

Land reform and access to land is a central issue throughout the Brazilian Amazon. The issues are hotly contested and bring landless workers, peasant farmers, and social activists into direct conflict with large landowners, especially ranchers, as well as loggers and miners. Frustrated by years of promises and no demonstrable results, landless workers and others, including liberal elements of the Catholic Church, have advocated tirelessly for equitable access to land and forest resources. A major consequence of these efforts has been the development of a broad coalition of groups under the banner of the Brazilian Landless Workers' Movement (Movimento dos Trabalhadores Rurais sem Terra, [MST]). Land invasions, forced explusions, violent confrontations, and political assassinations occur with alarming regularity across the Amazon Basin. Despite the election of presidents representing left and center-left political parties since 1995, who claim to be sympathetic to these demands, government efforts at land reform here have been tepid at best. In many areas of the basin the government seems unable or simply unwilling even to enforce the country's laws. Political assassinations and violence often go uninvestigated by government authorities.

Land Colonization

In most of Latin America resistance to land reform programs based on land redistribution stiffened throughout the 1960s. The receding threat of communist insurgencies supported by Cuba, the early expropriation of the largest and most identifiable latifundia, the equivocal policies of the United States, and the continued resistance of conservative politicians and wealthy landowners all worked to stall or at least significantly slow the pace of land reform. Government policymakers urgently sought other solutions to the problem of the *minifundia*, inequitable land tenure patterns, increasing rural populations, and rural poverty.

Beginning in the 1960s, Latin American governments increasingly turned to land colonization as a solution to the dilemma posed by redistribution agrarian reform programs. Programs of land colonization permitted governments to utilize vast land resources, usually in unsettled or sparsely settled tropical regions, to ease land pressure in densely settled highland regions. Property ownership in these regions was often ill-defined or only vaguely asserted by indigenous peoples who had long lived beyond the effective national territory. Government ownership of land was easily asserted and property rights could be assigned to colonists with little or no resistance and few problems. Land colonization represented a feasible strategy, particularly in countries lying largely within the Tropics and with appreciable tropical rain forests that had never been colonized. In some instances, land colonization proved possible in arid environments like northwest Mexico, where the construction of reservoirs and irrigation systems permitted the cultivation of arid lands that could not previously be farmed. However, the high cost of infrastructure limited the extent to which this strategy could be pursued. In one way or another, almost all Latin American countries attempted to implement land colonization programs. Chile, Argentina, and Uruguay are the only notable exceptions to this pattern.

Spontaneous Colonization

Tropical land colonization experiences in Latin America typically followed one of three approaches. One is spontaneous colonization, where, in many areas, colonization occurred spontaneously as settlers moved into the frontier with little or no direction from government authorities. Such settlement frequently occurred at the margins of settled regions as colonists pushed beyond the government road network, staking claims to farmsteads and clearing forests. This pushed back the frontier incrementally. In the tropical lowlands, spontaneous settlement has often followed river courses that have provided a ready means of penetration of tropical forests in the absence of roads. Government support and participation in efforts of spontaneous settlement has essentially been nil, at least in the initial phases of the process. As the settlement frontier has filled in, roads have been constructed, while the extension of effective government control and social services like the presence of the national police and basic public schools have followed in a tentative fashion. Spontaneous settlement requires little or no outlays of government monies or support initially. Investment from the central government to support spontaneous settlement invariably comes many years later when settlement is established and population numbers make it easier to justify.

Directed Colonization

The second approach, directed colonization, is another matter. The central government, often through a special agency or legislative program created to promote tropical land colonization, is involved in the colonization process from the beginning. The government itself selects the colonization zone, ostensibly with the intention of identifying fertile soils and other favorable environmental conditions, surveys the chosen zone, and chooses sites for roads, towns, and land parcels. Government planners attempt to identify potential colonists with the experiences or backgrounds that suggest they could successfully settle on and farm the new lands. The government invests heavily in the construction of roads, the laying out of town sites, and the provision of essential social infrastructure, especially schools and health facilities. Land titling is a key element in the early process of settlement for it guarantees settlers land security. Technical assistance, agricultural extension, and even programs to facilitate the marketing of agricultural products may form part of directed land colonization programs. In some of the most ambitious and well-financed programs, government agencies have even offered initial financial subsidies during the first 6 months of settlement and access to emergency food supplies.

Despite considerable government investment and official support for directed colonization schemes, results have often fallen far short of expectations. Several problems have proved difficult to overcome. Costs are high. Some estimates have placed the per-family costs to settle colonists at close to $10,000, many times the annual income peasant farmers in tropical rain forests might expect to reap even under the best conditions. Tropical soils have often proved far less fertile than anticipated. In addition settlers have used farming and cropping methods far more appropriate to temperate highlands or midlatitude regions rather than imitating or adapting indigenous tropical farming techniques. Within a few years after land clearance, crop yields tend to fall as soil nutrients are rapidly depleted.

Semidirected Colonization

Semidirected colonization is a third approach to land colonization. Efforts at semidirected land colonization date back well over 100 years, although the role of governments in promoting this sort of colonization has changed over that period. Mennonite colonization in the first half of the 20th century in northern Mexico, eastern Bolivia, and the Chaco region of Paraguay were an early form of semidirected colonization. In these instances, national governments eager to promote colonization in marginal regions permitted Mennonite settlers broad leeway over social and economic development in specific regional contexts, thereby allowing the establishment of semiautonomous ethnic enclaves. By the end of the 20th century, however, the success of these enclaves as agricultural colonies attracted additional settlement by non-Mennonites and prompted national governments to try to exert greater control over the social and economic activities of the Mennonite settlers. Mennonite enclaves in the Chihuahua Desert of Mexico and in the Gran Chaco of Paraguay have recently come under these pressures.

In the face of mixed results at best from planned colonization programs in tropical regions during the last half of the 20th century, national governments have sometimes attempted to promote other forms of semidirected colonization programs. In such cases, governments provide only the most basic infrastructure support. Frequently, this consists of little more than the construction of roads into tropical rain forest regions believed to be suitable for settlement and farming. Secondary support may be limited to land subdivision, distribution, and land titling. Government support and subsidies end here, with the exception of the standard rural infrastructure governments provide throughout their national territory, notably schools and health facilities. This approach has great appeal for many governments since costs and project duration are comparatively finite, and the responsibility for success or failure lies largely with the colonists themselves and not with government development agencies.

Tropical land colonization has led to increases in total amount of cropland in the region as well as even larger increases in unimproved pasture. Mexico, the Central American republics, the Andean nations, and Brazil have all experienced extensive tropical land colonization. It is not clear, however, whether this expansion of settlement and of the agricultural frontier has resulted in a more economically or socially equitable rural sector or led to increased agricultural productivity. The independent yeoman farmer and a strong rural middle class have not evolved in most colonization zones, and the minifundia and latifundia have appeared again as well as their attendant problems.

TABLE 12.6. Changes in the Rural Population of Latin America

	Percentage of labor force in agriculture		Rural population (as % of total)
Country	1965	1990–1992	1992
Argentina	18	13	13
Belize	—	—	49
Bolivia	54	47	48
Brazil	49	25	23
Chile	27	19	15
Colombia	45	10	29
Costa Rica	47	25	52
Ecuador	55	33	42
El Salvador	58	11	55
Guatemala	64	50	60
Guyana	—	27	66
Honduras	68	38	55
Mexico	49	23	26
Nicaragua	56	46	39
Panama	46	27	46
Paraguay	54	48	51
Peru	49	35	29
Uruguay	20	5	11
Venezuela	30	1	9

Source: United Nations Development Program (1994), cited in Bebbington (1996), p. 142.

Modern Transformations of the Agrarian Sector

The most striking transformation in the rural sector has been the steadily diminishing proportion of Latin America's population that is employed in agriculture (Table 12.6). As recently as the mid-1960s, the percentage of the labor force employed in agriculture exceeded one-half in all but a handful of nations and in a few cases reached nearly two-thirds of the labor force. In the intervening 30 years, this percentage dropped precipitously in many countries. In Brazil and Mexico, the region's largest and most populous nations, the proportion dropped from one-half of the labor force to just one-quarter. The Central American nations of Panama and Costa Rica experienced changes of a similar magnitude, while in Colombia and El Salvador the proportion dropped from around one-half to only one-tenth! A few countries experienced declines of less than 10 percent, but these were the exception and typically represented the region's poorest and most traditional societies with the highest proportion of indigenous peoples: Bolivia, Guatemala, and Paraguay.

At the same time, rural populations have continued to grow in absolute terms as many nations' populations increased at rates between 2.5 and 3.5 percent during the 1970s, 1980s, and then slowed somewhat in the 1990s. In the early 1990s the percentage of rural population still exceeded 50 percent in many of the region's smaller nations. Even in the most urbanized nations, rural populations stood at about 10 percent.

Agricultural and trade policies have increasingly followed neoliberal economic models. During the 1980s and 1990s these models have had a dramatic effect on the agricultural sector. An emphasis on maximizing export trade and generating foreign currency revenues has led to agricultural strategies that exploit the particular comparative advantages of each nation's or region's agricultural sector. As a consequence, there is a tendency for agricultural production to focus on a handful of crops that find ready markets and the best prices on the world market.

Chile pioneered this strategy in Latin America in the last decades of the 20th century. The military dictatorship of Augusto Pinochet that seized power in 1973 adopted an economic development model that emphasized neoliberal policies. In the agricultural sector, this meant encouraging agricultural production that capitalized on the region's Southern Hemisphere location and the burgeoning demand in Northern Hemisphere countries, particularly in North America and Europe, for fresh fruit during the winter months. The export of fresh table grapes exploded during the 1980s, becoming one of Chile's top agricultural export commodities. Other fresh fruits, especially peaches, nectarines, and cherries, also found ready customers in the Northern Hemisphere during the winter months. Improvements in postharvest processing and transportation allowed these highly perishable fruits to reach these markets in excellent condition. In addition to exporting fresh fruit, Chileans also sought ways to process fresh fruits in Chile and export food products. The production of dried fruit, fruit juices, and wine also grew dramatically, creating a boom in agro-industrial employment and export earnings.

Chile's successes contributed to an increased regard for neoliberal policies in many Latin American nations. At the same time, the World Bank and the International Monetary Fund (IMF) insisted on the adoption of these policies as preconditions for the receipt of critical foreign assistance. This too contributed to a widespread adoption of neoliberal strategies in other countries. In the western portion of the Brazilian state of Paraná and to a lesser extent in eastern Paraguay, the pursuit of agro-exports resulted in a crop monoculture that transformed the region. Soybean cultivation skyrocketed as voracious world demand fueled a rapid expansion of forest clearing and small-farm consolidations. As a result, a soybean crop monoculture evolved that is dominated by large commercial farms. In Paraná, this process accelerated and displaced sharecroppers and small farmers, fueling a dramatic migration to Amazonian colonization areas to the north in the state of Rondônia beginning in the 1980s. Despite modernization in some aspects of the agricultural sector, in many countries abusive and exploitative forms of labor extraction still persist. The existence of forced labor, slavery, in Brazil is a shocking example (Vignette 12.1).

Multinational corporations are playing an increasingly significant role in the agricultural sector in most Latin American countries. This has often meant that small farmers and middlemen are squeezed. In the southern Brazilian state of Rio Grande do Sul, tobacco has long been a major crop. In recent years, multinational tobacco conglomerates have moved into the region, bought out small tobacco wholesalers and processors, and enforced tough marketing conditions on the region's 160,000 growers. Here, the huge conglomerates Philip Morris and Souza Cruz, a Brazilian firm, have received nearly $1 billion in state tax breaks. At the same time, growers complain that the companies' monopoly control over marketing has allowed them to drive down prices. Many growers fear that they will go broke and have to sell their lands.

Multinational corporations have again entered the agricultural land market. In Argen-

VIGNETTE 12.1. SLAVERY IN BRAZIL ~~IN~~ THE 1990s

It is sometimes difficult for many Europeans, North Americans, and even Latin Americans to believe that many of the forms of labor exploitation that characterized the development of the *latifundia*, specifically *haciendas* and plantations, during the colonial and republican periods could still exist. Slavery and debt peonage, for instance, must have surely disappeared during the 20th century. This, however, is not the case. Throughout Latin America extreme forms of labor exploitation can still be found, usually in isolated rural contexts. The situation in some parts of Brazil, which has Latin America's largest economy, most numerous population, and greatest geographical extent, is shocking.

Slavery was officially outlawed in Brazil in 1888. Yet, over 100 years later, the vestiges of the institution still exist in some parts of the country, especially on isolated *fazendas* and plantations on the north coast and in the interior. In recent years the Brazilian Ministry of Labor's antislavery mobile enforcement team has freed hundreds of slave laborers. Almost 600 were freed in 2000 and more than 1,400 in 2001.

Extreme poverty, geographical isolation, powerful connections, and near impunity from prosecution provide the conditions that permit this system of quasi-slavery to exist. Typically poor landless peasants are recruited with offers of paid labor on large farms. They are transported to immense, isolated agricultural estates. Here they are often compelled to purchase their tools, food, and other basic commodities with credit from company stores at grossly inflated prices. They work at a variety of basic agricultural tasks, including land clearance, crop cultivation, harvesting, and even the primary processing of agricultural commodities, during workdays that can last up to 15 hours. Armed guards watch over them during working hours and patrol the estates at nights to guarantee their continued presence on the plantation. Those who flee are pursued by armed posses and forcibly returned. Some landowners even engage in the sale of individuals they hold captive on their estates, selling them to other estate owners when their own labor demands no longer justify holding them. Sadly, substandard living conditions, inadequate food, and oppressive labor requirements are not the most significant threats to those individuals held as slaves. Brazilian government sources estimate that almost 20 percent of those who are held as slaves are killed for one reason or another by armed guards or estate owners and are buried on the estates.

Landowners and their employees face few challenges. A weak legal system is largely ineffective in rural areas. The political and economic power of the owners of large estates is immense, and it allows them to continue these practices with relative impunity. Even if convicted, the penalties violators face under current law are surprisingly mild, a maximum of 2 years imprisonment. The government of President Fernando Cardoso declared its concern with these practices and proposed more aggressive legislation to address these problems. However, even with legislative change, the absence of an effective and independent local police and judiciary makes it seem unlikely these practices will end any time soon. For example, a survey in 2000 by a Roman Catholic Church group, the Pastoral Lord Commission, estimated that as many as 25,000 individuals were being held as forced laborers in Brazil.

tina, entrepreneurs with financial backing and assistance from international financiers have moved aggressively to consolidate massive agricultural landholdings during the late 1990s. As the prospects for increased globalization of food production increase, the low cost of land in Argentina and its high productivity make it an exceptional investment opportunity. A dramatic example is the Argentine farm company Cresud which, with the assistance of international financial backing, has amassed landholdings of over 1.1 million acres.

In many ways, however, these trends reflect a reassertion of several of the historic patterns typical of Latin America's agricultural economy: large landholding (latifundia), agricultural monocultures (plantations), the concentration of wealth and power in the hands of a few (rural oligarchies), and the marginalization of small landholders and peasants.

Summary

Indigenous peoples had varying understandings of land, none of which coincided with European ideas of land ownership. Chiefdoms had land-use rights, and the royal families in advanced civilizations claimed all the land they ruled and granted portions of it to their retainers. Europeans introduced the concept of land ownership. Spanish land ownership followed a feudal structure, with multigenerational family-owned manors. These latifundia included haciendas that raised horses and cattle and plantations on the coastal plains that raised sugarcane, cotton, tobacco, and other crops and used slaves for labor. At the other end of the land spectrum were minifundia, small, subsistence farms owned by peasants.

By the early 20th century, the social injustice of this system brought calls for reform and fueled a revolution in 1910 in Mexico. Land reform became a major issue in many countries. The Cuban revolution of 1959 transformed large estates to state-owned farms. The United States-backed Alliance for Progress supported land reforms in the 1960s. Those land reforms brought mixed results, however. Although successful in Mexico and Bolivia, in other Latin American countries they often missed the most marginalized peasants. In the 1960s land colonization of sparsely populated areas, usually tropical rain forest lowlands, received political support and occurred in many countries. With no entrenched landowning interests to block them, such programs relieved some of the pressure for land reform in settled agricultural areas. Land colonization occurred spontaneously as roads were built, but in other instances efforts at colonization were directed by government agencies. With the beginning of the 21st century, land reform remains a contentious issue in Central America, Venezuela, and especially in the Brazilian Amazon. Nevertheless, throughout the region the trend is an overall decline in numbers participating in agriculture and a move to large-scale commercial agricultural production for national and export markets.

Further Reading

Alston, L. J., et al. (1999). *Titles, conflict, and land use: The development of property rights and land reform tenure on the Brazilian frontier.* Ann Arbor: University of Michigan Press.

Barraclough, S., and Domike, A. (1966). Agrarian structure in seven Latin American countries. *Land Economics, 42*(4), 391–424.

Breton, B. L. (2003). *Modern-day slavery in the Brazilian Amazon*. Broomfield, CT: Kumarian Press.

Brockett, C. (1988). *Land, power, and poverty: Agrarian transformation and political conflict in Central America*. Boston: Unwin Hyman.

de Almeida, L. F., and Sanchez, F. R. (2000). The landless workers' movement and social struggles against neoliberalism. *Latin American Perspectives, 27*(5), pp 11–32.

Denevan, W. M. (2001). *Cultivated landscapes of Native Amazonia and the Andes: Triumph over the soil*. Oxford, UK: Oxford University Press.

Dobyns, H. (Ed.). (1971). *Peasants, power, and applied so-*

cial change: Vicos as a model. Beverly Hills, CA: Sage.

Doolittle, W. E. (1990). *Canal irrigation in prehistoric Mexico: The sequence of technological change*. Austin: University of Texas Press.

Hecht, S., and Cockburn, A. (1990). *Fate of the forest*. New York: Harper Perennial.

McReynolds, S. A. (2002). Land reform in El Salvador and the Chapultepec Peace Accord. *Journal of Peasant Studies, 30*(1), 135–169.

Meszaros, G. (2000). No ordinary revolution: Brazil's landless workers' movement. *Race and Class, 42*(2), 1–18.

Prem, H. J. (1992). Spanish colonization and Indian property in Central Mexico. *Annals of the Association of American Geographers, 82*, 444–459.

Seligson, M. A. (1995). Thirty years of transformation in the agrarian structure of El Salvador, 1961–1991. *Latin American Research Review, 30*, 43–74.

Sempat, C. (1992). The colonial economy: The transfer of the European system of production to New Spain and Peru. *Journal of Latin American Studies, 24*, 55–68.

Simmons, C. S. (2004). The political economy of land conflict in the eastern Brazilian Amazon. *Annals of the Association of American Geographers, 94*, 183–206.

Soustelle, J. (1964). *Daily life of the Aztecs*. Harmondsworth, UK: Penguin Books.

Thiesenhusen, W. C. (1995). *Broken promises: Agrarian reform and the Latin American campesino*. Boulder, CO: Westview Press.

Whitmore, T. M., and Turner, B. L., II (2001). *Cultivated landscapes of Middle America on the eve of conquest*. New York: Oxford University Press.

Wolf, E., and Hansen, E. C. (1972). *The human condition in Latin America*. New York: Oxford University Press.

Wolford, W. (2004). This land is ours now: Spatial imaginaries and the struggle for land in Brazil. *Annals of the Association of American Geographers, 94*, 409–424.

13 Brazil's Atlantic Coastal Plain

Portuguese exploration of Brazil began on the Atlantic Coastal Plain in the early 1500s. This region servFed as the focus of colonization and settlement until the end of the colonial period in 1822. Sugar cultivation and the plantation system that evolved on the plain was the economic backbone of Brazil throughout most of the colonial period. The country's first two capital cities, Salvador and Rio de Janeiro, respectively, were located on the plain. The region remained the center of the nation's economic, social, cultural, and political life until the beginning of the 20th century when the city of São Paulo and the Paraná Plateau region began to overtake it economically and politically. Today the Atlantic Coastal Plain is home to approximately 45 million of Brazil's citizens. It is also the location of the nation's second largest metropolitan region, Rio de Janeiro, with a population of about 10 million. The region retains much of its preeminence in the nation's social and cultural life, but its economic and political significance has been eclipsed by São Paulo.

Geography and Environment

Cabo San Roque (Cape Saint Roque in English) marks the northeasternmost point of both the South American continent and the country of Brazil. The cape, which sits some 100 km north of the coastal city of Natal, marks the beginning of Brazil's Atlantic Coastal Plain. The coastal plain stretches south and then southwest for over 2,300 km, terminating south of Rio de Janeiro at a latitude close to the tropic of Capricorn at 23.5°S (Figure 13.1). Within Brazil, most of the region is known as the *Zona de Mata*, although typically Brazilians themselves do not include the southern one-quarter of the coastal plain within this zone. The coastal plain is wedged between the coast and the mountainous escarpment, which runs parallel to the plain over much of its course. It is widest along the northeast coast and narrowest in the south. The plain reaches its greatest width along a 300-km stretch of coast north of the colonial capital, Salvador; here it stretches inland for up to 100 km. But over most of its extent its width rarely exceeds 50 km and it is often even narrower. This is especially so in the south; at Rio de Janeiro, its breadth is less than 10 km. It ends just to the south of Rio.

Despite a latitudinal range running from 5°S to the tropic of Capricorn, the climate of the Atlantic Coastal Plain varies little. It is warm and humid, with nearly imperceptible changes in mean temperatures between the summer and winter months. It is, in a word, tropical. The plain's proximity to the Atlantic Ocean strongly influences its climatic patterns. The South Equatorial Current and its extension the Brazil Current bring warm humid conditions to the coast, as well as slight seasonal temperature changes. Despite the tropi-

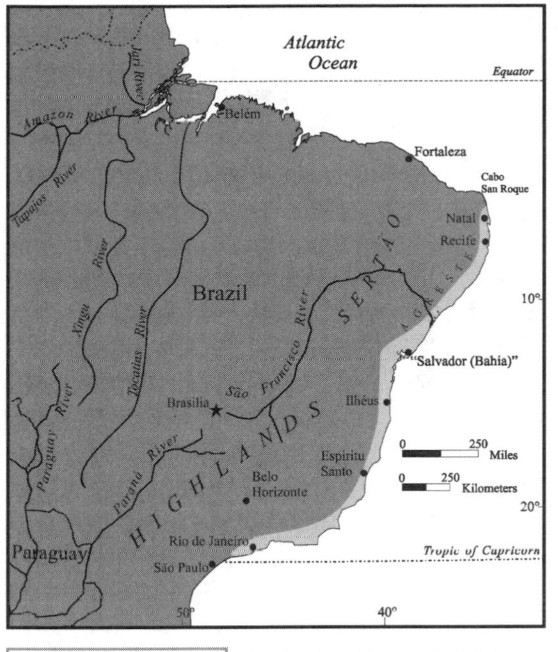

FIGURE 13.1. Brazil's Atlantic Coastal Plain.

cal location and a warm offshore current, a regular diurnal pattern of land and sea breezes constantly refreshes much of the coastal plain. The continuity in temperatures is demonstrated by the statistics for Natal in the far north and by those for Rio de Janeiro in the south. The temperature mean in the warmest months, January/February, is 27°C in Natal and 26°C in Rio de Janeiro. Mean temperatures on the coastal plain in the winter's coolest month, July, vary slightly, standing at 24°C in Natal and 20°C in Rio de Janeiro (Figure 13.2).

Precipitation displays a more complex pattern, particularly seasonally. The coastal plain receives moderate amounts of precipitation in most years, typically ranging from 1,300 to 1,600 mm, which is sufficient to support rain-fed agriculture. Annual precipitation totals 1,400 mm at Natal, 1,600 mm at Recife, and 1,100 mm at Rio de Janeiro; nearby mountains block some of the precipitation Rio might otherwise receive. Precipitation occurs throughout the year—there is no real dry season on the coast. A rainier season, however, can be readily identified. Along the central and northeastern portions of the coast, precipitation peaks between April and July. In the south this pattern is reversed; at Rio de Janeiro, for example, the period between April and July is marked by lower-than-average rainfall.

Historical Geography and Economic Development

At the time of contact with Europeans, mangroves cloaked much of the coastal margin, while broadleaf tropical forests covered the coastal plain. Known as the *mata Atlantica* in Brazil, these tropical broadleaf forests, and especially the brazilwood tree, or *pau brasil* (*Caesalpinia echinata*), were extensively logged by the early colonists during the first half of the 16th century. The brazilwood tree, from which a rich red dye can be produced, proved to be the colony's first significant export. Its economic importance was such that, subsequently, the colony dropped its early name, Terra de Vera Cruz (Land of the True Cross) and became simply Brasil (Brazil is the English spelling) in recognition of the economic importance of this tree. By the end of the colonial period, the forest cover had largely disappeared, giving way to agriculture, especially sugarcane, cacao, and pasture. The mangroves still cling to some portions of the coast, but urban and industrial encroachment, as well as water pollution, have greatly reduced their extent.

Portuguese colonization and settlement of Brazil began on the Atlantic Coastal Plain. This region evolved into the Brazilian heartland during the three centuries of Portuguese colonial rule. Initially, the Portuguese found little that attracted them to Brazil. The exploitation of brazilwood provided an early economic boom during the first decades of the

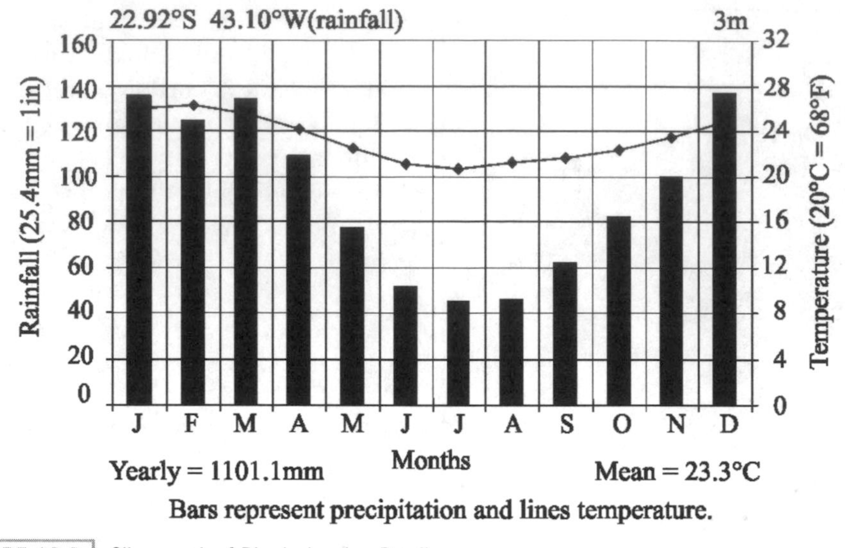

FIGURE 13.2. Climograph of Rio de Janeiro, Brazil.

16th century, but few Portuguese settled on the coast. Fearing the incursions of other European colonial powers into Brazil, the Portuguese Crown began serious efforts at establishing permanent settlements here in the 1530s.

In the beginning, the Crown established 12 "captaincies" along the coast, granting these territories to loyal subjects (*donatarios*) directly accountable to the Crown. These *donatarios* were granted broad rights to govern, but were also held responsible for economic development in their domains. The captaincy system of colonization proved unsatisfactory, as most *donatarios* lacked the financial means, administrative skills, or vision to promote and direct colonial economic development. Thus, in 1549, the Crown consolidated all power into the hands of a governor-general and established the capital of the Brazilian colony at Bahia (later renamed Salvador) on the spectacular anchorage at Todos Santos Bay.

Sugarcane cultivation transformed Brazil from a forgotten backwater of Portugal's colonial empire to a dynamic and productive colony. An Old World crop, sugarcane was probably introduced to Brazil from the Azores. The warm, moist, tropical climate and the fertile alluvial soils on the coastal plain provided an ideal environment for its cultivation. A seemingly insatiable demand for sugar in Europe provided a ready market for this production. Eventually, much of the coastal plain from Bahia northward was dedicated to the growing of sugarcane.

The regional economy, social structure, and culture were intimately linked with sugarcane growing, processing, and export. Between 1550 and 1650, the coastal plain was transformed as sugarcane cultivation became the dominant land use. This is reflected in the proliferation of sugar mills, which grew in number from 70 in 1550 to 230 by 1627. The regional economy boomed, producing vast fortunes for the Portuguese Crown and the local planters, but at the price of immense human suffering.

The economic viability of sugarcane production for export was predicated on the use of large tracts of fertile land and accessibility to a dependable and inexpensive labor supply. Land became available as the humid tropical forest, the *mata Atlantica*, that had blanketed the coastal plain quickly disappeared and was replaced by sugarcane (Figure 13.3).

Labor proved to be a more difficult problem to resolve. The Portuguese first cajoled and then enslaved Indians to toil in the fields. The Indians, however, resisted enslavement by disappearing into the forests, or, if enslaved, simply died from exposure to exotic diseases or the multitude of abuses suffered at the hands of the Portuguese. Soon it became evident that Indian labor could not support the intense demands created by the sugarcane economy.

African slaves provided a ready solution to the labor problem on Brazil's sugar plantations. The Portuguese had used African slaves for sugarcane cultivation and production in their colonies on the Madeira Islands and São Tomé. These experiences provided a model for the system of plantation slavery established in Brazil.

The Portuguese had developed an extensive trading network along the West African coast during the previous century. This network facilitated the procuring, transporting, and marketing of slaves. The first shipments of slaves entered Brazil around 1540, but it was not until the last quarter of the century that substantial numbers arrived in the colony. By 1600, between 10,000 and 15,000 slaves were imported annually, establishing a pattern that persisted in Brazil for more than 200 years. Thus, by 1620, Indians had all but disappeared from the sugarcane fields of the Atlantic Coastal Plain. Africans proved far superior to Indians because of their resistance to Old World diseases, superior physical strength, and, in many cases, superior knowledge of farming systems and animal husbandry.

FIGURE 13.3. *Mata Atlantica*, natural forest vegetation in Tijuca National Park, Rio de Janeiro, Brazil, 2004.

The sugarcane plantation, known as the *engenho* in Portuguese, became the basic social and economic unit early in the colonial period. The plantation was structured much like a feudal medieval manor. Typically, plantations occupied extensive tracts of flat, cultivable land, although only rarely did these exceed 1,000 ha. The infrastructure of the plantation consisted of the owner's home, the "big house," or *casa grande*; a chapel with a resident priest; the sugarmill, or *engenho*; a complex of workshops and storage sheds; living quarters for white overseers; and sheds or barracks for slaves. The plantation owner and his family ruled supreme over the plantation and enjoyed nearly absolute autonomy, and often over neighboring farmers and planters of lesser economic and political standing.

Planters and their families quickly evolved into a rural aristocracy that dominated the social, political, and economic life of the Atlantic Coastal Plain. Contrary to the situation in most other parts of Latin America, urban society here was subordinate to and largely controlled by the rural landed gentry. As an institution, the family-owned plantation endured until the end of the 19th century, leaving an indelible imprint on the economic and social geography of the Atlantic Coastal Plain.

The abolition of slavery in 1888 marked the beginning of dramatic transformations in the sugarcane economy. Technological advances in sugar processing required more capital-intensive investment and encouraged increased corporate ownership and the consolidation of smaller properties into larger, more efficient units. The social prestige and economic power of the region's rural aristocracy declined and urban elites gradually replaced them at the top of the regional power structure.

The economic history of the Atlantic Coastal Plain has been characterized by cycles of great prosperity followed by economic depression. Much of this has been due to a heavy dependence on one or two crops destined for export markets, but the exhaustion of fertile lands and poor agricultural practices have also contributed to the down part of these cycles.

The sugarcane industry boomed until about 1650. Then other European colonial powers began to cultivate sugarcane in their Caribbean colonies. Utilizing many elements of the Brazilian system, especially slavery, and applying new technology, the sugar industry in the Caribbean flourished. Virgin soils and proximity to European markets favored Brazil's competitors. Sugarcane, nevertheless, has remained a key crop along the coast of Brazil up to the present. Its fortunes have waxed and waned as world market conditions have changed—for example, falling when Dutch entrepreneurs began to till virgin lands in Surinam to grow the crop (late 1600s) and rising when a slave rebellion in Haiti ended sugar production in the French colony (1791).

Cotton enjoyed a brief boom in the 1860s when the Civil War in the United States disrupted world supply. Cotton, now destined for the domestic Brazilian market, continues to be grown in the interior and in transition areas between the coastal plain and the uplands of the Sertão.

At the end of the 19th century, cocoa cultivation flourished in the tropical forests of the *mata Atlantica* in southern Bahia and northern Espiritú Santo (Figure 13.4). Cocoa bean production expanded until after World War I. Then competition from Ghana and careless agricultural practices resulted in an increasingly inferior product. Still, the lavish turn-of-the-century architecture that adorns the city of Ilheus, the "chocolate capital," in southern Bahia is testimony to the fortunes brought by chocolate.

The discovery of mineral wealth in the Central Plateau region in the late 17th century and the transfer of the colony's capital from Salvador in the north to Rio de Janeiro in the south in 1763 marked the ascendancy of the southern portion of the Atlantic Coastal Plain. This process was accelerated in 1808 when the Portuguese king, fleeing Napoleon's armies in Lisbon, transferred his entire royal court to Rio de Janeiro. When Napoleon's forces were defeated in Europe and the Portuguese royal court returned to Portugal, the king's son remained in Brazil. Subsequently he declared Brazil's independence and was installed as the country's first emperor. With the country's independence from Portugal in 1822, Rio de Janeiro's role as the nation's capital further consolidated its economic, social, and political power and the importance of the southern quarter of the Atlantic Coastal Plain.

Contemporary Economic and Social Geography

The social and economic geography of the Atlantic Coastal Plain at the end of the 20th century reflects many elements of the region's historical geography. The legacy of the plantation system is clearly marked by the extreme differences in wealth and social class that persist to the present. The racial makeup of the population also reflects the legacy of the plantation:

FIGURE 13.4. Brazil's states and territories.

African Brazilians (*negros* and *pardos*) comprise about three-quarters of the region's total population. Much of the region's population is rural and its economy is still closely tied to agriculture. Sugarcane, cacao, cotton, and tobacco are all major commercial crops, although much of this production, especially of cotton and tobacco, is destined for Brazil's internal market rather than export markets, as was the pattern in the past. Mineral resources are scarce; modest petroleum reserves along the coast north of Salvador are the most significant.

In the smaller urban centers, agro-industrial processing and the production of basic consumer goods and textiles are the principal manufacturing activities. Most manufacturing, however, is concentrated in the major urban centers: Recife, Salvador, and especially Rio de Janeiro, the nation's second most important industrial center. Petrochemicals, iron and steel production, and the manufacturing of capital goods and consumer durables concentrate around Rio. The service sector, including government employment, is a major employer in all the region's urban centers. Tourism has been important in Rio for decades. Increasingly, however, other urban centers, particularly in the far northeastern portion of the Atlantic Coastal Plain, have begun to capitalize on their tropical climate, attractive beaches, and colonial architecture and heritage to cultivate tourism, especially from western and northern Europe.

The Atlantic Coastal Plain is one of Brazil's most densely settled and populous regions. Over 40 million people inhabit the coastal plain, representing approximately one-quarter of Brazil's total population. The population, however, is unevenly distributed, with the states of Rio de Janeiro and Espiritú Santo in the far south accounting for almost one-third of the total. Much of the population is also concentrated in a few metropolitan centers. Rio de Janeiro dominates the region with a population over 10 million, followed by Salvador and Recife with 3.1 million and 1.3 million, respectively. Nearly 75 percent of the population is classified as urban, but urban versus rural populations vary considerably from state to state. Over 95 percent of the state of Rio de Janeiro is classified as urban, while the northeastern states of Bahia and Alagoas post the lowest rates of urbanization on the coast, about 60 percent. The racial composition of the population reflects the region's historic links to Africa: approximately three-quarters of its inhabitants are African Brazilians. The regional culture reflects this heritage: African cultural elements are evident in the regional cuisine, religious practices, folk medicine, women's dress (in Bahia), and even martial arts. Levels of economic development vary dramatically along the coastal plain as well. In the south, Rio de Janeiro ranks among the most developed of Brazil's states. To the north, however, development levels are much lower and these coastal states rank among the nation's least developed (Figure 13.5).

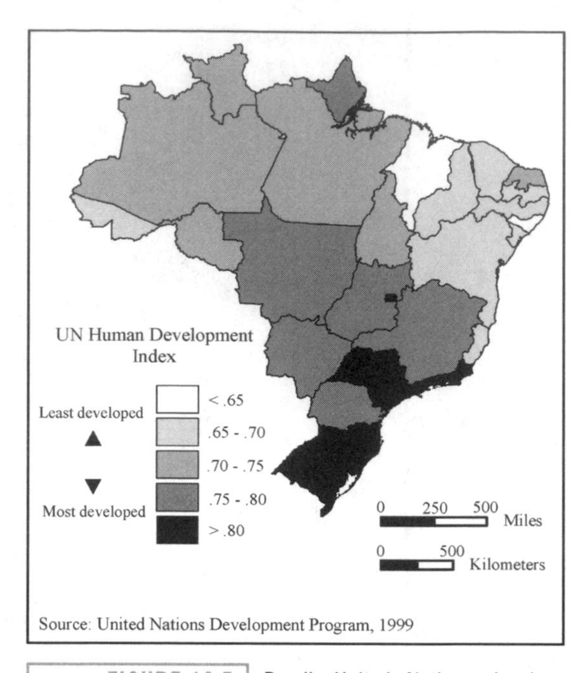

FIGURE 13.5. Brazil, United Nations development levels by state, 1999.

Rio de Janeiro is a major world metropolitan area and one of the largest cities in the world. Its metropolitan population accounts for about one-quarter of the total population of the Atlantic Coastal Plain. As the long-time national capital (1763–1960), the key social and cultural center of Brazil, and the country's second most important economic center, Rio's influence over the rest of region is immense. The city's residents, and to some extent all residents of the Atlantic Coastal Plain, are known colloquially as *cariocas* in Brazil. This is a sympathetic regional label that evokes the image of a fun-loving, sensual people. The city's economy is gigantic and diversified. Transportation, shipping, communications, and finance are key industries in the metropolitan area. Heavy and light manufacturing are also important. One of the country's principal iron and steel mills is located at Volta Redonda, a planned industrial city on the periphery of Rio's metropolitan area. Service industries, including tourism, employ a large percentage of the city's workforce.

The city is a study in contrasts. Its geographical setting is stunningly beautiful: lush green mountains rise sharply behind it and beautiful Guanabara Bay stretches in front of it. However, with a second glance at the same scene, one would be hard-pressed not to notice the provisional and sometimes ramshackle quality of the housing built on the steepest slopes. Perched on steep slopes poorly suited for construction, these neighborhoods or shantytowns, known as *favelas*, are home to about 20 percent of the city's residents (Figure 13.6). Nevertheless, in the city's core and upscale districts, stunning historic and modern architecture is commonplace (Figure 13.7).

Basic public services like potable water and sewers are absent from most *favelas*, although electrical service is often available. Crime can be a problem in some, and for a time in the 1990s criminal gangs took over entire favelas and ruled them like fiefdoms until police authorities eventually intervened decisively. The squalid conditions in the *favelas* contrast with the lavish homes and businesses in its upscale neighborhoods and trendy tourist centers.

Salvador and Recife are the two other major metropolitan centers on the coastal plain. Salvador, capital of the Brazilian colony from 1549 to 1763, is the larger of the two cities, with a metropolitan population of about 3.1 million. The city's rich colonial history is evident in the multitude of forts and colonial churches that form a key part of its architectural heritage (Figure 13.8). For much of the colonial period the city was the principal port of entry for the African slave trade. Even today people of African descent form a majority of the city's population. As a result, today the city is clearly the capital of African Brazilian culture. The city's port serves an extensive region that reaches deep into the interior. A broad range of activities characterizes its manufacturing base, ranging from cigar to petrochemical manufacture. Recife, meaning "reef" in Portuguese, prospered during the early colonial period because of the safe anchorage its offshore reefs provided. Its favorable site characteristics led the Dutch to establish their capital in the city during the quarter century (1630–1655) they occupied much of the northeast coast. Its port continues to play a key role in the regional economy. The industrial sector includes sugar refining, cotton milling, textiles manufacturing, agro-industrial processing, and leather goods production. The metropolitan population reaches 1.3 million.

Summary

Brazil's Atlantic Coastal Plain was colonized by the Portuguese in the early 1500s. The region was the center of Brazil's economic, social, cultural, and political life until the early 20th century and home to major cities and the

FIGURE 13.6. *Favelas*, or shantytowns, on the slopes overlooking Rio de Janeiro, Brazil, 2004.

FIGURE 13.7. 19th-century ferry terminal in Rio de Janeiro, Brazil, 2004.

country's first two national capitals, first Salvador and then Rio de Janeiro. The tropical climate is influenced and moderated by the region's proximity to the Atlantic Ocean. Early colonists harvested the timber from trees known as brazilwood, and initiated a thriving colonial trade based on its exploitation. Subsequently, an economy based on plantation agriculture evolved. Sugarcane was the dominant crop. The Portuguese brought in African slaves to work on the plantations. The plantation was the central unit throughout the colonial period, although the crops changed to include cotton and cocoa. At the beginning of the 21st century, this region still manifests the effects of its plantation heritage, with sharp economic and social disparities in its population and a large population of African Brazilians. Rio de Janeiro, although no longer Brazil's capital, has a population of about 11 million, but nationally ranks second to São Paulo.

FIGURE 13.8. An example of the colonial baroque style: the Church of the Third Order of Saint Francis, Salvador, Brazil, 2004.

Further Reading

Boxer, C. R. (1969). *The golden age of Brazil, 1695–1750: Growing pains of a colonial society*. Berkeley and Los Angeles: University of California Press.

Dean, W. (1995). *With broadax and firebrand: The destruction of the Brazilian Atlantic forest*. Berkeley and Los Angeles: University of California Press.

Eakin, M. (1998). *Brazil: The once and future country*. New York: St. Martin's Press.

Eisenberg, P. (1974). *The sugar industry in Pernambuco: Modernization without change, 1840–1910*. Berkeley and Los Angeles: University of California Press.

Freyre, G. (1986). *The masters and the slaves: A study in the development of Brazilian civilization*. Berkeley and Los Angeles: University of California Press.

Gay, R. (1984). *Popular organization and democracy in Rio de Janeiro: A tale of two favelas*. Philadelphia: Temple University Press.

Prado, C. (1967). *The colonial background of modern Brazil*. Berkeley and Los Angeles: University of California Press.

Riley, E., et al. (2001). Favela Bairro and a new generation of housing programmes for the urban poor. *Geoforum, 34*(4), 521–531.

Schwartz, S. B. (1985). *Sugar plantations in the formation of Brazilian society: Bahia, 1550–1835*. New York: Cambridge University Press.

Smyth, C. G., and Royle, S. A. (2000). Urban landslide hazards: Incidence and causative factors in Niteroi, Rio de Janeiro State, Brazil. *Applied Geography, 20*(2), 95–117.

Wolford, W. (2004). Of land and labor: Agrarian reform on sugarcane plantations of northeast Brazil. *Latin American Perspectives, 31*(2), 147–170.

World Bank. (2004). *Brazil: Equitable, competitive, sustainable: Contributions for debate*. Washington, DC: Author.

14 Contemporary Cities and Urban Patterns

At the beginning of the 20th century, the vast majority of Latin Americans lived in rural environments, either dispersed in the countryside or nucleated in agricultural villages and small towns. In Mexico, for example, just over one-quarter of the population resided in urban places at the turn of the century. Mexico was typical of the pattern that prevailed in most other Latin American countries until the 1930s and 1940s.

However, this distribution varied considerably among the region's nations. The proportion of urban population in some countries, especially Argentina and Uruguay, grew rapidly in the early decades of the 20th century. Immigration and industrialization fueled rapid urban growth during these decades, and by 1940 well over 50 percent of each nation's population lived in urban centers, with Buenos Aires and Montevideo accounting for most. Many other countries, especially smaller nations and those more peripheral to the world economy, urbanized more slowly. For instance, the proportion of urban population in Bolivia, Paraguay, and Guatemala in 1940 did not exceed 20 percent.

Broad structural changes in the economy and society during the 20th century have essentially transformed Latin America from the predominately rural agricultural society it was at the end of the 19th century to the urban, semi-industrialized society it is at the dawn of the 21st century. While barely one-third of all Latin Americans lived in cities in the 1940s, the following decades saw a steady increase in the proportion of urban population in all Latin American countries. By 1980 nearly two-thirds of the region's population resided in urban places; by 1990 nearly three-quarters of the population was urban.

Despite a high regional rate of urbanization—comparable to Anglo-America and Western Europe—there are still dramatic contrasts in the level of urbanization among individual Latin American nations. For example, in South America's Southern Cone, the percent of urban population is very high, exceeding 80 percent in Uruguay, Argentina, and Chile. Urban population exceeds 70 percent in the region's most populous and economically powerful nations, Brazil and Mexico, as it does in the Andean nations of Peru and Colombia and the United States's Caribbean island colony of Puerto Rico. At the opposite end of the spectrum, a majority of Central America's population is still rural. Only in Nicaragua and Panama does the proportion of urban population exceed 50 percent, while in the remainder of the region's countries it stands between 40 and 50 percent (Table 14.1, Figure 14.1).

The rapid rate of urbanization of Latin America's population during the 20th century

TABLE 14.1. Urban and Rural Population Numbers and Percentages for Latin American Countries, 2001

Country	Total population (1000s)	Urban population (1000s)	Rural population (1000s)	Percentage urban (%)	Percentage rural (%)
Argentina	37,488	33,119	4396	88.3	11.7
Bolivia	8,516	5,358	3,159	62.9	37.1
Brazil	172,559	141,041	31,518	81.7	18.3
Chile	15,401	13,254	2,147	86.1	13.9
Colombia	42,803	32,319	10,484	75.5	24.5
Costa Rica	4,112	2,448	1,664	59.5	40.5
Cuba	11,237	8,482	2,755	75.5	24.5
Dominican Republic	8,507	5,615	2,892	66.0	34.0
Ecuador	12,878	8,171	4,707	63.4	36.6
El Salvador	6,400	3,935	2,465	61.5	38.5
Guatemala	11,687	4,668	7,019	39.9	60.1
Honduras	6,574	3,531	3,043	53.7	46.3
Mexico	100,367	74,846	25,521	74.6	25.4
Nicaragua	5,208	2,943	2,265	56.5	43.5
Panama	2,899	1,639	1,260	56.5	43.5
Paraguay	5,637	3,194	2,443	56.7	43.3
Peru	26,093	19,084	7,009	73.1	26.9
Puerto Rico	3,952	2,987	965	75.6	24.4
Uruguay	3,361	3,097	264	92.1	7.9
Venezuela	24,632	21,475	3,157	87.2	12.8
Total	510,311	391,206	119,106		

Source: United Nations, Population Division, Department of Economic and Social Affairs (2002a).

can be traced to improvements in public health, high population growth, and a stagnant rural sector. The 20th century witnessed dramatic advances in the understanding of diseases and their transmission, as well as the development of vaccines, the creation of immunization programs, and the widespread adoption of other public health improvements. Cuban and U.S. medical researchers uncovered the causes of yellow fever and developed a vaccine for it at the turn of the century. An understanding of mosquitoes as the vector for the transmission of malaria led to the implementation of malarial control programs, which began to eliminate that disease as a major threat to public health during the same period. The widespread availability of vaccines and the implementation of immunization programs for measles, diphtheria, and typhoid effectively combated these common and often deadly diseases. Similarly, advances in the supply of potable water as well as progress in sanitation and the disposal of solid wastes, particularly in urban areas, contributed to vastly improved public health conditions. As a result, population grew rapidly and life expectancy

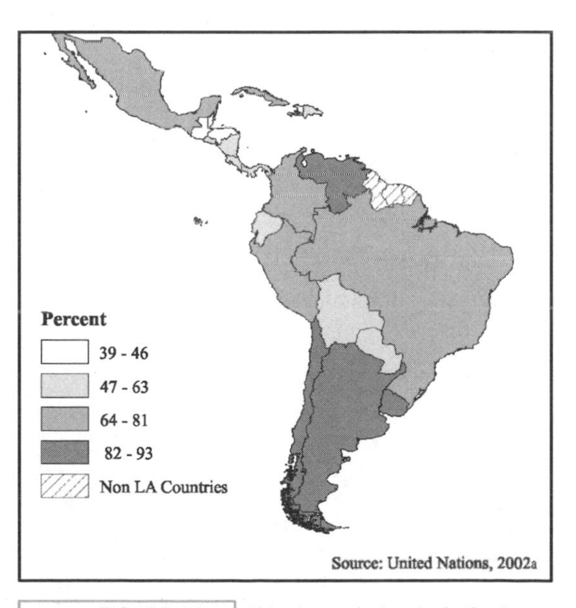

FIGURE 14.1. Urban population in Latin America, 2000.

increased. Between 1930 and the end of the century Latin America's population increased from slightly more than 100 million to approximately 480 million, nearly a fivefold increase in just 70 years. On an individual level, this meant that, on average, life expectancy increased sharply. Regionally, life expectancy stood at about 35 years in 1930, but it had increased to 70 years by the end of the 1990s.

Finally, a stagnant rural sector provided few opportunities for this rapidly increasing population. Most of the population increase in rural areas was quickly siphoned off to urban centers where opportunities in the expanding industrial and service sectors offered poor migrants some hope of employment. Rural population numbers as a consequence remained comparatively stable, while urban populations grew steadily.

Economic Development and Urbanization

The economic structure of Latin American countries began to diversify beginning in the early decades of the 20th century. While the export of agricultural products and raw materials continued to play a central role in many economies, manufacturing and other industrial activities began to account for a larger share of national incomes. Latin America's nations began to industrialize, first through primary processing of raw materials and the manufacture of consumer products, and then later with the fabrication of consumer durables and durable goods. Manufacturing and industrial enterprises concentrated in cities, creating new employment opportunities and fueling growth in retailing, services, and the public sector. The complexity of Latin America's major cities increased markedly during this period as a more diverse range of economic activities—from industrial fabrication to international financial services—began to appear in these cities.

After 1930, rural migrants swelled the population of urban places large and small. The highest rates of growth and the greatest population numbers, however, concentrated in just a few cities within each nation's urban system. Usually the national capital figured as the fastest growing among these cities. With the exception of Buenos Aires and Montevideo, whose growth was due to massive foreign migration in the early decades of the century, most major Latin American cities experienced extremely high annual growth rates during the 1960s (Table 14.2). During the 1960s, annual population growth rates exceeded 5 percent in Mexico City, Bogotá, São Paulo, and Lima. Population growth rates of this magnitude produce a doubling of a population within 15 years. The population of other major cities increased swiftly as well, although not quite as

TABLE 14.2. Average Annual Population Growth in Selected Latin American Cities, 1960-2005

	Growth-Rate Percentage					
City	1960–1965	1970–1975	1980–1985	1990–1995	1995–2000	2000–2005
Mexico City	5.7	4.0	1.6	1.8	1.5	0.9
Bogotá	6.4	5.0	3.5	2.8	3.4	2.3
Caracas	4.8	2.6	1.2	1.0	1.0	0.7
Santiago	3.8	2.8	2.2	1.9	1.7	1.4
Buenos Aires	2.2	1.7	1.2	0.8	0.7	0.7
São Paulo	5.4	4.4	1.7	1.7	1.7	1.7
Rio de Janeiro	3.6	2.1	1.0	1.0	1.0	1.0
Lima	5.5	4.4	2.9	2.7	2.2	1.9

Source: United Nations, Population Division, Department of Economic and Social Affairs (2002a).

precipitously. Rio de Janeiro, Santiago, and Caracas all had annual growth rates in excess of 3.3 percent, a rate that would lead to a doubling of population in 20 years or less.

But in the following decades the rate of population increase moderated. For instance, Lima's growth rate dropped to 4.4 percent in the early 1970s and fell to about 2.9 percent in the early 1980s. By 2005, its growth rate is estimated to have fallen to 1.9 percent. Progressive reductions of similar magnitude occurred for almost all major urban centers by that time, slowing the pace of population growth in these metropolitan areas at least slightly (Table 14.2).

Nevertheless, the annual population growth rates of Latin America's largest cities, and many of its secondary cities, still range between 2 and 3 percent. Increases of this size indicate a steady, consistent rate of urban population increase and a doubling of population in 25–35 years. Growth rates of this magnitude far exceed those experienced by major urban centers in the United States or Western Europe during the last 150 years.

Specific examples are useful in illustrating the true magnitude of population change. Lima, Peru, for example, had a population of slightly less than 650,000 in 1940. Fifty years later, the city's population had increased nearly tenfold to 6.3 million! Mexico City experienced even more vertiginous growth, increasing in population from about 1.5 million in 1940 to about 20 million inhabitants by 1990. In contrast, Havana is essentially the only major metropolitan center in Latin America that has not grown by leaps and bounds during the last 50 years. Its population has grown very slowly ever since the success of the Cuban revolution in 1959 when the government implemented a wide range of measures to restrict migration to the city and to make residence in the city difficult for those who were not "legal" residents. These measures have been successful in the case of Havana because a totalitarian national government can enforce them.

Latin America's Metropolitan Centers

The urban landscape of Latin America boasts some of the world's most populous metropolises. Although precise population numbers for world metropolitan areas often range widely depending upon the source (these numbers are frequently debated), São Paulo and Mexico City are consistently rated among the top 10. United Nations (U.N.) demographers have estimated that in 2010 these cities will rank as the second and fourth most populous urban agglomerations in the world, with populations of 20.5 million and 19.7 million, respectively. The same U.N. demographers have projected that both Buenos Aires and Rio de Janeiro will have populations in excess of 10 million by 2010, ranking 14th and 16th in the world. In comparison, in the United States, only the New York City metropolitan area and Los Angeles rank among the top 10 world cities; they rank seventh and 10th, with 17.5 million and 14.1 million, respectively.

Besides having some of the world's most populous urban centers, Latin America had 50 metropolitan areas with populations in excess of 1 million in 2000 (Figure 14.2). With the exception of Honduras, every country has at least one city with 1 million or more inhabitants. Brazil and Mexico accounted for about half of this total, with 14 and 9, respectively. The names of many of these million-plus cities are widely recognized in the English-speaking world because of their roles as national capitals and/or prominence as economic or historic centers (e.g., Lima, Quito, Havana, Caracas, La Paz, or Recife). But many other cities among this group are practically unknown beyond their nation's borders (Valencia, Barranquilla, Rosario, Curitiba, Campinas, Maceió, and Goiânia). A coastal orientation predominates among the region's million-plus cities, although there are exceptions, and the urban systems of Mexico and Colombia defy

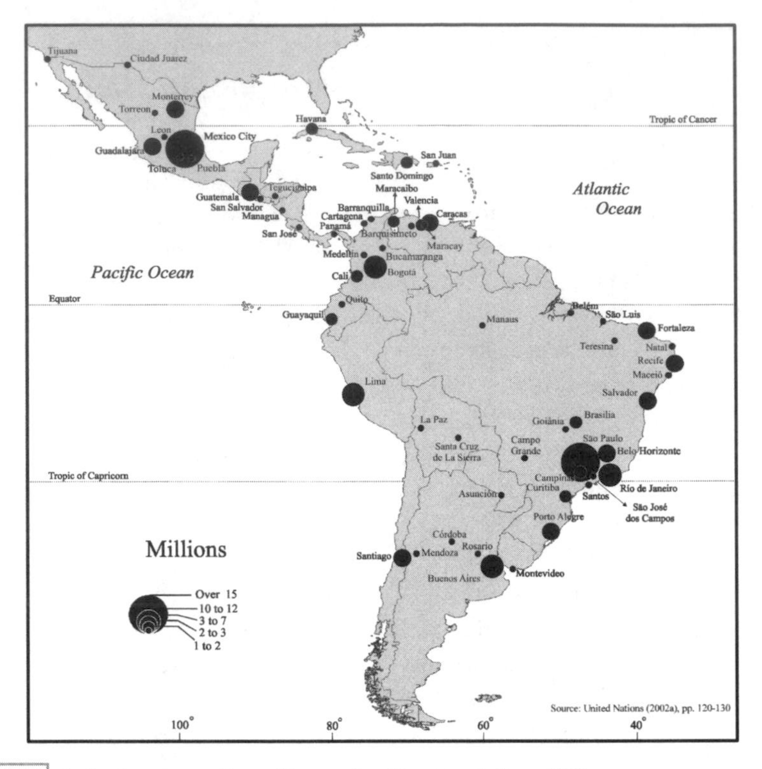

FIGURE 14.2. Latin American cities with over 1 million population, 2000.

this generalization almost completely. The U.N. demographers have projected that this pattern of urban growth will continue and that in 2010 there will be 65 cities in Latin America with populations over 1 million. Nearly one-third of the total will be in Brazil.

Characteristics of Million-Plus Cities

What are these "1 million-plus" cities like? The urban landscape and social and economic geography of these cities reflect a combination of influences but also reveal historic and environmental differences. While individual circumstances create distinct urban landscapes, there are commonalities shared by most metropolitan centers in Latin America. In the first instance, these are thoroughly modern places. Modern office towers characterize the city centers and suburban business districts of these cities. The upper classes, including professionals and successful business entrepreneurs and executives, share part of a world culture with similar social groups in other Latin American and Western countries. They follow national and international news; own handheld video camcorders; talk on cellphones; use satellite dishes to view movies, sports, and 24-hour news programming; and own home computers and surf the Internet (Figure 14.3).

The most affluent of the citys' residents live in spacious houses and apartments in "smart" (upscale) residential and commercial districts. Their homes are well appointed and feature the latest in modern appliances and conveniences. A significant proportion of the middle and upper classes hold university degrees, and many have received undergraduate or graduate degrees in the United States or

Europe. Vacation destinations for the well-to-do include major urban areas like New York City or Rio de Janeiro, theme parks like Disney World in Orlando, Florida, and international beach resorts like Cancún, Mexico, or Viña del Mar, Chile. Multiple automobile ownership, including luxury sedans and four-wheel-drive recreational vehicles, is a given for members of the upper class.

Stark contrasts characterize life in the modern Latin American city. A small proportion of the population, perhaps 10 percent, lives in a world as thoroughly up-to-date as anyone anywhere. The middle class, whose proportion of the population does not typically exceed one-quarter of the total, lives in reasonable comfort and dignity in most metropolitan regions. The accoutrements of contemporary living in an electronic age are employed in the homes of many, although these are often less varied, older, and fewer than among the upper classes. Refrigerators and televisions are standard household appliances, and computers are increasingly common. Living quarters are usually ample, with little need to double up to share bedrooms with other family members (Figure 14.4). Indoor plumbing, hot and cold running water, and often multiple bathrooms characterize these modest, functional homes and apartments. Many middle-class adults, perhaps as many as one-third, have university degrees or have some college or university education. Financial constraints usually limit their holidays and vacations to national destinations, but international attractions are well known, if not often visited. Automobile ownership is common among the middle class and is generally seen as a necessity. Car maintenance and operating costs often stretch individual or family budgets beyond acceptable limits, and cars may be sold or simply left to sit until better times make it feasible to operate them regularly.

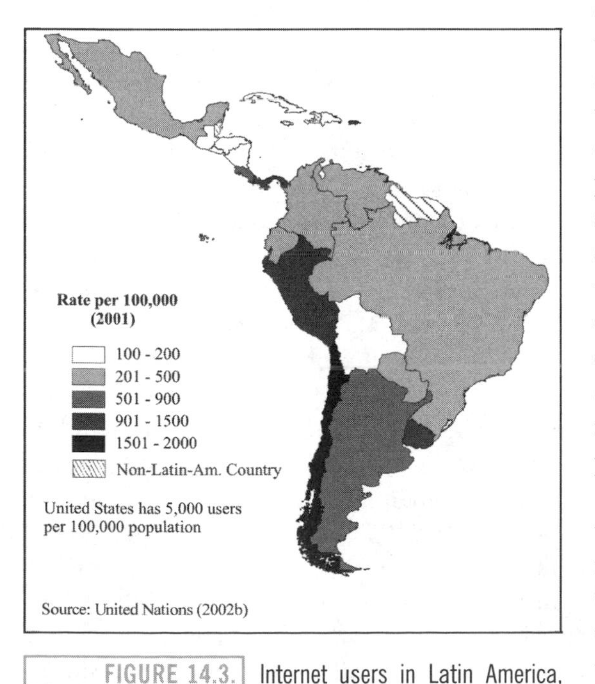

FIGURE 14.3. Internet users in Latin America, 2001.

The remainder of the population of major metropolitan centers, the working class and the poor, lives in more marginal conditions. A large proportion, one-third to one-half, is estimated to live in poverty by the standards set by local and national agencies (not by foreign countries or international organizations). Service and domestic work, semiskilled and unskilled labor, and self-employment through petty trading or artisan skills are typical of the employment options available. Part-time employment and underemployment are endemic among the working class and poor. Some work at regular wage labor in the formal sector of the economy where workers enjoy the benefits of social security and pension programs. However, the majority are engaged in the "informal sector" where wages are irregular and workers fall outside the umbrella of state or private pension programs.

The characteristics of residential quarters vary. The working class and the poor occupy many of the neighborhoods in the central districts of these metropolitan centers. In some urban centers, the urban poor occupy squalid tenements known as *conventillos* in Lima and Buenos Aires and *vecindades* in Mexico City.

FIGURE 14.4. Upper-middle-class suburban housing in San Juan, Argentina, 1993.

These generally consist of a series of single rooms or sets of interconnected rooms opening onto a common patio. They typically have a common water supply and share cooking and sanitary facilities located in or just off the patio.

The congestion of the central city and increasing population densities have led upper- and middle-class residents to view the center of the city as a less desirable residential location. Thus, they move to the peripheral residential districts, or suburbs. As a result, housing succession has occurred as buildings originally utilized by well-to-do and middle-class residents have been subdivided and transformed to accommodate much higher occupation densities. Shared sleeping quarters and multifamily or household cooking and toilet facilities are common in these circumstances. Access to electricity, although sometimes provisional and pirated, is almost universal, but access to piped water and sanitary facilities are not. While an important segment of the working class and poor live in the central neighborhoods of the metropolis, the majority live in vast peripheral settlements or towns that ring most of Latin America's million-plus cities.

Shantytowns

Shantytowns, or slums, emerge on the edges of the principal urban centers in almost all Latin American countries. They are called *favelas* in Brazil, *pueblos jovenes* in Peru, *barrios de emergencia* or *villas de miserias* in Argentina, and *callampas* in Chile (Figure 14.5). Regardless of the particular term, shantytowns have evolved as the basic means of meeting the housing needs of the majority of residents in Latin American cities over the last half century.

The poor rural migrants who inundated the urban centers of the region found few housing alternatives available in the city. As urban populations grew at rates of 4–5 percent annually, the deteriorated low-rent residential housing stock of the urban core quickly reached capacity. The private sector found inadequate financial rewards in the construction or management of low-cost housing, and built none. In the 1950s and 1960s, national government programs underwrote the construction of public housing blocks in many of the principal urban centers. While an appreciable number of housing units were constructed, the resources dedicated to the task were never sufficient to address the magnitude of the demand for low-rent housing. Public housing projects, with the exception of those in Havana, rarely housed a significant percentage of the urban population. By the 1980s, such projects had ceased to figure in the strategies urban planners proposed to solve urban housing problems.

On the urban fringe, open space, agricultural land, undeveloped land parcels, and marginal terrain like ravines, swamps, and steep hillsides proved critical in meeting the housing needs of the urban working class and poor (Figure 14.6). These vacant spaces offered the urban poor a means of resolving their housing needs by occupying the land through legal or, more commonly, illegal methods including squatting, planned land invasions, and land subdivision. Marginal lands and difficult terrain, bypassed by conventional urban development and with a low market value, proved attractive for squatting. In this way, settlers encountered little sustained or effective resistance from municipal authorities.

The settlement and urbanization of the more valuable agricultural lands and open space on the periphery of the city proved more contentious. Landowners and municipal authorities frequently resisted the occupation of these parcels, sometimes resorting to the forced eviction of squatters by police. Resistance and eviction led to greater organization and strategic skills on the part of the poor, who often employed massive land invasions under the cover of darkness as a common settlement strategy. The preliminary plotting of the town into lots and the construction of provisional homes of cardboard, sheets of tin, and straw mats occupied the hours of the first night as the hundreds and often thousands of the new settlement's residents worked frantically to build their new town. Daybreak invariably found the new settlement bedecked with the national flag and christened with a name likely to elicit sympathy from officials and the public. At sites where confrontation was likely, cordons of the new settlement's men, women, and children often surrounded the parcel in a protective ring. Police confrontations often ensued and the rapid clearance of the new settlement frequently occurred.

Nevertheless, a dogged persistence on the part of the poor sometimes brought them success. This came as either the resolve of municipal authorities to resist waned, as landowners settled for a sale price out of fear of receiv-

FIGURE 14.5. A shantytown, known as a *barrio de emergencia* in Argentina, located adjacent to a railroad right-of-way in San Juan, Argentina, 1993.

FIGURE 14.6. A low-income neighborhood near the waterfront in Iquitos, Peru. When the river rises, the simple latrines in the middle of the picture are inundated and canoes are a key form of local transportation, 1993.

ing no compensation for the loss of their land, or the offer of some other plot of land by municipal authorities at an alternative site made the invasion unnecessary. Squatting on marginal terrain and small parcels often occurs spontaneously by individuals, families, or small groups. However, community associations, political organizations, and religious groups provide the logistical and political skills required for successful large-scale land invasions. In these cases, the influence and leadership of these organizations provide an immediate social and political structure for the community, contributing the organization, vision, and stability often missing in marginal settlements.

If the settlers manage to resist the efforts of municipal officials or landowners to dislodge them, an uneasy truce often ensues with the settlement assuming a kind of de facto legitimacy due to its existence, even if the settlement is technically illegal and hence ineligible for the provision of municipal services like water, sewer, and electricity. Despite the absence of official municipal approval, streets and alleys take shape, and taxis, jitneys, and eventually buses provide access to the main public transportation system. Soon cistern trucks ply streets of the settlement or adjacent thoroughfares, selling potable water, while petty entrepreneurs in trucks sell steel canisters of natural gas for cooking. A trickle of electricity, albeit pirated, begins to reach the homes of some, as precariously and often dangerously strung lines tap the urban electric network.

The housing solutions created by marginal settlements for poor urban immigrants are often very positive. Property ownership is clearly an essential motivating factor for settlers. However, "ownership" is not really the correct term since settlers do not "own" anything in a strict legal sense. Nevertheless, the act of squatting or occupying a land parcel effectively establishes a certain de facto ownership, and each day that passes without challenge increases the security of those who occupy a property and their willingness to invest in it.

Settlers immediately begin the process of constructing a shelter on their land parcel. Initially, these dwellings are flimsy and provisional, but as individual and family fortunes improve or diligent saving pays off, hard mud floors are surfaced with concrete and tin walls are replaced with cinder blocks. An incremental process of residential construction and upgrading begins that can continue over decades. Family members plan and carry out the construction themselves, often employing outside assistance only for specialized tasks like plumbing and electricity. In time, the unadorned single-story cinderblock houses on dusty streets can be transformed into two-story structures on paved streets where some households even own automobiles.

De facto recognition of the legitimacy of the settlement often comes when municipal or other governmental authorities finally relent and permit the provision of basic public services like water, electricity, and sewers for the settlement. Land and property titles often remain murky, and many property owners may have no real documentation of ownership even after decades of residence. The conveyance of property—that is, its purchase, its sale, and the development of an active housing market—is severely impeded in the shantytowns of Latin American cities because of the widespread absence of legal property titles.

For much of the 20th century, most of Latin America's municipal authorities, national government officials, urban planners, and academics viewed the city's marginal settlements as a form of urban blight. Beginning in the late 1960s, however, a broad range of professionals began to appreciate the truth that marginal settlement and self-directed housing solutions could effectively satisfy much of the demand for urban housing and provide a number of positive benefits—especially de facto home ownership among a significant proportion of the urban working class and poor. Many subsequent public policy initiatives concerning the supply of urban housing have emphasized the provision of basic municipal services, property titling, and subsidies for self-help housing construction and improvement.

The growth of these squatter settlements and the proportion of the urban population housed in them has grown dramatically in the last decades of the 20th century (Table 14.3). The Latin American city has been truly transformed by their appearance. In most cities marginal settlements did not begin to appear until the 1930s and 1940s, their numbers were few, and they accounted for a very small proportion of the housing stock. The growth of *pueblos jovenes* or *barriadas* in Lima, Peru, is illustrative of that growth. In 1940, just five *barriadas* existed, but this number jumped to over 150 in 1960. Twenty-five years later, in 1965, the Lima metropolitan area had almost 800 marginal settlements! Today, with over one-third of its population living in self-help housing, Lima's situation is not extreme in the Latin American context. Indeed, it is probably just about average for a major urban center.

The growth in marginal settlements has been widespread. Today, in some major urban centers, it accounts for a majority of residential settlement. In 1950, for instance, about 15 percent of Mexico City's population was housed in marginal housing; this figure grew steadily in the ensuing decades, reaching about 60 percent in the early 1990s. Even in prosperous oil-rich Venezuela, similar growth occurred. In 1960, about 20 percent of the population lived in the *rancherías*, marginal settlements, and this grew to close to 60 percent in 30 years. Even in Argentina, traditionally one of the region's wealthiest nations, marginal settlement accounts for an appreciable component of the urban population. In Buenos Aires, 5–10 percent of the population now lives in these settlements on the edge of the metropolis. The growth of marginal settlements has been most notable where migration has been most intense, at the top of the settlement hierarchy in national capitals and in the most populous urban centers. But marginal settlements have appeared in all manner of cities and towns in the urban hierarchy. In Tucúman, Argentina, a

TABLE 14.3. The Growth of Self-Help Housing in Selected Latin American Cities, 1950s–1990s

City	Year	City population (thousands)	Population in squatter settlements (thousands)	Percentage
Rio de Janeiro	1947	2,050	400	20
	1957	2,940	650	22
	1961	3,326	900	27
	1970	4,252	1,276	30
	1991	9,696	921	10
Mexico City	1952	2,372	330	14
	1966	3,287	1,500	46
	1970	7,314	3,438	47
	1976	11,312	5,656	50
	1990	15,783	9,470	60
Lima	1956	1,397	112	8
	1961	1,846	347	17
	1972	3,303	805	24
	1981	4,608	1,455	32
	1989	6,234	2,338	38
Buenos Aires	1956	6,054	10	2
	1970	8,353	434	5
	1980	9,766	957	10
	1991	10,911	659	5
Caracas	1961	1,330	280	21
	1964	1,590	556	35
	1971	2,200	867	39
	1985	2,742	1,673	61
	1991	2,966	1,238	42
São Paulo	1973	6,561	72	1
	1980	8,493	321	5
	1985	8,929	440	6
	1987	9,109	813	8
	1991	9,483	1,050	9
	1989	10,436	3,238	31
Bogotá	1955	917	367	40
	1965	1,780	766	43
	1975	3,069	921	30
	1985	4,123	1,278	31
	1991	4,824	1,254	2

Source: Gilbert (1996), pp. 74–75.

major regional center, marginal settlements spread southward for several kilometers along the main north–south highway that skirts the Andean piedmont. In coastal Peru, these settlements appear at the edges of a long string of intermediate cities and towns that punctuate the coastline. Even cities with populations as low as 100,000 have experienced the addition of marginal settlements to their urban landscapes.

These approaches to urban housing provision prove cost-effective from a societal perspective. As a consequence, both municipal and national governments have found it convenient to ignore building codes, land subdivision requirements, public utility service provision standards, and other legislation and administrative policies governing marginal settlement, thereby providing tacit approval to the existence of these settlements.

Settlement policy is political, and, in Latin America's major metropolitan areas, the poor have some political power. Not much, but enough that eventually marginal settlements acquire basic municipal services and even a few urban amenities. Governmental public policy initiatives on housing have focused increasingly on providing assistance for individual homeowners to upgrade and expand housing units and on providing a range of community improvements within marginal settlements.

Surprisingly, Latin American cities are distinguished by a high rate of home ownership, reaching as much as two-thirds in large urban places like Mexico City, Caracas, and Santiago, and close to one-half in many other metropolitan centers. These rates far exceed the rates of home ownership in many other world regions. In Latin America, home ownership contributes an important element of social and political equilibrium to societies in which the discrepancies in living standards and wealth between the rich and the poor are far more extreme than in Western Europe or in the United States.

Marginal settlements are dynamic places. This fact is sometimes overlooked in efforts to summarize the significance and magnitude of marginal settlements in the modern Latin American city. Many evolve over a period of several decades from ramshackle collections of provisional huts and shelters into prosperous lower-middle-class suburbs. Today many of these neighborhoods bear no resemblance whatsoever to their appearance two or three decades ago. For instance, the municipal district of San Martín de Porras in Lima was created initially as a marginal settlement in the 1950s, but by the 1990s it had evolved into a well-established working-class neighborhood. In Lima and other cities, marginal settlements frequently develop in this fashion. However, all are not necessarily so fortunate. In some cases, the marginal settlements of the 1950s and 1960s have become the fetid slums of the 1980s and 1990s. The *favelas* that cloak many of the steep slopes surrounding Rio de Janeiro remained marginal settlements throughout most of four decades. It was not until the 1990s that municipal authorities and international lending agencies began programs to provide basic services to many of these *favelas*.

Urban Development and Form

Until the 1950s, most of Latin America's cities, even its largest, were essentially pedestrian cities, geographically compact and managed with comparative ease on foot. Horse-drawn trolleys and then, later, electric-powered trams and trolleys contributed to the expansion of some of the region's largest cities along major transportation axes in a pattern reminiscent of the development of urban centers in the United States at the end of the 19th century. Buenos Aires, São Paulo, Rio de Janeiro, and Lima all experienced this kind of development to varying degrees. Most grew slowly, however, so the process of suburbanization occurred gradually in most major urban areas. With the exception of Buenos Aires, where the first subway line was completed in 1914, subway lines made a very late appearance in Latin American cities. Mexico City, São Paulo, Rio de Janeiro, Caracas, and Santiago all boast subway lines of varying degrees of complexity constructed between the late 1960s and late 1980s (Figure 14.7). In general, these subways have had little influence on the overall pattern of urban growth.

Rather, it was the increasing availability of

FIGURE 14.7. Subway passengers in Santiago, Chile, 1993.

automobiles to members of the upper class and the use of buses as the principal means of public transportation for the middle and working classes after 1950 that determined the sprawling pattern of subsequent urban development in Latin American cities. Initially, broad avenues, boulevards, arterial streets, and access roads and highways provided the principal means of ingress and egress from the city center. In the 1960s the largest metropolitan centers began the construction of multilane limited-access highways following, or at least similar to, the pattern of urban expressway development in the United States (Figure 14.8). In most cities, however, the expense and political difficulties inherent in their construction has severely restricted the development of these highways, and they reach only a small portion of the urban area.

While large metropolitan centers are common in Latin America and almost all nations have at least one city whose population exceeds 1 million (Figure 14.2), as a rule, most Latin America cities exhibit low-density development that is geographically extensive. Lima, Mexico City, and São Paulo are especially notable examples; these cities sprawl over hundreds of square kilometers. High-rise development is focused in city centers, along upscale boulevards, and at a few nodes in the wealthy suburban districts. The high-rise buildings are primarily used for commercial office space and government offices; few urban dwellers live in high-density apartments. Medium-density residential dwellings, two- to four-story apartment buildings and the like, account for most of the residential living space in the central city and in the immediately adjacent districts. One- and two-story dwellings, usually built immediately adjacent to each other, with abutting buildings or property lot lines on all sides, dominate the residential housing stock in the metropolitan area's suburban districts. Higher density housing, often two- and three-story apartment buildings, concentrate along the principal boulevards and avenues.

In the most prosperous and well-heeled suburbs, lower density, free-standing (detached) single-family dwelling units predominate. Most of these homes have been constructed since 1950, and many within the last two decades. These dwelling units, and to a significant measure the residential suburbs in which they are found, clearly reflect the importance of U.S. cultural and economic influence in Latin America during the 20th century.

FIGURE 14.8. A multilane limited-access highway running through the center of São Paulo, Brazil, 2004.

The widespread adoption of the bus and automobile as the principal means of urban transportation by the middle of the century permitted the geographical expansion of the city over a large area, a pattern typical of most U.S. cities. In Latin America, this pattern of urban development benefited both the rich and the poor. The rich built lovely suburban districts, studded with spacious and sometimes even palatial detached single-family homes surrounded by well-groomed lawns and attractive gardens. And they used private automobiles to get there. The poor, on the other hand, could ride buses to the edge of the urban fringe where they found ample space to establish marginal settlements and build their homes. The result is the sprawling cities found in most of Latin America today.

Urban Primacy and Primate Cities

The urban systems of many Latin America nations are characterized by a strong tendency for the major urban center, often the capital, to exert an overwhelming influence over the country's economic, social, and political life. *Urban primacy* is the term used to describe this phenomenon, and the cities that demonstrate this characteristic are called *primate cities*.

Precisely what are primate cities and how is urban primacy identified? A primate city exhibits an overwhelming role in the national system of a country; indeed, often the city is viewed as almost synonymous with the country. It is, for instance, inconceivable to imagine Mexico without Mexico City, Argentina without Buenos Aires, or Peru without Lima. Like many other Latin American capitals, all are primate cities. When a primate city dominates the urban system, the nation is often viewed as being divided into just two parts: the capital (primate city) and the rest of the country. For example, a geographer of the author's acquaintance who lives in Peru divided his address book into two parts: "Lima" and "Outside Lima."

Population data are most commonly used to test or demonstrate urban primacy. A variety of measures are employed to illustrate or suggest primacy, as well as to test it more rigorously. The percent of the total national population living in the urban area provides one of the most straightforward means of assessing the relative dominance of a particular city in the nation and its urban system. Frequently, primate cities account for at least one-fifth of the nation's population. This figure can reach close to one-third of the total, as in the cases of Buenos Aires, Lima, and Santiago.

One of the most satisfactory, and common, approaches to examining primate cities is to simply divide the population of the largest city in a nation by the population of the next largest city. This is known as the "two-city primacy ratio." It provides a statistic that reveals how many times larger the top city in the urban hierarchy is than the next largest city. Urban systems are considered "balanced" when the resulting ratio approximates what is often called the "rank-size rule." Thus, in urban systems where the largest city demonstrates little or no primacy, the population of the second largest city will be about one-half of the largest, the third largest will be about one-third the size of the largest, and so on. In larger nations, more than one city might occupy the second and subsequent tiers. Although there is no specific ratio that is used as a benchmark for identifying urban systems dominated by a primate city, once the largest city is over five times as large as its nearest competitor, urban systems are generally considered to demonstrate primacy. Using this criterion, many Latin American nations have urban systems that exhibit tremendous primacy (Table 14.4). The capital cities of Paraguay, Chile, Guatemala, Uruguay, Costa Rica, Nicaragua, Argentina, and Peru all have primacy ratios that ex-

TABLE 14.4. Two-City Primacy Ratios, Latin America, ca. 1990

Country	Primacy	Country percent urban	First city (capital) percent urban
Dominant Primacy (6.00 or over)			
Paraguay	25.8	62	44
Chile	15.2	81	44
Guatemala	14.7	39	36
Uruguay	14.6	84	52
Costa Rica	11.3	46	64
Nicaragua	10.5	53	47
Argentina	10.4	83	45
Peru	10.4	65	39
Strong primacy (4.00–6.00)			
Panama	5.7	51	66
Cuba	5.4	68	38
Weak primacy (2.00–4.00)			
Dominican Republic	3.6	51	54
Venezuela	3.3	84	26
Bolivia	2.8	44	44
Colombia	2.8	64	26
Honduras	2.4	37	33
El Salvador	2	39	22
Little or no primacy (below 2.00)			
Ecuador	1.5	55	29
Brazil	1.4	66	1

Source: Davis 1970 and U.N. 1980, as cited in C. E. Browning (1990), p. 74.

ceed 10, while Mexico, Cuba, and Panama have primacy rates that exceed 5.

Population data provide a consistent and comparable means of identifying primacy, but they do not give much sense of the immense impact and role a primate city plays in all aspects of a nation. In many ways the saying "All roads lead to Rome" aptly describes the relationship between a primate city and its hinterland—which is typically the entire national territory! The transportation and communication networks focus sharply on the primate city. The primate city is the hub for the national highway system, as it is also for nearly all bus routes, train lines, and air services. The hub-and-spoke network of railroad lines that emerged in Argentina in the early decades of the 20th century is an oft-cited example of this pattern.

These patterns are not simply historic relics; rather, they characterize the transportation systems of almost all countries with a primate city at the end of the 20th century. In the 1990s, in Argentina, a national hub-and-spoke network characterized air travel. A trip between almost all provincial capitals, including a number with populations exceeding 1 million, required a flight from the city of origin to Buenos Aires, and then a second flight outbound from Buenos Aires to the final destination.

Primate cities account for a far greater proportion of their nation's economic activity than their population numbers might suggest. Slightly more than 20 percent of Mexico's population resided in Mexico City during the mid-1980s in a geographical area representing just 0.01 percent of the national territory. Yet at the same time the city's economy generated 38 percent of the gross domestic product, 45 percent of commercial activity, 48 percent of manufacturing, 52 percent of services, 60 percent of transport services, and almost 70 percent of bank assets. The preeminence of Mexico City in Mexico's economy is similar to that of Lima in the economy of Peru. In the late 1980s, Lima accounted for an astounding 70 percent of the gross domestic product, 87 percent of tax collections, 83 percent of bank deposits, 98 percent of private investment, 73 percent of physicians, 48 percent of hospital beds, and 62 percent of university students.

Population and economy, however, represent just two of the spheres in which primate cities play a dominant role. Despite some efforts at government decentralization, essentially all government decisions of any consequence are made or ultimately approved in the primate city, which is also invariably the national capital. Local and regional officials travel frequently to the primate city to lobby national government officials and to expedite the approval of budgets and infrastructure projects. The presence of central government offices in the primate city encourages many business and industrial enterprises, especially larger companies and those oriented toward exterior markets, to locate their headquarters in the primate city. Access to transportation facilities, a superior labor pool, dependable and higher quality public services—especially electricity—and a range of other agglomeration factors favor business and industrial location in the primate city. Finally, the primate city will be the center of a nation's cultural and social life. The most prestigious universities, the best museums, the most important artistic institutions, and the most respected scholarly societies invariably concentrate in the city. The city also serves as the center of the social world for the nation's business, governmental, and cultural elite.

Urban primacy and primate cities are not unique to Latin America. A number of European countries have urban systems that exhibit a high degree of primacy. These include Great Britain (London), France (Paris), Austria (Vienna), Denmark (Copenhagen), Hungary (Budapest), and Ireland (Dublin). Over half of Latin America's nations have urban systems dominated by a primate city, while comparable percentages for Europe and Asia stand at just about one-quarter of the regions' nations.

Urban primacy in Latin America has strong colonial roots. Almost without exception, the cities that have evolved into primate cities were capitals of one or another colonial administrative division: viceroyalties (Mexico City, Lima, and Buenos Aires), audiencias (Santo Domingo), and capitancy generals (Santiago and Guatemala City). The aftermath of political independence from Spain tended to reinforce the dominance of these cities as their nations' burgeoning agro-export economies continued to focus economic growth and political power in the national capital.

Primate city systems did not evolve in all Latin American countries. Indeed, several have urban hierarchies whose city populations come close to fitting the "ideal" model of the rank-size rule. While no direct link between geographical size and primacy can be demonstrated, larger nations, like Brazil, tend not to demonstrate high levels of urban primacy. Logically, urban primacy is less likely to occur when a nation's political and economic development is more evenly balanced among a series of regions and their respective urban centers. In South America, Brazil, Bolivia, Colombia, Venezuela, and Ecuador all exhibit low rates of urban primacy, while in Central America a similar situation prevails in El Salvador and Honduras.

Is urban primacy important, or is it simply an academic curiosity? On the one hand, many urban planners and academics had lamented the growth of primate cities, suggesting that the concentration of power and resources in the capital chokes off and stunts economic and social development in the balance of the nation, especially in secondary cities. In the primate city, the multiple effects of agglomeration economies in the labor market, the transportation sector, financial markets, and governmental and private decision making are almost impossible to overcome elsewhere.

Other observers have worried that primate cities, especially very large ones, generate a disproportional share of serious urban problems. Obviously, scale economies in service provision can reduce the marginal costs of the provision of many public and private services. But many argue that the largest primate cities no longer generate economies of scale but rather diseconomies of scale in which the marginal costs of service provision increase disproportionately. In Mexico City, for instance, garbage collection is totally inadequate, and in the city's poorer neighborhoods where garbage collectors' opportunities for tips are few and the garbage provides little to recycle or resell, garbage collection is irregular, and refuse and rotting garbage collects on street corners and vacant spaces. The transportation system of many primate cities strains under the intense demands of the public to commute from home to work and back again. The working poor often pay a tremendous cost in their time for inadequate transportation systems. In Mexico City and Buenos Aires, for example, the working poor often spend 4–5 hours a day commuting. The air pollution problems of Mexico City are legendary, but they would be serious even if the city were half its size. However, these problems are not unique to primate cities. Rio de Janeiro, São Paulo, and Bogotá all suffer from serious problems of pollution, public sanitation, and transportation gridlock, but none are primate cities.

Urban Economy: Formal and Informal Sectors

The urban economies of Latin America's cities are complex, but some broad similarities help us to understand their present characteristics. One useful approach at conceptualizing the urban economy suggests that it can be best understood if it is envisioned as being divided into two distinct but interrelated components. This dichotomy is useful in creating a rough conceptual model of the urban economy (Table 14.5).

One sector, often called the *formal sector*, or the upper circuit, is comprised of commercial, industrial, financial, transportation, and other enterprises that participate legally in the official economic system. These businesses produce manufactured products; provide financial, banking, and insurance services; and provide modern transportation and communications services and technology. Their factories, products, brand names, property, and vehicles are legally registered and comply, at least nominally, with official regulations and standards for health, safety, social security, and pension plans (Figure 14.9). The business methods of the industrial firms, financial institutions, and commercial enterprises that comprise the formal sector reflect modern capitalist business practices. These are capital-intensive enterprises, which make use of advanced technology in production and administration and produce standardized products at

TABLE 14.5. The Characteristics of the Upper and Lower Circuits of the Urban Economy

	Upper circuit	Lower circuit
Technology	Capital-intensive	Labor-intensive
Organization	Bureaucratic	Primitive
Capital	Abundant	Limited
Labor	Limited	Abundant
Regular wages	Prevalent	Exceptional
Inventories	Large quantities, and/or high quality	Small quantities, poor quality
Prices	Generally fixed	Negotiable between buyer and seller (haggling)
Credit	From banks, institutional	Personal, noninstitutional
Profit margin	Small per unit, but large turnover	Large per unit, but small turnover
Relations with customers	Impersonal and/or on paper	Direct,personal
Fixed costs	Substantial	Negligible
Advertisement	Necessary	None
Reuse of goods	None (waste)	Frequent
Overhead capital	Essential	Not essential
Government aid	Extensive	None or almost none
Direct dependence on foreign countries	Great, externally orientated	Small or none

Source: Santos (1979), cited in Bromley and Bromley (1988), p. 85.

FIGURE 14.9. Formal-sector activities in São Paulo, Brazil. A grocery store, Barateiro, and an armored car from a security firm, Protege, 2004.

fixed prices. The size of these business enterprises varies from small to very large, although medium-sized and large firms dominate and employ most of the labor in this sector. Overall, this sector generates only a modest percentage of all urban employment, perhaps one-quarter of the total, but it produces as much as three-fourths of the "gross regional product" of the urban economy. The government sector—national, regional, and local—while often not quite as modern or efficient as the private sector, is also an integral part of the formal sector. Working in the formal sector in Latin America is similar in most respects to the experience of the working population in the United States, Canada, and Western European countries.

The *informal sector*, or the lower circuit, is the term used to conceptualize the other major component of the urban economy. The informal sector produces a wide range of goods and services. Characteristically, individual artisans, small-scale entrepreneurs, and petty capitalists provide these products and services. Retailing in the informal sector typically consists of stationary or ambulatory street vending (Figures 14.10 and 14.11). Petty merchants will spread their goods out on a sidewalk, in a park, or in the street on the margins of a busy market or commercial district. They sell a wide variety of goods ranging from CD players to books, magazines, lottery tickets, and toiletries. Ambulatory street venders carry bags or small trays of merchandise hung around their neck, proffering clothes, cassettes, candy, cigarettes, chewing gum, and a myriad of inexpensive merchandise to passersby.

Informal-sector artisans work from small shops or makeshift shelters in markets, on street corners, or in other nooks and crannies of the urban landscape. They tailor clothing, repair shoes, and produce a plethora of simple products bought by the working class and the poor. Entrepreneurs play a role in the informal sector as well, although their financial resources and scale of operations are smaller than those of entrepreneurs in the formal sector. Modest workshops and industrial firms employing a small number of workers prosper in the informal sector. Unlicensed taxis, jitneys, and minibuses provide much of the available public transportation. The vast majority of the domestic service workers who are found in most upper- and middle-class homes—maids, cooks, housekeepers, nannies, gardeners, and the like—are part of the informal sector. Similarly, many day laborers, unskilled construction workers, and even many skilled workers toil in the informal sector.

By definition, the business operations and

FIGURE 14.10. A market vendor sells yucca and other tubers under a tarp in a street market in Cajamarca, Peru, 1985.

FIGURE 14.11. Happy customers pose with a smiling street vendor selling ice cream in Potosí, Bolivia, 1997.

the economic and financial activities of the informal sector occur on the margins of the legal or official economy. Informal-sector microbusinesses are generally not licensed or registered with municipal or national authorities, nor is it probable that their products or services meet official health, safety, or sanitary conditions. These enterprises are not likely to pay business taxes to local or national authorities, nor are they likely to charge sales or value-added taxes to their customers. Financial transactions take place almost exclusively on a cash basis. Financial tools like business loans, lines of credit, credit cards, checks, and bank accounts are alien to business operations in the informal sector. Employees, if they exist, are paid in cash and no employer contributions are made for government pension or social security programs. Business enterprises in the informal sector employ simple technology, involve little capital, and are labor-intensive. Products are not standardized, nor are prices fixed. Business hours and days of operation are flexible, but tend to be longer than those of businesses in the formal sector.

The terms and conditions of work also vary dramatically between those in the formal sector and those in the informal sector. Workers in the formal sector usually work regular hours or some kind of standard shift. They are paid regularly, usually with checks and increasingly through electronic deposits to their bank accounts. Income taxes, pension contributions, and social security payments are deducted directly from their pay. These workers are often organized into some form of labor union, whether they are governmental employees or industrial workers. The economic status of these workers and their relationship to the formal economy is similar to that of most members of the middle class in other modern capitalist countries at the end of the 20th century.

In the informal sector, a large percentage of all workers are essentially self-employed. This includes almost all petty retailers, street-corner artisans, and a whole set of individuals providing personal services: bootblacks, car park attendants and "guards," and prostitutes. Others—domestic servants, unskilled laborers, and skilled workers—are employed by others and paid cash wages, but the customary terms of employment are limited to hourly or daily compensation or pay at piece rates. Typically there is no other form of compensation whatsoever and no participation in government pension and social security programs.

The informal sector employs as much as one-half to two-thirds of the population in some Latin American urban centers, especially in those countries that have lower per capita GNPs and standards of living. In the metropolitan centers of the more developed nations, employment in the informal sector may be as low as 10 percent of the labor force. Part-time employment and underemployment are widespread in the informal sector. Many individuals may pursue more than one informal-sector employment option while also working in the formal sector. The informal sector plays a critical role in the operation of metropolitan and national economies.

It provides employment and a livelihood for millions for whom there is no place in the formal capitalist economy.

In Latin American cities the relationship between the formal and the informal sector is symbiotic; each is dependent upon the other for its survival. For example, a large proportion of the consumer goods manufactured by industrial firms in the urban economy are sold by informal-sector merchants in markets, from street stands, and from pushcarts. The informal-sector ice-cream vendor sells ice-cream bars on the street corner produced in factories by formal-sector food-processing firms. The factory-made cigarettes, candy, clothes, watches, radios, cassette tapes, and the like produced in the manufacturing enterprises of the formal sector find markets in the hands of informal-sector vendors. Similarly, the workers who earn their living in the informal sector use immense quantities of consumer goods ranging from toothpaste and processed foods to small home appliances—all produced in formal-sector factories. Domestic service provides another key linkage between the two sectors. The upper and middle classes entrust their automobiles, kitchens, homes, and even children to a veritable army of informal-sector domestic workers. Thus, some of the income from their formal sector salaries is cycled into the informal sector, providing employment for many millions of men and women in metropolitan centers across Latin America.

This pattern is most pronounced in the metropolitan areas of countries with lower per capita incomes or where extreme income contrasts exist between metropolitan areas and their hinterlands. The ability of the informal sector to foster innovation, utilize scarce resources, and absorb large numbers of workers, albeit at wage levels and under working conditions that are uniformly inadequate, has increasingly led government planners and economists to seek ways of promoting informal-sector activities and integrating them into the formal sector.

Models of Latin American Urban Structure

Latin America's major metropolitan centers are gigantic urban places that extend over vast areas. On one level they are chaotic. Often they are characterized by a pattern of land use showing sharp contrast and even conflicting land uses in close proximity. Nevertheless, urban centers in Latin America share commonalties in land-use patterns, internal structure, and morphology. In Spanish America, the central plaza and the rectilinear-grid street system imposed by the colonial Laws of the Indies on new town establishment in the Americas still define the majority of urban places.

At the present, in cities with populations of less than 150,000, land use usually still follows the patterns established during the colonial period, with the central plaza serving as the urban center. The plaza is the religious, political, and social center of the town. The church or cathedral is located on the plaza, as are the offices of the representatives of the central government, the national police, and the local government (e.g., *municipalidad* or *ayuntamiento*). The plaza, often graced by a bandstand, gazebo, fountain, or statue, may also be planted with flowers, shrubs, and other ornamental plantings. In almost all cases, however, park benches will be distributed throughout the park. As such, the plaza constitutes perhaps the single most significant public social space in a small- to medium-sized Spanish American city today.

The principal business district typically runs off one of the main streets connecting to the plaza or closely adjacent to it. Residential patterns still retain some of the characteristics of the colonial period, although rapid urban growth and the mobility provided by the bus and automobile have transformed residential patterns. Some upper-class and elite residence still centers in the downtown, sometimes on or closely adjacent to the central plaza; at the same time, and in the same vicinity, many upper-class residences have been converted into apartments and boarding houses. Centrality, nevertheless, traditionally ranks high in decision making regarding residential location. As a consequence, increasingly peripheral residential locations in the city often reflect lower social status.

However, several factors have created numerous exceptions to this rule in the last 50 years, especially in large metropolitan areas. While squatter settlements and the modest dwellings of recent arrivals to the urban center tend to dominate the urban periphery, upper-class suburban housing developments and gated compounds for the wealthy are also found on the urban fringe. State-financed housing for the middle and lower classes is also often constructed on the fringe where land prices are lower and development costs less.

At the beginning of the 21st century, the urban land-use patterns, internal structure, and morphology of the major urban centers in Spanish America—those with populations that exceed 150,000—only vaguely reflect the heritage of the central plaza or the rectilinear street grid mandated in colonial times. In Portuguese America, urban centers with populations exceeding 150,000 have also outgrown their early colonial centers that were typified by small plazas, colonial churches, and narrow winding streets. These cities have expanded far beyond such limits, often giving rise to urban chaos. While broad-brush statements about the patterns of land use and urban morphology in Latin America's metropolitan centers are somewhat difficult to make, it is possible to identify some generalizations that hold across national frontiers and are consistent throughout the region.

Perhaps the most widely accepted model of land use, structure, and form for Latin American cities, at least among Anglo-American geographers, appeared in the early 1980s. The model, originally proposed by U.S. geographers Griffin and Ford, conceptualizes the Latin American city as being divided into six major zones (Figure 14.12); a subsequent revision of the model by Ford added three additional land-use categories. The first of these, located at the center of the urban area, is the *central business district* (CBD) that is focused on the city's traditional colonial center and incorporates the city's original street grid. Often an extension of this area, the *spine*, radiates outward from the core along a key transportation arterial, usually a major boulevard. A wide range of activities including wholesaling, retailing, industry, and government services all occur in the CBD and the Spine. A wedge of elite residential land use parallels the Spine from the center of the city and typically extends to the urban periphery. Modern business and shopping districts, often competitive with the CBD and replete with modern shopping malls and office complexes, often grow up near the end of this wedge on the urban periphery. Miraflores in Lima, Peru, Sopocachi in La Paz, Bolivia, and Los Condes in Santiago, Chile, are all examples of this pattern (Figure 14.13).

While a wedge of upper-class residences follows an axis in one direction in most large metropolitan centers, one or two wedges of contrasting land use also radiate from near the CBD in opposing directions. Polluted river courses, garbage-filled ravines, and busy, dirty, exhaust-choked highways frequently characterize these land-use wedges. The range of land uses runs from squalid housing to heavy manufacturing. At the city edge, these *mixed land-use wedges* grade imperceptibly into the squatter settlements that ring most of Latin America's largest metropolitan areas. In recent decades, efforts at industrial planning and development have given rise to industrial parks on the peripheries of many centers. Similarly, freeways, toll roads, or highway bypasses that circumvent the urban core often run through these peripheral squatter settlements and ring the

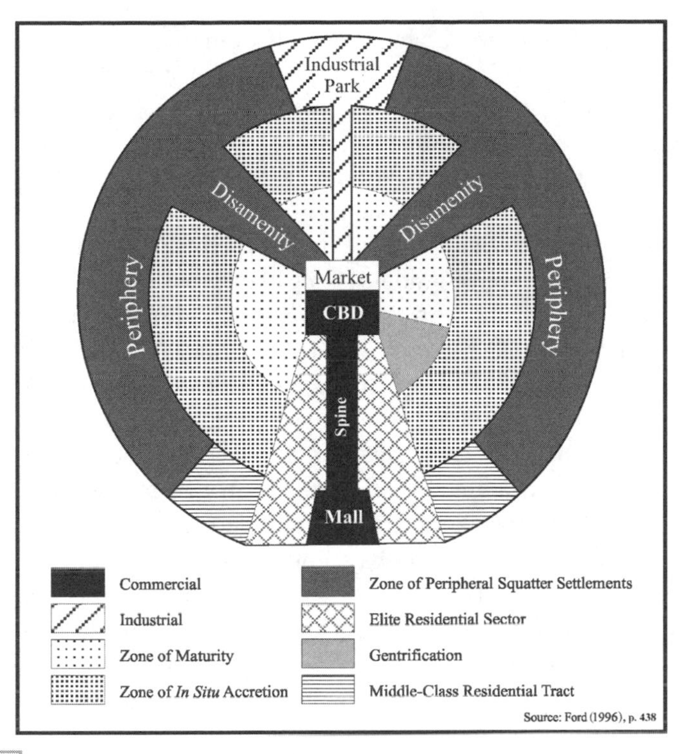

FIGURE 14.12. A model of Latin American city structure. Reproduced from *Geographical Review* with the permission of the American Geographical Society.

city. Besides this ring, two other roughly circular land-use zones surround the CBD.

Immediately adjacent to the CBD lies the *zone of maturity*. Once the site of middle-class and upscale residences, as well as commercial activity, this area has been transformed as the metropolitan center has grown. Accessibility to the CBD is the attraction of this area, which generally tends to be marked by low-quality housing occupied by the poor

FIGURE 14.13. Avenida Centroamerica, an urban "spine" in Tegucigalpa, Honduras, 2002.

and working class. Former large homes have often been subdivided into flats, and in some instances high-density low-cost housing has been constructed. Often these neighborhoods have evolved into full-fledged inner-city slums. Formal businesses, offices, informal street vendors, transportation services, and a myriad of small manufacturing establishments all coexist, although not necessarily harmoniously, in the zone of maturity that surrounds the CBD. In some cities, a small segment of the middle and upper classes have found the accessibility of the area attractive and have sought housing here, setting off a small-scale gentrification process. In Buenos Aires, for instance, in the zone of maturity, warehouses along the city's waterfront have been converted into luxury apartments, fashionable restaurants, and upscale boutiques and specialty shops.

A broad zone roughly resembling a concentric circle lies between the zone of maturity and the peripheral ring of squatter settlements, industrial development, and transportation infrastructure on the edge of the city. In the model proposed by Griffin and Ford this zone, called the *zone of in situ accretion*, is dominated by residential neighborhoods of varying qualities and inhabited by different social classes.

Typically, these include lower-working-class neighborhoods, or *barrios*. Originally many were squatter settlements, but they have been upgraded and transformed into substantial urban neighborhoods over the course of several decades as the metropolitan region has expanded. Middle- and upper-middle-class neighborhoods are also found here. In contrast to working-class neighborhoods that are typically unplanned and developed somewhat haphazardly, these residential areas have evolved from a series of planned suburban developments, similar in some respects to those found in urban areas in the United States and Canada.

Commercial activity, largely services and retail establishments, concentrate along main transportation arteries, but neighborhood stores, especially small establishments selling foodstuffs and other necessities, are frequently found on nearly every city block. In many larger metropolitan areas, the last two to three decades have also seen the construction of nodes of high-rise buildings in this zone of in situ accretion. These high-rise complexes are usually located at or near the intersection of major traffic routes and include upscale apartments, office space, and commercial establishments on the ground floor.

Although common patterns of land use and urban morphology can be identified among Latin America's principal urban centers, it is important to underscore the fact that no single model can adequately describe the range of land-use patterns and urban structures found in the region's cities. The model proposed by Griffin and Ford provides a quick and easily understood thumbnail sketch of urban structure. Other scholars have focused on the land-use patterns and morphology for cities in specific countries, in geographic subregions, or for specific functional types—for example, port cities. Thus, for example, models have been proposed to explain the urban structure of Argentine cities, Mexican cities, and Mexican border cities.

Planned Cities

Planned cities have appeared throughout Latin America since the 1880s. Characteristically, they have been built from scratch according to a master plan, often of grandiose dimensions. Most commonly, these planned cities have been built to serve as national or provincial (state) capitals. Brasília, which became Brazil's national capital in 1960 when its construction was completed, is undoubtedly the best known and the most elegant of these planned metropolises. Grand avenues, monumental architecture, parks, promenades, and impressive ceremonial spaces typify Brasília, as they do other planned cities.

Brasília, however, was not the first of these planned capitals. Brazilians have established a good number of planned cities, most of which are found in the nation's interior due to the fact that much of its early development focused on the coast, and the settlement of much of the interior occurred centuries later. Belo Horizonte, constructed to replace Ouro Prêto as the capital of Minas Gerais, was completed in 1897. Others followed in subsequent decades: Goiânia, the capital of Goiás, was finished in 1933 and Boa Vista, the capital of Roraima, was founded in 1943. Outside of Brazil, few Latin American nations have demonstrated much enthusiasm, or experienced much need, to plan new national or state capitals. La Plata, the capital of the Argentine province of Buenos Aires, is one of the few exceptions. Established in 1882 after the city of Buenos Aires became the nation's capital, the city's plan is strikingly similar to Washington, D.C., and Paris.

Economic development policies and growth-pole strategies gave rise to a number of efforts to develop planned industrial cities in the second half of the 20th century. Volta Redonda, a city of about 250,000 in the state of Rio de Janeiro, Brazil, was planned and constructed as a steel production center during the late 1940s. Other planned industrial cities include Ciudad Guayana in Venezuela, which has a population of nearly 500,000. This city, situated along the Orinoco River in Bolivar state, was developed as a steel and aluminum production center in the 1960s. Albeit smaller, with a population slightly more than 100,000, Ciudad Lázaro Cárdenas in Michoacán, Mexico, is another example of a planned industrial city specializing in industrial metals. Built around a government-financed iron and steel plant and a modern port, the city took form in the early 1970s.

Although as yet there are few examples, the latest round of planned city development in Latin America has taken a distinct twist, focusing on a different sort of economic development: foreign tourism. Mexico's government has led the way by constructing the tourist resort and city of Cancún on the Caribbean coast of the Yucatan Peninsula in the early 1970s. Cancún has since become one of the most popular tourist destinations in Mexico. Cancún's success spurred a similar effort by the Mexican government some years later, although on a smaller scale. Along the country's southern Pacific coast, in the state of Michoacán, the government planned and developed the tourist town and resort of Ixtapa in the late 1970s.

Summary

In the early decades of the 20th century, most Latin Americans were rural. Structural changes in the 20th century have transformed Latin American society to a largely urban, quasi-industrialized society in the first decade of the 21st century. The rapid rate of urbanization can be attributed to improvements in public health, population growth, and a stagnant rural sector. The economy diversified in the early 20th century, with industry growing in importance. Urban centers have grown rapidly, and by the mid-1990s Latin America had some of the most highly populated cities in the world. Because Latin American cities typically cover large areas, urban population densities are generally low.

The people of these large cities live widely contrasting lives. The 10 percent who are wealthy and educated live a cosmopolitan, high-tech existence. The middle class, some 15–20 percent, also live fairly comfortable lives. But the great bulk of urban residents live in poverty. While the affluent tend to live in suburban areas, some of the poor live in the central city. Most of the poor, however, live on the urban periphery in shantytown settlements. The growth of shantytowns is most intense in those cities where migration from rural areas and small towns is greatest. Home ownership, however, occurs more often in Latin American urban areas than it does in other areas of the globe.

Until the 1950s, cities were largely pedestrian with some public transportation. The increasing availability of the automobile introduced suburban sprawl in the second half of the 20th century. Often one major city, usually the national capital, dominates the country's economic, social, and political life. These cities are known as primate cities. The current urban economy can be understood as two sectors, one formal that is a legal participant in the economic system, and the other informal, comprised of day laborers, street vendors, and small-scale artisans and entrepreneurs who operate at the margins of the formal economy.

Further Reading

Arreola, D. D., and Curtis, J. R. (1993). *The Mexican border cities: Landscape anatomy and place personality*. Tucson: University of Arizona Press.

Bromley, R. D. F., and Bromley, R. (1988). *South American development: A geographical introduction*. Cambridge, UK: Cambridge University Press.

Browning, C. E. (1990). Urban primacy in Latin America. *Yearbook, Conference of Latin American Geographers, 15*, 71–78.

Crowley, W. K. (1998). Modeling the Latin American city. *Geographical Review, 88*, 127–131.

Crowley, W. K. (1995). Order and disorder: A model of Latin American urban land use. *Association of Pacific Coast Geographers, 57*, 9–31.

Curtis, J. R. (2000). Praças, place, and public life in urban Brazil. *Geographical Review, 90*, 475–492.

Eckstein, S. (1990). Urbanization revisited: Inner-city slum of hope and squatter settlement of despair. *World Development, 18*, 165–181.

Epstein, D. (1973). *Brasilia: Plan and reality. A study in planned and spontaneous settlement*. Berkeley and Los Angeles: University of California Press.

Ford, L. R. (1996). A new and improved model of Latin American city structure. *Geographical Review, 86*, 437–440.

Gilbert, A. (1994). *The Latin American city*. London: Latin American Bureau.

Gilbert, A. (1996). *The mega-city in Latin America*. Tokyo: United Nations University Press.

Griffin, E. C., and Ford, L. R. (1980). A model of Latin American city structure. *Geographical Review, 70*, 397–422.

Gwynne, R. N. (1986). *Industrialization and urbanization in Latin America*. Baltimore: Johns Hopkins University Press.

Hardoy, J. E. (Ed.). (1975). *Urbanization in Latin American: Approaches and issues*. Garden City, NY: Anchor Books.

Keeling, D. J. (1996). *Buenos Aires: Global dreams, local crises*. Chichester, UK: Wiley.

Mangin, W. (1967). Latin American squatter settlements: A problem and a solution. *Latin American Research Review, 2*, 65–98.

Mitchell, J. (2000). Political decentralization, municipal fragmentation, and the geography of real estate investment in Caracas, Venezuela. *Urban Geography, 21*(2), 148–169.

Pick, J. B., Butler, E. W. (1997). *Mexico megacity*. Boulder, CO: Westview Press.

Pineo, R., and Baer, J. (Eds.). (1998). *Cities of hope: People, protests, and progress in urbanizing Latin America, 1870–1930* . Boulder, CO: Westview Press.

Ryder, R., and Brown, L. A. (2000). Urban-system evolution on the frontier of the Ecuadorian Amazon. *Geographical Review, 90*, 511–535.

Scarpaci, J. L. (2004). *Plazas and barrios: Heritage tourism and globalization in the Latin American centro histórico* . Tucson: The University of Arizona Press.

Scobie, J. (1971). *Argentina: A city and a nation* (2nd ed.). New York: Oxford University Press.

15 The Southern Brazilian Highlands

The Brazilian Highland are a vast and complex region. The highlands lie roughly parallel to Brazil's eastern coastline, running from the northeast toward the southwest from about 5°S to 30°S latitude. They extend inland from the Atlantic coast approximately 1,000 km, although in the far south their width shrinks to between 500 and 600 km (Figure 15.1). The Southern Brazilian Highlands encompass about two-thirds of this area, while the Sertão, a populous but impoverished and peripheral economic region in the northeast, occupies the remainder (Figure 15.1). The Southern Brazilian Highlands are divided into two large sub-regions, the Paraná Plateau, and the Central Plateau.

Nearly two-thirds of the latitudinal extent of the Brazilian Highlands lie within the Tropics (3°S–23°S), while the remainder falls into the subtropics (23°S–30°S). This latitudinal range combined with altitudinal variation and the differing influences of coastal and interior location produce distinctive environmental patterns over the extent of the highlands region.

Structurally, the highlands are more akin to a plateau than a mountain chain, and sometimes the term "Brazilian Plateau" is used to describe this region. The highlands rise sharply in the form of a steep escarpment along much of the Atlantic coast where they abut the Atlantic Coastal Plain. In some sections, especially between Rio de Janeiro and São Paulo, the mountains that form the escarpment reach elevations as high as 2,400 m within sight of the ocean. Such is the case for the Serra do Mar lying between the Atlantic coast and São Paulo. Typically, the highlands reach their greatest height along this eastern escarpment, with the plateau sloping downward toward the west.

The highlands are composed of ancient rocks of Precambrian origin that have been highly eroded. Beyond the escarpment, there are few areas that can be considered truly mountainous. Rather, the topography is more typically "rounded" and characterized by low hills and broad valleys.

The Brazilian Highlands end abruptly in the south. At about 30°S latitude, a low escarpment, a continuation of the coastal escarpment, runs due west for about 500 km to the Uruguay River. South of this line the landscape changes quickly, giving way to the gentle rolling terrain of the northern Pampa, broken only occasionally by hilly outliers of the highlands.

The western and northern boundaries of the highlands region are not as easily defined as those on the south and east, where steep escarpments and coastlines facilitate the identification of clear regional boundaries. In general terms, however, the western limits of the highlands can be conceptualized as corresponding to a northeast–southwest line formed by the lower course of the Paraná River. Here, however, as along the region's northern margin, broad transition zones often define the boundary between the highlands and neighboring lowland regions.

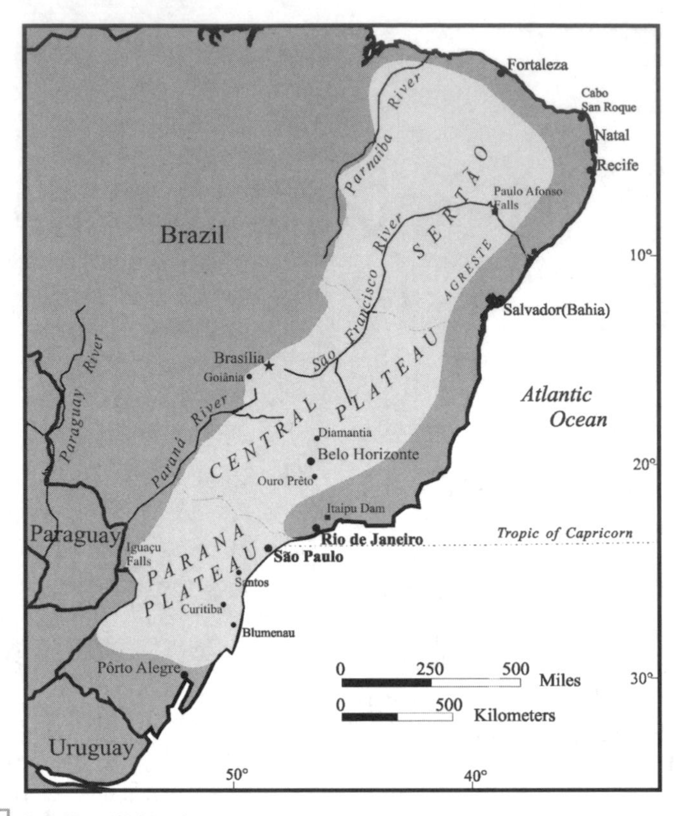

FIGURE 15.1. Brazilian Highlands.

The Brazilian Highlands are characterized by a complex hydrographic pattern, but two major river systems drain the northern and southern sections of the region and are worthy of note. The headwaters of the São Francisco River rise in the central section of the highlands near the city of Belo Horizonte. From here the river runs in a northeasterly direction for nearly 1,500 km before its course turns sharply to the east; it disgorges into the Atlantic Ocean some 400 km further downstream. The São Francisco is the only navigable river to penetrate the highlands, but it is only accessible to oceangoing vessels for 250 km. At this point, the Paulo Alfonso Falls block fluvial transport. River traffic further upstream is restricted to shallow-draft boats. Despite its limitations, the São Francisco played a key role in the economic development of the highlands as a critical transportation route well into the second half of the 20th century. By then, the construction of an extensive all-weather road network significantly reduced the importance of the river as a transportation route.

The Paraná River and its tributaries drain the southwestern section of the highlands. The river's headwaters also rise in the central plateau region of Minas Gerais near Belo Horizonte, but the river follows a southwesterly course for over 1,500 km before it leaves the highlands and enters the humid and often swampy lowlands of the Paraguay–Paraná river system. The shallow and shifting nature of the river along the lower portion of its course and frequent falls and rapids along its upper reaches have severely restricted its use as a transport route. The spectacular Iguaçu Falls on one of its lower tributaries, the Iguaçu River, is an excellent example. On the other hand, the river is a tremendous resource for the generation of hydroelectric power. Itaipu Dam, built jointly by Brazil and Paraguay on the Paraná River just 20 km upstream from Iguaçu Falls, is one of the largest hydroelectric-generating plants in the world.

Indigenous peoples, known as Tupians or Tupinambas, occupied the length and breadth of the Brazilian Highlands at the time of contact with Portuguese explorers and colonists in the early 16th century. Perhaps as many as 2 million natives occupied the region, but this number is only a rough approximation. For these sedentary peoples, large village societies with populations reaching several thousand formed the basic unit of social organization. Tribal groups and chiefdoms were the highest level of social development. However, almost all the tribes spoke closely related languages belonging to the linguistic group Tupi–Guaraní, providing an element of cultural cohesion over the region. Agriculture was the mainstay of the natives' economies. Principal food crops included cassava, peanuts, beans, and sweet potatoes. The exploitation of aquatic resources figured prominently for some groups for whom turtles and turtle eggs were especially important. Indian populations declined rapidly after contact with Europeans as introduced diseases and slavery took a severe toll. On the northeastern coastal plain where Portuguese settlement first focused, most indigenous peoples had disappeared by the end of the 16th century, either dying, fleeing, or being absorbed. In more remote areas in the central and southern highlands indigenous populations survived intact much longer. For instance, on the southwestern margins of the highlands on the Paraná Plateau tens of thousands of Guaraní Indians lived on Jesuit missions until the last half of the 18th century. Nevertheless, by the time of independence from Portugal in the early 19th century, indigenous peoples had ceased to be a numerically significant element in the population geography of the Brazilian Highlands.

The Paraná Plateau

The Paraná Plateau is the industrial and population heartland of the Southern Highlands of Brazil and the home of approximately 50 million people. It is also Brazil's most prosperous region. Nearly 40 million of its residents live in just two Brazilian states, São Paulo and Paraná, which are the core of the region. The city of São Paulo, Latin America's most extensive and populous metropolitan area and one of the five most populated urban areas in the world, with a population that exceeds 20 million residents, is located here. The region also has the greatest concentration of industrial production in all of Latin America. Finally, it is a rich agricultural area. In the 19th century, the Paraná Plateau became the world's most important producer of coffee, a niche it continues to occupy. Other crops, namely, citrus fruits and soybeans, are also major export commodities at the present.

GEOGRAPHY AND ENVIRONMENT

The Paraná Plateau is the southernmost extension of the Brazilian Highlands. In the north, it grades into the Central Plateau across a broad transition zone that is roughly coincident with the modern border between the Brazilian states of Minas Gerais in the north and São Paulo in the south. The differences between the two regions are defined along this boundary more by climate and vegetation than by topography. Save for a narrow coastal strip, the Paraná Plateau extends almost to the Atlantic coast in the east. A series of parallel mountain ridges form this eastern boundary. A steep escarpment rising almost 1,000 m in places, known as the Serra do Mar in the north and as the Serra Geral in the south, marks the beginning of the plateau. Further inland, parallel ranges like the Serra da Mantiqueira and the Serra Paranápiacaba rise sharply to elevations in excess of 2,000 m, the highest elevations on the Paraná Plateau.

In stark contrast to the steep slopes that characterize its eastern margin, the plateau descends gently toward the west into the central geographical feature of the plateau: the upper drainage basin of the Paraná River. The western margin of the plateau is marked by a series of low mountain ranges, notably the Serra de Maracaju in the southwest and the Serra dos Caiapos, whose eastern slopes are drained by tributaries of the Paraná River.

The same structural escarpment that defines the eastern boundary of the region does so in the south as well. At about 30°S latitude, the escarpment formed by the eastern slopes of the Serra Geral turns sharply westward. Marking the southern boundary of the plateau, the escarpment continues westward, declining in elevation until the plateau landscape disappears altogether on the eastern banks of the Uruguay River.

The highlands' limits correspond roughly to the geographical area formed by the following Brazilian states: São Paulo in the north, Paraná in the center, and Santa Catarina in the south. In the west, the plateau includes parts of the Brazilian state of Mato Grosso du Sul.

The region lies squarely within the subtropics, falling almost precisely between the tropic of Capricorn (23°S) and 30°S latitude. Thus, while the climate over most of the Brazilian Highlands can be classified as tropical, the climate of the Paraná Plateau ranges from subtropical in the north to almost temperate in the far south. During the winter months occasional incursions of cold air masses from the South Atlantic can subject most of the southern and central sections of the plateau to killing frosts; the northern and western sections are typically frost-free. Seasonal temperature variations are more extreme here than in the other two sections of the Brazilian Highlands. During the summer months, December to February, mean monthly temperatures stand in the low 20s°C—for example, 22°C at São Paulo, 21°C at Curitiba, and 24°C at Blumenau. Temperatures dip modestly in the winter, May to September. In the coolest month, the temperature mean at São Paulo is 15°C, at Curitiba it is 12°C, and at Blumenau it is 15°C. The plateau receives moderate rainfall, with annual precipitation averaging between 1,250 mm and 2,000 mm. Precipitation is highest in the south along the southern escarpment, and tends to decline toward the north and west. Rainfall occurs throughout the year. In the winter precipitation is usually associated with the northward incursion of midlatitude storm fronts from the South Atlantic, while summer precipitation is characteristically due to convectional heating on the plateau that brings afternoon and evening thundershowers (Figure 15.2).

The natural vegetation of these humid subtropical uplands falls into three principal formations: Araucarian conifer forests, humid broadleaf forests of the Paraná River Basin,

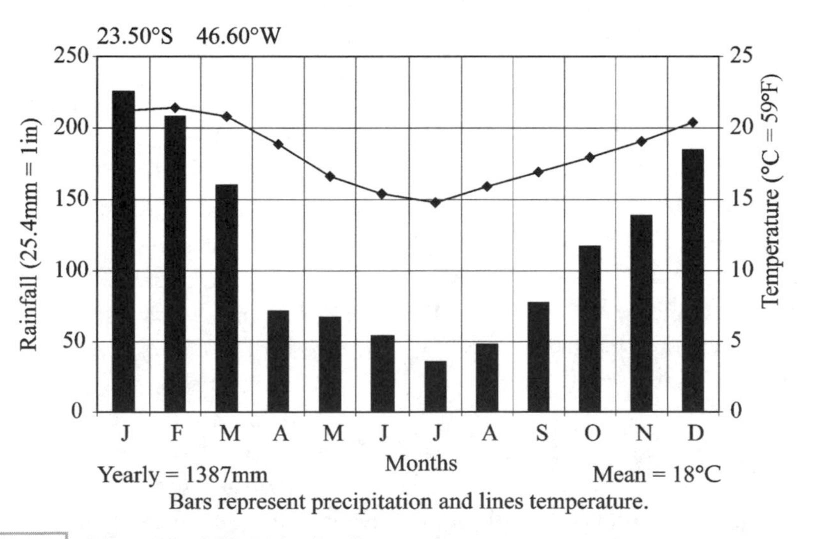

FIGURE 15.2. Climograph of São Paulo, Brazil.

and subtropical broadleaf forests. The first and most extensive are the Araucarian forests that occupy the central portion of the plateau. These unique-looking evergreens grow to heights of 30 m or more and are characterized by straight trunks almost absent of limbs, but with large dense canopies. Araucarian forests grow over a range of ecological conditions at elevations between 850 m and 1,300 m. In the cool uplands where frosts occur, the trees form a dense homogenous forest with a closed canopy. In the highland's more subtropical environments to the north and to the west, the Aruacaria tree is common, but is just one species in a more heterogeneous forest community. At lower elevations in the west and along the basin of the Paraná River and its adjacent tributaries, humid broadleaf forests and open gallery forests comprised of a more diverse species mix predominate. Humid broadleaf dominates on the northern portion of the plateau where it eventually grades into the *cerrado* forests of the Central Plateau. Finally, subtropical broadleaf forests comprise the natural vegetation of the highland escarpment on the coast. These subtropical broadleaf forests, similar to the *mata Atlantica* found further north along the coast, take on a more tropical aspect than those in the interior of the plateau because of more precipitation and milder temperatures.

All three of these natural vegetation formations have been severely reduced by human land use during the last century. Only isolated pockets of the subtropical broadleaf forests along the coastal margin remain, amounting to no more than 5 percent of their original extent. The destruction of the Araucarian forests, which began in earnest in the latter half of the 19th century as colonists cleared land for coffee and other crops, has been less complete. Up to 20 percent of these forests remain, and some reforestation efforts have been successful. The broadleaf forests of the Paraná River basin have largely disappeared during the last half of the 20th century as highly mechanized commercial agricultural enterprises have cleared these lands for the production of export crops, notably soybeans.

HISTORICAL GEOGRAPHY AND ECONOMIC DEVELOPMENT

Early in the colonial period the Captaincy General of São Vicente, which included the port of Santos and the small inland settlement of São Paulo, was one of the few captaincies that was economically viable. Despite this early development, the area remained very much at the periphery of the nation's economy. The early colonial history of the Paraná Plateau provided little indication of the role it would eventually play in the development of the Brazilian nation.

Eclipsed by the sugar economy of the Atlantic coast in the 16th and 17th centuries and by the mining boom in the Central Highlands during the 18th and early 19th centuries, the region developed slowly. Livestock raising provided the catalyst for the massive market that developed at Sorocaba on the edge of the escarpment. Here, some 200,000 livestock reared on the Paraná Plateau and the Pampa further to the south were traded to buyers from the north at twice-a-year livestock fairs during the 1700s. While sugarcane was grown as a commercial crop along the narrow coastal plain that separates the plateau from the Atlantic, subsistence farming and the cultivation of food crops for local and regional markets predominated inland.

One of the most enduring colonial legacies of this region is the exploration and discovery of the Brazilian interior by roving groups of adventurers, the *bandeirantes*. These men from the state of São Paulo engaged in extended expeditions into the interior. Both culturally and racially, they represented the mixing of European and Indian genetic stock and customs, giving rise to the *mameluco*, the Brazilian equivalent of the Spanish American mestizo.

These expeditions spanned thousands of kilometers and lasted many months, if not years.

The search for Indian slaves was the primary motive of these adventurers during the 17th century. These slaving expeditions pushed the limits of geographical knowledge inland and established clear Brazilian claims to extensive territories in the north, west, and south. In the southwest, along the eastern banks of the Paraná and Uruguay Rivers, in a region known as "Missiones," bandeirantes waged a near war on the system of Jesuit missions and their Guaraní Indian charges as they sought to enslave the Indians as laborers. Brazilian colonial authorities tacitly approved of these actions, while Spanish colonial authorities often remained passive in the face of such aggression as long as it was directed at the Jesuits, whose presence they found objectionable. The Spanish colonial authorities viewed the Jesuits as a threat to the colonial establishment because of the economic success of the mission communities and their teaching and training of the Indian converts who lived under their protection in the missions. A fictionalized account of the bandeirantes' assaults on the Jesuit mission system is artfully portrayed in the movie *The Mission* (1986), directed by Roland Joffé.

In the short term, the bandeirantes filled the demand for Indian slaves in the environs of São Paulo and on coastal sugar plantations. Of greater consequence for Brazil was the long-term effect of pushing Brazil's western border far beyond the Line of Tordesillas, greatly expanding the nation's geographical limits. The Line of Tordesillas, set by a treaty between Spain and Portugal in 1494 demarcated the spheres of influence of each nation's colonial interests. West of the line, which corresponded roughly with 46°W longitude, was acknowledged as Spain's sphere of influence, while Portugal's extended eastward from the line.

As the 17th century drew to a close and the demand for indigenous slaves slackened, the bandeirantes increasingly extended their searches to include precious stones and metals. Their travels and explorations played a significant role in the discovery and initial exploitation of major mineral lodes in the Central Plateau during the 18th century. For example, bandeirantes discovered rich diamond deposits in the vicinity of what would become the town of Diamantina, setting off a major mining boom there.

The bandeirantes disappeared as an important social and economic force as the 18th century progressed and the effective settlement frontier was pushed inland throughout the Brazilian Highlands by cattlemen, subsistence farmers, and miners, creating a network of settlements. The bandeirante passed from reality to legend, taking on a kind of mythic stature as the embodiment of a range of positive attributes, characteristic of the prototypical person from São Paulo, a *Paulista*, daring, innovative, and hardworking.

In the middle of the 19th century the region's economy and social structures were radically transformed by the introduction of coffee. The economic boom it created affected not only the Paraná Plateau, but much of the Brazilian Highlands as well. Coffee's cultivation began in Brazil in the early 1700s; the country's first coffee exports date from the 1730s. By the end of the 18th century coffee growing became widespread in the environs of Rio de Janeiro, especially in the Paraíba Valley. The valley lies roughly to the north of the city and runs in an east–west direction for several hundred kilometers. Coffee grew well on the cool upper slopes of the valley, and by the 1840s the valley had become one of the country's principal areas of production. Between 1830 and 1840 coffee accounted for nearly 44 percent of the nation's total exports, supplanting sugar, which continued to decline in economic importance throughout the remainder of the century.

As poor agricultural practices exhausted the virgin soils of the coffee lands of the Paraíba Valley and yields declined, growers sought new areas. Coffee growing spread north into the uplands of the Central Highlands and westward toward São Paulo and the Paraná Plateau by the 1860s (Figure 15.3). The fertile *terra rosa* (red earth) soils on the cool

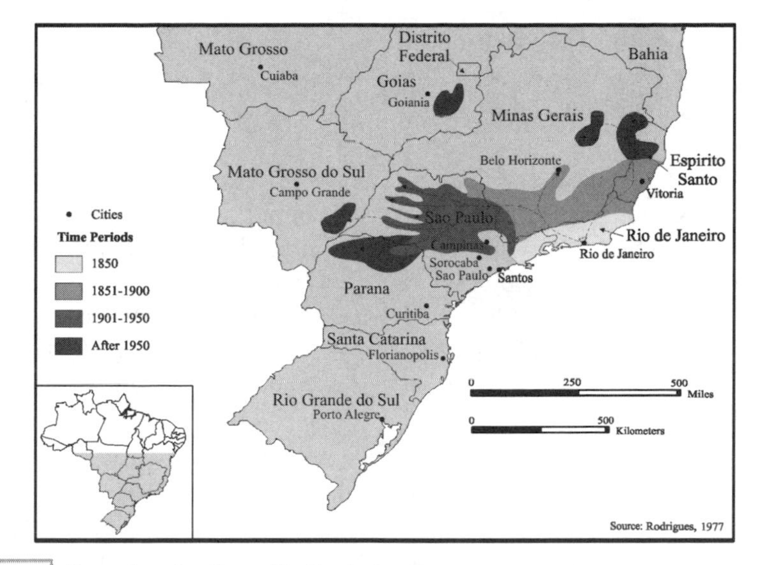

FIGURE 15.3. Expansion of coffee cultivation in Brazil.

rolling uplands of the plateau produced well, and by 1885 the area around Campinas, to the west of São Paulo, had surpassed the Paraíba Valley as the principal coffee-growing region. In subsequent decades, this region expanded westward for several hundred kilometers and production exploded. By the end of the 19th century coffee accounted for two-thirds of the value of all Brazilian exports. Brazil dominated the world trade in coffee, and coffee brought the nation, especially São Paulo and the Paraná Plateau, immense wealth.

The coffee and the economic bonanza it created brought many of the same dangers that had characterized Brazil's previous export economies focused on sugarcane and mineral wealth. An export economy in which two-thirds of all export value is represented by a single commodity begs disaster. First, cool winter weather, including occasional frosts, occurs in the Paraná Plateau. These infrequent frosts can damage or destroy a year's coffee crop, while a hard freeze can kill the coffee trees themselves. Second, high coffee yields in other growing regions in Latin America and Africa can increase supplies and drive down prices, just as a bumper crop in Brazil itself could do. Third, the changing tastes of consumers can affect market demand and hence the health of the regional economy. Fourth, poor land-use practices can lead to significant soil loss via erosion or can simply reduce soil fertility and cause yields to decline. Finally, changing economic conditions in consuming countries can reduce the disposable income of consumers and stifle demand. While Brazil's coffee industry experienced many of these structural difficulties that typically plague export economies, coffee and the profits it generated transformed southern Brazil as well as the nation.

This transformation occurred largely because of the characteristics and vision of the coffee planters and entrepreneurs from São Paulo. Initially, the agricultural economy that developed around coffee growing shared many characteristics with the sugarcane plantation economy that had prospered on the Atlantic Coastal Plain during the colonial period. Ownership of much of the coffee acreage was concentrated in the hands of a small elite class of growers. Furthermore, most production came from large plantations, *fazendas*, which covered hundreds of hectares. Typically, the physical infrastructure of these fazendas bore a strong resemblance to the sugar plantations of the coast and included the "big house" or owner's mansion, large paved areas for drying the coffee beans, warehouses to store them, and workers' barracks.

But in São Paulo the behavior of the coffee elite varied notably from that of the sugar growers on the coast and brought vastly different consequences in the long run. While sugar growers remained steeped in tradition and were unwilling to adopt new technologies, coffee plantation owners and entrepreneurs in São Paulo understood that change was inevitable and were characterized by a willingness to adopt new technologies and methodologies. For example, coffee planters foresaw the end of slavery in Brazil (1888) and took strong measures to ensure a stable supply of agricultural laborers by promoting large-scale foreign immigration. They supported liberal trade policies that ensured open markets for coffee exports and encouraged the importation of machinery and technology critical for the region's and the nation's industrial development. São Paulo's coffee and business elite also realized that coffee might not always reign supreme in Brazil as it did at the end of the 19th century. As a consequence, they diversified their investments and business interests, contributing in large measure to the creation of a broad-based regional economy focused on the city and state of São Paulo, but eventually encompassing much of the Paraná Plateau.

Foreign immigration had a tremendous impact on the region. During the middle decades of the 19th century, tens of thousands of German immigrants colonized the southern reaches of the Paraná Plateau, clearing virgin forests and establishing a thriving agricultural economy based on crops and livestock. The coffee boom in São Paulo created a strong demand for rural labor, which in turn led to the encouragement of further foreign immigration to that state. Between the 1880s and the 1920s, about 1.5 million foreign immigrants arrived from Western Europe, with Italians constituting the vast majority. Not all immigrants to the region hailed from Europe, however: in the first decades of the 20th century about 200,000 Japanese settled around São Paulo, most coming to Brazil under labor contracts to work on coffee plantations.

Immigrants played a key role in the establishment of new coffee plantations and in the workforce of existing operations. Labor contracts often bound the immigrant to terms of 3–5 years during which time they were to clear the forest and plant and tend the young coffee trees. While they waited for the trees to begin production, the colonists could raise subsistence and cash crops in the open spaces between the rows of coffee trees. Once the trees began production, colonists either moved on to clear and plant new coffee plantations or they remained as wage laborers to tend the coffee trees and harvest the crop. A growing class of rural laborers, earning regular wages, provided the beginnings of a regional consumer market that stimulated industrial development in São Paulo. This in turn provided attractive investment opportunities for the vast wealth coffee had brought. The demand for rural labor declined after the 1930s as the world depression, overproduction of coffee, and dropping world prices for the bean drastically reduced returns. Excess rural workers then filled the increasing demands for urban labor by the rapidly growing industrial sector in the city of São Paulo. As their incomes rose and a sizeable percentage began to attain a middle-class life style, they became a critical market for the region's industrial production.

CONTEMPORARY ECONOMIC AND SOCIAL GEOGRAPHY

With a total population of nearly 50 million, the Paraná Plateau accounts for almost one-third of Brazil's entire population. Over 85 percent of the population in the region is urban, and the vast majority is located in a few large urban centers.

A mild subtropical climate and an environment well endowed with fertile, arable land supports an intense and varied agricultural economy. These lands are among the most productive in Brazil. Agricultural crops destined for export markets still play a key role in the economy. Most of Brazil's coffee crop is harvested here and to the north in the Central Plateau. In recent years, however, increased world production and falling prices have undercut coffee's traditional role as a dependable agricultural mainstay in this region. Farmers have diversified production as a result. Among other crops, orchards of oranges and other citrus fruits now cover immense acreages, producing frozen juice concentrate and other products for markets in the United States and elsewhere overseas.

During the last 30 years hundreds of thousands of hectares of forest have been cleared on the western portions of the plateau. In the agricultural equivalent of a gold rush, large-scale, highly capitalized agricultural production companies have planted soybeans. Soybeans enjoy a brisk demand on world markets. Large-scale commercial soybean production on the plateau extends from western Paraná state in Brazil into eastern Paraguay.

Cotton, sugarcane, and cattle are also key agricultural commodities produced on the plateau, although typically this production is destined for national, rather than international, markets. Local demand for agricultural products has also transformed the agricultural landscape around São Paulo's metropolitan area. Here, independent farmers operating on small plots of land have created a vast and productive truck-farming (market-gardening) belt around the metropolis, capitalizing on access to its market of nearly 20 million inhabitants.

In the century between 1870 and 1970, the Paraná Plateau changed from a sparsely populated agricultural hinterland to Brazil's industrial, financial, and commercial core. Today it is the site of the greatest concentration of industrial activity in Latin America. Industrialization focused initially on São Paulo's metropolitan area. It had become so dominant by the middle of the 20th century that it accounted for about two-thirds of the nation's industrial output. By the 1970s, however, industry had begun to disperse throughout the state of São Paulo. The metropolitan area of São Paulo accounts for 50 percent of the nation's industrial output, while the remainder of the state accounts for an additional 25 percent of the total. A well-developed hydroelectric power supply spurred the region's early industrial development and contributed to the growth of a diversified industrial base producing every imaginable consumer good, as well as processed foods, textiles, pharmaceuticals, electronics, automobiles, and heavy machinery. Significant industrial activity in the Paraná Plateau is limited to a few population centers, notably Santos, Curitiba, and Pôrto Alegre.

São Paulo is the region's preeminent urban place. While precise statistics are elusive, the city is likely to have a population of about 10 million, while its surrounding suburban districts add close to another 10 million, bringing the metropolitan population to nearly 20 million. The diverse ethnic mix of the city's inhabitants reflects the role of immigration in the metropolitan area's history and population growth. A large proportion of the population can trace its origins to southern Europe, from where large numbers of Italians and Spaniards immigrated in the late 1800s. Nevertheless, many other ethnic groups have contributed to the population mix. Notable among these are other Europeans: Germans and Russians; Middle Easterners: Lebanese, Arabs, and Armenians; and Asians: Japanese, Chinese, and Koreans. Jews too proved a minor component in the city's population mix (Figure 15.4). African-Brazilians comprise a small proportion of the population.

São Paulo is the nation's most important single center of financial institutions and commercial activity. Brazil's, and arguably Latin America's, most significant stock exchange,

FIGURE 15.4. Jewish synagogue in São Paulo, Brazil, 2004.

BOVESPA, is located here. The city is the headquarters for a very large proportion of Brazil's banks, insurance companies, and other financial institutions. Commercial activity is intense and varied. The region's technological advantage over the rest of the country is reflected in recent patterns of personal computer ownership. With the exception of the state of Rio de Janiero and the Federal District of Brasília, no other states in the nation have such a high proportion of personal computer ownership (Figure 15.5).

Despite the superlatives that characterize its size, as well as its industrial, financial, commercial, cultural, and educational importance within Brazil and in South America generally, the city suffers from a host of vexing urban problems. Living conditions for the millions of its inhabitants who live in shantytowns on the urban periphery are abysmal, while traffic congestion and the air pollution fed by automobiles degrade the quality of life for all of its residents (Figure 15.6).

São Paulo dwarfs other urban centers on the Paraná Plateau, but several cities are significant. Santos, with a population of about 1.4 million, is located approximately 70 km to the east of São Paulo on the Atlantic coast. It is one of the region's principal ports, serving São Paulo as well as an immense agricultural hinterland. Curitiba, the capital of the state of

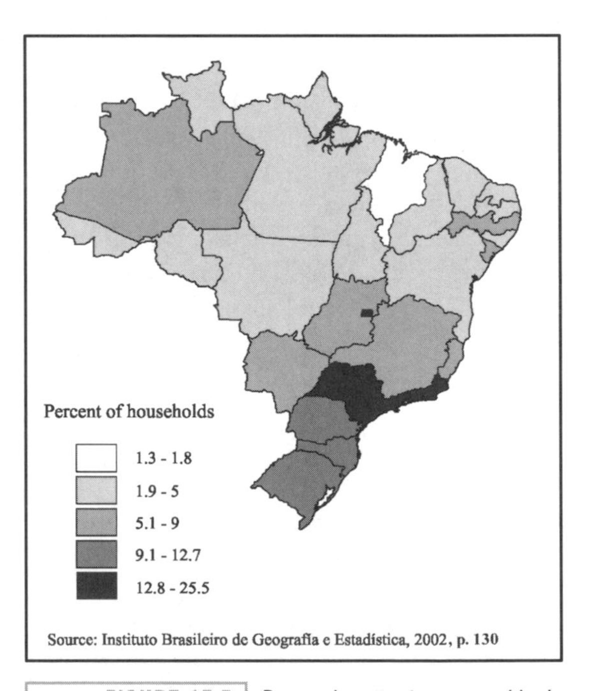

FIGURE 15.5. Personal computer ownership in Brazil, 2000. It is highest in the states of São Paulo, Rio de Janeiro, and the Federal District of Brasília.

Paraná, has a population of close to 3.0 million and is a diversified commercial and industrial center. The city has gained considerable renown for its progressive efforts to improve urban transportation and the quality of life for its residents. Of nearly equal size, Pôrto Alegre, on the southern margins of the Paraná Plateau, is an industrial center as well as a major Atlantic port with a population of almost 4.0 million.

FIGURE 15.6. A homeless camp under a pedestrian overpass in São Paulo, Brazil, 2004.

Brazil's Central Plateau

Although its economic clout and population numbers consign it to a secondary role in the Southern Highlands, the Central Plateau has played a key role in the economic development of the Brazilian nation since the middle of the colonial period. Initially the remote inaccessible highlands remained beyond the colony's settlement frontier. But the discovery of rich lodes of gold and diamonds beginning at the end of the 17th century set off a massive population migration to the plateau, enriched the local economy and populace, and filled the treasury of the Portuguese Crown. When the boom ended about 100 years later, the region settled into a pastoral economy supported initially by livestock raising and later by a more diversified agricultural economy.

Mining and agriculture continue to play a prominent role in the area's modern economy; however, commerce, manufacturing, and administrative services are increasingly important. The regional population totals nearly 25 million. Several of Brazil's million-plus metropolitan centers and the nation's capital, Brasília, are located here.

GEOGRAPHY AND ENVIRONMENT

As its name implies, the Central Plateau, or *Planalto Central* in Portuguese, occupies the central section of the Brazilian Highlands (Figure 15.1). This region is situated astride the highlands' two major drainage basins, the São Francisco in the northeast and the Paraná in the southwest. The tributaries of the Amazon drain its northwestern and western portions. Much of the plateau lies at about 1,000-m elevation, but its terrain varies from comparatively flat plateaus to rolling hills and deeply incised valleys. In some areas, low mountain ranges rise 500–600 m above the surrounding topography. The region's core lays squarely in the modern Brazilian state of Minas Gerais and includes the Federal District of Brasília, much of the state of Goiás to the west, and portions of southern Bahia and eastern Mato Grosso.

The region exhibits a climate typical of dry tropical highlands. Temperature means in the warmest summer months (November–February) stand in the mid-20°s C, while mean winter temperatures (May–September) run from the high teens to low 20°s C. Local relief produces much of this variation. At Belo Horizonte in the south-central area of the plateau, monthly temperature means in the summer and winter months stand at about 23°C and 19°C, respectively (Figure 15.7). Annual precipitation on the plateau varies from about 1,100 mm in the north, where it borders the Sertão, to about 1,750 mm in the southwest. Precipitation increases in the southern and western portions of the plateau.

Rainy wet summers and dry winters typ-

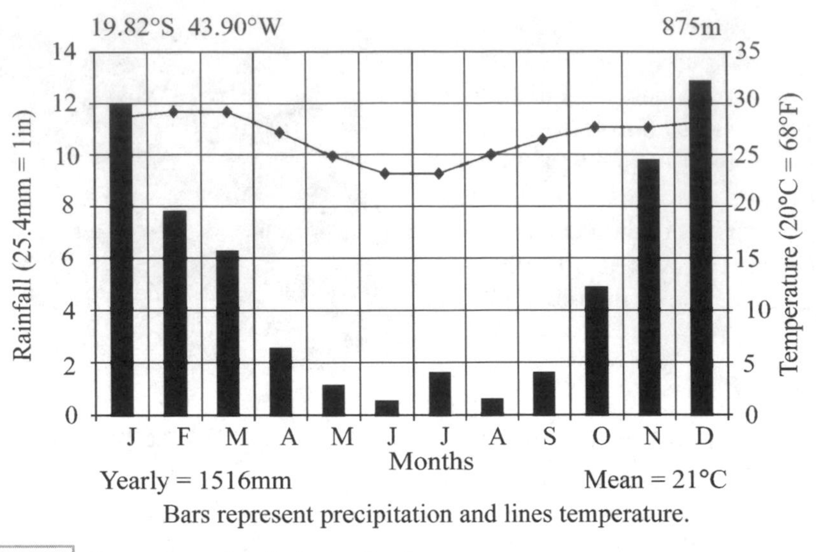

FIGURE 15.7. Climograph of Belo Horizonte, Brazil.

ify the Central Plateau. This pattern is attributable to two mechanisms, one global and the other regional. In the first instance, the ITCZ (Inter-Tropical Convergence Zone) moves over the plateau in the Southern Hemisphere's summer, bringing unstable air and precipitation. The reverse occurs in the winter months when the ITCZ moves north and the highlands fall under the influence of the global air pressure belt known as the "subtropical high" that brings stable descending air and no rainfall.

The second mechanism involves the formation of a low-pressure cell over the plateau region during the hot summer months. A monsoon-type circulation pattern develops in which moist, comparatively higher pressure air is pulled inland from the Atlantic, bringing with it seasonal rains. In a reversal of this pattern in the winter months, high pressure develops over the plateau and precipitation is scant.

Three principal vegetation formations dominate the plateau. Much of the region is covered by *cerrado*, a forested savanna. It is composed of two vegetation strata: an open canopy of widely spaced trees reaching heights of 20–30 m and a lower strata of grasses and herbaceous plants with a maximum height of perhaps 0.5 m. Where precipitation is less, trees are more widely dispersed and the upper canopy of the cerrado opens, producing a savanna landscape called *campo cerrado* (Figure 15.8). In the driest areas of the central plateau, campo cerrado grades into open, almost treeless, grasslands. Land-use practices, especially during the last 300 years, have reduced the extent and quality of these vegetation formations. Livestock raising, agriculture, and the demand for wood and wood products like charcoal are the main culprits in this decline. Much of what was cerrado and campo cerrado has been deforested and transformed into grass savanna. Much of it is eroded, barren, and severely degraded as a result of overgrazing.

HISTORICAL GEOGRAPHY AND ECONOMIC ENVIRONMENT

The Planalto Central remained essentially beyond the frontier during the first two centuries of the colonial period. In 1698, however, bandierantes discovered gold in the southern portion of the mountains of the Serra do Espinhaco.

The discovery of readily exploitable placer deposits set off a gold rush that lasted almost a century and left an imprint on the placenames of the region that endures to the present. A prominent example is the name of the Brazilian state that sits at the core of the region, Minas Gerais, which in Portuguese means "general mines." Ouro Prêto, a colonial mining center and the former state capital, means "black gold," while the name of the city Diamantina recalls the immense wealth in diamonds extracted by miners from placer deposits surrounding the city.

The impacts of this gold rush were dramatic and far reaching. Scholars estimate that the Central Plateau produced between 30 and 40 percent of the world's gold supply during the 18th century. This generated immense wealth for Portugal, as all gold produced was subject to a 20 percent royal tax, although much escaped this levy. This wealth allowed Portugal the luxury of purchasing manufactured products from Great Britain, but at the same time it retarded the process of industrialization in Portugal. Population growth and demand for foodstuffs and beasts of burden in the gold-mining districts stimulated livestock raising, especially of cattle and mules, over a broad geographical area including the Central Highlands, the Sertão in the north, and the Pampa of Rio Grande do Sul in the south. The interaction between the country's major regions as a result of the livestock business, which involved breeding, fattening, transporting, and marketing livestock, proved to be a key contributing factor in the long-term integration of the Brazilian nation.

FIGURE 15.8. *Campo cerrado* vegetation in the Parque Nacional de Brasília in the Central Highlands, 2004.

The gold rush, and the subsequent discovery of diamonds in 1729, generated tremendous in-migration from the coast and immigration from Portugal. One estimate places the number of Portuguese immigrants to Brazil during the 18th century between 300,000 and 500,000. Most were destined for the highlands. As a result, the population of the region is said to be the most "Portuguese" in all Brazil, and Catholicism is stronger here than in other parts of the nation. The population, also known as *mineiros*, or miners, in the rest of Brazil, is often viewed as conservative and traditional.

During the economic boom created by gold and diamonds towns sprung up throughout the highlands wherever placer deposits proved rich enough to support them. A few urban centers grew to impressive proportions. Perhaps the most famous is Ouro Prêto. During the mid-18th century the city's population exceeded 100,000. By the early 19th century, when the gold boom was finally ending, the city's population had dwindled to just 20,000. Nevertheless, much wealth had poured into the city, leaving a colonial architectural legacy that includes baroque churches and religious art, opulent homes, and picturesque cobblestone streets and plazas. Today, Ouro Prêto is a national historic monument.

The picaresque Brazilian movie *Xica* (1976), directed by Carlos Diegues, recounts the fictionalized tale of a beautiful black slave who becomes the rich and powerful mistress of the royal diamond contractor. Filmed on location in Minas Gerais and in Diamantia, it provides excellent views of the plateau's landscape and the colonial architecture of urban centers.

CONTEMPORARY ECONOMIC AND SOCIAL GEOGRAPHY

The present-day economy of the Central Plateau reflects many of its historic antecedents.

Mining remains a central economic activity, especially in Minas Gerais. A rich lode of mineral resources supports the regional economy. These include chromium, tungsten, titanium, manganese, zinc, gold, and iron ore. Diamonds and other precious stones are still mined commercially. The region's iron ore reserves are extensive; thus iron and steel production and metal fabrication are important industries. A broad range of industrial activities including agro-industries, textiles, and equipment production characterize the region's industrial mix.

Agriculture, however, continues to play a major role in the regional economy. Livestock raising endures as a widely practiced and economically viable livelihood. Cattle raising is most significant; dairy products, meat, and hides are commercialized on an immense scale. Where soil and water resources permit, a wide range of agricultural products prospers in the tropical to subtropical climate. For example, coffee and oranges are exported to foreign markets and basic food commodities for the national and regional market are also grown. Commercial production of sugarcane, manioc, beans, and potatoes is notable.

The region's population exceeds 25 million. The racial composition of the regional population reflects the strong tradition of Portuguese immigration created during the century-long gold rush in Minas Gerais. People categorized as "whites" comprise a narrow majority of the population, while mestizos, mulattos, and blacks account for the balance. About two-thirds of the population is concentrated in its southeast corner, in the state of Minas Gerais. Elsewhere, the Central Plateau is thinly populated, with population densities declining as one moves west through the states of Goiás and then Mato Grosso. The urbanization rate here is higher than one might expect, with a regional average of 77 percent. This elevated rate is due largely to the dense urban network of small cities, towns, and villages that covers the eastern portion of the region.

The urban system of the plateau is closely linked and subordinated to the nation's principal urban centers of São Paulo and Rio de Janeiro, which lie just to the south. Nevertheless, the nation's third largest city, Belo Horizonte, and the national capital, Brasília, are located here. The populations of the region's three principal cities exceed 2 million—Brasília (2.3 million), Goiânia (3.6 million), Belo Horizonte (4.8 million). These metropolitan centers share several characteristics. All are governmental administrative centers: Brasília is the national capital, while Belo Horizonte and Goiânia are the capitals of their respective states. All were founded in the modern period: Belo Horizonte was established in 1897, Goiânia in 1942, and Brasília in 1960. Finally, all were planned capital cities, conceived, designed, and built as completely new capital cities.

The most populous and complex of these urban centers is Belo Horizonte. Its metropolitan population, encompassing the city proper and 14 contiguous municipalities, reaches 4.8 million. Conceived in the late 19th century, Belo Horizonte (which means "beautiful horizon" in Portuguese) is sited in a broad upland basin (800 m) with a pleasant climate. The city's urban design shares the broad streets and diagonal boulevards of Washington, D.C. and La Plata, Argentina, the cities whose designs are said to have inspired Belo Horizonte's plan. Inaugurated officially as the state capital in 1897, the city grew slowly during the first decades of its existence, fulfilling mainly administrative functions. Capitalizing on the region's mineral wealth, an aggressive industrialization program in the 1940s began to transform the city's economic structure. The population reached 350,000 in 1950 and 1.2 million in 1970. Belo Horizonte's steady growth, much of it industrial, since then is linked to vast and accessible reserves of iron ore, the development of hydroelectric power supplies, and the establishment of Brasília as the national capital in 1960. The metropolitan region is Brazil's third most important manufacturing center after São Paulo and Rio de Janeiro and produces metals, textiles, processed food, electrical and construction materials, and automobiles. The city has experienced considerable in-migration in recent decades, straining its infrastructure and housing stock. Over one-quarter of the city's residents live in abject poverty. This is less than in most coastal cities, but far more than in the urban centers to the south in the Paraná Plateau.

Despite the fact that its construction was completed and the city inaugurated only in 1960, Brasília, Brazil's national capital, is one of the most famous and well-known cities in Latin America. As a measure of the city's national and international significance, UNESCO declared the city a "cultural patrimony of humanity" in 1987, less than 30 years after it was founded.

Brasília, in its design and execution, is a thoroughly monumental and modernistic city. The city is laid out on an immense scale, with its streets and principal boulevards forming what appears like an airplane to some or a bird with extended wings to others. Massive futuristic designs, characterized by flowing lines and arcs, typify the principal government buildings designed by the world-class Brazilian architect Oscar Neimeyer. These buildings occupy the higher ground at the key symbolic places in the city's layout at the head of the "bird." In contrast, the gigantic superblocks that occupy the peripheral spaces—for government ministries and offices, residences, and commercial spaces—are monotonous and far from inspiring.

Designed for the automobile, the city lacks much of the dynamic street life that characterizes other Brazilian cities (Figure 15.9). Many residents and visitors find the urban environment sterile. The design of the city made essentially no allowance for housing for the poor and the working class, those who built it and toil in its service industries. These people live in a series of six satellite cities that ring Brasília. All lie within the Federal District and form part of the city's metropolitan region. The metropolitan population is about 2.3 million.

Goiânia is the third major metropolitan center. Planned as the state capital of Goiás, the city's physical layout exhibits a certain grandeur, including broad streets, diagonal boulevards, circular avenues, and many public open spaces. Although the city has been occupied only since 1937 and has served as the state capital only since 1942, it has grown very rapidly and its population today stands at approximately 3.6 million. The establishment of Brasília, only 200 km to the north, contributed tremendously to the city's development. The

FIGURE 15.9. Brasília's central transportation axis with government buildings in the background, 2004.

city's key economic activities are services, especially government services, and the commercialization and marketing of agricultural products, including cattle. Industry plays a very limited role in the local economy, and that which is found in Goiânia concentrates on the production of basic consumer goods and agroindustrial processing.

The Sertão: A Populous Periphery

The Sertão is a densely populated peripheral region that occupies the northeastern one-third of the Brazilian Highlands. The term *Sertão* translates roughly into English from Brazilian Portuguese as "backwoods" or "bush" and is used as a generic term to refer to a broad swath of the northeastern highlands, or simply the "Northeast." The region is also sometimes referred to as Brazil's "drought polygon" because of the erratic nature of the region's precipitation regime and the regular droughts that affect it.

Subsistence agriculture, commercial gathering of forest products, and cattle raising have been the principal economic activities in the Sertão since colonial times. High fertility rates and adverse economic and social conditions have encouraged out-migration at least during the last 150 years. Although no major urban centers are located in the Sertão, its population exceeds 20 million.

GEOGRAPHY AND ENVIRONMENT

The Sertão includes the upland interior region of the states of Piauí, Ceará, Rio Grande do Norte, Paraíba, Pernambuco, and Bahia. It also includes the northern quarter of the vast interior state Minas Gerais. The Parnaíba River defines its western boundary, while the mountainous escarpment that forms the margins of the Atlantic Coastal Plain defines its northern and eastern boundaries.

The region is characterized by a varied upland topography including rolling hills, tablelands, mountainous areas, and severely eroded badlands. Nevertheless, most definitions of the Sertão consider that in the north it extends to the Atlantic coast, interdigitated with areas exhibiting the more humid characteristics of the Atlantic Coastal Plain. This somewhat confusing situation arises because the region is defined principally by its climatic characteristics rather than by its topography.

The region is semiarid. Rainfall generally averages less than 1,100 mm. In the region's core, however, roughly coincident with the middle course of the São Francisco River, mean annual precipitation is less than 750 mm. At Petrolina, for example, mean annual precipitation is even less, about 440 mm (Figure 15.10). Precipitation variability is high. This high degree of variability is tied to the effects of the El Niño Southern Oscillation (ENSO) which occurs periodically in the eastern Pacific with profound effects on precipitation patterns. Extended periods of comparative drought, at times several years in length, occur about every decade. In other years, torrential rains fall and run off quickly. Those rains that do fall normally occur during the summer months, roughly from December to March, with little rainfall during the remainder of the year.

The combination of tropical latitudes, an interior location, and modest elevation produce high temperatures over most of the year in the Sertão. Mean summer temperatures (November–March) range between 25°C and 30°C, and daytime highs often reach into the upper 30°sC. The temperatures moderate slightly in the winter (May–August) when mean temperatures fall into the mid-20°sC. "Hot and dry" aptly describe the prevailing climatic conditions.

The vegetation cover of the area is typified by a mix of species, known collectively in Brazil as the *caatinga*. The term originates from an indigenous word meaning "white forest." This vegetation community is drought-resistant. Its plants have evolved a wide variety

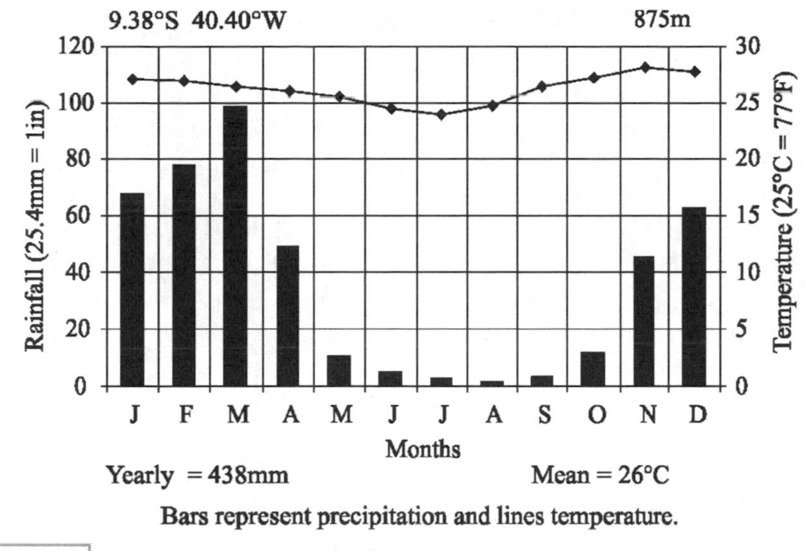

FIGURE 15.10. Climograph of Petrolina, Brazil.

of adaptive strategies that include deep taproots, seasonal loss of foliage, thick waxy leaves, a small leaf size, protective spines and thorns, succulent growth for water storage, and short life cycles for many annual plants. In the more humid areas of the region, a thick, sometimes impenetrable, thorn forest reaching heights of up to 10 m or more covers the landscape. These woods are comprised mostly of spiny leguminous species, especially from the genera *Prosopis* and *Mimosa*. In those areas where precipitation is less abundant, the *caatinga* opens to a discontinuous tree cover where tree growth is somewhat stunted. Cacti, xerophytic shrubs, and seasonal grasses occupy the open ground. Land-use practices have had severe negative impacts on the *caatinga*. Overgrazing, cultivation of fragile lands, and woodcutting have reduced the extent of the thorn forest. Severe soil erosion and even desertification have occurred in some areas. On its eastern and northern margins, where the Sertão borders the Atlantic Coastal Plain and precipitation is higher, a narrow transition zone called the *Agreste* supports a more luxuriant plant cover and a diverse and productive agricultural economy.

HISTORICAL GEOGRAPHY AND ECONOMIC ENVIRONMENT

The Sertão, with its semiarid climate and dense thorny vegetation, did not attract the immediate attention of the Portuguese. However, as the plantation economy of the Atlantic Coastal Plain grew and flourished, it created a demand for meat and animal products that encouraged livestock raising and the movement of cattlemen into the Sertão. The river valleys that punctuate the coastal plain provided the initial avenues for penetration of the interior. By the 1650s cattle raising was well established along the middle and upper courses of the São Francisco River and by the 1680s cattlemen had pushed deep into the interior of Ceará and Paraíba. As early as the beginning of the 1700s Brazilian cowboys, *vaqueiros*, drove cattle hundreds of kilometers through the Sertão eastward to reach plantation and urban markets on the Atlantic Coastal Plain as well as southward to mining centers in the Central Plateau.

High fertility rates, limited economic opportunities, and a drought-prone environment have produced a consistent historical pattern of emigration as residents have sought improved living conditions elsewhere. This tradition of migration takes several forms. The abolition of slavery in the 1880s changed the dynamics of the labor market. Modest opportunities for seasonal wage-labor employment on sugarcane and cacao plantations on the coast developed and attracted the impoverished *flagelados* (literally the "whipped ones" in Portuguese) of the Sertão. These early migrants initiated a pattern of migration that persists to the present as landless peasants, subsistence farmers, and unskilled laborers travel to the coast to meet these labor demands, and then return when the harvest ends. Longer sojourns, of several years' duration, often leading to permanent relocation, are also common.

The periodic droughts that affect the region frequently provide the immediate incentive for migrants to leave. This occurred, for instance, in the late 1870s when an extended drought coincided with the Amazonian rubber boom, and tens of thousands of *sertanejos* flooded into the rain forests of the Acre. When the boom ended, many remained in Amazonia. Similarly, in the middle of the 20th century, demand for unskilled laborers for the construction of Brasília attracted migrants from the Sertão at the same time a drought ravaged the region. The booming economic center of São Paulo on the Paraná Plateau has generated tremendous demands for unskilled labor in both the rural and the urban sectors, and has drawn hundreds of thousands of *sertanejos* in search of work.

The Brazilian government has long recognized the precarious economic condition of the region, which droughts and famine have underscored. The government has made periodic efforts to address these problems. One of the first was the establishment in 1909 of a federal government agency, the Inspectoria Federal de Obras Contra as Secas, charged with the construction of dams, wells, and irrigation works. In 1959 the government undertook a more comprehensive effort to promote regional development in the entire northeast, including the Sertão and the northeast coastal plain, and established SUDENE, the Superintendencia para o Desenvolvimiento do Nordeste, a regional development agency with a broad mandate. Not only did the agency promote rural development, but it also employed a range of fiscal incentives to encourage industrialization in the urban centers of Recife and Salvador on the coast, theoretically creating employment opportunities in these centers for migrants from the region. Despite the investment of huge sums of government money and significant fiscal concessions to investors intended to improve conditions in the Sertão, it remains Brazil's "backwoods" whose principal export continues to be its people, the *flagelados*. *Central Station* (*Central do Brasil*; 1998), a superb movie directed by Walter Salles, recounts the travails of an orphaned immigrant in Rio de Janiero's urban core. The film traces the journey of the boy and the old woman who helps him from Rio to the backlands of the Sertão in search of the boy's father. The movie provides an insightful look at the rural and urban landscapes of the area set against the broad pattern of rural-to-urban migration from the rural hinterland to the country's urban centers.

The region's poverty and marginal position in the national context has spawned a range of social movements. Fanatic religious movements have arisen here, as messianic leaders have capitalized on the hopelessness and despair that is endemic in the Sertão. Perhaps the most well known of these were the followers of Antonio Conselheiro (Anthony the Counselor), who numbered many thousands during the early 1890s. Rejecting the evolving capitalist social order that was rapidly undoing traditional economic and social relations in Brazil's interior, Conselheiro and his adherents established a large and successful folk community called Canudos in the backlands of Bahia. It was viewed with great suspicion and concern by the central government and the elite of the coastal urban centers. A dispute about lumber provided the spark for a violent confrontation between the community and the government. Beginning in 1896, the Brazilian army made three unsuccessful and humiliating attempts to destroy the community. Finally, with a heavily armored force of over 8,000 and after a siege of several months, the army destroyed the community and killed its leader and many of his followers in October 1897. A contemporary account of the Canudos War, published originally in Portuguese in 1902, is given by Euclides da Cunha in *Os Sertós* (*Rebellion in the Backlands*). An engaging novel, based on the same historical events, is *Guerra del Fin del Mundo* (*The War of the End of the World*), published by the modern Latin American author Mario Vargas Llosa.

Outlaw bands have also been part of the social fabric of the Sertão. Such bands continued to operate in the region well into the 20th century. Lampião, one of the region's most infamous bandits, and a small cadre of followers operated at will until the late 1930s when military police finally tracked them down, ambushed them, and killed them. In the early 1960s, landless rural workers who rented land from absentee landlords formed the Peasant Leagues, aggressive organizations demanding land reform and challenging the traditional rural social order. The Peasant Leagues enjoyed modest success, but were short lived and fell apart after the military coup in 1964 when its leaders were arrested and exiled.

CONTEMPORARY ECONOMIC AND SOCIAL GEOGRAPHY

The Sertão is the home to over 20 million Brazilians despite its poorly endowed resource base. The urbanization rate here is one of the lowest in Brazil. Most of the population lives in the countryside, small hamlets, and villages and is dependent on agriculture, herding, or the gathering of forest products.

Indigenous peoples survived longer here than on the coast. Miscegenation between indigenous women and Portuguese men was frequent. Individuals of mixed European and indigenous origin, often called *mamelucos* in Brazil, are dominant in the Sertão, accounting for about two-thirds of the total population. Blacks, mulattos, and whites comprise the balance.

A pastoral economy based largely on cattle, but also including goats, sheep, and donkeys, has dominated the region since colonial times. Open-range grazing, a lack of adequate forage, an absence of veterinary care, and unimproved stock have characterized the livestock-raising methods of the Sertão in the past and continue to do so at the present.

Subsistence farming, commercial agriculture, and harvesting forest products supplement livestock raising in the rural economy. Subsistence crops like manioc, corn, and beans are cultivated in favorable sites where moisture is available, often in the beds of ephemeral streams and rivers. Commercial agriculture is practiced in some areas, known as *jardims*, where water is more plentiful. These oasis-like agricultural areas produce cotton, sisal, and other crops that can prosper in warm semiarid conditions. The commercial gathering of wild forest products has been, and continues to be, significant. The carnauba palm (*Copernicia prunifera*) is perhaps the most important of several products of economic consequence that are gathered in the wild. The palm produces an edible fruit and a fine hardwood used for veneers. It also produces a wax, carnauba wax, from its leaves that is widely used as a lubricant, polish, and wax. The Sertão dominates world production of carnauba wax.

No metropolitan centers are located in the Sertão proper. The urban areas sited in the adjacent Atlantic Coastal Plain play a dominant role in the region's urban system. The hinterlands of the coastal cities of Salvador and Recife reach far into the interior, encompassing essentially all of the southern and central portions of the region. Fortaleza, although located on the northern coast, with a metropolitan population of over 3.6 million, is widely regarded as the Sertão's largest urban center. The city's economy and population growth have been, and continue to be, closely tied to the interior. Cotton, carnauba wax, and rum are among the principal exports from the city's port. The city is an important educational and cultural center and boasts three universities. Industrial development in Fortaleza was retarded until after the 1950s because of the absence of a dependable electrical supply. Energy from the Paulo Afonso hydroelectric complex on the São Francisco River now supplies the city, where agro-industrial processing, notably of cotton and the carnauba palm, dominates the industrial sector. Tourism became a significant component of the urban economy during the last two decades of the 20th century: local entrepreneurs have capitalized on the city's splendid beaches, pleasant tropical climate, and colonial heritage to draw foreign visitors, mostly from Europe, to the city. The region's periodic droughts have sent tens of thousands of migrants to the city in search of work or simply the means to survive. The urban fabric shows the effects of this constant migratory pressure. Substandard housing is commonplace, unemployment and underemployment are widespread, and one-third of the city's population lives below the poverty line as defined by Brazilian authorities.

Other urban centers of lesser size and importance locate on the periphery of the region, characteristically in the Agreste, between the Sertão and the coastal plain. Trading, warehousing, and shipping of natural resource commodities and agricultural products predominate in these cities. Feira de Santana has grown from a regional cattle market in the middle of 18th century to an urban center of some 500,000. Although it is located just 80 km from the coastal center of Salvador, its strategic location at the border of the coastal plain and the interior uplands promoted its growth. The city continues to be a major livestock center complemented by some industrial activity, particularly the manufacture of leather goods and construction materials. Campiña Grande, a city of some 325,000 inhabitants, sits on the edge of the Borborema Plateau in the tiny northeastern state of Paraíba just 80 km from the Atlantic coast. It is located at the juncture of three regions: the Sertão, the Agreste, and the Atlantic Coastal Plain. The city draws raw materials from all three regions: cotton from the Agreste, leather from the Sertão, and sugarcane from the Atlantic Coastal Plain. The city's industrial sector includes textiles, cottonseed oil production, sugar refining, and leather goods manufacture. Teresina is a city of over half a million. The city is sited on the eastern banks of the Parnaíba and situated in the transition zone between the Sertão and the coastal plain in the northwest. Founded in 1850, it has grown rapidly as a service center for an immense hinterland.

Despite the Sertão's lack of large cities, the region is dotted with a dense network of villages, small towns, and minor urban centers. Most are marketing and service centers for a small tributary area, and few exceed 50,000 inhabitants. Two of the largest urban centers in the interior are Juazeiro (150,000), located at the head of navigation along the middle course of the São Francisco River, and Juazeiro do Norte (215,000), a regional commercial center and pilgrimage site.

Summary

The Southern Highlands is a large, complex region that extends over the southernmost section of the Brazilian Highlands. The Paraná Plateau is the core of this region with its population centered in the states of São Paulo and Paraná. During the early colonial period, the region was known for its aggressive explorers and adventurers (*bandierantes*) who led expeditions into the interior seeking Indian slaves and riches. During the mid-19th century the region's economy and social structures were altered by the introduction of coffee. Reliance on a single crop posed many financial risks; however, the region's coffee growers were willing to change, to adopt new technologies, and to invest in other nonagricultural pursuits, thereby contributing to the rapid industrialization there. Immigrants from Germany, Italy, and Japan invigorated the region's cultural and economy. The Paraná Plateau continues to be highly productive. Today São Paulo is the center of Brazil's manufacturing, commercial, and financial activity.

The Central Plateau encompasses the northern half of the Southern Highlands. It became important in the 17th century when gold and diamonds were discovered across a broad swath of the area. These discoveries led to a massive population movement from the coast to the interior, as well as an influx of Portuguese immigrants. Mining remains important to the region, as does livestock raising and agriculture. In 1960, in an effort to promote the development of the country's interior, the national capital was moved from coastal Rio de Janeiro to Brasília on the Central Plateau.

The Sertão is a peripheral region that occupies the northeast section of the Brazilian Highlands, know for its erratic rainfall and periods of drought. During the colonial period livestock raising dominated the regional economy. Steady population growth, an inadequate resource base, and a lack of economic opportunities have resulted in out-migration. Its poverty and an oppressive social system have contributed to social movements by peasants and other impoverished and disenfranchised groups. The region has no major urban centers, but is dotted with villages, towns, and small urban areas.

Further Reading

Biswas, A. (1999). *Management of Latin American river basins: Amazon, Plata, and Sãō Francisco*. Tokyo: United Nations University Press.

Brannstrom, C. (2000). Coffee labor regimes and deforestation on a Brazilian frontier, 1915–1965. *Economic Geography, 76*(4), 326–346.

Caldeira, T. P. R. (2000). *City of walls: Crime, segregation, and citizenship in Sãō Paulo*. Berkeley and Los Angeles: University of California Press.

Da Cunha, E. (1975). *Rebellion in the backlands*. Chicago: University of Chicago Press.

Dean, W. (1969). *The industrialization of Sãō Paulo, 1880–1945*. Austin: University of Texas Press.

Eakin, M. (2002). *Tropical capitalism: The industrialization of Belo Horizonte, Brazil*. New York: Palgrave.

Epstein, D. G. (1973). *Brasília, plan and reality: A study of planned and spontaneous urban development*. Berkeley and Los Angeles: University of California Press.

Evanson, N. (1973). *Two Brazilian capitals: Architecture and urbanism in Rio de Janeiro and Brasília*. New Haven, CT: Yale University Press.

Font, M. (1990). *Coffee, contention, and change in the making of modern Brazil*. Cambridge, UK: Blackwell.

Geld, E. (2003). *View from the fazenda: A tale of the Brazilian heartlands*. Athens: Ohio University Press.

Graham, L., and Wilson, R. H. (1990). *The political economy of Brazil*. Austin: University of Texas Press.

Lucier, R. L. (1988). *The international political economy of coffee*. New York: Praeger.

Perlman, J. E. (1976). *The myth of marginality: Urban poverty in Rio de Janeiro*. Berkeley and Los Angeles: University of California Press.

Shoumatoff, A. (1987). *The capital of hope: Brasília and its people*. Albuquerque: University of New Mexico Press.

Vargas Llosa, M. (1997). *The war of the end of the world*. New York: Penguin Books.

Wilkening, E. (1969). *Some problems of development in the central plateau of Brazil*. Madison: University of Wisconsin Press.

16 Latin American Development in Perspective

In order to understand Latin America's regions and peoples, it is essential to place them and their experience in a broader context. This chapter seeks to provide that perspective. First, it compares Latin America's historical experience, and especially its long colonial history, with the experiences of other world regions with significant colonial legacies. Then it introduces the concept of "development" and identifies a variety of definitions and measures of that phenomenon, and explores some elements of the debate about the term and its application. What does "development" really mean? In addition, it documents the current state of economic and social development in Latin America and compares it to that of other developing world regions and specific countries. The chapter details how development levels in individual countries can vary tremendously—so much so that the statistical averages that are often cited create distorted impressions and lead to false conclusions about development levels in many Latin American countries.

The Colonial Experience in Perspective

For many of the world's countries, especially those in the Americas and Africa, and to a lesser extent in Asia and the Middle East, the colonial experience left an indelible mark on the country's history, economy, political system, and even culture. Although by no means unique, the Latin American experience is distinct from that of most other countries and regions in several ways.

In many ways, the colonial experience in Latin America was more profound and left a far deeper and more lasting imprint on the region than did colonial experiences in other regions. First, the colonial period in both Spanish and Portuguese America lasted far longer than that in other places. In Spanish America, the most densely populated and richest regions had been conquered and brought under Spanish colonial administration by 1540. While many peripheral or sparsely populated regions were never effectively integrated into Spain's American empire, Spanish colonial rule was largely uncontested for almost three centuries until the wave of independence wars between 1810 and 1825 when most countries achieved political independence. However, Spain did manage to hold onto two of its American colonies, Cuba and Puerto Rico, until the very end of the 19th century. The situation in Brazil, Portugal's American colony, was much the same, although early colonization lagged behind that of Spain's colonies and effective control of the colony was largely limited to coastal regions. Independence, however, occurred at about the same time. By comparison, British colonial rule in what is now the United States lasted barely 150 years, and only slightly longer in Canada. European colonies in Africa, Asia, and the Middle East were short-lived by comparison, with colonial rule often lasting less than 100 years.

In Latin America, the Portuguese and the Spanish introduced new agricultural production systems that radically transformed preexisting agricultural practices and land-use patterns. With the introduction of livestock by early colonists, *haciendas*, or immense livestock ranches, dominated much of the interior countryside in Mexico, Colombia, Cuba, Argentina, Brazil, and other Latin American colonies. In coastal regions, Europeans established plantations, specialized farms that produced sugar and other high-value agricultural crops for export. Long-distance trade predominated here as the specialized and high-value products from these plantations linked the Latin American colonies to the world economic system from the initial decades of the colonial period.

The colonial period profoundly transformed the region's settlement geography, especially in the Spanish colonies. Insofar as possible, Spanish civil authorities pursued a policy of concentrating indigenous peoples in village and towns rather than allowing them to live in dispersed dwellings or small hamlets in rural areas. Concentrated urban settlement facilitated social control, taxation, and labor conscription of native peoples. Similarly, ecclesiastical authorities also pursued similar policies, concentrating natives in *reducciones*, or mission settlements, to ease their evangelization efforts and to promote native conversion to Catholicism, as well as to facilitate the agricultural and craft production that the missions attempted to instill in their converts (Figure 16.1).

In no other world region did Europe's colonizing nations so effectively impose their religion, Christianity, and more specifically Catholicism, than in Latin America. Close to 80 percent of Latin Americans are at least nominally Roman Catholic today. The Roman Catholic Church is a well-established and powerful force in most of Latin America's countries, and while certainly less powerful than in the colonial period, the church has tremendous influence on political and social policy in most countries. Christianity's success in Latin America has made it one of the largest concentrations of Christians in the world at the beginning of the 21st century (Figure 16.2). Among Roman Catholics, Latin Ameri-

FIGURE 16.1. A modern-day mission settlement (a *reducción*) in Beni, Bolivia, 1987.

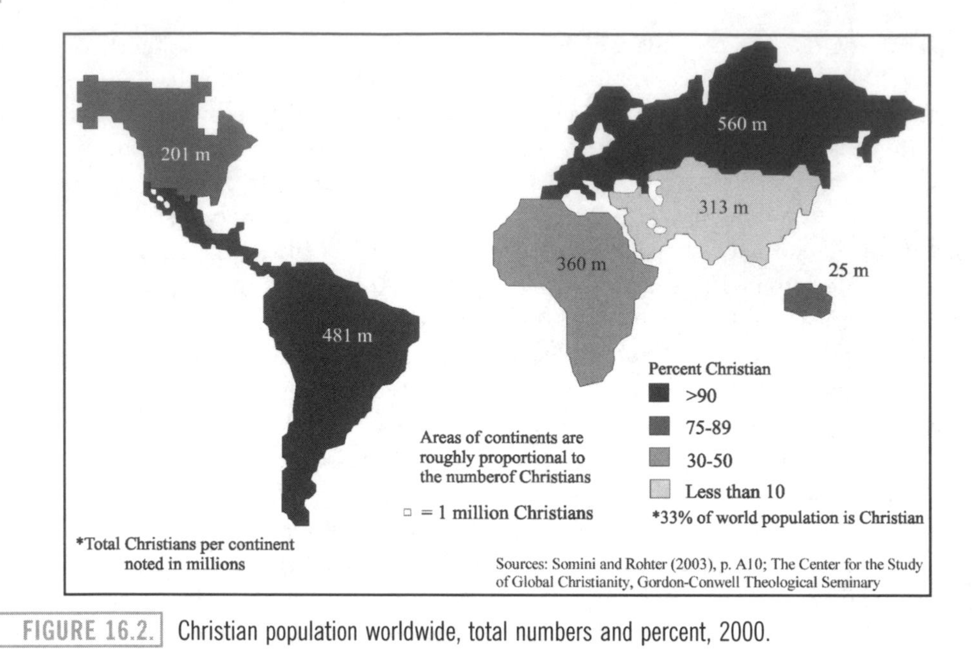

FIGURE 16.2. Christian population worldwide, total numbers and percent, 2000.

cans are the most numerous group. Evangelical Christians have become increasingly numerous in Latin America since the 1950s (Figure 16.3). Finally, religion has left an almost ubiquitous stamp on the urban geography of nearly every town in Latin America from south of the Rio Grande to the Central Valley of Chile. The Roman Catholic church, whether a chapel or a cathedral, always occupies a prominent location adjacent to each town's central plaza, regardless of whether the town is the smallest forgotten municipality or the national capital (Figure 16.4). While European colonists evangelized in other regions, and indeed enjoyed some successes in religious conversion in Africa, and to a lesser degree in Asia, in no instance has the influence been as deep and as long lived as in Latin America.

Another lasting legacy of the colonial period in Latin America is the framework for local government: the town and the town council. Drawing on the earlier influence of Roman governance on the Iberian Peninsula, both the Spanish and the Portuguese authorities established towns whose control and administration fell to local town councils, often appointed by the central government, but sometimes selected through quasi-democratic processes. These municipal governments, focused on urban centers, but also encompassing the surrounding countryside, form the backbone of local governance and grassroots democracy across Latin America today. While English administrative and governance structures are deeply woven into the fabric of American, Canadian, and Australian society, they were imposed in a landscape almost devoid of precontact occupants. But this was not at all the case in Latin America, Africa, Asia, or the Middle East. And in none of these regions was an administrative and governance structure imposed on an existing population (on a continental or even a subcontinental scale) as enduringly and completely as occurred in Latin America. For instance, the British ruled the Indian subcontinent through a system of local proxies. Other colonies, like Hong Kong (British) and Macau (Portuguese) in China, and Goa (Portuguese) in India, were tiny trading enclaves, powerful and rich, but nonetheless exerting little or no control over the hinterlands. In Africa, colonial control was often limited to coastal regions, mining districts, and major urban centers. Interior regions remained largely beyond the control or direct interest of the colonial powers.

Development, Levels of Living, and Quality of Life

Development, its definition and subsequent measurement, has been the subject of considerable discussion, debate, and study. In some ways there is little agreement on its definition and how it should be measured. While a precise, or even a generic, definition, is far from accepted, most social scientists and politicians, and indeed the educated public, is likely to at least agree that the term *development* carries with it the idea of progress, advancement, improvement, or betterment for a people, society, or country.

Many argue that development is economic growth. Measuring growth hinged on the ability of economists to measure national income in one form or another and then to standardize it to a per capita measure. A common measure is the GDP, the gross national product, a monetary measure of all goods and services produced inside a country. A related and more comprehensive measure is the GNI, the gross national income, which measures the total monetary value of all a country's production including domestic production and net income sources from individuals and corporations located outside the country but that is repatriated. This statistic provides a convenient monetary measure in U.S. dollars to compare total and per capita national incomes. Nevertheless, critics have noted that the purchasing power of a U.S. dollar varies widely from country to country and that per capita measures of the GNI fail to accurately reflect real differences in living standards and purchasing power.

In an effort to address these concerns, a new measure has been developed to more fairly portray economic conditions among the world's countries and allow meaningful comparisons. This measure is called *purchasing power parity* (PPP). The measure standardizes

FIGURE 16.3. An evangelical bookstore at the bus station in Brasília, Brazil, 2004.

FIGURE 16.4. In a pattern typical of most Latin American towns, the Roman Catholic church sits on the central plaza in Punata, Bolivia, 1991.

country-by-country comparisons by calculating what the annual per capita GNI (more or less the average annual per capita income) in U.S. dollars for a particular country would buy in goods and services in the United States. For example, if $1.00 will buy a 25-lb. bag of potatoes in Bolivia but it would cost $7.00 to buy the same bag of potatoes in the United States, then the index of PPP would credit Bolivia with $7.00, the monetary value of the bag of potatoes, and not $1.00.

Specific country-by-country comparisons help us to understand this measure and suggest why it may be a more telling statistic than simply comparing monetary income among nations (Table 16.1). For the United States, both the per capita GNI and the per capita GNI-PPP are the same, reflecting the fact that the United States is used as the benchmark for the GNI-PPP. Switzerland and Japan present interesting contrasts between the two measures. In both countries the per capita GNI is actually higher than in the United States, but the per capita GNI-PPP is 20–30 percent lower. This means that the $38,000 per capita GNI in Switzerland will only buy about $31,000 of equivalent goods and services in the United States. Or put another way, the same basket of products and services in Switzerland costs more than it costs in the United States. The reverse can also be true. Consider Bolivia and Brazil, for example. Per capita GNI in Bolivia is just $950, but that cash income would buy the equivalent of $2,240 worth of goods and services in the United States. This occurs, because as is noted above, the $1.00 bag of potatoes in Bolivia would cost $7.00 in the United States. Both income and purchasing power are greater in Brazil than in Bolivia, but the pattern is the same. In Brazil per capita GNI is about $3,000 and the per capita GNI-PPP is just over $7,000.

Assessing the significance of economic statistics can be tricky. Whenever doing so, one should remember that statistics may obscure as much as they reveal. While a rising tide lifts all boats, such is not always the case with increases in measures of economic well-being. Is it possible for GNI and GNI-PPP to increase substantially in a country, and yet have increases in poverty rates and a declining standard of living for the majority of the population? The answer is yes. One may reasonably wonder just how that might be true. Simply put, population growth may outpace economic growth, and, as a consequence, once the increase in population numbers is considered, then per capita GNI and per capita GNI-PPP could actually decrease. Many observers argue that this is precisely what happened in many Latin American countries during the past two decades as neoliberal economic reforms and trade liberalization have created increases in GNI and GNI-PPP. However, steady increases in population combined with a pattern of income distribution heavily concentrated in a small proportion of the population have actually resulted in more people, as well as a higher percentage of people, living at or below the poverty level than previously.

TABLE 16.1. Gross National Income (GNI) Per Capita and GNI Purchasing Power Parity (PPP) Per Capita among Latin American and Selected Countries, 2001

Country	GNI-PPP per capita, 2001 (US$)	GNI per capita, 2001 (US$)
United States	34,280	34,400
Switzerland	**30,970**	**38,330**
Canada	**26,530**	**21,930**
Japan	**25,550**	**35,610**
United Kingdom	**24,340**	**25,120**
Finland	**24,030**	**23,780**
Spain	19,860	14,300
Argentina	10,980	6,940
Costa Rica	9,260	3,970
Chile	8,840	4,590
Uruguay	8,250	5,710
Mexico	8,240	5,530
Brazil	7,070	3,060
Colombia	6,790	1,890
Dominican Republic	6,650	2,230
Venezuela	5,590	4,760
Panama	5,440	3,260
Paraguay	5,180	1,350
El Salvador	5,160	2,040
Peru	4,470	2,000
Guatemala	4,380	1,680
China	**3,950**	**890**
Ecuador	2,960	1,080
India	**2,820**	**470**
Honduras	2,760	900
Bolivia	2,240	950
Ethiopia	**800**	**100**
Nigeria	**790**	**290**
Cuba	N/A	*a*
Nicaragua	N/A	*b*

*Non-Latin American countries are in bold typeface.
[a]Estimated to be lower middle income ($746–$2,975).
[b]Estimated to be low income ($745 or less).
Source: World Bank (2002).

While economic measures are widely used to assess levels of development despite their limitations, other approaches to the study of development emphasize measuring societal and individual welfare and quality-of-life issues. Proponents of these approaches argue that using economic statistics often presents an incomplete picture of an individual's or society's living conditions. They contend that measuring health, living conditions, political freedom, personal security, or working conditions are as important as measuring per capita income or per capita GNI. Common quality-of-life measures include statistical measures of infant mortality (deaths per 1,000 live births), life expectancy at birth (in years), literacy rates (in percent of population), and daily caloric intake (in number of calories). These kinds of quality-of-life metrics are often easy to generate from existing data or survey sources, many compiled by government agencies in each country.

Other quality-of-life measures are less straightforward in terms of the data upon which they are dependent and include a much greater element of subjectivity. Nonetheless, such rating systems provide another useful perspective on development. Freedom House, a nonprofit agency, assesses the state of "freedom" in the world's countries by examining the state of political rights and civil liberties in each country and assigning index rankings ranging from 1 to 7 for each factor. Assessments of political rights and civil liberties are based on a series of questions addressing issues related to representative governance, political expression, legal and judicial protections, and other factors. Using a rating system for these factors, index values are created and countries are classified as "free" (1–2.5), "partly free" (3–5.5), and "not free" (6–7). Transparency International compiles another country index called the "Corruption Perception Index" (CPI) which compares perceived levels of corruption among politicians and public officials in the world's nations. This index, produced annually since 1995, draws on the surveys of other institutions, individuals in business and academia, and professional risk analysts inside and outside the country. Some 13 different institutions contribute survey data to Transparency International, which then rates the world's countries on a scale from 1 to 10, with 1 representing widespread and deep corruption, and 10 representing the lowest level of perceived corruption.

Latin American Development in World Perspective

How does Latin America compare to other world regions with respect to its development? It depends in part on what measures one chooses to employ to assess the region's comparative development.

Since economic "yardsticks" are among the most commonly used, it makes sense to first place Latin America in world perspective by focusing on GNI-PPP. Perhaps surprisingly to many, when compared to aggregate statistics on an international stage, Latin America's GNI-PPP is just about at the world average (Figure 16.5). Per capita GNI-PPP in Latin America stands at $6,820, not much below the world average of $7,160. Asia lags behind with an average of $4,290, while Africa and sub-Saharan Africa both have per capita GNI-PPPs that are less than one-third those for Latin America, $2,120 and $1,710, respectively.

An examination of quality-of-life measures reveals that the region is often far better off than world averages and in some instances approximates conditions in Western Europe, the United States, and Canada (Table 16.2). In terms of the infant mortality rate, a widely accepted measure of quality of life, Latin America falls below world averages by a substantial margin. While the world's average is 55 deaths per 1,000 live births, Latin America's rate is just 29 deaths per 1,000 live births. However, Latin America's rates do seem high when compared to 7 per 1,000 for the United States and Canada or 8 per 1,000 for Western Europe. On the other hand, Latin Americans tend to live long lives, with the life expectancy at birth being 71 years, while the world's average is 67 years and that in the United States and Canada is 77 years. Literacy rates in Latin America also average considerably higher than comparable figures for the world. Female literacy, for those 15 years and older, in Latin America exceeds world averages by nearly 20 percent. Male literacy rates are similar, about 89 percent of those 15 years and older, and also exceed world averages, but only by about 6 percent.

Other quality-of-life measures provide useful comparisons between Latin America and other world regions. Freedom House's annual assessments of the state of political rights and civil liberties and the subsequent categorization of individual countries as free, partly

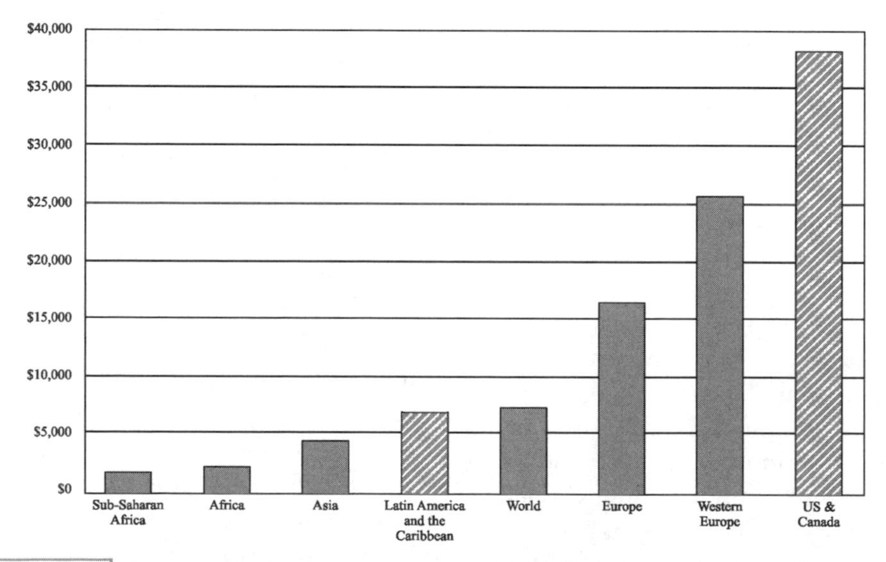

FIGURE 16.5. Gross national income-purchasing power parity (GNI-PPP) in major world regions, 2003.

TABLE 16.2. Development Indicators for Major World Regions, ca. 2003

	World	Africa	Sub-Saharan Africa	U.S. and Canada	Latin America and the Caribbean	Asia	Europe	Western Europe
Population mid-2003 (millions)	6,314	861	711	323	540	3,830	727	185
Birthrate/1,000 population	22	38	40	14	23	20	10	11
Death rate/1,000 population	9	14	16	8	6	7	12	10
Rate of natural increase[1]	1.3	2.4	2.5	0.5	1.7	1.3	–0.2	0.1
Infant mortality[2]	55	88	93	7	29	54	8	4
Life expectancy (years)	67	52	48	77	71	67	74	79
Urban population (%)	47	33	30	79	75	38	73	78
GNI-PPP per capita, 2001 (US$)[3]	7,160	2,120	1,710	33,510	6,820	4,290	16,270	25,430
Literacy rate, female (%)[4]	69	52	53	—	88	65	—	—
Literacy rate, male (%)[4]	83	70	69	—	89	82	—	—

[1]Rate of natural increase of population (Birthrate - death rate).
[2]Deaths per 1,000 live births.
[3]Gross national income (GNI) adjusted for purchasing power parity (PPP).
[4]Literacy rate for ages 15 and over (2000).
Source: Population Reference Bureau (2004)

free, and not free provides at least an idea of the strength of democracy and freedom worldwide. While many countries in Africa, the Middle East, and across Asia are categorized as "not free," only one Latin American nation, Cuba, falls in that category (Figure 16.6). Many of the region's countries, however, are classified as only "partly free" because civil and political rights are restricted, corruption is common, and the rule of law is weak. Freedom House includes Guatemala, Nicaragua, Honduras, Colombia, Venezuela, Ecuador, Paraguay, and Argentina in this group. The remaining Latin American countries are classified as "free." Conditions in many countries, however, are fluid, and rankings can change from year to year based on local and national political events. For example, Mexico, Brazil, and Peru were classified as only "partly free" in 2000, but as "free" in 2002.

Transparency International's CPI provides another perspective on Latin America and how it compares to other countries in the world community (Table 16.3). Countries are ranked on a scale from 1 to 10 on the CPI, with a 1 representing the most extreme level of perceived corruption and a 10 representing the lowest level of perceived corruption. Countries are also placed in rank order from the least corrupt country, 1, to the most corrupt country, 133. Countries with the same score fall within the same rank. On a world scale, Finland has a CPI of 9.7 and ranks No. 1, indicating that it is perceived as the least corrupt country on the planet. By way of comparison, Canada's and the United Kingdom's CPIs are 8.7, the United States's CPI is 7.5, Japan's CPI is 7.0, and Spain's CPI is 6.9. By and large, however, Latin America's countries rank poorly with respect to both the CPI and country rankings. Chile, with a CPI ranking of 7.4 and a country ranking of 20, is perceived to be the most corruption-free country in Latin America. The CPI rankings for most Latin American countries are much lower, with the majority of the CPI rankings falling between 2.0 and 4.0. There is a clear sense that business and government activities in these countries are seriously hampered by endemic corruption. Six Latin America countries have country rankings below 100, with Paraguay ranking the lowest in Latin America with a

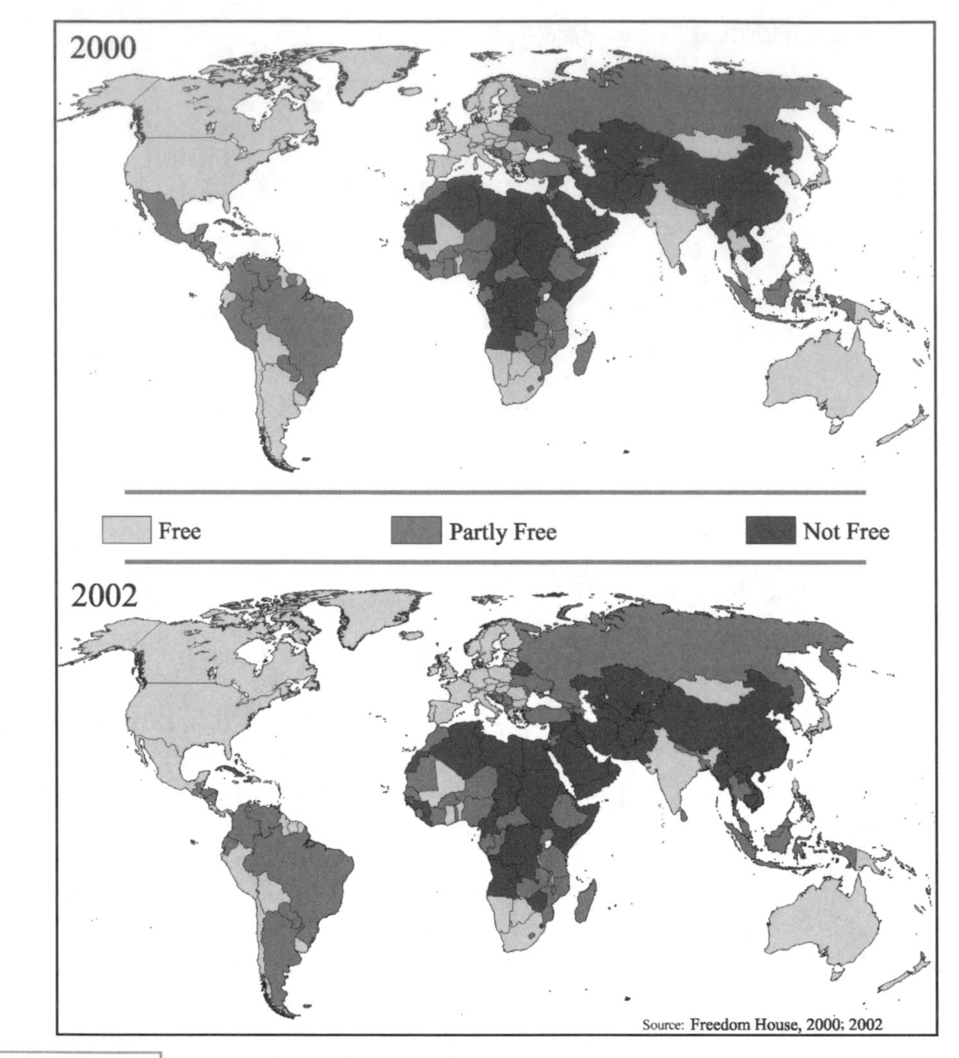

FIGURE 16.6. World freedom, 2000 and 2002: Latin America in perspective.

rank of 129th and a CPI score of only 1.6! These data suggest that the citizens of many Latin American countries face significant difficulties in their struggles to establish honest government, open and transparent markets, and effective and efficient business practices.

Inside Latin America: Comparative Development

As noted earlier in this chapter, development and quality-of-life indicators for Latin America suggest that when compared to other developing world regions, especially Africa and Asia, the region is relatively well off. Yet it is important to understand that the region is not uniform and that an examination of the development and quality-of-life indicators at a national level within Latin America show considerable differences among individual countries. The comparative purchasing power (GNI-PPP), for instance, averages $6,820 in Latin America, but regional patterns and individual country statistics vary widely (Table 16.4, Figure 16.7). The Southern Cone of South America, including Argentina, Chile, and Uruguay, stands out as a relatively prosperous region within Latin America: the annual GNI-PPP in all these countries exceeds $8,000. Argentina, historically the region's most affluent nation, but buffeted in recent decades by economic hardships, still ranks at the top of the heap regionally with a GNI-PPP of nearly $11,000. Other comparatively prosperous countries include Costa Rica and Mexico, both of whose GNI-PPPs exceed $8,000. At the other end of the spectrum, most Central American countries as well as most Andean republics are characterized by comparably lower levels of PPP. The most impoverished, with annual GNI-PPPs below $3,000, include Honduras, Ecuador, and Bolivia. The middle ground is occupied by Brazil, Colombia, and the Dominican Republic, with average annual GNI-PPPs between $6,000 and $7,000. A quick look at comparative statistics for a handful of industrialized nations in North America and Western Europe provides additional perspective on these data (Table 16.5). In the United States the average annual GNI-PPP is nearly $35,000, but falls to about $27,000 in Canada. In Western Europe, Spain and the United Kingdom stand at about $20,000 and $24,000, respectively.

Infant mortality is often used as a key

TABLE 16.3. Perceptions of Freedom and Corruption among Latin American Countries and Selected Other Countries, 2003

	Freedom			Corruption	
Country Name	Political Rights	Civil Liberties	Rating	Perception Index	Country Rank
Finland	1	1	Free	**9.7**	**1**
Switzerland	**1**	**1**	**Free**	**8.8**	**8**
Canada	**1**	**1**	**Free**	**8.7**	**11**
United Kingdom	**1**	**2**	**Free**	**8.7**	**11**
United States	**1**	**1**	**Free**	**7.5**	**18**
Chile	1	1	Free	7.4	20
Japan	**1**	**2**	**Free**	**7.0**	**21**
Spain	**1**	**2**	**Free**	**6.9**	**23**
Uruguay	1	1	Free	5.5	33
Cuba	7	7	Not free	4.6	43
Costa Rica	1	2	Free	4.3	50
Brazil	2	2	Free	3.9	54
Colombia	4	4	Partially free	3.7	59
El Salvador	2	3	Free	3.7	59
Peru	2	3	Free	3.7	59
Mexico	2	3	Free	3.6	64
China	**7**	**6**	**Not free**	**3.4**	**66**
Panama	1	2	Free	3.4	66
Dominican Republic	3	2	Free	3.3	70
India	**2**	**3**	**Free**	**2.8**	**83**
Nicaragua	3	3	Partially free	2.6	88
Argentina	2	2	Free	2.5	92
Ethiopia	**5**	**5**	**Partially free**	**2.5**	**92**
Guatemala	4	4	Partially free	2.4	100
Venezuela	3	4	Partially free	2.4	100
Bolivia	3	3	Partially free	2.3	106
Honduras	3	3	Partially free	2.3	106
Ecuador	3	3	Partially free	2.2	113
Paraguay	3	3	Partially free	1.6	129
Nigeria	**4**	**4**	**Partially free**	**1.3**	**132**

*Non-Latin American countries are in bold typeface.
Sources: Transparency International (2003); Freedom House (2003).

TABLE 16.4. World Development Indicators: Socioeconomic Statistics for Latin America, ca. 2003

Country	Life expectancy (years)	Urban population (%)	GNI-PPP per capita, 2001 (US$)[1]	Literacy rate, female (%)[2]	Literacy rate, male (%)[2]
Costa Rica	79	59	9,260	96	
Puerto Rico	77	71	18,090		
Chile	76	87	8,840	96	96
Cuba	76	75	—	97	97
Mexico	75	75	8,240	89	93
Uruguay	75	93	8,250	98	97
Argentina	74	89	10,980	97	97
Panama	74	62	5,440	91	93
Venezuela	73	87	5,590	92	93
Colombia	71	71	6,790	92	92
Ecuador	71	61	2,960	90	93
Honduras	71	46	2,760	75	74
Paraguay	71	54	5,180	92	94
El Salvador	70	58	5,160	76	82
Brazil	69	81	7,070	85	85
Dominican Repbulic	69	61	6,650	84	84
Nicaragua	69	57	—	70	67
Peru	69	72	4,470	85	95
Guatemala	66	39	4,380	61	76
Bolivia	63	63	2,240	79	92

[1]Gross national income (GNI) adjusted for purchasing power parity (PPP).
[2]Literacy rate for ages 15 and over (2000).
Source: Population Reference Bureau (2004).

quality-of-life indicator, reflecting to some degree a range of factors including maternal prenatal care, quality and quantity of the food supply, and the availability and accessibility of quality medical care. The regional average is 29 per 1,000 live births, but Latin America's countries show tremendous variation (Figure 16.8, Table 16.6). In this regard, some of the region's countries are comparable to developed industrialized countries in North America and Western Europe whose infant mortality rates per 1,000 range between 3 and 7. Cuba's infant mortality rate of 6 per 1,000 is the lowest in Latin America by a significant margin—it is actually less than that for the United States, which is 7 per 1,000! Low infant mortality rates are also characteristic of the Southern Cone countries and Costa Rica. On the other hand, many Central American nations, as well as those in the Andes, report high infant mortality rates, usually in excess of 30 per 1,000. Bolivia is an extreme case: its in-

FIGURE 16.7. Gross national income-purchasing power parity (GNI-PPP) in Latin America in U.S. dollars, 2003.

TABLE 16.5. World Development Indicators for Canada, Spain, the United Kingdom, and the United States, ca. 2003

Country	Canada	Spain	United Kingdom	United States
Population mid-2003	31,600,000	41,300,000	59,200,000	291,5000,000
Birthrate/1,000 population	11	10	11	14
Death rate/1,000 population	7	9	10	9
Rate of natural increase[1]	0.3	0.1	0.1	0.6
Infant mortality[2]	5.3	3.5	5.4	6.9
Life expectancy (years)	79	79	78	77
Urban population (%)	79	64	90	79
GNI-PPP per capita, 2001 (US$)	$26,530	$19,860	$24,340	$34,280
Literacy rate, female (%)[3]	—	97	—	—
Literacy rate, male (%)[3]	—	99	—	—

[1]Rate of natural increase of population (Birthrate - death rate).
[2]Deaths per 1,000 live births.
[3]Literacy rate for ages 15 and over (2000).
Source: Population Reference Bureau (2004).

fant mortality rate is the highest in Latin America at 61 per 1,000.

Other key quality-of-life indicators reflect similar regional patterns across Latin America, although regional differences are less. Life expectancy is shortest in the Andes, especially in Bolivia (63 years), and in some Central American countries. In contrast, the longevity of citizens of Southern Cone countries, Costa Rica, and Cuba extends into their mid- to late 70s, nearly as long as it does for those in many industrialized countries of North America and Western Europe (Figure 16.9, Tables 16.4 and 16.5).

Literacy rates too show similar patterns. Rates are lowest, 60–80 percent, in the Andes and some Central America countries. Often they are the lowest in those countries like Bolivia, Peru, and Guatemala with the largest proportions of indigenous peoples, many of whom still speak native languages like Maya or Quechua. Male literacy rates exceed female literacy rates in these countries by as much as 10–15 percent. On the other hand, the data suggest that many Latin American countries have come close to achieving fully literate societies with no differences in literacy between males and females. Literacy rates in all of the Southern Cone countries (Chile, Argentina, Uruguay), Costa Rica, and Cuba exceed 95 percent, rates commensurate with industrial countries in North America and Western Europe (Figure 16.10, Tables 16.4 and 16.5).

It is important to underscore the point that the use of arithmetic averages over large areas and large numbers of people can obscure as much as it reveals. In Latin America, it might be reasonably argued, the greatest differences in development and levels of living are not so much *between countries* as *within*

FIGURE 16.8. Infant mortality in Latin America, 2003.

TABLE 16.6. World Development Indicators: Selected Population Variables for Latin America, ca. 2003

Country	Population mid-2003	Birthrate/1,000 population	Death rate/1,000 population	Rate of natural increase[2]	Infant mortality[1]
Cuba	11,300,000	12	7	0.5	6
Chile	15,800,000	18	6	1.2	10
Costa Rica	4,200,000	18	4	1.4	10
Puerto Rico	3,900,000	16	7	0.8	11
Uruguay	3,400,000	16	9	0.6	14
Argentina	36,900,000	19	8	1.2	17
Venezuela	25,700,000	24	5	1.9	20
Panama	3,000,000	23	5	1.8	21
Mexico	104,900,000	29	5	2.4	25
Colombia	44,200,000	23	6	1.8	28
El Salvador	6,600,000	29	6	2.3	30
Dominican Republic	8,700,000	25	6	1.9	31
Nicaragua	5,500,000	32	5	2.7	31
Brazil	176,500,000	20	7	1.3	33
Peru	27,100,000	26	7	2.0	33
Honduras	6,900,000	34	5	2.9	34
Ecuador	12,600,000	27	6	2.1	35
Paraguay	6,200,000	31	5	2.7	37
Guatemala	12,400,000	33	7	2.6	41
Bolivia	8,600,000	32	9	2.3	61

[1]Deaths per 1,000 live births.
[2]Rate of natural increase of population (Birthrate - death rate).
Source: Population Reference Bureau (2004).

the national borders of individual countries. Rural/urban disparities and core/periphery differences are convenient shorthand terms that can be used to summarize and explain in broad brush strokes the genesis of these contrasts. Disparities in levels of development and quality of life between the million-plus national capitals and metropolitan centers, and provincial towns, villages, and rural areas can be dizzying.

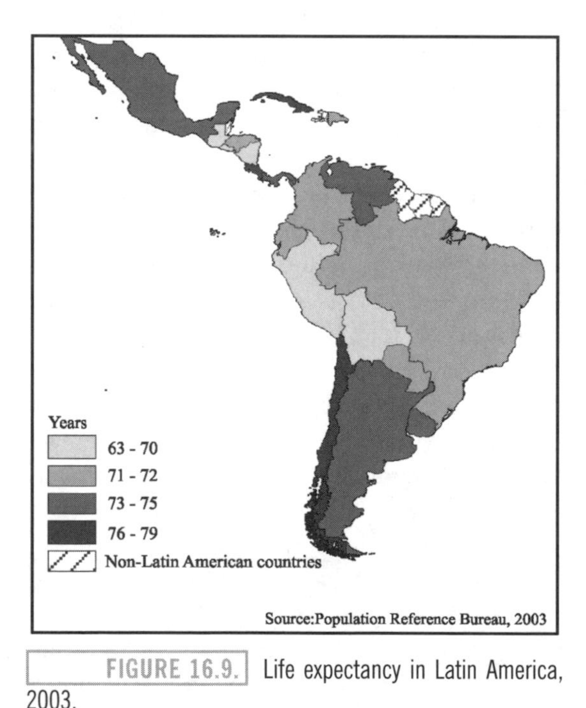

FIGURE 16.9. Life expectancy in Latin America, 2003.

The Mexican Example

An examination of several key statistics at the country level illustrates how extreme these disparities can be. While Mexico provides the examples in this case, it is only illustrative of the disparities that are common in almost all Latin American countries. Infant mortality shows considerable variation throughout Mexico. It ranges from as few as 7 deaths per 1,000 live births in some states, a rate comparable to

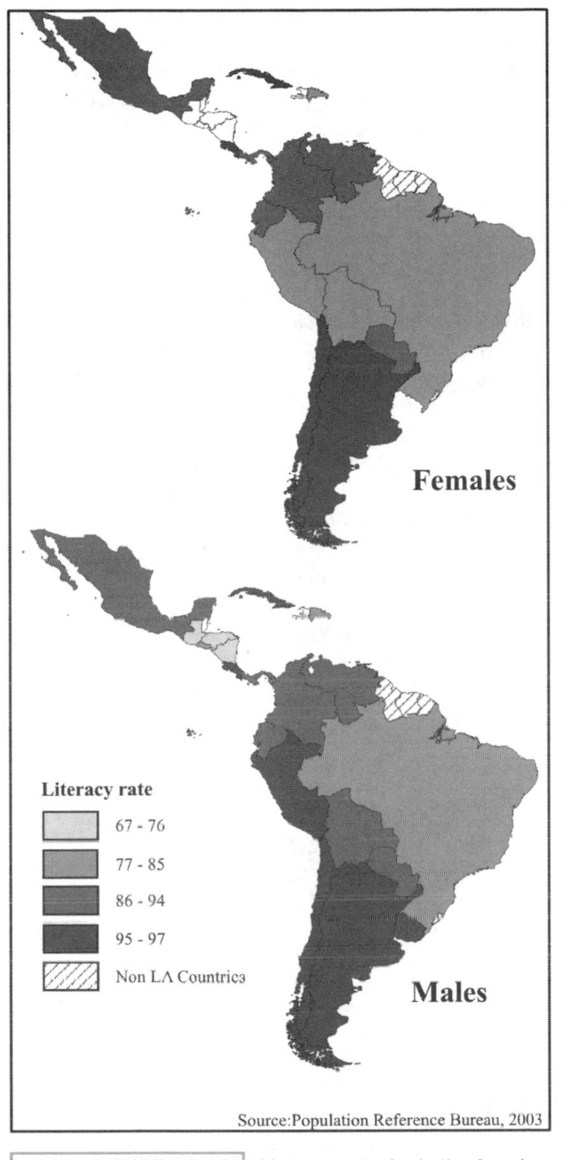

FIGURE 16.10. Literacy rate in Latin America, 2003.

some Western European nations, to as many as 24 deaths per 1,000 live births in others (Figure 16.11). Wages and wage rates are often another useful measure of development and quality of life. Mexico, like many other countries, sets a minimum wage rate for workers, but, as in many other Latin American countries, not all workers actually receive it. In early 2001, the daily minimum wage in Mexico was 40 pesos, or approximately $4.25 per day. Geographically, a striking pattern is evident (Figure 16.12). In the states along the country's northern border, Baja California, Sonora, Chihuahua, Coahuila, and Nuevo Leon, almost 90 percent of the labor force was paid at least the minimum daily wage. However, as one moves south, the percent decreases, such that as little as 44 percent of the labor force in the southern states of Oaxaca and Chiapas earns the minimum wage. Literacy rates for both males and females also show strong regional variation in the nation. Female literacy rates, for example, are less than 76 percent in the south, but reach 95 percent in the north (Figure 16.13).

Summary

The unique character of the colonial experience has had a profound impact on the development of Latin America. The duration of the colonial rule of Spain and Portugal in the region far outstrips the length of European colonial rule in other parts of the world. Spanish and Portuguese rule established settlement

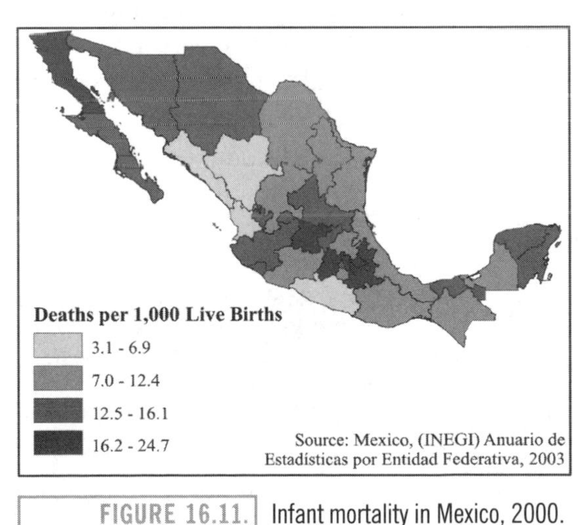

FIGURE 16.11. Infant mortality in Mexico, 2000.

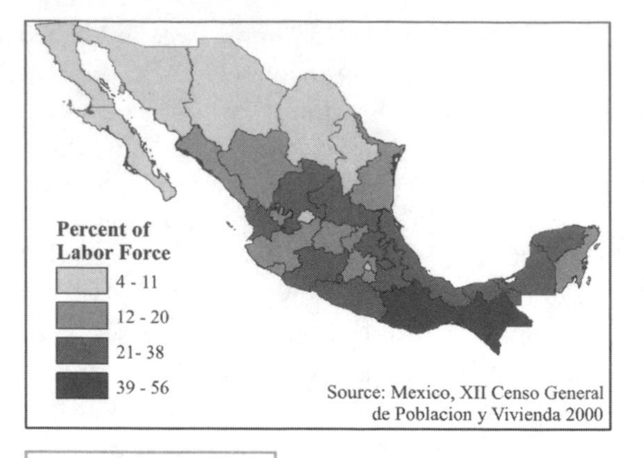

FIGURE 16.12. Labor force earning minimum salary or less in Mexico, 2000.

patterns, Catholicism as the predominant faith, and the framework for local government.

Development in Latin America can be examined using many measures, including those that measure economic as well as social well-being. In world perspective, the region's development compares favorably with that of Africa, the Middle East, and much of Asia, but it lags behind that of North America and Western Europe. Within Latin America significant differences can be observed between the levels of development among the region's countries. Striking differences in levels of development also occur within most of the region's countries, with core regions and national capitals generally being most developed and peripheral regions and secondary cities being considerably less developed.

Further Reading

Abel, C., and Lewis, C. (Eds.). (2002). *Exclusion and engagement: Social policy in Latin America*. London: Institute of Latin American Studies, University of London.

Bethell, L. (Ed.). (1989). *Latin America: Economy and society, 1870–1930* . New York: Cambridge University Press.

Browder, J., and Godfrey, B. (1997). *Rainforest cities, urbanization, development, and globalization of the Brazilian Amazon*. New York: Columbia University Press.

Brundenius, C., and Lundahl, M. (1982). *Development strategies and basic needs in Latin America: Challenges for the 1980s*. Boulder, CO: Westview Press.

Cardoso, F. H., and Faletto, E. (1979). *Dependency and development in Latin America*. Berkeley and Los Angeles: University of California Press.

Foders, F., and Feldsieper, M. (Eds.). (2000). *The transformation of Latin America: Economic development in the early 1990s*. Northampton, MA: Elgar.

Furtado, C. (1970). *Economic development of Latin America*. Cambridge, UK: Cambridge University Press.

Gallup, J. L., et al. (2003). *Is geography density?: Lessons from Latin America*. Palo Alto, CA: Stanford University Press.

Green, D. (2003). *Silent revolution: The rise and crisis of market economics in Latin America*. London: Latin American Bureau.

Hofman, A. (2000). *The economic development of Latin America in the twentieth century*. Northampton, MA: Elgar.

Inter-American Development Bank. (2002). *Beyond borders: The new regionalism in Latin America*. Baltimore: distributed by Johns Hopkins University Press.

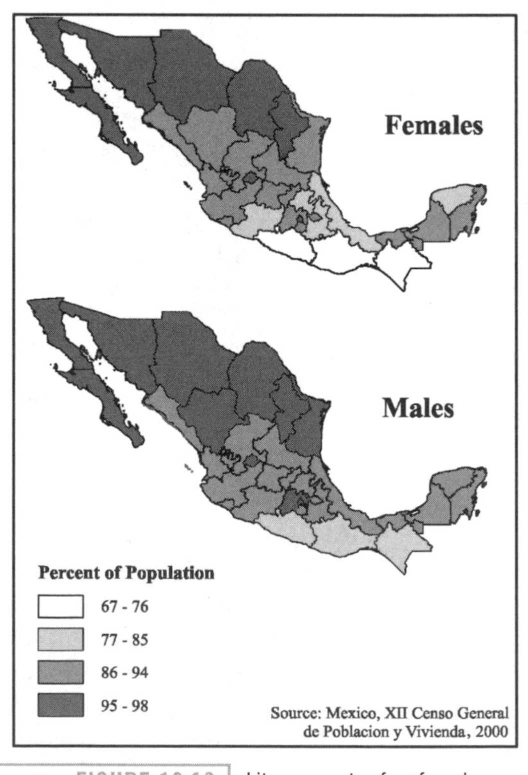

FIGURE 16.13. Literacy rate for females and males in Mexico, 2000.

Kitching, G. (Ed.). (1990). *Development and underdevelopment in historical perspective*. New York: Routledge.

Little, W., and Posada-Carbó, E. (Eds.). (1996). *Political corruption in Europe and Latin America*. New York: St. Martin's Press.

Mahon, J. (1996). *Mobile capital and Latin American development*. University Park: Pennsylvania State University Press.

Nagel, S. (Ed.). (1994). *Latin America, development and public policy*. New York: St. Martin's Press.

Preston, D. A. (Ed). (1996). *Latin American development: Geographical perspectives* (2nd ed.). Harlow, UK: Longman.

Tapinos, G., Mason, A., and Bravo, J. (1997). *Demographic responses to economic adjustment in Latin America*. Oxford, UK: Clarendon Press.

Thorp, R. (1998). *Progress, poverty and exclusion: An economic history of Latin America in the twentieth century*. Baltimore: Johns Hopkins University Press.

Ward, J. (1997). *Latin America: Development and conflict since 1945*. London: Routledge.

Wirth, J. D. (1991). *The politics of Brazilian development, 1930-1954* . Palo Alto, CA: Stanford University Press.

17 The Pampa

The Pampa is the dominant settled geographical region of the temperate Southern Cone of South America. It is a vast plain with fertile soils and a climate that ranges from humid to semiarid. The Pampa extends inland for 600 km in all directions from the city of Buenos Aires. Although often associated exclusively with the country of Argentina, it also encompasses all of the country of Uruguay and much of Brazil's southernmost state, Rio Grande do Sul (Figure 17.1).

The Pampa is one of the world's richest agricultural regions. It is one of the foremost producers of premium beef cattle, as well as a major supplier of wheat, soybeans, and other grain crops to world markets. Its agricultural significance is comparable to the Corn and Wheat Belts of the U.S. Midwest, the Ukraine in eastern Europe, and Australia's grain belt.

Approximately 37 million people inhabit the Pampa. About two-thirds of this total locates within the Argentine portion of the region, with the remainder in Uruguay and Rio Grande do Sul, Brazil. Surprisingly, about 80 percent of the population is urban. The Buenos Aires metropolitan region accounts for about one-third of the regional population total. While agriculture was critical to the region's economic development and still plays a significant role in the regional economy, manufacturing, commerce, and service industries account for most of its economic base today.

Geography and Environment

The Pampa is often conceptualized as being composed of two major parts. The first is humid natural grassland, centered on the estuary of the Río de la Plata. This tall-grass prairie, similar in appearance to the moist eastern portions of the Great Plains in the United States, grades into the second part, a drier short-grass prairie. In the south, this drier region is abutted by the windswept Patagonian steppe, while in the west and northwest the short-grass prairie gives way to increasingly sparse and xerophytic vegetation. Desert conditions prevail along the western and northwestern front of the Andes. These two regions, known as the Cuyo and the Northwest, respectively, are subject to a strong rain shadow effect created by the Andean mountain chain. Northward from the core of the Pampa on the estuary of the Río de la Plata and along the course of its principal tributary, the Rio Paraná, the ecology is dominated by the presence of the river system, and the Pampa conditions disappear. This humid, marshy lowland, dissected by the Paraná and its tributaries, is known locally as Mesopotamia. This humid area disappears upstream and grades into the Gran Chaco, semiarid interior plains dominated by scrub and thorn forest vegetation. In the northeast, across the estuary, the Pampa continues into Uruguay and the southern Brazilian state of Rio Grande do Sul. Low hills,

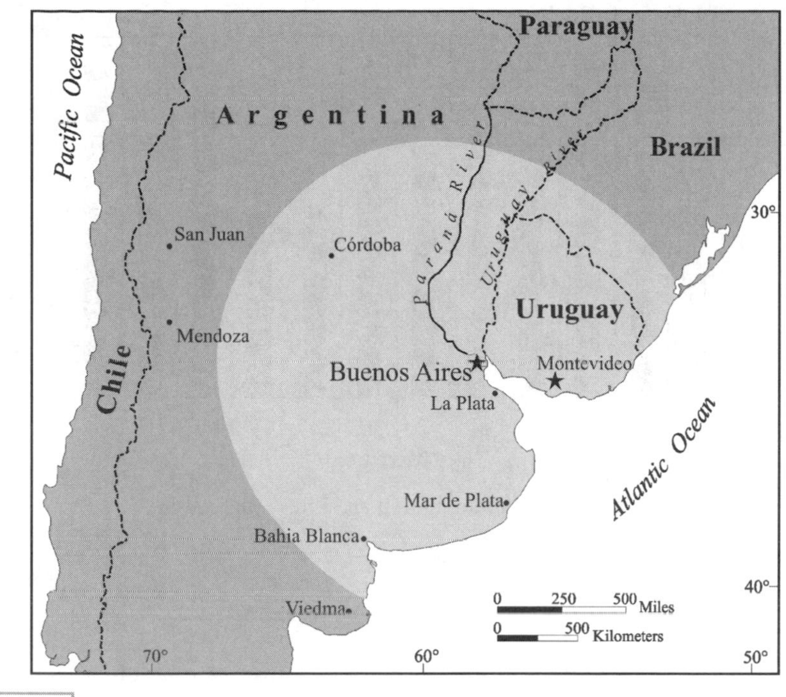

FIGURE 17.1. The Pampa.

plateaus, and uplands of modest elevations become frequent as the Pampa grades into the other southern Brazilian states of Santa Catarina and Paraná, marking the beginnings of the Brazilian Highlands.

The climate of the Pampa is temperate and is described as humid subtropical (Figure 17.2). The proximity of much of the region to the coast moderates the climate, so temperature variation is less than might be expected at that latitude (28°–40°S). Mean summer temperatures in the warmest summer months, December and January, at coastal locations usually average in the low 20°sC and are a few degrees higher further inland. Average temperatures in the coolest winter months, June and July, range from 8° to 10°C and are slightly lower in the interior. Periods of extended cold are uncommon, snowfall is rare, and hard freezes are infrequent over most of the Pampa. Rainfall occurs throughout the year, but more precipitation occurs during the warm summer months, when thundershowers are a frequent occurrence. Rainfall is most plentiful in the north and east and declines toward the south and west. In southern Brazil and Uruguay, annual precipitation totals range between 1,000 and 1,500 mm, while less than 1,000 mm characterizes most of the Argentine Pampa. On the dry southern and western margins of the Pampa, precipitation declines to below 500 mm annually. Rain-fed crop agriculture is possible over most of the Pampa, except in the south and west where precipitation is lower.

Historical Geography and Economic Development

The region's indigenous occupants eked out a subsistence existence as seminomadic hunters and gatherers. They specialized in hunting animals native to the Pampa. Among these were the rhea, a large flightless bird related to the ostrich and the emu, and the guanaco, a member of the Camilidae family related to the al-

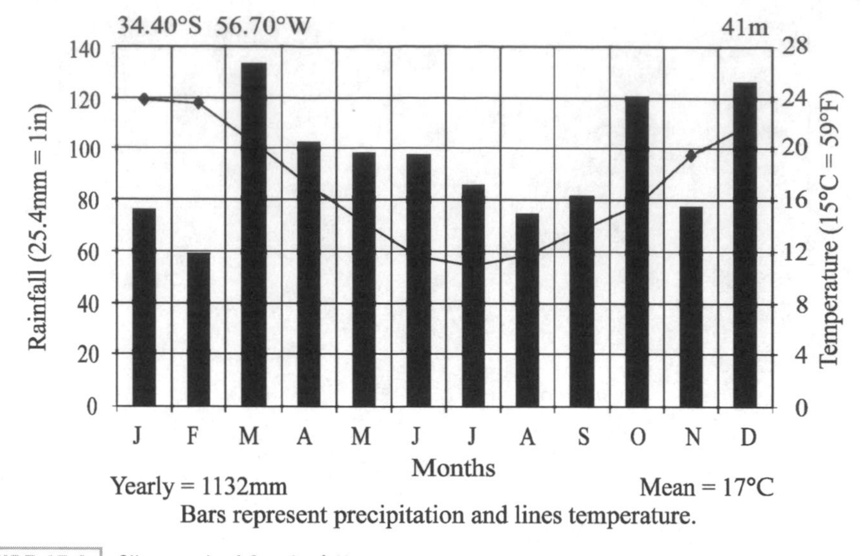

FIGURE 17.2. Climograph of San José, Uruguay.

paca and the llama of the Andes. In addition to employing the lance and the bow and arrow in their hunting, these peoples invented the use of the *bolas*, a unique hunting tool comprised of weighted balls attached to short lengths of rope or cord. When thrown accurately, this hunting tool can entangle and capture the hunter's prey. The social organization, material culture, and technological level of these peoples were rudimentary. Clan and tribal forms of social organization dominated. They had no permanent settlements or dwellings, and they did not make pottery, textiles, or metalwork.

The region's low indigenous population numbers, the natives' rudimentary subsistence hunting economies, and their tenacious independence and willingness to aggressively confront the Spanish conquistadores and colonists helped keep the Pampa on the periphery of Spain's New World colonies for much of the colonial period. The first Spaniards who attempted any settlement in the Pampa found it inhospitable. The Spanish noble and explorer Juan de Mendoza established the city of Buenos Aires early in the colonial period, in 1536. Some 1,500 men comprised the original population of the settlement. Easily exploitable resources were few, the indigenous population was sparse and disinclined to aid the Spaniards, and early attempts at agriculture were inadequate and unsuccessful. The Spaniards alienated the local Querandi Indians, who responded with repeated attacks on the settlement. Within 5 years, the Spanish had abandoned their settlement in favor of more hospitable local peoples and a more forgiving environment nearly 1,000 km up the Paraná River at Asunción (Paraguay). In 1580, a second effort at establishing Buenos Aires succeeded.

The settlement and economic exploitation of the extensive grasslands of the Pampa evolved slowly (Figure 17.3). Buenos Aires remained a small, remote urban outpost of the Spanish colonial empire, and settlement of the Pampa was sparse. Early exploitation of the grasslands focused largely on the hunting of feral cattle and horses for their hides by *gauchos*, fiercely independent roving horsemen of mestizo descent (Vignette 17.1). By the beginning of the 1700s, the hunting of feral livestock waned as entrepreneurs began to establish herds of semidomesticated cattle that were easier to manage and harvest. Eventually, they began the process of establishing *estancias*, vast livestock ranches. Hides and tallow were the principal products in this economy, although a trade in salted meat destined for the plantations of coastal Brazil became increasingly important. By the end of the 1700s the technological basis for improved meat drying and salting was available, and the first meat-drying and meat-salting plants, *saladeros*, were constructed. Demand for dried and salted meat in Europe and the Americas initiated a minor industrial revolution on the banks of the Río de la Plata as more *saladeros* were constructed. This led to increased production of livestock and deeper penetration into the vast Pampa. By the early decades of the 1800s, the annual cattle slaughter for the *saladeros* had grown to between 350,000 and 500,000 head.

The fortunes of Buenos Aires changed considerably during the last decades of Spanish colonial rule. In 1776 the Spanish Crown divided the Viceroyalty of Peru in two, creating the Viceroyalty of the Río Plata and making Buenos Aires its capital. Prior to this time colonial regulations mandated that all trade pass through the distant viceregal capital of Lima. Although illegal trade had been a component of the local economy for decades, viceregal status permitted direct trade with Spain, and hence indirect trade with Europe, and buoyed the city's fortunes. Argentine independence from Spain in 1810 served to further bolster the growth of the city; by 1850 the city boasted a population of over 1 million.

The growth of Buenos Aires into a major metropolitan center mirrored the increasing pace and intensity of development on the vast prairies of the Pampa during the second half of the 19th century. Sheep raising, fueled by intense demand in Europe for wool and tallow, began to be practiced on a vast scale after 1850. The best pasturelands for sheep were found in the humid Pampa, and here flocks of sheep began to displace cattle herds. Cattle herding shifted into the drier portions of the Pampa further inland. Sheep raising was widespread throughout most of the Pampa. As an example of the magnitude of their impact, in the 25 years between 1850 and 1875 the number of sheep in the province of Buenos Aires increased nearly tenfold from 4 million to over 40 million head.

The boom in sheep raising brought new changes to the Pampa. Europeans, principally

FIGURE 17.3. The vast open expanses of the prairie landscape of the humid Pampa, 1993.

VIGNETTE 17.1. THE *GAUCHO*

Who were the gauchos and who are they today? *Gauchos* were fiercely independent, seminomadic horsemen, often of mestizo origin, who eked out a subsistence living on the Pampa during most of the colonial period. They have been described as "frontiersmen," roughly akin to the American "cowboy" or the Mexican "*vaquero*," although neither characterization is entirely complete. Exploiting feral cattle herds, these expert horsemen hunted the cattle for their hides, leaving the carcasses to rot on the open Pampa. They lived on the margins of the formal economy and society-shunning towns and subsisting on a diet that consisted largely of beef, corn, and *yerba mate*. *Yerba mate* is the dried leaves of a native shrub native to the Gran Chaco, *Ilex paraguariensis*. Guaraní peoples used the leaves of the plant to make a tea and European settlers adopted the habit from them. Many Argentines and Paraguayans consume this tea, or *mate*, in the same way many other peoples drink tea or coffee.

The gaucho's key possessions included a train of horses that might number as many as 20, a saddle, and a bedroll. His distinctive dress included a broad hat, or *sombrero*; loose, pleated, baggy pants gathered at the ankles called *bombachas*; ankle-length leather boots; a broad belt, the *chiripa*, encircling the waist; and a woolen poncho. A lasso, dagger, and *bolas* were all tools of the trade for the gaucho.

As the patterns of economic development on the Pampa changed, so did the role of the gaucho. Gauchos played a critical role as fighters in the armies of independence against Spain; subsequently, during the early republican period, gauchos were well represented in the armies of rival *caudillos* who vied for power and control in the civil wars that plagued Argentina. The establishment of private property claims and the eventual fencing of the Pampa into vast cattle estates and farms spelled the end of the gaucho as an independent, seminomadic occupant of the Pampa. Most gauchos became cowhands and *peones* (peasants) working on the immense cattle ranches called *estancias*. As the gaucho began to disappear from the Pampa at the end of the 19th century, he became the stuff of legend. He was immortalized and remembered in romantic literature epitomized by the epic poem *El gaucho Martín Fierro* (1872) by José Hernández and the novel *Don Segundo Sombra* (1926) by Ricardo Güiraldes. The gaucho remains a key folk hero and legendary figure in both Argentina and Uruguay. Other Latin Americans often refer to Argentines as "gauchos." In Brazil, the residents of its southernmost state, Rio Grande do Sul, which falls mostly in the Pampa, are nicknamed "gauchos" as well.

Irish and Scots, immigrated to the Pampa by the thousands to work as shepherds. Cattle ranchers, on the other hand, pushed toward the outer limits of the Pampa, which put increasing pressure on the Araucanian Indians who had previously managed to keep European influences and control confined to areas north of the 37th parallel. This ended, however, with the War of the Desert (1879–1883) in which the Argentine army defeated the Indians, and thereafter essentially eliminated them from the Pampa as well as Patagonia, allowing a major expansion of the settlement frontier.

The elimination of the Indians coincided with the beginning of a series of processes that transformed the Pampa from an unsettled and semiwild frontier to a settled, agricultural region. Technological innovations in transportation and in food storage played critical roles in this transformation. Beginning in the early 1880s, railroad construction (financed almost exclusively by British investors) produced a network radiating outward from Buenos Aires. The principal objective of the railroads, which the British owned and operated until the 1940s, was to move agricultural production from rural areas to Buenos Aires. From the city's port, products could be shipped directly to overseas markets. There were few interconnections between the "spokes of the wheel." This situation reinforced the dominant position of Buenos Aires in the national economy and the marginalization of provincial cities and regional capitals.

Improvements in refrigeration technology, permitting the shipment of fresh chilled meat to export markets, revolutionized agriculture on the Pampa. Initially, fresh mutton was shipped to Europe on refrigerated ships in the 1880s. Later improvements in refrigeration technology permitted the shipment of the larger and more unwieldy beef carcasses. These new export products and markets placed increasing pressure on the Pampa and land use intensified. The demand for fresh meat in Europe, and especially for beef in Great Britain, was astronomical. By 1889 the annual export of beef to Europe exceeded 1 million cattle. Supplying these quantities of beef transformed the economy and landscape of the Pampa.

The British preferred beef that was lean and tender, shunned the stringy, tough meat characteristic of the traditional *criollo* cattle. These herds had grazed on the Pampa for over 300 years and had been the source of the salted meat processed by the *saladeros*. British demand led to the introduction of improved cattle breeds whose meat pleased the pallet of the British public: shorthorns, Angus, and Herefords. But these new breeds changed the nature of cattle raising on the Pampa as well as the landscape. The increasing value of cattle and the use of improved breeds corresponded with a breakthrough in fencing technology: the invention of barbed wire and its increasing availability and decreasing cost. The endless unbroken vistas of the prairies more and more often gave way to straight fence lines disappearing into the distance. Fences, however, were crucial in maintaining cattle-breeding programs as well as for pasture rotation and improvement.

The unimproved pastures of the Pampa proved inadequate for maximizing the potential of the pure-breed cattle. Ranchers turned to an ingenious mechanism to secure adequate acreages of improved pasture. Using promoters, pamphlets, and newspaper advertisements, the ranchers recruited men from rural areas of Italy and Spain to immigrate to Argentina as tenant farmers. The tenancy model permitted the tenants to work small plots under contract for 3–5 years. They broke the prairie sod and could plant crops of their choosing all but the last year of the contract, when the entire acreage was to be planted in alfalfa or other high-grade forage for cattle. Landowners transformed the Pampa in this way. The tenant farmers moved on to other contracts or rural settings, but eventually many migrated to the city, most frequently Buenos Aires.

Agriculture on the Pampa diversified, moving from basically a grazing economy to one based on a mixed system of cropping and livestock raising. Wheat was first exported from Argentina in 1875; by the end of the 19th century, Argentina became and has remained one of a handful of major wheat-exporting nations in the world.

Wheat cultivation also contributed to a migratory flow that was transforming the human geography of the Pampa. During the harvest season for wheat labor demands were high, but there were few laborers available in Argentina. Again, southern Europe, in this case Italy, provided the necessary workers. Poor rural laborers harvested Italy's wheat crop during the Northern Hemisphere summer, and then later traveled to Argentina to harvest wheat again during the Southern Hemisphere's summer. Dubbed the *golondrinas*, or "swallows," these migratory workers continued this pattern of intercontinental commuting well into the first decades of the 1900s. Like the tenant farmers, many of the golondrinas eventually settled permanently in Argentina. Between 1880 and 1910 immigration exceeded 250,000 in some years, and totaled over 2 million by the end of the period, transforming the Pampa, as well as Argentina, into a place whose population was principally European or of European descent and whose social customs reflected those origins.

Contemporary Economic and Social Geography

The agricultural economy of the Pampa prospered throughout most of the 20th century. Both World War I and World War II produced long booms for the economies of Argentina and Uruguay, as civil unrest and war disrupted and often paralyzed the economies and agricultural production of most European nations. The rich soil and climatic resources of the Pampa produced abundant agricultural exports, jobs, and tax revenues that, in turn, contributed to the attainment of a standard of living in both countries comparable to that of Europe and the United States. By the 1950s, however, crippling export taxes and an overdependence on agriculture to fill government coffers and to fuel growth created strong disincentives in the agricultural sector.

By the 1990s, however, economic reforms, especially in Argentina, as well as increasing world demand for basic agricultural commodities, provided the impetus for increasing agricultural productivity, total production, and crop diversification, with corn, sorghum, sunflower oil, rice, and soybeans joining wheat as export commodities. As pressure has built for increased agricultural efficiency, farm parcels have been consolidated, smaller operators have been forced out, population densities have declined in rural areas, and many small towns have withered. However, since agriculture on the Pampa has always been characterized by large farm parcels, the changes here have not been as dramatic as on the Great Plains of the United States where smallholders had traditionally predominated (Figure 17.4).

The immense wealth of the Pampa and the success of its farmers and ranchers propelled the growth of Buenos Aires and Montevideo, as well as that of other secondary urban centers. Both capital cities completely dominate the urban hierarchies of their countries. As primate cities, they account for nearly one-third of the total population of each country and are the focus of social, cultural, economic, and governmental life in each nation.

Buenos Aires grew by leaps and bounds, and by the early decades of the 20th century was clearly a world-class city in terms of both population and economic importance. By 1920 its population reached close to 2 million as the urban area expanded beyond the traditional urban core and a process of suburban development began that has continued to the present. An extensive system of trolley lines crisscrossed the urban area by the end of the 19th century. Horse-drawn trams, and subsequently those powered by electricity, provided access to a wide geographical area and offered a range of residential options for many of the city's inhabitants. The first commuter rail service and a subway system were completed in 1914. Expanded subsequently, this urban transportation system allowed easy access to the city core. Subsequently, the metropolitan area sprawled into the Pampa.

The city prospered as trade and an increasingly diversified economy offered employment for tens of thousands of Italian and Spanish immigrants and Argentines alike during the first half of the 20th century. Both World War I and World War II provided an immense economic stimulus for Argentina and particularly Buenos Aires. Demand for food exports soared during and immediately after

FIGURE 17.4. The wheat harvest on the Pampa is dependent on mechanized equipment like these combines near Bahia Blanca, Argentina, 1993.

both wars. Also, because of the wars, the flow of European manufactured goods was cut off, stimulating domestic manufacturing. These years marked a period of unprecedented prosperity for the city and its metropolitan area. Its population continued to grow, reaching nearly 7 million by 1960.

The downtown and central districts of the city have an urbane, cosmopolitan air characterized by a dense urban population, a fine opera house, wide boulevards, public monuments, a modern subway, and some stunning architecture reminiscent of western European cities (Figures 17.5 and 17.6). More mundane architecture and public spaces typify the outlying residential districts and industrial and commercial areas. The increasing availability of the automobile to the middle and working

FIGURE 17.5. Calle Florida, a crowded pedestrian street in downtown Buenos Aires, 1994.

FIGURE 17.6. A municipal office building off the Plaza de Mayo in downtown Buenos Aires imitates late-19th-century French architectural designs, 1994.

classes after 1960 encouraged the construction of limited-access highways and the growth of more peripheral suburban districts.

Buenos Aires remains a world-class city, but the last decades of the 20th century saw its fortunes wane and the quality of life for many of its residents decline. Its population has continued to grow; as of 2000, estimates of the population of the metropolitan region stood at close to 13 million. Argentina experienced a deepening crisis during the late 1990s. This culminated in a massive economic meltdown in 2002 when the country defaulted on its international loans, devalued its currency, froze bank assets, and liquidated millions of depositors' savings accounts denominated in U.S. dollars and then repaid their owners only a portion of the accounts' values in Argentine currency.

This crisis put increasing pressure on the city's housing stock and public services. Most of the city's residents have adequate housing and receive basic services, and many enjoy a standard of living comparable to Western Europeans and North Americans. However, an increasingly large proportion, perhaps as much as 20 percent, live in poverty. Many of these people live in century-old inner-city slum dwellings known as *conventillos*, while others inhabit shantytowns, known as *barrios de emergencia*, on the urban periphery.

The city enjoys a diverse economic base. It has one of the world's busiest ports, serving a vast agricultural hinterland encompassing the Pampa and beyond. A diverse range of agricultural products is shipped from the city, including wheat, wool, meat, dairy products, and hides. Its industrial base encompasses an immense range of activity—notably agro-industry, consumer goods, textiles, metalworking, machine tools, motor vehicles, and petrochemicals.

Montevideo, Uruguay, and Córdoba, Argentina, are the two other metropolitan centers of significance in the Pampa region. Both cities have populations between 1.2 and 1.4 million. As Uruguay's capital, Montevideo is arguably the more important of the two. It is the seat of government and of economic, social, and political life in Uruguay and accounts for over one-third of the nation's total population. It boasts an active harbor that is home to a large fishing fleet and also serves as the nation's only important port. Montevideo is also the undisputed leader of the country's industrial and manufacturing sectors. Córdoba lies far inland on the western margins of the Pampa and in the heart of a hinterland rich in agricultural and mineral resources. The city supports a small industrial sector, but most industrial activity is oriented toward primary processing, agro-industry, or consumer goods.

Patagonia: A Region on the Margins of the Good Earth

South of Bahia Blanca, the Pampa grades into a region called Patagonia. Until the 1880s, indigenous people, the Mapuche, controlled the region. However, an expanding settlement and a growing regional economy on the Pampa created a demand for more grazing lands and the indigenous inhabitants were driven from the region or exterminated by the Argentine military. Agricultural products, wool, mutton, and some crop agriculture have been the economic mainstays of the region since the early 1990s. Extensive petroleum reserves in the region are exploited for oil and natural gas; in 2000 these accounted for a major share of the regional economy.

In terms of its geographical extent, Patagonia is immense. Its northern limit is the Río Colorado, corresponding roughly with a line running along 39°S latitude. The region extends southward for over 1,600 km to the terminus of the island of Tierra del Fuego at about 55°S latitude. For most of its north–south extent, Patagonia is bordered on the west by the Andes and falls exclusively into Argentine national territory. However, at about 50°S latitude, the Andes effectively disappear as a mountain chain and the Patagonian region extends westward into Chile, becoming the extreme south of that country (Figure 17.7).

Patagonia is a vast semiarid steppe. Steppes are cool plains dominated by grasses, although they sometimes also have trees.

FIGURE 17.7. Patagonia.

Within this framework of uniformity, regional variation occurs. In the north, a series of broad plateaus of volcanic origin rise in somewhat of a stair-step fashion from about an average 100-m elevation along the Atlantic coast to over 1,000 m near the base of the Andes. Southward this pattern becomes less regular. In the central portions of Patagonia, in the area north of the Río Chubut, upland areas composed of crystalline outcrops exceed elevations of 1,500 m. Along the Atlantic coast, the region's plateaus end abruptly with cliffs fronting directly on the ocean or separated from it by narrow coastal plains. These cliffs can range in height from 50 m in the north to as much as 500 m in the south. A series of rivers occupying deeply incised valleys traverse the region. Described by geomorphologists as "underfit," these rivers occupy wide flat river valleys created during the Pleistocene when glacial meltwater drained off of the Andean ice sheets in vast quantities. These protected flat valley floors, with fertile alluvial soils and access to water, offer the best opportunities for permanent settlement and crop agriculture in the region.

Climatic conditions in Patagonia vary. Generally, the region may be described as having a midlatitude temperate climate, although it is moderated by the marine influence of the Atlantic Ocean and, to a lesser extent, the Pacific Ocean. One of the most consistent climatic characteristics of Patagonia is the incessant and desiccating winds that blow across the open steppe from the west. The natural vegetation on the steppe is composed of tussock grasses and low shrubs and brush.

Over most of Patagonia precipitation regimes range from arid to semiarid (100–500 mm), with the lowest precipitation occurring in the northwest and the highest in the south (Figure 17.8). The Andes serve as an effective barrier to the moisture-laden winds (the Westerlies) that blow in from the Pacific. However, in the extreme south, where the mountain barrier disappears, precipitation is somewhat higher.

Temperatures are rarely extreme. Summer temperatures in the north can reach 30°C, but drop southward. Winters are cool, but freezing temperatures are rare and snow falls only occasionally on the steppe.

Settlement and Historical Geography

Hunting and gathering, often focused on marine resources, formed the basis of the econ-

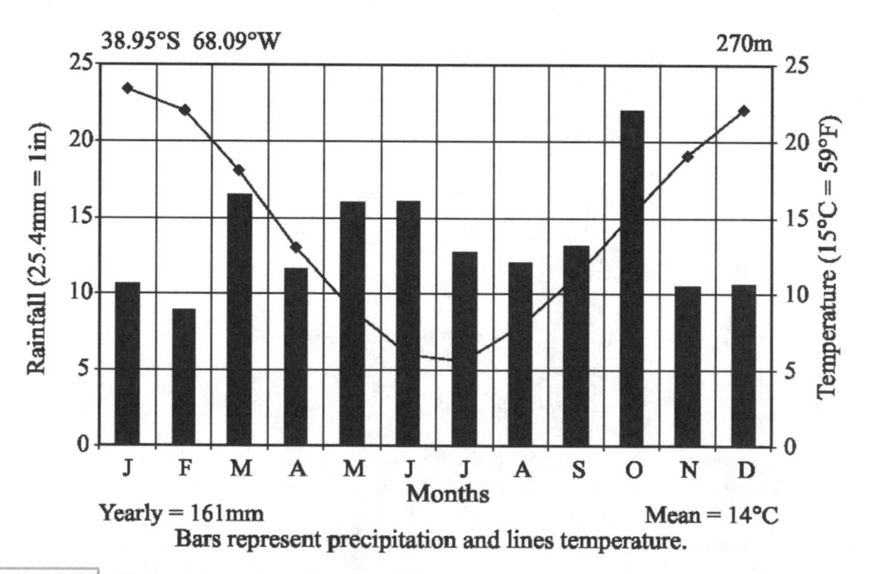

FIGURE 17.8. Climograph of Neuquén, Argentina.

omy of the aboriginal inhabitants of Patagonia, known as the Tehuelche. The sparse population of these peoples was reduced dramatically after contact with Europeans; introduced diseases for which they had no immunity killed many. During the late part of the colonial period, Araucanians, aggressive, highly mobile, specialized hunters who had adopted the horse, migrated into the northern portion of Patagonia from southern Chile. Aboriginal Patagonia remained largely beyond the effective control of Spanish authorities during the colonial period. Argentina exercised no control over the region until the late 1880s when a genocidal campaign, euphemistically known as the "War of the Desert," carried out by the Argentine army, fought on the southern Pampa and in Patagonia, eliminated most of the native inhabitants.

Settlement by Europeans and people of European descent began in Patagonia in the mid-1860s when Welsh settlers established a colony based on sheep raising on the lower reaches on the Río Chubut in the vicinity of the present-day cities of Rawson and Trelew. The increasingly important role of cattle in the economy of the Pampa in the 1860s and 1870s caused the displacement of sheep from the Pampa and provided the main reason for the settlement of Patagonia. Argentines, as well as immigrants, including Welsh, British, Spanish, and Chilean flows, established vast sheep *estancias* during the last two decades of the 19th century, effectively settling the region. However, because sheep raising requires little labor, inhabitants were few. British landowners and investors played a critical role in providing the investment capital necessary to stock the flocks with improved breeds as well as to develop an infrastructure to ensure effective export of their products.

The development of the transportation infrastructure followed the classic pattern of an extractive economy in a colony or dependent region. Along the Atlantic coast, and usually at the mouth of a major river, small port towns were established. The construction of railroads followed. These were almost all single-line systems running inland and upriver from a small port city into the herding hinterland, providing the most efficient means of moving wool and meat to export markets. The northernmost line connected to Bahia Blanca, a large grain-exporting city on the southern margin of the Pampa, but the other lines ran from the small port towns into the steppe; most did not even reach the Andes. This track pattern, and the lack of north–south linkages between any lines, prevented the railroads from serving a broader role in the region's development.

Contemporary Conditions

Early sheepherding concentrated on the production of wool, while the export of fresh, chilled and frozen mutton on refrigerated freighters, *frigoríficos*, became an additional component of the export mix during the early 20th century. In the last decades of the 20th century, the export of mutton fell dramatically as national and world demand lessened. Sheep raising once dominated the Patagonian economy. While this is no longer the case, livestock are still important in the regional economy (Figure 17.9). Limited crop agriculture in the north and the extraction of raw materials in the south are also now key components of the regional economy.

Crop agriculture concentrates in the northern half of Patagonia, where the growing season is long enough for commercial production. Here, crop agriculture is confined to the broad river valleys that cross the steppe. The width of these valleys is measured in kilometers. For instance, at its mouth, the valley of the Río Chubut measures 6 km wide. Irrigation is required for crop agriculture in Patagonia. Dams, reservoirs, and irrigation works have been constructed on the Río Chubut and the tributaries of the Río Negro. Orchards and vineyards are widespread in the

FIGURE 17.9. Livestock graze on the Patagonia steppe in southern Chile, 1996.

northern valleys, especially in those of the Río Colorado and the Río Negro. High-quality apples and pears are grown here for the export market in Brazil and western Europe, while lesser quantities of peaches, plums, and grapes are grown for the national market. Row crops are also important; nearly half of the national tomato crop is produced in the Río Negro. In the valley of the Río Chubut, alfalfa, grains, and vegetables are grown commercially. In more southerly valleys, options are increasingly limited and grazing begins to dominate.

Energy resources are abundant in Patagonia. The discovery of oil in Comodoro Rivadavia in 1907 marked the beginning of the exploitation of oil and natural gas here. The industry has continued to grow since then. Three principal oil and gas fields are found in Patagonia: the Néuquen Basin in the northwest in the Andean piedmont, the San Jorge Basin centered on Comodoro Rivadavia, and the Austral Basin centered on Río Gallegos but extending as far south as the island of Tierra del Fuego. The latter two basins have extensive offshore deposits. Nearly two-thirds of Argentina's petroleum production comes from Patagonia. Similarly, natural gas reserves in Patagonia represent about two-thirds of the nation's total. A system of gas pipelines and compression plants link production areas in these basins with the key consuming regions, cities on the Argentine Pampa and Buenos Aires. Patagonia also boasts limited coal deposits. These deposits are located in the extreme southern portion of the region along the Chile–Argentina border in an area known as Río Turbio. Almost all of Argentina's coal production and known reserves are found here.

Patagonia is sparsely populated. Total population for the region stands at around 2 million, with over three-fourths of these located in the Argentine portion of the region. Over most of Patagonia's extent population densities do not exceed one person per square kilometer, although in the river valleys of the north population densities can exceed 25 people per square kilometer. Much of the region's population is concentrated in urban areas.

Néuquen, with a population of about 200,000, is the region's largest urban center. Punta Arenas, in the Chilean portion of Patagonia, follows, with a population of some 125,000. Other significant urban centers include the small agricultural service centers of

Rawson and Trelew; the towns of Rio Gallegos (50,000) and Comodoro Rivadavia (100,000), both of which are closely linked to the petroleum and natural gas industries; and the port city of Ushuaia, often said to be the world's most southerly city, on the southern margins of the island of Tierra del Fuego. Patagonia has experienced considerable population growth since the 1970s, but this growth has been concentrated almost exclusively in urban areas and specifically in those towns linked to petroleum and natural gas exploitation. Tourism represents an increasingly important component in the region's economy; some population growth is attributable to its expanded role.

Summary

The Pampa is the paramount settled geographical region of the Southern Cone of South America. The vast open plains of the Pampa encompass much of Argentina, all of Uruguay, and the southernmost state of Brazil. The Pampa is one of the world's best agricultural areas, producing beef and grain crops on prairies similar to the Great Plains of the United States. The humid midlatitude climate is moderated by the region's proximity to the Atlantic Ocean. The indigenous people were seminomadic hunters with low population densities. This and their rudimentary economies limited Spanish interest in the Pampa during the colonial period. By the end of the 18th century, the development of improved methods of meat salting and drying led to expanded livestock production. In the late 19th century the Argentine army eliminated the Indians who had held sway on much of the Pampa. Demands from the British market for higher quality beef led to improved pastures, barbed-wire fencing, enclosed grazing, and cattle-breeding programs. The agricultural economy prospered throughout the early 20th century and again more recently in the 1990s. Agricultural wealth has contributed tremendously to the growth of the major urban centers of Buenos Aires and Montevideo.

Patagonia is a semiarid windswept steppe to the south of the Pampa. It was home to indigenous peoples until the 1880s, when they were exterminated by the Argentine army. Its sparse population is engaged principally in sheepherding, limited commercial agriculture, and petroleum exploitation.

Further Reading

Adelman, J. (1994). *Frontier development: Land, labour, and capital on the wheatlands of Argentina and Canada, 1890–1914*. Oxford, UK: Clarendon Press.

Amaral, S. (1998). *The rise of capitalism on the Pampas: The estancias of Buenos Aires, 1785–1870*. Cambridge, UK: Cambridge University Press.

Briones, C., and Lanata, J. L. (Eds.). (2002). *Archaeological and anthropological perspectives on the native peoples of Pampa, Patagonia, and Tierra del Fuego to the nineteenth century*. Westport, CT: Bergin & Garvey.

Chatwin, B. (2003). *In Patagonia*. New York: Penguin Books.

Hora, R. (2001). *The landowners of the Argentine pampas:A social and political history, 1860–1945*. Oxford, UK: Oxford University Press.

Jefferson, M. (1926). *Peopling of the Argentina Pampa*. New York: American Geographical Society.

Keeling, D. (1996). *Buenos Aires: Global dreams, local crises*. New York: Wiley.

Keeling, D. (1997). *Contemporary Argentina*. Boulder, CO: Westview Press.

McEwan, C., et al. (Eds.). (1997). *Patagonia: Natural history, prehistory, and ethnography at the uttermost end of the earth*. London: British Museum Press.

Moya, J. C. (1998). *Cousins and strangers: Spanish immigrants in Buenos Aires, 1850–1930*. Berkeley and Los Angeles: University of California Press.

Reding, N. (2001). *The last cowboys at the end of the world: The story of the gauchos of Patagonia*. New York: Crown.

Rock, D. (1985). *Argentina, 1516–1982: From Spanish colonization to the Falklands War*. Berkeley and Los Angeles: University of California Press.

Scobie, J. (1971). *Argentina: A city and a nation* (2nd ed.). New York: Oxford University Press.

Slatta, R. (1992). *Gauchos and the vanishing frontier*. Lincoln: University of Nebraska Press.

18 Latin America in the World Economy

Much of Latin America remained beyond the control of the Spanish and Portuguese after the Conquest. But in the coastal regions and wherever there were dense indigenous population centers the colonial economy took root, transforming the region's landscapes and peoples and linking them to the world economic system. The mainstays of these colonial economies included the privatization of land and other property resources, forced labor and labor tribute/taxation, the establishment of large estates like haciendas and plantations, and the extraction of valuable natural resources. Several of the key elements of the colonial economy have been examined in other chapters in this book, including land privatization, colonial tribute systems, the slave trade, and new agricultural systems.

This chapter examines Latin America's place in the world economy. It outlines the geographic and economic dimensions of the mercantile systems that connected the colonies' economies to those of Spain and Portugal and then to the rest of the world. It documents the realignment of the region's geopolitical relationships and economic linkages in the postindependence period, and the establishment of commodity export economies. The economic dislocations caused by World War I, the Great Depression, and World War II are chronicled and the inwardly oriented economic strategies of that period that promoted industrial development are explained. The concluding sections of the chapter explore early efforts at regional economic integration in the 1960s and 1970s and the rapidly evolving panorama of regional trade blocs and customs unions in Latin America at the beginning of the 21st century. A discussion of the shift to neoliberal economic models in the closing decades of the 20th century and the increasing globalization that is transforming the region concludes the chapter.

The Colonial Mercantile System

A system of mercantile colonialism that controlled the terms of trade between the New World colonies and their metropolitan centers was the glue that held these vast colonial economies of Latin America in place and provided immense financial wealth to the colonial powers that controlled them, Spain and Portugal. Mercantile systems of various types characterized the colonial economies of European powers throughout the world during the period of intense European colonialism that occurred between 1500 and 1900. At their most elemental, these systems used a range of mechanisms that established terms of trade that favored the mother country at the expense of the colony.

In Latin America, both Spain and Portugal developed mercantile systems. Arguably, the Spanish system was the more efficient of the two. Trade was strictly controlled and only trade between the colonies and Spain was permitted. Trade to foreign ports or with foreign merchants was prohibited. All legal trade came from Spain, and returned there as well. As the colonial period advanced, the Spanish Crown and Spanish colonial authorities devised a number of legal and administrative mechanisms that ensured that trade was conducted under the control of the colonial authorities and that Spanish traders, merchants, and craftsmen, as well as the Spanish Crown, profited from this trade.

The control of trade also included limiting the number of ports and cities where trade between Spain and the colonies could occur. In Spain, all trade with the colonies passed through just one of Spain's many ports, Cadiz, on the southwest coast of the Iberian Peninsula; in the New World, a handful of cities served a similar function. In the Caribbean Basin there were effectively four ports through which all trade passed: Veracruz in Mexico, Cartagena in Colombia, Portobelo (Colón) on the Isthmus of Panama, and Havana in Cuba.

Acapulco, on Mexico's Pacific Coast, served as the official trading port for trade with Manila in the Philippines—at the time another important Spanish colony. Portobelo served as the transit point for trade to Lima, from which all trade from the central and southern portions of Spanish South America was to pass until the late 1700s. While this proved to be an efficient system for extracting taxes for the Spanish Crown and profits for colonial merchants, it promoted tremendous economic inefficiencies. For instance, exports from Buenos Aires and other points in Argentina had to pass overland, and then over the Andes, to Lima, a journey that was both torturous and long; imports from Spain to Argentina followed the reverse route.

A fleet system of trade between Spain and its New World colonies also served to reinforce the control of the Spanish Crown over nearly all external trade and commerce (Figure 18.1). The wealth of Spain's new colonies naturally attracted the attention of foreign pirates and privateers, the latter of whom operated with the direct support of competing colonial powers, especially the English, but also the Dutch and French. The slow unarmed merchant ships that carried gold, silver, and a host of other valuable colonial products to Spain from the New World were easy prey for the maneuverable and well-armed vessels used by foreign pirates, including the likes of Sir Francis Drake, John Hawkins, and Piet Heyn. In response to these attacks, the Spanish instituted a convoy system for colonial trade that began in the mid-1500s and continued to operate until the late 1700s. Operating from its principal ports in the Caribbean, the two main Spanish fleets, called the "Panama Galleon" and the "Mexican Flota," made one round-trip to Spain and back each year. The two fleets arrived in the Caribbean in late spring and summer, respectively, and then returned to Spain during the winter months, usually January and February. The return route brought both fleets to Havana, where they sometimes joined forces for the journey to Cadiz in Spain. Somewhat later in the colonial period, the Portuguese too established a fleet system between Lisbon and its principal Brazilian ports: Recife, Salvador, Rio de Janeiro, and Santos. Initiated in 1650, it too ceased to operate in the late 1700s.

Colonial governments also attempted to control trade through the licensing of trading companies. Often these companies were licensed to operate from certain ports, like the Guipuzcoa Company in Caracas or the Havana Company, while others had broader territorial mandates, like the Maranhão and Pará Company in Brazil. Other companies, especially in Spain, were authorized to trade in specific commodities. Unlicensed trading was prohib-

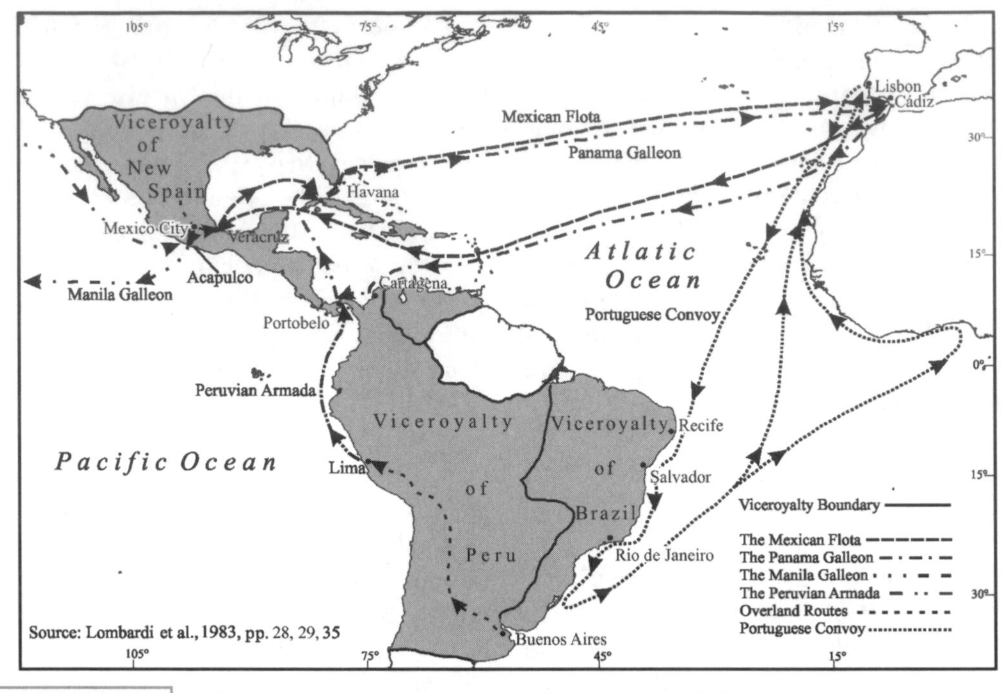

FIGURE 18.1. Trade routes, ports, and colonial administrations, circa 1650.

ited by colonial and royal authorities, but unlicensed trading as well as smuggling were common features of the colonial economy, especially in peripheral locations. In Spanish America, for example, the Atlantic coastal region around Buenos Aires and Montevideo proved to be a major center of illegal trade and smuggling. The small port town of Colonia, directly across the Río de la Plata from Buenos Aires, was the region's principal smuggling center during the 17th and 18th centuries. European manufactured goods from Great Britain and other countries entered Buenos Aires illegally from Colonia. The foreign vessels carrying this contraband were loaded with hides for their return trips (Figure 18.2). Colonia's importance waned when Spain finally established the Viceroyalty of Río de la Plata in 1777, permitting direct legal trade between Buenos Aires and Spain.

A variety of administrative rules and colonial policies served to reinforce the advantages of the mother country in trading relationships. Aside from immense quantities of silver and gold, both the Spanish and the Portuguese colonies provided an array of valuable raw materials to the Iberian Peninsula. These included commodities like sugar, cacao, leather, and tallow, as well as more exotic products like the cochineal bug and the leaves of indigo used respectively for making red and blue dyes. Manufactured goods came from Spain; if

FIGURE 18.2. A contemporary view of a colonial-period residential street in the historic core of the one-time smuggling center of Colonia, Uruguay, 1993.

not made there, they still passed through the hands of Spanish merchants and were shipped from Spanish ports. Colonial supplies of firearms, steel weapons, paper, fine textiles, books, soap, wine, olive oil, and other products passed through Cadiz to the New World. Spanish policies often prohibited the introduction to the colonies of manufacturing technologies or agricultural products that might compete with interests in Spain. For instance, in Spain's colonies the introduction of the honeybee, an Old World insect domesticate, was restricted for nearly 200 years in an effort to protect the interests of beekeepers who supplied beeswax for sacramental candles to Catholic churches throughout the colonies.

Taxes provided direct income sources to colonial authorities. These taxes came from a variety of sources. Municipal taxes represented an accessible, although sometimes difficult to collect, source of revenue for local governments (Figure 18.3). The greatest tax revenues, however, came from levies on mining and trade. In the Spanish colonies, taxes on gold and silver varied from a "royal one-fifth" to "royal one-tenth," known in Spanish as the *quinta real* and the *decimo real*. In the mid-1600s, as much as 40 percent of Spain's colonial revenue came from taxes on gold and silver. Taxes on other more mundane commodities, as well as other royal fees and levies, provided the balance, and often accounted for nearly three-quarters of all colonial tax collections. In the early 1700s the Crown established an even more lucrative source of revenue from its colonies: a monopoly on the tobacco trade in both Spain and its colonies.

The end of the colonial period saw a slow decline in the strength of the colonial mercantile system. The administrative reforms instituted by the Bourbons in the late 1700s allowed more open trading opportunities for the colonies. The fleet system disappeared in the late 1700s as changes in sailing technology and the geopolitics of the colonial period made it possible for individual vessels to make transoceanic voyages safely. But smuggling and illegal trade continued to flourish. The contraband trade in tobacco to Spain, for example, is estimated to have exceeded the legal trade by a factor of 2 or 3! Nevertheless, the colonial mercantile systems of Spain and Portugal in Latin America continued to provide immense financial benefits to the metropolitan centers on the Iberian Peninsula until the end of the colonial period.

Independence, Neocolonialism, and Export Economies

The first decades of the 1800s brought independence to most of Spain's and Portugal's

FIGURE 18.3. A Spanish-speaking municipal council officer responsible for tax collections in an Andean town receives payment from an indigenous Quechua speaker in this drawing from Felipe de Guamán Poma de Ayala's monumental illustrated book, *Nueva crónica y buen gobierno,* ca. 1550.

New World colonies. The only exceptions were Cuba and Puerto Rico, which remained Spanish colonies until the end of the 19th century. The disarticulation of the colonial mercantile system had been underway for a number of decades before actual independence, but political independence thrust the newly independent nations of Latin America into the capitalist world economy and new forms of economic relationships.

The stability of the colonial system and the mercantile economy that it supported was dependent on the imposition of control by Spain and Portugal. Decrees, laws, legal decisions, and administrative fiats backed up by direct military force maintained order and the structure of the economic system in the colonies. Independence brought the new nations into other types of relationships with the world economic system. In many ways, Great Britain replaced Spain and Portugal as the principal colonial power in Latin America during the 100 years after independence. However, it exercised political and economic control through indirect means: loans to foreign governments, investment in key export activities, construction of basic communication and transportation infrastructure, and trade concessions. If these more benign techniques failed to bring Great Britain the results it desired, the threat of military force always hung in the background. The effectiveness of these tools was such that military force was rarely applied. Often referred to as "neocolonialism," this type of relationship with Great Britain characterized almost all of Latin America's nations well into the 20th century. By the 1920s, however, the United States had supplanted Great Britain as the new neocolonial power in Latin America.

National economies dependent on the export of primary products to generate income for foreign exchange to pay for imports dominated Latin America's economic landscape during this period. As an example, nearly 100 years after independence, in 1914, a single commodity, usually an agricultural product or a mineral, accounted for over 50 percent of exports in at least 10 Latin American nations. In South America, for instance, these countries included Venezuela (coffee), Ecuador (cocao), Bolivia (tin), Chile (nitrates), and Brazil (coffee). In several Central American countries banana plantations and banana exports figured so prominently in the national economies, and the foreign companies that controlled them were so dominant in the politics of national life, that the term "banana republic" was coined to describe them. The term is now used more generally to describe any small nation controlled by outside economic interests (it has also been appropriated by a clothing company as a brand name) (Figure 18.4). Coffee played an extremely prominent role in this export economy during much of this period. It was the principal export crop in six Latin American countries; in many of these countries coffee accounted for an overwhelming percentage of the nation's export earnings. This occurred in small economies like Guatemala (85 percent), El Salvador (79 percent), and Nicaragua (65 percent), but also in large economies like Brazil (62 percent) and Venezuela (52 percent). Coffee also was the principal export crop in Colombia, but thanks to Colombia's slightly more diversified export sector, it only accounted for 37 percent of exports (Table 18.1).

Dependence on one or a handful of export commodities led to economic instability. Boom–bust economic cycles were characteristic of most nations. The health of these economies was highly dependent on a range of factors largely beyond their control. When international demand was great, higher commodity prices spelled good fortune for national economies and the collection of export taxes enriched government treasuries. However, the good times never lasted, and periods of economic decline and even depression always followed. The international market for tropical commodities drew from many world regions,

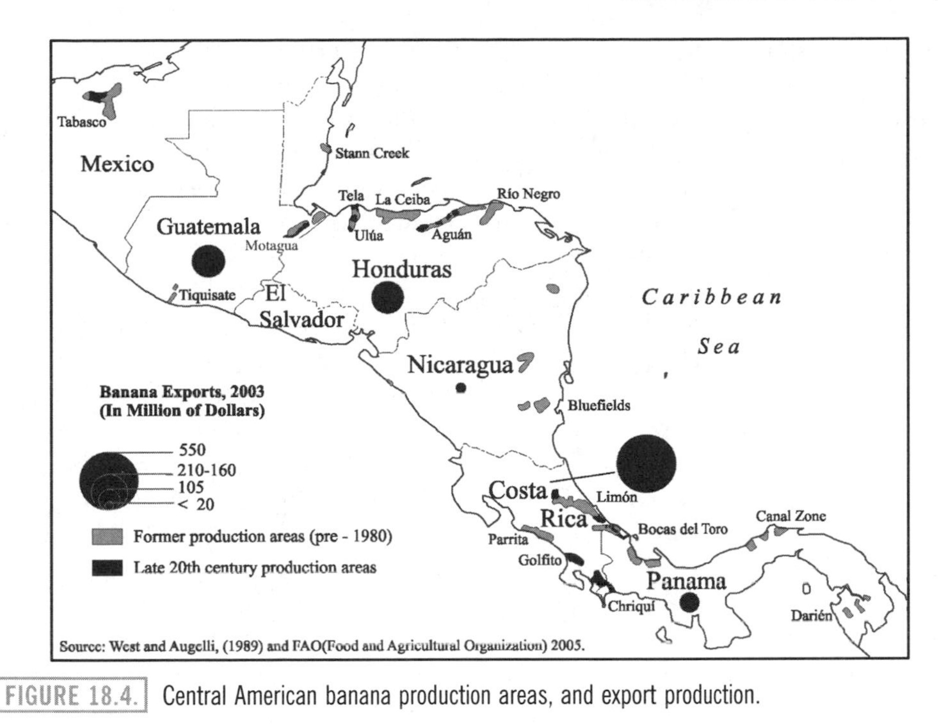

FIGURE 18.4. Central American banana production areas, and export production.

TABLE 18.1. Export Commodity Concentration Ratios, ca. 1913

Country	First product	Percentage	Second Product	Percentage	Total
Argentina	Maize	22.5	Wheat	20.7	43
Bolivia	Tin	72.3	Silver	4.3	77
Brazil	Coffee	62.3	Rubber	15.9	78
Chile	Nitrates	71.3	Copper	7.0	78
Colombia	Coffee	37.3	Gold	20.4	58
Costa Rica	Bananas	50.9	Coffee	35.2	86
Cuba	Sugar	72.0	Tobacco	19.5	92
Dominican Republic	Cacao	39.2	Sugar	34.8	74
Ecuador	Cacao	64.1	Coffee	5.4	70
El Salvador	Coffee	79.6	Precious metals	15.9	96
Guatemala	Coffee	84.8	Bananas	5.7	91
Haiti	Coffee	64.0	Cacao	6.8	71
Honduras	Bananas	50.1	Precious metals	25.9	76
Mexico	Silver	30.3	Copper	10.3	41
Nicaragua	Coffee	64.9	Precious metals	13.8	79
Panama	Bananas	65.0	Coconuts	7.0	72
Paraguay	Yerba maté	32.1	Tobacco	15.8	48
Peru	Copper	22.0	Sugar	15.4	37
Puerto Rico	Sugar	47.0	Coffee	19.0	66
Uruguay	Wool	42.0	Meat	24.0	66
Venezuela	Coffee	52.0	Cacao	21.4	73

Source: Bulmer-Thomas (2003), p. 58.

and higher production in other areas could force down prices and spell catastrophe for Latin American economies. Coffee, for instance, was a key export commodity in Colombia, Brazil, Venezuela, Guatemala, and Costa Rica, so regional competition played a role in prices, as did overproduction. But coffee was also produced in the highlands of Africa and Asia and increased production there affected Latin American markets negatively.

Agricultural export commodities were also susceptible to the development of new production areas in other world regions. Rubber and cacao are classic examples. Natural rubber comes from the sap of a wild tree, *Hevea brasiliensis*, native to the Amazon Basin. During the late 1800s and the beginning of the 20th century, the Amazon region experienced a tremendous economic boom as demand for rubber on world markets surged. The rubber boom brought immense wealth to the region. Manaus in Brazil and Iquitos in Peru blossomed into major urban centers. Despite Brazilian government prohibitions against the exportation of rubber trees or seeds, British botanists secured seedlings and took them to the Kew Botanical Gardens in London, where they were successfully cultivated. Subsequently, these contraband seedlings led to the introduction of rubber trees into British colonies in Southeast Asia, where the British successfully established vast plantations and soon dominated world production. Rubber gathering in the Amazon Basin collapsed, leaving economic ruin for many in its wake.

Cacao, a New World domesticated plant, whose seeds are the principal ingredient in chocolate, had a similar history. Initially, Latin American cacao producers enjoyed a secure market for their product. However, enterprising foreign agriculturalists soon successfully introduced the tree to equatorial areas in Africa and Asia, where the climatic conditions were similar to those of its American homeland, and foreign production boomed. West Africa became a major production region. Cacao producers in Latin America experienced a significant reduction in market share and have since been subject to dramatic swings in demand and prices as a consequence of supply fluctuations in other production regions around the world.

Resource depletion, replacement, and product substitution also plagued the export economies of Latin American nations in the 19th and early 20th centuries. Mining economies were highly susceptible to the depletion of their resource base and subsequent economic collapse. Regions dependent on the mining of precious metals (gold and silver) often experienced the exhaustion of the ore lodes, but this occurred in the case of nonprecious metals as well. In some instances, newly discovered resources eliminated the demand for existing commodities exports or reduced it dramatically.

Guano, or bird dung, was mined along the dry coastal margins and small offshore islands of southern Peru beginning in the 1840s. High in nitrogen, the guano made superior fertilizer. It found a strong market in Europe and proved to be a major source of income for the Peruvian economy and government. However, the depletion of the guano supply and the discovery of vast nitrate deposits in the Atacama Desert led to the demise of the guano export industry. The guano boom ended in the 1870s, taking with it one of Peru's most remunerative export products. Peru's misfortune was Chile's gain. Nitrate mining expanded dramatically in the Atacama Desert, generating immense tax revenues for the nation as well as strong profits for its oligarchs. While demand for nitrates remained strong through the early 1900s, the development of synthetic chemical fertilizers in Europe eventually spelled the end of the nitrate boom.

The export economies that characterized almost all of Latin America's nations during the 19th century and much of the 20th century produced narrowly focused economic benefits, although they were often significant. The number of jobs created was often limited. While unskilled jobs went to locals, the technical, supervisory, and managerial positions typically went to foreigners. While some profits were reinvested in the productive equipment and infrastructure of the export enterprise, often little found its way into the national economy. Most of the products were repatriated to the home countries of the foreign firms and reinvested there or elsewhere. Investment in infrastructure naturally focused on improving the means of efficiently exporting commodities. Port facilities, roads, and railroads linking production areas with export nodes (ports), and telecommunication infrastructure—first the telegraph and subsequently the telephone—absorbed most infrastructure investment. While these investments clearly created some positive spin-off effects for the national economies, the support of export industries remained the prinicipal objective of these investments.

The pattern of railroad construction in many Latin America countries illustrates this phenomenon. Railroad construction in Peru focused almost exclusively on east–west routes that ran from mining centers down narrow Andean valleys to small ports on the Pacific coast from which the ore was shipped. While many of these train lines also carried passengers and other cargo, the mining operations provided the reason for their existence. When the ores were exhausted or if changes in the terms of trade made their extraction unprofitable, the railroads ceased to operate. In Argentina, railroad construction prior to 1885 focused on the principal entrepot, Buenos Aires, and its port facilities on the Río de la Plata. Over the next three to four decades, during the heyday of railroad construction in Argentina, the pattern of railroad line construction mirrored early trends. While some north–south lines were constructed and a few provincial cities did have direct rail connections, the overwhelming pattern of rail lines as well as passenger and freight flows moved to Buenos Aires (Figure 18.5).

The commodity export strategy pursued by almost all Latin America nations for nearly 100 years proved to be a failure in many key respects. While some social sectors, usually the elites, and specific geographical regions did benefit from the commodity export trade, overall most Latin American economies showed only negligible economic growth during the century-long export boom. Even more problematic was the fact that in some countries living standards actually declined during the period, leaving much of the population worse off than 100 years earlier. Because growth was negligible and economic benefits were narrowly focused, the commodity export trade did little to stimulate internal demand and consequently encourage local manufacturing and industry.

Economic Retrenchment in the 20th Century: World Wars and Depression

The outbreak of World War I upset world trading patterns and had profound effects on the predominant model of economic development in Latin America: the commodity export economy. The changing demands of the war economy on the region's principal trading partners, Europe and the United States, reverberated through the commodity export sector. Strategic commodities (those critical for the war economy) enjoyed strong demand. The export of industrial metals (tin, copper, lead), petroleum, and basic foodstuffs, like grains and meat, boomed. On the other hand, demand for nonessential commodities, for instance, cacao and coffee, dropped precipitously. Although the fate of individual nations varied depending upon their mix of export commodities, overall the upshot of these changes was a reduction in export earnings and foreign exchange. The war

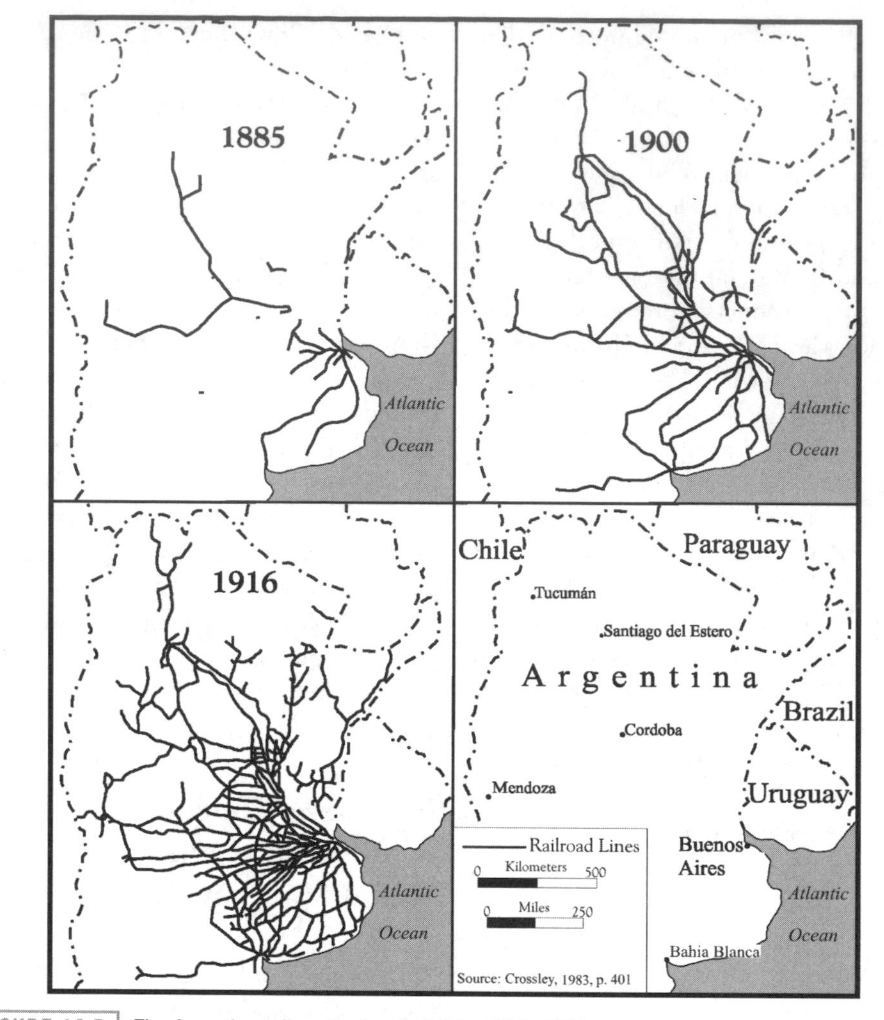

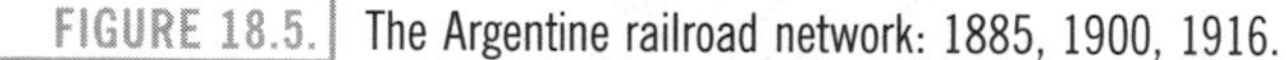
FIGURE 18.5. The Argentine railroad network: 1885, 1900, 1916.

also brought a structural reorientation of Latin America's economy. Great Britain lost its traditional role as the region's preeminent neocolonial power and was replaced by the United States.

At the outbreak of World War I, Latin America was one of the world's least industrialized regions. In most countries, little or no effort had been directed at developing industrial production to capitalize on the needs of the export sector, nor had the commodity trade generated a sufficiently broad increase in wages and living standards to provide much of a national market for consumer goods. In most nations, the commodity export economy of the 19th century generated sufficient foreign exchange earnings to permit the importation of most manufactured products, whether these were consumer goods, like textiles, or capital goods, like transportation equipment or manufacturing machinery.

However, the outbreak of war rapidly transformed the industrial complexes of Europe and the United States, where production shifted to military supplies, armaments, munitions, and kindred needs. The production of consumer and capital goods for export markets declined dramatically. By the war's end there were few manufactured goods available for Latin American countries to import, even if they had the foreign exchange to do so. The worldwide economic depression that began in the late 1920s and continued through the 1930s only deepened Latin America's economic crisis and reinforced the need for a new approach to economic growth and development.

The economic shocks ushered in by the outbreak of World War I in 1914, followed by the depression in the 1930s, and then by World War II in the 1940s, contributed to a shift in the economic strategies pursued by Latin American nations. This period has been characterized by many scholars as a period of "inward orientation." It began around 1914 and lasted until about 1980. The first half of this period, running from the beginning of World War I until the end of World War II, was typified by increasingly widespread national economic protectionism. Countries erected formidable trade barriers to limit the entry of manufactured products in order to protect their own manufacturing enterprises from external competition. Initially these efforts centered on the protection of industries producing simple manufactured products destined for consumers' use, like soap, toothpaste, textiles, clothing, and shoes. As manufacturing sophistication increased, and local firms developed the capacity to produce increasingly complex products, duties on other products that could be produced nationally were increased substantially. This tariff subsidy favored national producers to the detriment of foreign producers.

After World War II, many national governments articulated a more focused inward development strategy. This strategy, known as "import-substitution industrialization," sought to promote manufacturing and economic development by substituting locally manufactured products for imported goods. Initially, basic consumer goods, for example, clothing, shoes, and soap, were obvious targets for this kind of industrialization. This had occurred well before the end of World War II in many countries and especially in the region's larger economies, like Brazil, Mexico, and Argentina. While small countries, like those in Central America, struggled to promote manufacturing of basic consumer goods after the war, the region's more advanced economies promoted the manufacturing of consumer durables. These more complicated and expensive consumer products require more sophisticated manufacturing skills. During the 1950s and 1960s, behind the protective walls of high tariff barriers, the manufacturing of consumer durables like radios, televisions, refrigerators, and other similar goods blossomed.

Multinational firms, like Ford, Volkswagen, Renault, and General Motors (GM) dominated motor vehicle manufacturing in Latin America during the phase of inward orientation. The largest nations—Brazil, Mexico, and Argentina—successfully moved from automobile assembly to automobile manufacturing. The smaller countries also made many efforts during the 1970s and 1980s, usually with strong government support, to assemble or manufacture motor vehicles. Peru, Venezuela, and Ecuador were among that group. One such effort was the production of the Andino, a small simple motor vehicle manufactured in Ecuador in the 1970s using GM motors imported from Brazil. A basic vehicle designed for the rigors of hard use, it—like most other such efforts—was not commercially successful, although a few Andinos still plied Ecuador's highways well into the 1990s (Figure 18.6). By 2000, most of these efforts had failed and motor vehicle manufacture and assembly in the region was dominated by Mexico and Brazil. Argentina ranked a distant third, while Colombia and Chile produced a miniscule number of vehicles (Figure 18.7).

Import-substitution industrialization also sought to promote the development of basic industrial infrastructure: steelmaking, chemical production, and petroleum refining, for example. While tariff barriers provided some

FIGURE 18.6. An Andino pickup truck, an Ecuadorian effort at motor vehicle manufacturing as part of an import substitution strategy, Quito, Ecuador, 1979.

protection for these kinds of industries, the capital investment and sophistication required to establish and operate such industries meant that entry into enterprises was difficult and beyond the capabilities of most private capitalists and entrepreneurs. As a consequence, national governments often promoted such industrial developments by creating state-owned industries and in some cases state-owned industrial complexes. In Brazil, the construction of the massive Volta Redonda steel plant in the Paraíba Valley just to the east of Rio de Janeiro is one prominent example of this kind of state-sponsored industrialization. Initiating production in 1946, the Volta Redonda plant marked the beginning of an impressive steel industry in Brazil, promoted and supported by strong state intervention.

In the 1970s, based on the earlier development of basic industries like steel and allied manufacturing processes, some countries made active efforts to promote the manufacture of capital goods. *Capital goods* are those products that are used to manufacture other products. A lathe, a stamping mill (for molding and forming steel), and a drill press are all examples of capital goods. Few countries experienced any success with these efforts to extend import-substitution industrialization to capital goods. Most national economies were simply too small to generate sufficient internal demand for these capital goods. Moreover, since production was typically protected by stiff tariff barriers, the export of these products was impractical.

Only in Brazil, which benefited from a vast national market as well as a strongly developed and technically sophisticated industrial sector, did capital goods production succeed. One of Brazil's most stunning successes has been its entry into the global aerospace industry. In 1969, under government auspices, EMBRAER, the Empresa Brasileira de Aeronáutica S.A. (the Brazilian Aeronautical Company), was founded. Building initially on internal demand for military and commercial aircraft, EMBRAER built up its technical and production capacity by focusing on specialized market niches it was especially suited to serve. By 2005, the firm had grown into a substantial manufacturing enterprise with about 13,000 employees and had become one of the nation's top export companies. Its regional jets have been especially successful in the export market—for example, they are widely used in the United States where almost 400 were flying in 2000.

With the beginning of the 1980s the period of inward orientation and import-substitution industrialization drew to a close throughout Latin America. The results of these economic development strategies had been mixed. But the strategy had brought modern industry, in one form or another (rudimentary in some cases), to all Latin American countries.

The import-substitution strategy tended to work best in the countries with the region's largest economies. Brazil and Mexico accounted for over 60 percent of all Latin American manufactures, with most manufacturing located on the Paraná Plateau of southern Brazil and the Mesa Central of Mexico.

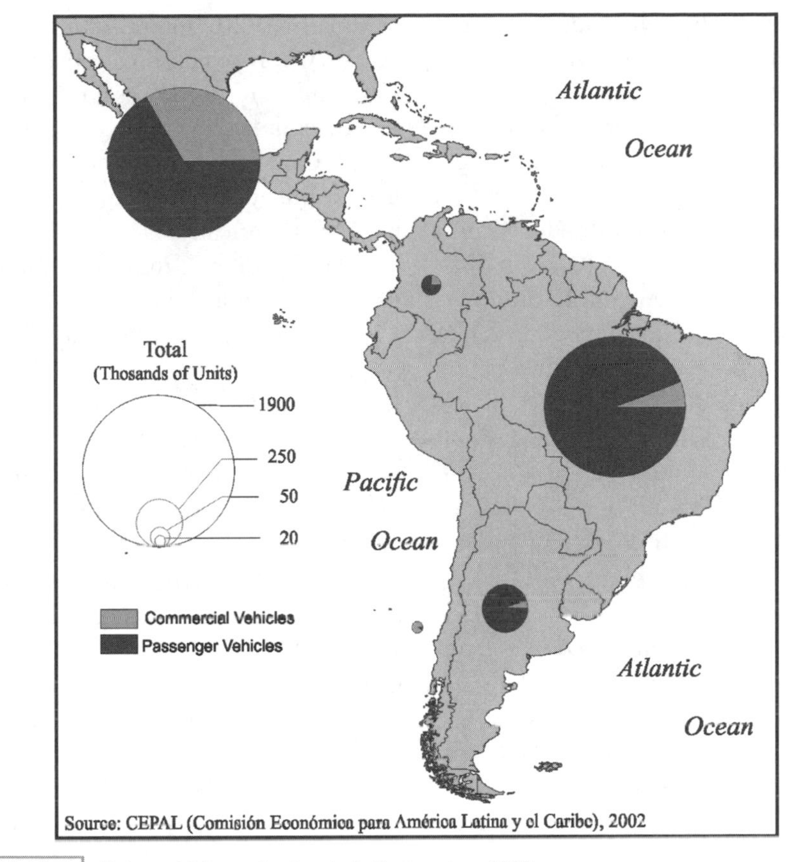

FIGURE 18.7. Motor vehicle production in Latin America, 2000.

Brazil's manufacturing sector had established itself not only on a regional scale in Latin America, but had evolved into a major exporter of aircraft and military weaponry. Its home-grown steel industry had achieved the status of a major world producer and in 2004 was poised for a massive expansion of capacity to fulfill rapidly growing demand in China over the next decade. Peru, Venezuela, and Puerto Rico also all experienced appreciable success in industrializing their economies during the inward-looking period.

Other nations, many which had done well exporting commodities to Europe and the United States during the 19th and early 20th centuries, like Cuba, Argentina, Chile, and Uruguay, did comparatively poorly during the period of inward orientation that ended in the 1970s and 1980s. In smaller nations, like those in Central America, import-substitution industrialization and other strategies of the phase of inward orientation were even less successful. With very small internal markets and little demand, there were few economies of scale in production, so manufacturing was often limited to the most basic consumer products in countries like Nicaragua and Honduras.

Halting Steps at Regional Integration: 1960s–1980s

The limitations of the strategy of inward orientation in promoting industrialization and manufacturing across much of Latin America had become increasingly apparent in the decades leading up to the 1980s. The model had not proved successful in many countries and the limitations of small markets had become obvious to most observers. During the period, Latin America's participation in the world export trade declined sharply. At the end of World War II, the region accounted for 25 percent of world exports, but by 1975 this had fallen to just 8 percent. Among all Latin American nations, only Brazil developed sufficient internal industrial capacity and sophistication to successfully break into the export of advanced industrial products, including commercial jet aircraft, computer electronics, and military armaments and munitions.

Regional economic integration in the form of multinational trade blocs with few restrictions on trade between member states became an increasingly popular concept as economists and politicians scrambled, with only the most limited success, to promote economic development and to improve living standards. The promise of larger markets would allow efficiencies in production that could not be realized in small national markets. It was reasoned that regional specialization, capitalizing on comparative advantages in the factors of production, would lead to the greatest efficiencies over larger geographic and population areas.

In 1960 Guatemala, Honduras, El Salvador, and Nicaragua joined forces to establish the Central America Common Market (CACM). Costa Rica joined in 1962. This customs union brought together again the same nations that had formed the United Provinces of Central America (1824–1836) immediately after independence from Spain (Figure 18.8). The CACM reduced tariffs for member states on a wide range of goods. During the decade of the 1960s intraregional trade rose sevenfold. However, the so-called Soccer War between El Salvador and Honduras at the end of the 1960s severely impacted the CACM's success when Honduras disrupted land transportation on the Pan-American Highway and imposed customs duties on goods from other CACM nations. CACM limped along until the early 1980s, when a combination of regional civil wars and continuing disagreements over customs tariffs led to its suspension. In the mid-1990s CACM was reorganized and reestablished, but its recent impact has not been especially significant.

The Andean Group, a multifaceted international organization formed in 1969 under what is known as the Cartagena Agreement, brought together Colombia, Ecuador, Peru, Bolivia, and Chile in an ambitious effort to improve regional integration across a broad socioeconomic spectrum, including the establishment of a customs union. Membership proved somewhat unstable, with Venezuela joining in 1973, Chile dropping out in 1977, and Peru suspending its membership between 1992 and 1997. Known as the Andean Community (*Comunidad Andina* [CAN] in Spanish) since 1997, the organization has heightened regional appreciation of economic integration, although it has been ineffectual at producing concrete results. While both the CAN and the CACM delivered less than expected, the two organizations did succeed in raising the consciousness of politicians and the public to the concept of broader regional economic cooperation. It seems likely that in the first decade of the 21st century these organizations are likely to disappear or to be absorbed by new, more open, and more comprehensive free trade agreements.

Economic Shock and Realignment: 1980–2000

Heavy borrowing, ill-advised investment programs and small returns, overdependence on a few export commodities for foreign exchange earnings, and financial mismanagement and irresponsibility brought most Latin American governments to near-bankruptcy in the 1980s. Many have called the decade that followed the

FIGURE 18.8. Latin American trade blocs, circa 2005.

"Lost Decade" because economic progress stalled and social and economic conditions for many of the region's populations declined appreciably. Poorly designed fiscal policies provided inadequate tax revenues desperately needed to provide basic social services as well as to service immense public debts. Governments often usually responded by printing more currency, creating inflationary spirals—at times reaching 1,000s of percent per year in Argentina, Bolivia, and Peru. In Argentina new banknotes often had to be overprinted with additional sets of three zeros before they went into public circulation because the rate of inflation was so great. In late 1985, in Bolivia, inflation was so rampant that $100 bought a grocery bag of Bolivian currency! While very few circulated in the streets, notes denominated at 5 million pesos had a value of less than $5!

Excessive public debt in almost all countries eventually led to default on international loan obligations and their subsequent renegotiation, although some countries like Peru under President Alan Garcia defiantly suspended international debt payments for a time in the late 1980s. In most countries these bilateral and multilateral negotiations addressed debt payments and loan rescheduling between Latin American nations and powerful debt-holding nations like the United States, Japan, Germany, and Great Britain, as well as debts and payments to international lending organizations like the International Monetary Fund (IMF) and the Inter-American Development Bank (IDB).

In addition to protecting their loans as much as possible, the holders of these debt obligations, especially the IMF and the IDB, used the negotiations to impose broad economic policies designed to reform and invigorate the economies of the debtor Latin American nations. These paradigms, often summarized under the catch-all term of "neoliberal economic reforms," emphasized the critical need to rationalize government activity, make markets more efficient and more open, and reform the system of public finance. Specifically, these reforms sought to cut government expenditures, to reduce the government's role in the national economy, to deregulate and to open markets, to promote free trade, and to capitalize on local, regional, and national comparative advantages in export economies. These policies cut public expenditures by shrinking the government payroll and privatizing many state companies and activities. The participation of the state in the exploitation, extraction, and export of natural resources, in service sectors like public transportation, and in basic industries like steelmaking, petroleum refining, and chemical production was significantly reduced, although not necessarily eliminated, in most Latin America countries. However, it is important to note that in the oil-rich nations of Mexico and Venezuela the state still owns the petroleum industry; any suggestions that these industries be privatized is met with intense political resistance.

Within Latin America, Chile, beginning in the mid-1970s, under the dictatorial leadership of General Augusto Pinochet, was the first country to wholeheartedly pursue these neoliberal reforms. In Chile, one of the central objectives of these reforms was to diversify the export economy, heretofore heavily dependent on the export of copper. As in other nations that subsequently followed the same strategy, Chile sought to generate a diversified export sector by focusing on the production of commodities in which it had a comparative advantage. Fruit growing and the export of both fresh and processed fruit products proved central to that strategy, although other abundant natural resources, like timber and seafood, also played a significant role in the diversification strategy (Figure 18.9).

Chile brought a number of comparative advantages to temperate and subtropical fruit production for the export market. The seasonal reversal of its Southern Hemisphere location proved critical, since this meant that when Chilean orchards and vineyards were in full production, it was winter in the major markets of the Northern Hemisphere. In addition, Chile's Central Valley offered a range of optimal climates for temperate and subtropical fruit growing, as well as sufficient land and water resources to support expanding production.

In the 20 years between the mid-1970s and the mid-1990s, fruit production and fruit exports boomed and helped lead Chile's strong economic growth over the period. The acreage in fruit tripled during this 20-year period, growing from about 65,000 ha in 1977 to 190,000 ha in 1996, while the value of fruit exports increased from just $30 million in 1974 to a whopping $1,146 million in 1995!

Chile has shown remarkable economic growth and a major expansion in its exports by following this model of promoting nontraditional exports. Foreign trade expanded dramatically, growing from just $2 billion in 1975 to over $15 billion in 1996, while dependence on traditional exports like metals (e.g., copper) have fallen notably. At the beginning of the period, metals accounted for nearly 80 percent of all exports; metals now account for around 50 percent of all exports. An increasingly diversified range of products—especially agricultural products, forestry products, and fish products—now account for nearly half of all exports. While Chile's traditional export partners were the United States and Western European countries, the pursuit of nontraditional exports has

FIGURE 18.9. This commercial lumber mill in southern Chile near Concepción is an example of one of the elements in Chile's successful nontraditional export strategy, 1993.

also helped Chile to diversify its trading partners to include a broad range of industrial and industrializing countries in Asia: Japan, Taiwan, Hong Kong, China, and South Korea. During the 1990s Chile had the highest annual rate of economic growth of any nation in the region, in excess of 7 percent, nearly double that of many other Latin American countries during the same time period (Table 18.2).

Neoliberal economic reforms have produced encouraging macroeconomic results. Aggregate economic indicators at the national level have been positive. Average annual growth rates, a key macroeconomic indicator used by economists and government officials, showed steady growth through the decade of the 1990s in many of the region's countries (Table 18.2). Most observers agree that the initial dislocations and economic disruptions occasioned by neoliberal economic restructuring hit the middle class, the working class, and the poor hard. Indeed, the adverse effects of these reforms typically hit the poorest sectors of the population the hardest. However, in most instances, after this initial period of economic restructuring and its effects occur, macroeconomic indicators do improve and economic growth is stimulated.

Nevertheless, there is considerable debate among neoliberalism's proponents and detractors about its long-term efficacy and the extent to which neoliberal economic policies have led, or will lead, to an improvement in the living standards of a particular country or the region generally. It is argued by some observers that one of the most troubling effects of the neoliberal economic model is that in some countries it has led to increasing income and social inequity. Critics suggest that the benefits of these economic reforms tend to be most concentrated among those social groups that already have had access to education and capital, while few benefits have accrued to the poor and those with limited education.

Argentina in the early years of the 2000s is an excellent example. It pursued a neoliberal economic model throughout the 1990s with some success, although unemployment hovered at about 20 percent for most of the decade. However, in 2002 the nation's economic system collapsed: wages fell, unemployment levels climbed as high as 50 percent, and the number living below the poverty line increased sharply. Argentina represents an ex-

TABLE 18.2. A Comparison of Inflation Rates in the 1980s and 1990s, and Growth Rates in the 1990s for Selected Latin American Countries

Country	Annual inflation average, 1984–1993	Annual inflation rate, mid-1998	Average annual GDP growth, 1990–1997
Chile	19.5	5.4	7.3
Argentina	811.5	1.2	5.9
Peru	1,283.7	8.4	5.2
Uruguay	75.5	12.3	4.3
Colombia	24.8	20.7	4.1
Bolivia	1,051.6	9.4	3.9
Ecuador	44.5	33.6	3.3
Brazil	944.8	4.1	3.0
Paraguay	24.5	6.6	2.8
Venzuela	34	39.6	2.4
Mexico	52.9	15.1	1.9

Source: Gwynne (1999), pp. 74 and 84.

treme case. However, critics of neoliberal policies suggest that the medium-term (10 years) social effects of neoliberal economic reforms in most countries can be summarized as follows: for the top 20 percent of the population, their share of the national income has actually increased; for the next 40 percent, the middle class and the working class, incomes have been relatively stable or perhaps have even declined slightly; and for the lowest 40 percent, the poorest and most vulnerable, incomes have definitely declined.

It is not surprising that the benefits of these neoliberal reforms designed to promote economic efficiency and market economics should benefit capitalists and entrepreneurs more than the working class and the poor. While "a rising tide lifts all boats," this has not been the case in many Latin American nations. They have not implemented fiscal and social policies (e.g., progressive income taxation and a basic social safety net) that would redistribute some of the economic benefits of neoliberal reforms to the nearly 80 percent of the population that has largely been left behind.

Data regarding the population living in poverty in Latin America's countries during the decade of the 1990s provide a useful perspective on issues (Table 18.3). For the region as a whole, poverty rates fell slightly during the decade, but at the end of the millennium about 45 percent of all Latin Americans lived in poverty. The situation in rural areas was more extreme, with about 65 percent living in poverty. However, the fates of individual countries varied considerably. Chile, considered by many to be a superb example of the benefits of neoliberal reforms, saw poverty rates cut in half, from almost 40 percent to just 20 percent. Argentina, on the other hand, which also aggressively pursued neoliberal policies, saw poverty rates in its cities climb. Some countries saw slight declines in poverty rates, but poverty remained deeply ingrained in the economies of nearly all Latin American countries, with poverty rates exceeding 50 percent of the population in Bolivia, Colombia, Guatemala, Honduras, Nicaragua, and Paraguay!

Regional Integration: 1990 and Beyond

The 1990s brought increasing emphasis on neoliberal economic policies and strategies in Latin America. Inspired and challenged by the success of the European Union (EU), in the 1990s Latin American nations pursued new efforts at regional integration to capitalize on regional comparative advantages. These efforts promised to bring more concrete benefits than the halting efforts at regional economic integration that characterized the 1960s and 1970s.

The "Common Market of the South" (*Mercado Común del Sur* in Spanish), or MERCOSUR as it is most commonly known, is an ambitious effort to establish a common market in Latin America. MERCOSUR evolved out of earlier efforts at regional economic integration in the region—notably, the

TABLE 18.3. Rural, Urban, and Countrywide Poverty Rates in Latin America, 1990–2000

Country	Year	Urban total	Rural total	Country total
Argentina	1994	16	—	—
	1999	24	—	—
Bolivia	1989	53	—	—
	1994	52	—	—
	1997	52	—	—
	1999	49	81	61
Brazil	1990	41	71	48
	1993	40	63	45
	1996	31	56	36
	1999	33	55	38
Chile	1990	38	40	39
	1994	27	31	28
	1996	22	31	23
	1998	21	28	22
	2000	20	24	21
Colombia	1991	53	61	56
	1994	45	62	53
	1997	45	60	51
	1999	51	62	55
Costa Rica	1990	25	27	26
	1994	21	25	23
	1997	19	25	23
	1999	18	22	20
Ecuador	1990	62	—	—
	1994	58	—	—
	1997	56	—	—
	1999	64	—	—
El Salvador	1995	46	64	54
	1997	44	69	56
	1999	39	65	50
Guatemala	1998	46	70	61
Honduras	1990	70	88	81
	1994	75	81	78
	1997	73	84	79
	1999	72	86	80
Mexico	1989	42	57	48
	1994	37	57	45
	1996	45	63	52
	1998	39	59	47
	2000	32	55	41
Nicaragua	1993	66	83	74
	1998	57	77	64
Panama	1991	40	51	43
	1994	31	49	36
	1997	30	42	33
	1999	26	42	30
Paraguay	1994	50	—	—
	1996	46	—	—
	1999	49	74	61
Peru	1997	34	73	48
	1999	36	73	49
Dominican Republic	1997	36	39	37
Uruguay	1990	18	—	—
	1994	10	—	—
	1997	10	—	—
	1999	9	—	—
Venezuela	1990	39	47	40
	1994	47	56	49
	1997	—	—	48
	1999	—	—	49
Latin America	1990	41	65	48
	1994	39	65	46
	1997	37	63	44
	1999	37	64	44

Source: United Nations, Comisión Económica para América Latina y el Caribe (2002), pp. 11–12.

Latin American Free Trade Association (LAFTA) (1960) and the organization that followed it, the Latin American Integration Association (1980). MERCOSUR evolved from a bilateral accord signed by Argentina and Brazil in 1985 to begin examining the integration of their economies. In 1991 these efforts culminated in the signing of a formal treaty creating MERCOSUR, with Argentina, Brazil, Paraguay, and Uruguay as its founding members. Subsequently, both Chile and Bolivia joined MERCOSUR as associated members. MERCOSUR has been somewhat successful at establishing a free trade area among its member states, and also at establishing a common system of tariffs, a customs union, for all goods brought into its member countries.

The ultimate goal is for the creation of a common market, a form of economic integration that not only eliminates customs duties between nations, but also allows for the free movement of both capital and labor among member states. As the United States presses its own version of regional economic integration through the Free Trade Area of the Americas (FTAA), MERCOSUR has moved aggressively to link itself with a broader range of countries and other trade associations. In 2003 it signed a free trade agreement with the Andean Community and it has been aggressively pursuing a trade accord with the EU, much to the consternation of the United States.

Including the countries of Mexico, the United States, and Canada, the North America Free Trade Agreement (NAFTA) is a trade association that breaks new ground by bringing together the highly industrialized and advanced economies of the United States and Canada with that of a developing and recently industrializing nation, Mexico. Authorized originally in an agreement signed in 1992, NAFTA officially came into being on 1 January 1994, eliminating tariffs on many products. Nevertheless, NAFTA did not immediately eliminate all tariffs, but rather will phase them out gradually over the first 15 years of the agreement.

The effects of NAFTA have been broad and have impacted all its members. Trade among the three nations has increased substantially over the decade of the accord, confirming in part the rationale for NAFTA's establishment. In the United States and Canada, employment losses have occurred in some sectors and in some regions. Manufacturing industries especially hard hit by NAFTA have included original and replacement automobile parts, consumer electronics (e.g., televisions), and a wide range of light manufacturing and assembly processes. Automobile manufacturing in Mexico has also grown substantially since the signing of the NAFTA treaty, with many cars destined for export to the United States. For example, of the over 1.3 million motor vehicles manufactured in Mexico in 1997, almost 1 million were exported principally to the United States and Canada.

NAFTA has also caused employment and economic dislocations in Mexico. Overall, real wage rates fell during the first decade of the agreement. The agreement's detractors point to this as a major shortcoming of the pact's promises, while its supporters acknowledge some loss of real wages, but contend that Mexicans would have been much worse off without the beneficial effects of the agreement.

The border region is one area where the effects of NAFTA are especially apparent. In the 1970s and 1980s, prior to NAFTA, Mexico had promoted industrialization along the U.S. border, eliminating customs duties on imports as long as those imports were used in the manufacture of products that were subsequently exported. The program was a resounding success. Hundreds of factories, called *maquilas* or *maquiladoras*, sprung up in all of the border's principal cities, but especially in the larger urban areas like Tijuana, Mexicali, Ciudad Juárez, Nuevo Laredo, Reynosa, and Matamoros. The number of maquiladoras grew steadily, as did employment in them. During the 1990s the number of maquilas along the border grew from 2,000 to about 3,700 and employment grew from about 500,000 to over 1.2 million. However, beginning in 2000, both the number of maquiladoras and the number of manufacturing employees in them began to decline. By 2004, over 500 factories had closed and total maquiladora employment slumped by 300,000.

With the advent of NAFTA one of the key advantages that the maquiladoras enjoyed, the ability to locate in a region with inexpensive labor, was extended to the entire nation. As a result, some manufactures have moved to interior locations in Mexico, usually in the Mesa Central, where wage rates are considerably lower than in the border region (Figure 16.12). Whereas nearly 90 percent of all maquiladoras were located in the border region in the mid-1990s, this Figure stood at only 60 percent a decade later. Wage pressure comes not only from other regions in Mexico. Factories have relocated further south in Central America, and even offshore. The wage differential is immense. At the beginning of the 21st century, average Mexican manufacturing employees earned $2.08 per hour, while their counterparts in China earned only $0.61.

NAFTA has also transformed the social and economic geography of Mexico's rural sector. Although the trade agreement did not immediately remove all tariffs on basic agricultural commodities, Mexico subsequently decided to reduce them more rapidly. Small marginal agricultural producers and subsistence farmers who produce basic grains, especially corn, and other food staples have been heavily impacted by these forces as their production costs often far exceed those of the heavily mechanized and highly subsidized commercial farms in the United States. Many farms have been pushed from full-time agricultural employment. Some will eventually abandon the rural sector altogether as agricultural lands are consolidated and farming enterprises depend increasingly on mechanized equipment and technology in place of field hands. NAFTA also has contributed to the success of firms operating large-scale export-oriented farms, and indirectly to the much more widespread use of chemical fertilizers, herbicides, and pesticides typical of commercial farmers in the United States and most other industrialized nations. Large-scale tomato growers in Mexico, for example, have made strong inroads into the U.S. market. They have captured a large share of the winter tomato market and have forced down prices, much to the consternation of growers in Florida and other southerly locations in the United States.

For over a decade, and from both Democratic and Republican administrations, U.S. foreign policy has viewed NAFTA as just a first step in the establishment of a much broader free trade area: the Free Trade Area of the Americas (FTAA). The surprising success of MERCOSUR has presented a serious challenge to this U.S. policy initiative. But the United States has pushed ahead by promoting bilateral and other subregional free trade pacts. At the beginning of 2004, a free trade pact between Chile and the United States went into effect. A similar agreement, the Central American Free Trade Agreement (CAFTA), between the United States and the five Central American Republics (Guatemala, Honduras, Nicaragua, El Salvador, and Costa Rica) has been successfully negotiated and was approved by the U.S. Congress in 2005. The United States views these agreements as building blocks that will eventually be joined in one form or another as the FTAA, encompassing the vast majority of countries in the Americas.

It seems inevitable that over the next 20–30 years the countries of Latin America and the other countries in the Americas will be increasingly bound by regional trade agreements, and perhaps eventually customs unions and even common markets similar to the EU. Whether it will be a supraregional organization encompassing nearly all countries, along the lines of the FTAA envisaged by the United States; or two major trade blocks, one focused on North America and the other on South America; or some other configurations, remains to be seen. One possibility, advanced by Brazil, is an effort to integrate MERCOSUR and the EU into a single free trade area. Although negotiations bogged down in 2004, the idea has not been abandoned and will likely be pursued in the future.

Global Markets and Global Lives

For over 500 years the countries that comprise Latin America have been linked to the global economy. This has been the case since the early decades of the colonial period when the first shipments of gold and silver moved from the New World to Spain and the first boatloads of the tropical hardwood *pau brasil* were sent to Portugal from Brazil's northeastern coast. Later, these trade contacts expanded to include a much broader range of natural resources and agricultural products. During the colonial period the export trade in sugar promoted the slave trade and the forced migration of hundreds of thousands of Africans to Brazil, the Caribbean, and Hispanic America. Sugar also promoted the development of the plantation system, which strengthened the global economic linkages between Latin America and Europe, and subsequently other world regions.

In the 19th century these global linkages contributed to the transformation of Latin America into an export powerhouse. Argentina and Uruguay supplied European markets with meat and grain. Producing coffee for export to Europe and the United States dominated the agricultural economies of Guatemala, Costa Rica, Colombia, and Brazil. The international trade in mineral commodities likewise dominated the economies of Peru, Chile, and Bolivia throughout most of the 19th century and much of the 20th century.

So, in many respects, globalization is nothing new in Latin America. Nevertheless, the late 20th century and the early 21st century have ushered in a new era of global contact and interconnectedness never before experienced. The contacts are more frequent, more rapid, and penetrate the daily lives of Latin Americans in more ways than in the past. Over the last few decades, technical and commercial advances in civil aviation have reduced time and economic barriers between Latin America and other world regions. Since the 1960s, Latin Americans have migrated in increasing numbers from their countries of origin to other Latin American nations, and even beyond to the United States, Japan, and Western Europe. Tremendous advances in telecommunications during the same period brought improved telephone services and much more extensive geographical coverage. In the 1990s the spread of cell phones throughout Latin America brought millions into the telecommunications revolution sweeping the world. With improved telecommunications infrastructure, Latin Americans have been able to link themselves to the world beyond their community or country through the Internet and the World Wide Web. Some scholars argue that because of the ease with which individuals can bridge geographical distance today, some individuals, especially recent immigrants, are "transnational." These transnational peoples live between two worlds—the homeland and the new land—transforming both and creating new communities in their wake.

The telecommunications revolution also facilitated the ease with which money can be moved globally. The formerly cumbersome process of all manner of financial transfers from one country to another—cashier's checks, money orders, letters of credit, and so on—can now be done electronically in seconds. For emigrants from Latin America, most of whom immigrate to the United States or Western Europe to improve their socioeconomic standing and often that of their families back home, this has been a boon. The ease and relatively low cost of such operations has made sending money, called *remittances*, back home much easier.

This phenomenon of financial remittances has grown dramatically, linking labor markets in the United States with local communities and regions across Latin America, but especially in Mexico, Central America, and the Caribbean. The total magnitude of these remittances is astounding. In a study conducted in 2004, the IDB surveyed the financial remittances of Latin American immigrants in the United States. The study revealed that these immigrants would send over $30 billion back home in 2004. While the average remittance was comparatively small, between $235 and $250, over 60 percent of all Latin American immigrants sent remittances, and most did so each month. The states with the largest immigrant populations, like California, New York, Texas, and Florida, sent the most money, but even states with small immigrant populations, like Kentucky, with less than 30,000 immigrants, sent over $50 million (Figure 18.10). The study revealed significantly different patterns in the immigrants' practices of sending remittances (Figure 18.11). In those states where Latin American immigrants have most recently arrived, the percentage that sends remittances is very high, exceeding 75 percent. States like Alabama, Georgia, the Carolinas, Virginia, and Maryland stand out among this group. On the other hand, in those states with a long history of Latin American immigration—Texas, Arizona, New Mexico, and Florida, for example—the percentage of immigrants that sends remittances is less than half!

The magnitude of these remittances is staggering in many ways. In simple aggregate terms, it is useful to compare the $30 billion in remittances that are sent annually to Latin American from the United States with the total foreign aid provided to the developing world by the world's richest nations—just $52 billion. The amount of these remittances is so great that in some countries in Latin America they represent one of the greatest sources of foreign exchange earnings. Such is the case in Mexico, where remittances rank third in foreign exchange earnings after petroleum exports and tourism. It is estimated that the remittances to Mexico are about $15 billion annually. Remittances play a significant role in Mexico; nearly 20 percent of all Mexicans receive remittances from relatives in the United

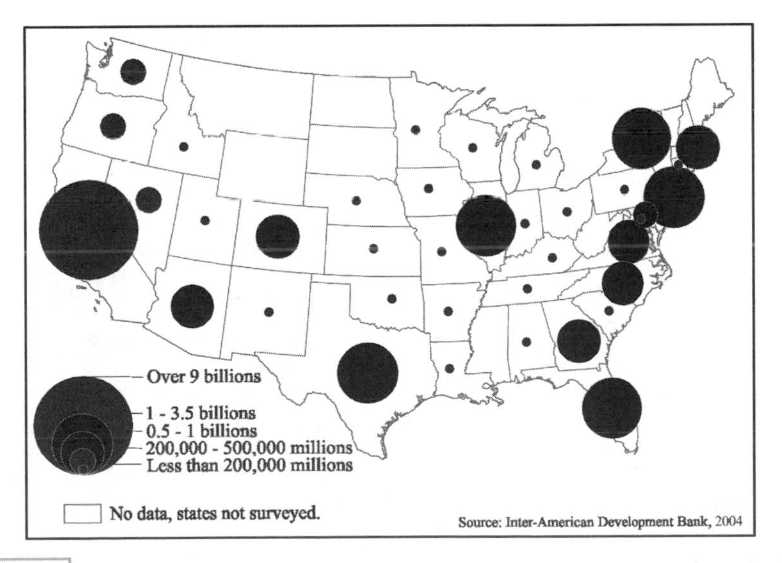

FIGURE 18.10. Remittances by Latin American immigrants in the United States, dollar value by state, 2004.

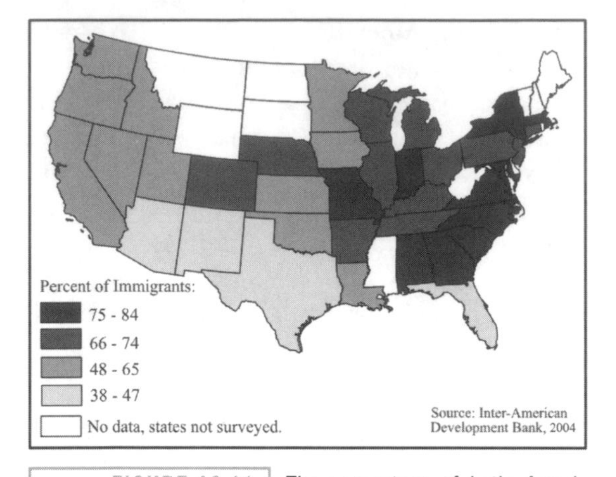

FIGURE 18.11. The percentage of Latin Americans immigrants in the United States who send remittances, by state, 2004.

States. Migrants send money to their families back home, but also often purchase homes and invest in farms and businesses back home. Not only do they send money to their families and invest for themselves, but they also contribute to civic projects and infrastructure investments through hometown associations formed in the United States. Although the Mexican example may be the most dramatic, similar patterns occur in other countries with large immigrant populations in the United States: the Dominican Republic, Cuba, El Salvador, and Guatemala, among others.

Another bellwether indicator of globalization in the region is the increasing penetration of foreign and multinational corporations into Latin America. While this process has been ongoing for decades, it has accelerated in recent years. Multinational firms have become more numerous and have penetrated more sectors of both local and national economies. In the 1950s the commercial penetration of multinational firms was most evident in sales of manufactured goods, both capital goods and consumer durables. At the beginning of the 21st century, multinational firms sell fast food and operate megastore retail chains across Latin America. These commercial developments have contributed to changes in Latin America retail forms as well as cultural habits. Urban forms characteristic of the United States and Canada are increasingly common across the region. Automobile-friendly commercial strip developments, shopping malls, and mall food courts are a fact of life for the middle and upper classes in the region's major cities and metropolitan centers (Figure 18.12).

The U.S. megaretailer Wal-Mart is one example of this more recent wave of multinational companies purveying fast food and retail goods to Latin Americans. Wal-Mart began its expansion into Mexico in 1991, with stores in Puerto Rico, Brazil, and Argentina following in subsequent years (Figure 18.13). In Mexico, Wal-Mart followed a two-pronged strategy. First, it bought up existing Mexican retail firms and operated them under their original Mexican brand names, and second, it established a series of stores under its flagship brands Wal-Mart and Sam's Club. After 12 years of operating in Mexico, Wal-Mart has a payroll of over 100,000, making it the single largest private employer in the nation! Some Mexicans chafe under Wal-Mart's presence, making many of the same complaints that its detractors raise in the United States. Cultural imperialism is often another complaint. The construction of a Wal-Mart store in 2004 in the shadows of the pre-Columbian ruins of the

FIGURE 18.12. A modern food court in a suburban shopping mall in Tegucigalpa, Honduras, 2002.

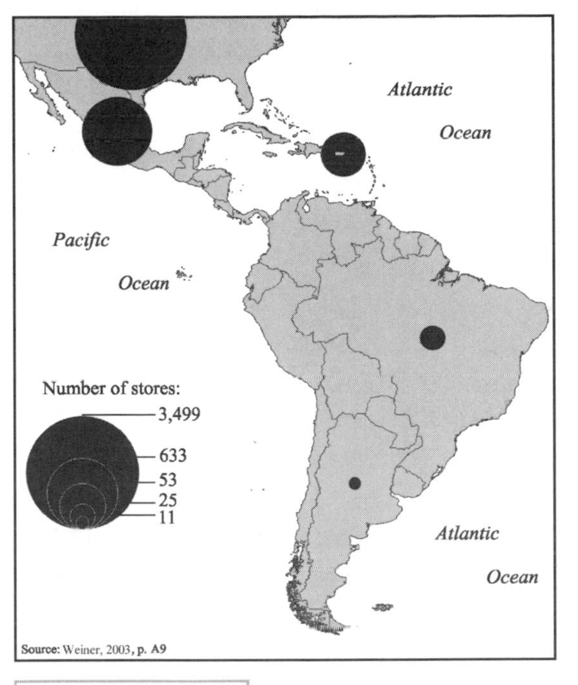

FIGURE 18.13. Wal-Mart stores in Latin America, 2003.

Pyramids of the Sun and the Moon in the Valley of Mexico outraged many Mexicans. At the same time, millions of Mexicans flock to Wal-Mart stores, making over $10 billion in annual purchases!

Globalization, of course, is not a one-way street. As interaction between Latin America and other world regions increases, Latin Americans also affect and transform other countries and economies. At the present, the international migration of hundreds of thousands of Latin Americans to the United States, Spain, and a handful of other countries may be the region's greatest effect. Few Latin American businesses have yet entered foreign markets, but some have. One Guatemalan fast-food chain, Pollo Campero, successfully entered the U.S. market in 2002 and looks to experience notable commercial success (Vignette 18.1).

Globalization is a complex process. How its evolution and specific effects will play out in Latin America over the next 20 to 30 years remains to be seen. While it is decried by many in Latin America as contributing to a loss of economic and cultural sovereignty, others readily drink Coca Cola, shop at Wal-Marts in Mexico or Brazil, eat hamburgers and fries at Burger King and McDonalds, and get on with their lives. It does seem clear that multinational companies and the products and services they provide will become even more widespread in the decades ahead. It seems reasonable to expect some kind of increasing economic and cultural convergence as not only products and services transform behavior, but as work habits and business processes are also adopted from multinational businesses and integrated into national economies as local firms struggle to compete. Hybridization will also occur, as it does already, when products, services, or even foods are transformed to satisfy local and regional markets. While the dominant flow of these changes has been from outside of Latin America, this will not necessarily always be the case, as the number of Latin Americans who emigrate abroad continues to grow. The immense popularity of "Mexican food" in the United States is but one example of how the United States has also been affected by this ongoing process of globalization.

Summary

Latin America has been part of the world economic system for nearly 500 years. Spain and Portugal established mercantile systems that controlled the terms of trade between the colonies and the mother countries. Raw materials and commodities were exported to Europe, while manufactured goods were imported by the colonies. The system ensured that vast wealth accrued to both Spain and Portugal, but economic development in the colonies was hindered.

Independence from Spain and Portugal

VIGNETTE 18.1. POLLO CAMPERO: A CENTRAL AMERICAN FAST-FOOD RESTAURANT FINDS SUCCESS IN THE UNITED STATES

Pollo Campero is an immensely popular and successful fast-food chain in Central America. The name translated into English means "country chicken." While its outlets resemble U.S. fast-food establishments in many respects, there are notable differences. One of these is that the company's restaurants not only provide food for take out, but also provide sit-down dining with table service for the same price. Founded in 1971 in Guatemala, the company developed a loyal customer base with a chicken-centered fast-food menu. Subsequently, it opened restaurants in other Central American countries with equal success and customer loyalty and today has close to 200 outlets. Pollo Campero has about 70 locations in Guatemala, and nearly 100 in other Central American nations, as well as restaurants in Mexico and Ecuador.

Its fried chicken inspires intense customer loyalty. As Guatemalan and Salvadoran migration to the United States grew during the 1990s, many who returned home for brief visits would bring back boxes of its signature chicken for friends and family back in the United States One customer made a take-out order for 1,600 pieces of chicken which she took back to the United States in two duffel bags, selling the chicken by the piece and paying for her flight with the profits. At one point, airline executives complained to the company that its planes were starting to smell like chicken and implored them to develop an odor-proof container. The company's executives declined to do so, but began studying the possibilities of entering the U.S. market.

The first store, opened in Los Angeles in 2002, was an immediate success. When the restaurant first opened customers waited up to 2 hours in line to buy its chicken and the number of pieces that could be purchased had to be rationed. Since then the company has opened additional stores in southern California, Houston, New York City, and Washington, D.C. While the stores tend to locate in neighborhoods with a significant population of Central American immigrants, the firm has adapted its menu to the United States in order to attract a broader range of customers. Table service is not an option in its U.S. stores because labor costs are too high. The company plans to expand into Europe too, beginning with Spain, Portugal, and Poland.

in the early 1800s brought significant changes to Latin America's trading and economic relationships. Agricultural commodities and other raw materials continued to be the region's main exports. However, Great Britain replaced Spain and Portugal as the region's principal economic power and principal trading partner for much of the 19th century. Despite Latin America's political independence, Great Britain exerted considerable indirect control over the region through its dominance of financial institutions, technology transfer, investment decisions, and trade. Often referred to as neocolonialism, this pattern of relationships between Latin America and outside economic powers has characterized the region since then. By the early 20th century, the United States had replaced Great Britain as the region's primary neocolonial power.

During much of the 20th century Latin American countries struggled to diversify their economies and to move away from dependence of one or two export commodities. Governments employed inward-oriented economic development polices to reduce dependence on foreign manufactured goods and encourage industrialization. The erection of formidable tariff barriers also protected manufacturing from import competition. For most countries, the small size of internal markets made success elusive. However, the largest economies, Mexico and especially Brazil, did experience some notable successes.

The late 20th century brought new ideas and policies to the region as the world economy became increasingly interconnected. These economic and policy reforms focused on emphasizing the benefits of economic comparative advantage and the reduction and eventual elimination of trade barriers. As a consequence, regional economic integration strategies have been pursed. Major trade pacts like NAFTA and MERCOSUR have emerged as a result, and efforts are underway to build a hemisphere-wide free trade zone. These policies, often referred to as neoliberal reforms, have also sought to reduce the role of government in business and financial markets and to ensure transparency and economic efficiency in markets. As the 21st century begins, the processes of globalization are bringing more of the outside world to Latin American as well as more of Latin America and more Latin Americans into the wider world.

Further Reading

Bulmer-Thomas, V. (2003). *The economic history of Latin America since independence* (2nd ed.). Cambridge, UK: Cambridge University Press.

Bulmer-Thomas, V. (2001). Regional integration in Latin America and the Caribbean. *Bulletin of Latin American Research, 20*(3), 360–369.

Chavez, J. (2004). Localized effects of globalization: The case of Ciudad Juárez, Chihuahua, Mexico. *Urban Geography, 25*(2), 120–138.

Clapp, R. A. (1995). Creating competitive advantage: Forest policy as industrial policy in Chile. *Economic Geography, 71*(3), 273–296.

Estevadeordal, A., et al. (2004). *Integrating the Americas: FTAA and beyond.* Cambridge, MA: David Rockefeller Center for Latin American Studies, Harvard University.

Folsom, R. H. (2004). *NAFTA and free trade in the Americas in a nutshell* (2nd ed.). Minneapolis, MN: West.

Gonzalez, D. (2002). Guatemala journal: Fried chicken takes flight, happily nesting in the U.S. *New York Times*, September 20, p. A4.

Gwynne, R. N. (1996). Direct foreign investment and non-traditional export growth in Chile: The case of the forestry sector. *Bulletin of Latin American Research, 15*(3), 341–357.

Gwynne, R. N., and Kay, C. (Eds.). (1999). *Latin America transformed: Globalization and modernity*. London, UK: Arnold.

Gwynne, R. N., et al. (2003). *Alternative capitalisms*. London: Arnold.

Harrison, L. E. (2000). *Underdevelopment as a state of mind: The Latin American case*. Lanham, MD: Madison Books and the Center for International Affairs, Harvard University.

Hiroka, L. S. (2001). *Global alliances in the motor vehicle industry*. Westport, CN: Quorum Books.

Keeling, D. J. (1999). Neoliberal reform and landscape change in Buenos Aires. *Conference on Latin Americanist Geographers Yearbook, 25*, 15–32.

Klak, T. (1999). Globalizaton, neoliberalism, and economic change in Central America and the Caribbean. In R. N. Gwynne and C. Kay (Eds.), *Latin America transformed: Globalization and modernity* (pp. 98–126). London: Arnold.

Kugel, S. (2004). A Guatemalan legend, fried and plucked, is headed here. *New York Times*, May 2, sec. 14, p. 6.

Orozco, M. (2002). Globalization and migration: The impact of family remittances in Latin America. *Latin American Politics and Society, 44*(2), 41–66.

Preston, D. A. (Ed.). (1996). *Latin American development: Geographical perspectives*. London, UK: Longman.

South, R. B. (1990). Transnational "maquiladora" location. *Annals of the Association of American Geographers, 80*(4), 549–570.

Throp, R. (1998). *Progress, poverty, and exclusion: An economic history of Latin America in the 20th century*. Baltimore: Johns Hopkins University Press.

Tuman, J. R., and Emmert, C. F. (2004). The political economy of U.S. foreign direct investment in Latin America: A reappraisal. *Latin American Research Review, 39*(3), 9–28.

Walton, M. (2004). Neoliberalism in Latin America: Good, bad, or incomplete? *Latin American Research Review, 39*(3), 165–183.

19 The Amazon Basin

The Amazon Basin is one of South America's megaregions. This lowland basin, the largest and wettest in the world, is focused on the Amazon River and its vast network of tributaries. It occupies the central portion of the South American continent, and encompasses in excess of 6 million sq. km (Figure 19.1). Expressed another way, it accounts for nearly half of the landmass of Brazil, nearly half of Peru, and significant portions of Colombia, Ecuador, and Bolivia. Superlatives often seem to be the only suitable terms to describe the immensity of the basin and the river system that defines it. The Amazon River, the "backbone" of the basin, is estimated to discharge nearly one-fifth of all the freshwater flowing into the Earth's oceans, exceeding the discharge of the Mississippi River by a factor of 10.

The Amazon Basin is a resource-rich region that is the focus of intense social and economic development. Ecologists note that half of the world's remaining tropical rain forests are found in the region and warn that rapid deforestation is a significant threat to the region's ecological balance and may also have worldwide ecological repercussions. Traditionally, the regional economy was based on the gathering of forest products like Brazil nuts, cinchona (quinine), and rubber which created a boom–bust pattern of economic development but had only limited long-term ecological impacts. In the last half of the 20th century, however, road building, logging and forest clearance, agriculture, cattle ranching, and mineral exploitation have caused what many believe is widespread environmental degradation. Social conflict, at times violent and even deadly, has also increased as competing economic interests contest the region's resources. The region's total population is comparatively small, 12–13 million, but it is growing rapidly and is increasingly urban. Its vast size and resource potential will make the Amazon Basin an increasingly important and conflicted region in the 21st century.

Geography and Environment

The headwaters of the Amazon River rise in the Peruvian Andes where the Ucayali and Marañón Rivers form two of the river's principal tributaries. On the northern side of the river's drainage basin other important tributaries include the Napo River, rising in Ecuador and joining the main current just downstream from Iquitos, Peru; the Caquetá River, known as the Japurá River in Brazil, draining southern Colombia and part of eastern Brazil; the Río Negro and its principal tributary the Branco River, which together drain much of the Guiana Highlands and a small portion of the Colombian and Venezuelan Amazon; and the Trombetas River, which drains the extreme

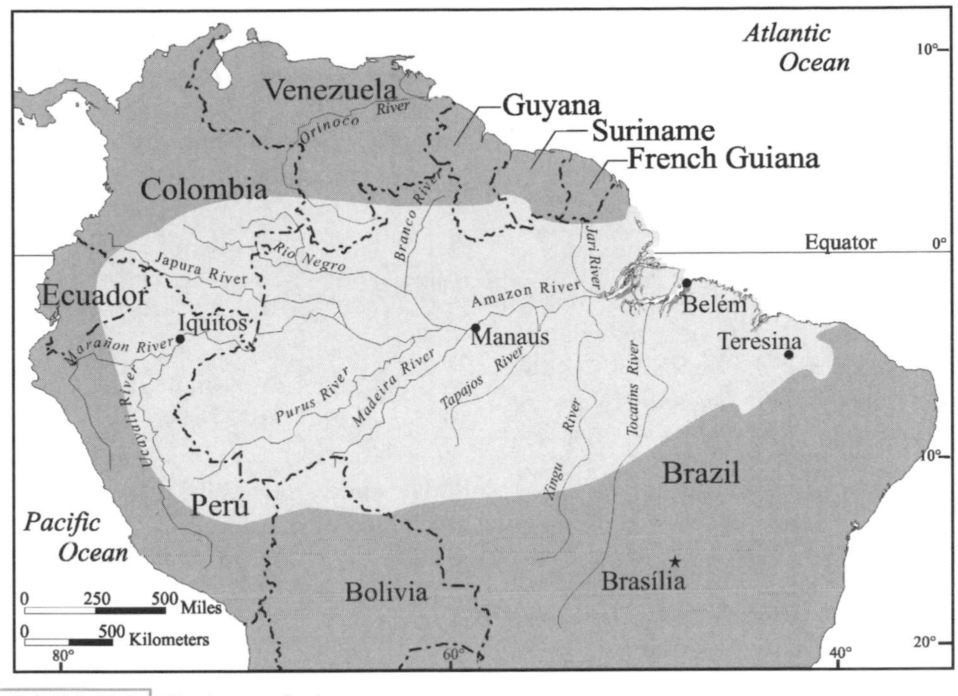

FIGURE 19.1. The Amazon Basin.

eastern portion of the Guiana Highlands. Major tributaries along the southern portion of the watershed include the Madeira and the Purus Rivers, whose tributaries flow northward from Bolivia's eastern Amazonian rain forests and savannas. To the east the Tapajós, the Xingu, and the Tocantins Rivers flow northward off the Brazilian Highlands.

At the Amazon River's confluence with the Atlantic Ocean, its main channel is some 80 km wide. The river is readily navigable by large oceangoing ships as far inland as Manaus, Brazil—1,600 km from the river's mouth; smaller oceangoing vessels can reach Iquitos, Peru, 3,700 km from the Atlantic Ocean. Nearly one-half of the world's remaining tropical rain forests are found in the Amazon Basin. This is one of the reasons environmentalists attempt to focus world attention on the issue of deforestation and its consequences in this region.

The topography of the Amazon Basin is varied. While the basin is a lowland area, there is considerably more relief here than is often assumed, and these variations in relief establish two of the most important geographic distinctions within the basin: the river floodplains and the interfluves.

Constituting less than 10 percent of the total land area of the basin, the floodplains, or *várezas* , as they are known in Brazil, are one of the basin's most significant resources. The periodic, usually annual, flooding that occurs along the basin's river courses deposits nutrient-rich silts in the *váreza* , creating a narrow band of fertile tropical soil along the river courses once the annual floods have receded. Traditionally, these floodplains have been the most important areas for indigenous agriculturalists, as they have been in recent decades as small-scale agriculturalists have colonized riverine areas in the basin.

The uplands, or interfluves, known in Brazil as *terra firme*, occupy the vast remainder (over 90 percent) of the basin's land area. Tropical rain forests dominate the natural vegetation cover of terra firme. However, in the extreme northern and southern margins of the Amazon Basin, *cerrado*, a broadly defined vegetation class in Brazil that ranges from grasslands with scattered trees and shrubs to scrub woodlands, dominates.

Situated astride the equator, the Amazon Basin is characterized by a warm, humid, tropical climate (Figure 19.2). Rainfall is plentiful and exceeds 2,000 mm in the central portion of the basin's interior and between 1,500 and 2,000 mm over the remainder of the region. Rain occurs in appreciable amounts in all months along the equatorial belt. This is caused by the passage of the tropical low pressure belt, the ITCZ, over the area twice each year, which brings abundant rainfall. To the north and south of this central region, precipitation is more markedly seasonal in its occurrence. Mean temperatures are relatively constant, although there is slight seasonal variation. Mean temperatures in January range between 25° and 30°C, and the pattern in July is nearly the same with mean temperatures falling off by 2°C or 3°C.

Historical Geography and Economic Development

European explorers began the exploration of the Amazon Basin during the earliest years of the colonial period, but knowledge of the region remained sketchy until well into the 20th century. Francisco de Orellana, a Spanish explorer, descended the Amazon from the Napo River in what is now Ecuador to its confluence with the Atlantic Ocean in 1540–1541. Nearly 100 years later (1637–1938) Pedro Teixeira ascended the river to its headwaters in the Ecuadorian Andes. Generally, however, the geography of the Amazon Basin remained largely unknown to the Spanish and Portuguese, who tended to focus their colonization and settlement along the coastal margins, in the highlands, and in areas of dense indigenous settlement. The Amazon Basin remained on the extreme periphery of the New World territories of both colonial powers throughout the entire colonial period. Nor did independence or the establishment of nation-states during the 19th century bring the region much attention, colonization, settlement, or economic ac-

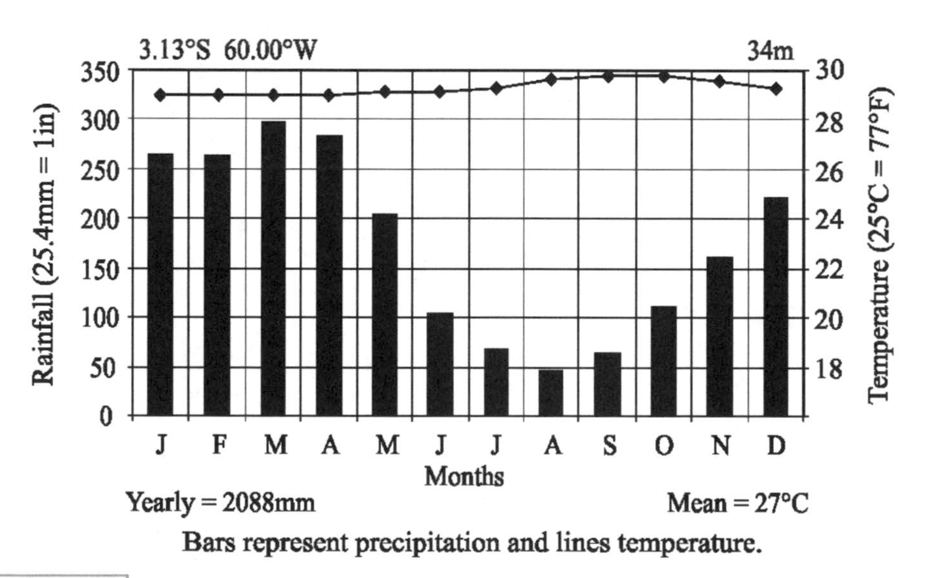

FIGURE 19.2. Climograph of Manaus, Brazil.

tivity. The region remained largely the domain of scattered indigenous tribes, missionaries, and small numbers of traders and settlers focused along the major watercourses of the region.

However, by the end of the 19th century, the Amazon Basin acquired prominence as the exploitation of forest products for export propelled Amazonia into the world economy.

The gathering of Brazil nuts, *Bertholletia excelsa*, which are rich in oil and protein and grow wild, began early in the Amazon Basin. This set a pattern that has endured to the present, especially in the Brazilian states of Pará and Amazonas. Brazil nuts are one of the most widely traded nuts in the international market. Traditionally, Brazil has dominated the world market but its production has fallen precipitously in recent years from about 19,000 metric tons in 2000 to barely 7,000 metric tons in 2003. Although the exact cause of the decline is hotly contested, many believe that the increasing pace of deforestation and the expansion of cattle raising in the Amazon Basin has contributed to this pattern. In the meantime, Bolivia has outstripped Brazil in production and is now the world's largest exporter of Brazil nuts.

Another early key forest product gathered for commercial exploitation and export during the 19th century was the bark of some 40 species of trees known generically as *cinchona*. These trees, which are native to the Amazon Basin and the Andean piedmont and foothills, are the source of quinine, an antimalarial drug. The collection of the bark of these trees began as early as the mid-1600s. Europeans successfully smuggled the seeds of several varieties of *cinchona* out of the Amazon Basin and established plantations in their colonies in Asia in the 1850s and 1860s. This occurred again nearly 50 years later with another tree that produced a valuable product, rubber.

The invention of a process to vulcanize (i.e., to treat chemically to make it elastic, strong, and stable) natural rubber, a white sap tapped from the bark of the rubber tree, *Hevea brasiliensis*, in 1839 set the stage for the first large-scale migration into the Amazon. As the Industrial Revolution produced increasing opportunities for the use of rubber as a component in manufactured goods, Brazilian entrepreneurs sought to find the means to provide it. Initially, indigenous peoples were recruited to tap the wild rubber trees and collect and concentrate the sap. Since indigenous peoples were not dependent on the market economy and could not be depended upon to work regularly, the entrepreneurs, called "rubber barons," who controlled large land concessions and marketed the crop to European and North American markets, sought another labor supply.

These entrepreneurs found the necessary labor force in the chronically drought-stricken mortheast of Brazil, known as the *Sertão* . Impoverished and landless rural workers from that region migrated into the Amazon Basin in large numbers. Between 1870 and 1890, some 200,000 northeasterners moved into the Amazon Basin, settling along river courses and playing a critical role in the rubber boom. These rubber tappers, known as *seringueiros*, provided the backbone of the rubber boom that lasted until about 1915, when Asian plantations, growing rubber trees descended from seedlings smuggled out of Brazil years earlier, began to outproduce Brazil.

The rubber boom transformed several small towns along the Amazon from muddy inhospitable places into wealthy cities; Belém and Manaus, in Brazil, and Iquitos, in Peru, are notable in this regard. Tremendous wealth was concentrated in these cities. In Manaus, for instance, the local elite built an opera house modeled on that found in Paris, and in Iquitos the local market was designed by Eiffel, the famous French architect for whom the Eiffel Tower is named (Figure 19.3). The compelling movie *Fitzcarraldo*

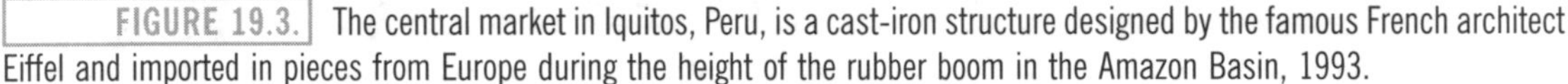

FIGURE 19.3. The central market in Iquitos, Peru, is a cast-iron structure designed by the famous French architect Eiffel and imported in pieces from Europe during the height of the rubber boom in the Amazon Basin, 1993.

(1982), directed by the German director Werner Herzog, provides a vivid fictionalized account of the rubber epoch in the region around Iquitos, Peru.

The Amazonian rubber economy did not die, however; it simply withered and stagnated. Plantations in Asia undercut Brazilian supplies, and the eventual development of synthetic rubber also reduced demand for the wild Brazilian product. The *seringuieros*, however, established themselves as a permanent part of the regional economy, occupying the upper reaches of the basin's tributaries and eventually becoming a potent force for ecological sanity and alternative forms of economic development in the rain forest during the 1980s and 1990s (Vignette 19.1).

Contemporary Economic and Social Geography

Settlement in the Amazon Basin was limited and of comparatively little consequence until the 1950s and 1960s. Then events in Brazil and the other countries of the Amazon Basin unleashed dramatic change. Social injustice and population growth in the rural areas of the Andean highlands of Colombia, Ecuador, Peru, and Bolivia created serious demands for agricultural land reform. Significant, although ultimately inadequate, reforms were undertaken in each country. Another option, pursued to varying degrees by each country, was the promotion of agricultural colonization and settlement in the Amazon Basin. These areas included regions in the Putumayo in Colombia, along the Río Napo in Ecuador, in the Huallaga Valley in Peru, and in the Chapare and the Beni in Bolivia. Settlers migrated into these areas, cut the forest cover, and established farms and ranches.

In Brazil, the military had taken power in 1964. It looked to the Amazon with a mixture of preoccupation and optimism. From a geopolitical perspective, the generals viewed the vastness of the Amazon and its sparse settlement, especially in border areas, with con-

VIGNETTE 19.1. CHICO MENDES AND THE RUBBER TAPPERS

In many ways the life and death of Chico Mendes (1944–1988) is a metaphor for the conflicts and struggles that typify the Amazon Basin at the beginning of the 21st century. Mendes grew up in the remote rain forest region in the state of Acre near Brazil's border with Bolivia. As the son of a rubber tapper and a rubber tapper himself, Mendes knew the forest and its ways intimately. There were no schools in the rain forest frontier—Mendes, although bright and inquisitive, learned to read only after he was 18 years old. As he expanded his worldview, he began to understand the political economy of rubber tapping, especially the exploitation of the tappers by the *seringalistas*.

Trading-post owners advanced basic supplies to the tappers on exorbitant credit terms, purchased their rubber at depressed prices, and kept the tappers in a perpetual state of debt peonage. Mendes spent much of his life at great personal risk organizing tappers to bypass these intermediaries in the sale of their rubber. In 1975 he began to work for the Confederaçao do Trabalhodores (the Confederation of Rural Workers Union) in an effort to defend the interests of the rubber tappers. As the pace of settlement in Acre quickened, the very livelihoods of the rubber tappers were increasingly threatened as loggers and ranchers began to usurp forestlands and clear them.

Mendes helped develop a strategy known as a stand-off, or *empate*. When gangs of loggers and armed gunmen hired by ranchers to clear forest moved into an area, the tappers and their families mobilized. Using nonviolent means, men, women, and children placed themselves between the loggers and the forests, acting as human shields. The strategy proved to be marginally successful. Frequently the destruction of forest tracts was stopped, although often the stoppage was just a postponement. Nevertheless, the strategy began to convince the rubber tappers of their own abilities to empower themselves. In 1985, the Conselho Nacional do Seringueiros (the National Rubber Tappers Council), often known informally as the "Rubber Tappers Union," was formed. Mendes was central to its establishment and became its leader. Two of the key premises of the organization were that (1) the destruction and colonization of forests used by extractors of natural resources (e.g., rubber tappers and Brazil nut gatherers) must stop, and (2) that the government must find ways to promote sustainable economic and regional development. The tappers began to develop a framework that called for the establishment of "extractive forest reserves" that would be a form of permanent, legally protected communal landholding reserves.

These were revolutionary ideas that challenged traditional power relationships. In 1987 legislation that permitted the establishment of extractive reserves was signed, and in early 1988 the first such reserve was established in Acre. Later the same year, the tappers, under Mendes's direction, pressed for the establishment of a reserve on contested forestland that local cattle ranchers also coveted. After a series of escalating confrontations, the rubber tappers prevailed when the state's governor, fearful of increasing conflict, declared the lands an extractive reserve. The victory, although significant, cost Chico Mendes his life only a few weeks later. In December 1988 two gunmen hired by the frustrated ranchers assassinated him as he stepped onto the back porch of his home.

Chico Mendes's murder reverberated worlds away. It focused increasing attention on the runaway and irresponsible destruction of the Amazon rain forest. It brought national and international pressure on the Brazilian government to take measures to promote sustainable development of the region. It galvanized rubber tappers and others to redouble their efforts at preserving their way of life and the tropical rain forest. Mendes's life and the conflicts over resources that continue to characterize the Amazon Basin are portrayed superbly in the Hollywood movie *The Burning Season* (1994), starring the popular Puerto Rican American actor, Raul Julia.

cern. They believed it to be vulnerable to penetration by neighboring countries. Although there were no indications that any other South American nation had designs on Brazil's Amazon territories, the military's worries were translated into plans for the long-term development of the basin.

The military's vision, however, was not new. Brazil's long-time president, Getúlio Vargas, had enunciated a vision for the development of Amazonia some 25 years earlier, under the rubric of the "March to the West." The construction of a road network, the promotion of vast livestock-raising schemes, agricultural colonization and settlement by smallholders, and incentives for a variety of economic development programs represented the key elements of the military's development program.

Major road construction in Amazonia began in 1960 with the initiation of the north–south route, the Brasília–Belém Highway, BR 14. When it was completed in 1964, this road opened up much of eastern Amazonia to colonization and settlement, and provided the first north–south land linkage between Amazonia and Brazil's core regions in the south. The construction of the equally ambitious Trans-Amazon Highway between 1970 and 1980 established the primary east–west vehicular artery across the central portion of the Amazon Basin. Beginning at Marabá in Pará state, the Trans-Amazon Highway runs some 3,100 miles westward to the city of Río Branco in the state of Acre close to the Peruvian border.

The construction of several other tremendously expensive major routes filled in the basic land transportation network (Figure 19.4). These include the well-known BR 364, running from the southern states of Mato Grosso and Goiás along the basin's western margin to Rondônia and Acre; BR 163, running from Cuiaba in Mato Grosso directly northward to Santarém, which is located at the confluence of the Río Tapajos and the Amazon; and BR

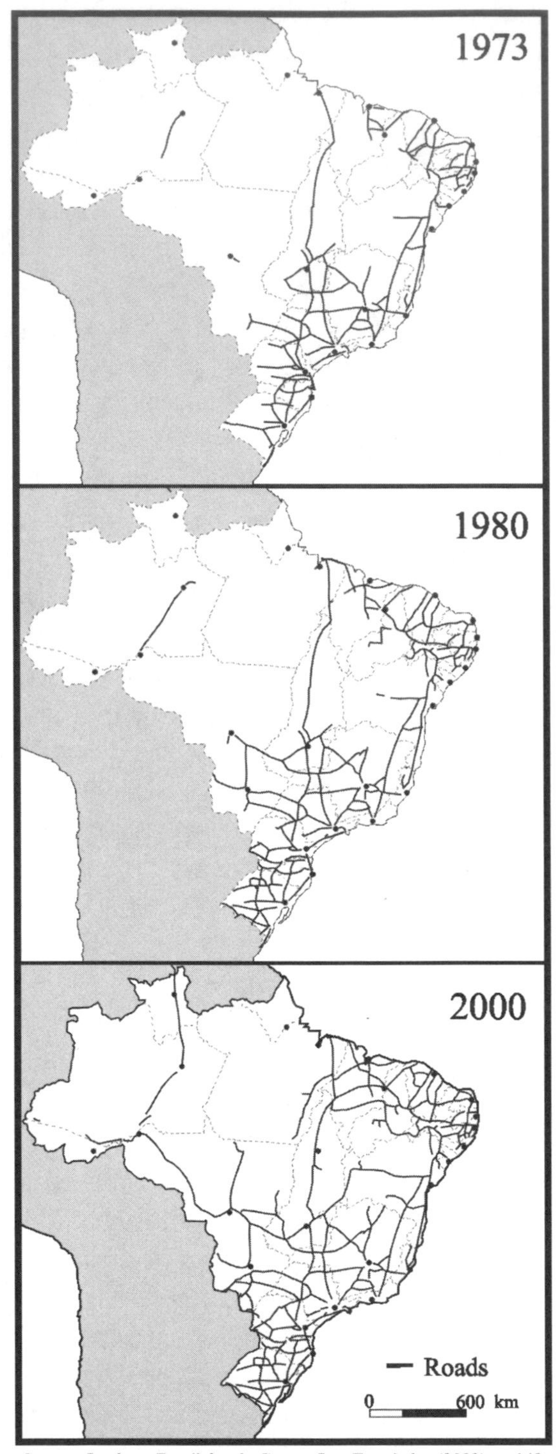

Source: Instituto Brasileiro de Geografia e Estatística (2002), p. 149

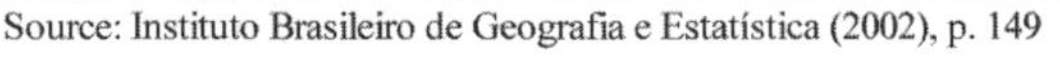

FIGURE 19.4. The growth of Brazil's network of paved roads, 1973-2000.

139, connecting Acre and Rondônia with Manaus and the northern side of the Amazon Basin watershed. This route near the border with Guyana connects with the partially completed Northern Perimeter Highway. While highway engineers have been relatively successful in building highways in Amazonia, no bridge crosses the Amazon River. Instead, the northern and southern portions of BR 139 are joined by a ferry at Manaus.

Brazil's military government created a range of programs to speed the process of settlement and economic development in the basin. These tended to favor the nation's wealthy elite at the expense of all others, and usually encouraged large projects promising rapid returns. Cattle ranching figured prominently in these early development schemes. In Brazil, hyperinflationary periods coincided with much of recent Amazonian development and encouraged many wealthy investors to view landownership, and to a lesser degree cattle ranching, as an excellent hedge against inflation. As a consequence of government subsidies and tax incentives that encouraged forest clearing as evidence of land improvement and development, hundreds of thousands of acres of tropical rain forest were cleared to create pasture for cattle herds. While cattle raising is widespread throughout the Amazon, the most significant concentration occurs in a crescent-shaped belt running along the southern margin of the region. This begins in the southwest in Rondônia, runs along BR 364 and then through the states of Mato Grosso and Goiás in the south, and finally arches northward into Tocantins and Pará.

Colonization and settlement by smallholders, or peasant farmers, has also figured prominently in the development of the Brazilian Amazon. Both planned and spontaneous colonization have occurred on a large scale.

Government-sponsored planned settlement played a key role in colonization along the Trans-Amazon Highway. The military government sponsored such settlement in an effort to head off the demands for land in the northeast from landless peasant groups known as the peasant leagues. Under the aegis of INCRA (Instituto Nacional de Colonizaçao e Reforma Agrário [National Institute for Colonization and Agrarian Reform]), the government sought to resettle as many as 100,000 families. Settlers along the Trans-Amazon Highway received what initially appeared to be a generous package. This included transportation to the Amazon, clear title to 100 ha, credit for planting basic food crops, an income subsidy for the first 6 months, and subsequent food subsidies.

The government planned a hierarchy of service centers, *agrovilas*, along the highway. These were established every 10 km. Basic health, education, and commercial and recreational services were offered in each agrovila. A second urban tier, regional centers, or *agrópolises* , was created to provide social and commercial services requiring higher threshold populations to support.

These efforts at planned settlement failed completely. Eventually fewer than 8,000 families participated in the program along the Trans-Amazon Highway. Despite generous benefits, most settlers abandoned the plan. By the early 1990s, along some sections of the highway, the farm abandonment rate among settlers ranged between 80 and 95 percent. Some remained only as long as the 6-month income subsidy lasted. The smallholders failed for a variety of reasons, including a lack of agricultural experience in a humid tropical environment, marginal lands, unrealistic assumptions about economic markets for smallholders, and inadequate credit and agricultural supplies. Some farmers succumbed to the quick profits to be made on the sale of land with a secure property title, while others fell victim to unscrupulous land barons who used violence and intimidation to usurp settlers' land.

Deforestation

The deforestation of the Amazon Basin proceeds at a rapid pace. In the last 50 years the area of tropical rain forests in the Amazon Basin has declined by about 25 percent. While precise estimates vary, since the late 1970s satellites and remote-sensing technology have permitted a careful monitoring of geographical outlines of deforestation trends. Such techniques documented the deforestation of a total of 17.2 million ha in the region between 1990 and 2000. This is equivalent to about one-half the land area of the Netherlands (also about half the land area of the state of Missouri in the United States). Driven by burgeoning demand for meat products in Europe and elsewhere, cattle raising has boomed in the last decade and with it additional pressure for forest clearance. In the state of Mato Grosso, on the southern margins of Amazonia, soybean cultivation has expanded massively in the last 10 years as demand from export markets, especially China, has soared. Both of these factors have played key roles in contributing to an increase in the rate of deforestation in recent years. Independent studies from the Brazilian government's Space Research Institute and the Center for International Forestry Research pegged the rate of forest loss in 2003 at about 2.4 million ha per year.

Ecologists decry the destructive environmental impact of forest loss, pointing out a range of negative environmental consequences. Soil erosion, loss of biological diversity, reduction in the long-term carrying capacity of the land, and even air pollution are among the effects of deforestation condemned by its opponents. Others denounce the development of a rural political economy in which a few control almost all land and assets, while huge numbers of poor landless peasants live in abject and often extreme conditions in the countryside and more commonly in urban centers.

Others see wholesale development of the region as an economic imperative and vigorously defend road building, forest clearance, mining, and the development of agro-industrial complexes raising cattle, soybeans, and other crops. They argue that the need to pursue economic development of this resource-rich region is essential if Brazil and the other nations that share the Amazon region are to develop to their full potentials. They note the needs of a steadily growing population and the constant migration of new settlers and job seekers into the region. Many acknowledge the potential dangers of environmental degradation, but argue that conservation and environmental management measures can be taken to mitigate the long-term negative effects of deforestation and development.

Land Degradation and Abandonment

Spontaneous settlement occurs throughout the Brazilian Amazon. Large numbers of colonists have occupied lands adjacent to the Trans-Amazon Highway and other major roads in the basin. One of the most well-known areas of spontaneous settlement has been along BR 364 in Rondônia where, after the completion of the road's paving, some 5,000 immigrants per month arrived. Here, as in other areas of Amazonia, a fairly common pattern of smallholder land use evolved. Settlers squat on available lands adjacent to the roads. Trees that are commercially valuable are logged for sale, and most of the remainder of the forest is cleared. The settlers employ agricultural techniques common throughout the Tropics, known variously as slash and burn, swidden, or—in Brazil—*roça* . Clearing is timed to coincide with the driest time of the year. The trees and brush are allowed to dry and then are burned in situ in what will become the farm's fields. The ashes of the burned forest provide critical nutrients and minerals to the soil for plant growth, usually contributing to adequate crop yields in the first year or two after clearing. Continuous cropping in the same fields leads to soil erosion, depletion of key minerals and nutrients, and compacting of the soil. Yields fall precipitously in subsequent years, and the economic position of the settlers becomes increasingly precarious, usually demanding off-farm employment to supplement farm income or encouraging the sale of the land or even its abandonment.

Displaced and unsuccessful colonists may follow a pattern of recurrent frontier settlement and land abandonment, moving through this cycle several times during their lives. These settlers are critical to the process of land clearance, land consolidation (latifundia formation), and large-scale cattle raising in much of Amazonia. A "hollow frontier" is created, with the most intense land use and population density at the frontier, and less intensive land use and low population densities in long-settled regions. A mix of crop farming, logging, and forest gathering typify the frontier, while a cattle monoculture predominates in the long-settled regions.

The use of the abandoned farmland in terra firme tends to follow one of two scenarios. Marginal pastureland may result. Then the smallholdings of early settlers are consolidated into large properties, latifundia, dedicated to cattle raising. In the long run, and even in the medium term, not all is well for cattle ranchers either. Extended use of these lands as pasture reveals their inherent fragility. For instance, along some sections of the Brasilia–Belém Highway, per hectare grazing capacities have been reduced by two-thirds in just 6 years.

In areas where even cattle raising proves to be a marginal proposition, abandoned croplands and pastures revert to secondary forest growth. The return of a forest canopy does serve to mitigate some of the more negative environmental effects of colonization and deforestation. Soil erosion is reduced and the leaching of nutrients and minerals slowed. However, often the secondary forest that colonizes these abandoned clearings is dominated almost exclusively by one species, and the resulting ecological niches lack diversity of species and habitats.

Growing world demand for soybeans in recent years, especially from an increasingly affluent population in China, has resulted in additional pressure on land resources in Amazonia. In the southern Amazonian states of Mato Grosso and Pará, land devoted to soybean cultivation has expanded dramatically since 2000 and continues to increase. While some previously deforested lands have been brought under cultivation, forest clearing for soybean cultivation has occurred at a rapid rate. This has brought calls of alarm from environmentalists and a skeptical response from farmers and agrobusiness interests.

Large-Scale Development Projects

The exploitation of natural resources and the promotion of large-scale projects represented important elements in the development strategy pursued for Amazonia by the nation's various military governments (1964–1988). Perhaps one of the most well known of the efforts at a large-scale development project in the early years of this period was the Jari Project, initiated in 1967. This grandiose project was the brainchild of a wealthy American, Daniel Ludwig, who purchased 1.2 million hectares (more than 13,000 sq. km) along the Jari River in northern Pará state.

Ludwig's plan envisioned a vast agro-industrial complex. The cornerstone of the complex was to be a wood-pulp plant that would process trees grown on plantations of *Gmelina arborea*, an imported East Asian species purported to grow rapidly under tropical conditions. Large-scale agricultural activities, specifically rice cultivation and cattle raising, were included as key components in the project. Kaolin, a commercially valuable clay, was subsequently discovered on the property and its mining became a key component in the project's operation. A massive wood-pulp plant constructed on two barges in Japan and towed across the Pacific, around Cape Horn, up the Atlantic coast to the Amazon River, and eventually to Jari focused worldwide attention on the ambitious plan. While Ludwig benefited initially from the goodwill of the Brazilian government, loans from development banks, and his own deep pockets, the project never realized its anticipated potential. Jari failed because of a lack of understanding of the ecology of the region as well as unrealistic expectations. In 1982 Ludwig sold the property to a Brazilian consortium backed by the government for close to $500 million.

Another large-scale project was initiated by government planners. They used regional development planning theory to pursue a development strategy that identified "growth poles." *Growth poles* are existing or future urban centers targeted for development of key economic and social infrastructure to promote regional economic development (e.g., roads, electric grids, schools, markets, and medical facilities). Outlined under the aegis of the Polo Amazonia program, some 15 growth poles were identified where investment and incentives would be focused. Planners hoped that these centers would concentrate, process, and eventually export primary products or their derivatives from Amazonia to markets in the Brazilian heartland or overseas.

A third example, the epitome of the large-scale development project in Amazonia, is Gran Carajás. In the late 1960s geologists discovered rich iron ore deposits in the Serra dos Carajás in the northeastern portion of Pará state. The richness and immense extent of these deposits are spectacular. The iron ore is 67 percent pure, and it is predicted that it will take well over a century to exhaust the ore in this region. Open-pit mining of the iron ore deposits and ore processing began in 1967 under the auspices of a consortium of the U.S. Steel Company and the Brazilian state-owned mining company, the Vale do Rio Doce Company (CVRD). The Brazilian government constructed key infrastructure projects to ensure the success of the project and to contribute toward a grander vision of a regional development project. One of the most critical among these was a 900 km railroad, completed in 1985, that runs east from the mine to the capital of Maranhão state, São Luis, on the Atlantic, where port facilities for iron ore export were constructed. To provide reliable energy sources, the government constructed the Tucuruí dam and hydroelectric plant about 240 km north of the Serra dos Carajás on the Tocantins River. The dam, inaugurated in 1984, impounds a reservoir over 150 km in length. It is one of the largest hydroelectric complexes in the world and cost nearly $5 billion to construct.

The Serra dos Carajás subsequently proved to be a treasure trove not only of iron ore, but of a broad range of other valuable minerals, including nickel, copper, manganese, and gold. The government has encouraged small-scale iron ore processing operations along the railroad to São Luis, where it has licensed over 20 pig-iron smelters. The use of local trees, the most economical means to produce charcoal used in the smelters, has caused widespread deforestation along the railroad route and negative environmental impacts. Despite the environmental costs, the government sought to encourage other primary economic activities in the project region, notably forestry, farming, and cattle raising. As in other areas, the environmental impact of such activities has been mixed. Finally, the Gran Carajás project required immense financial resources. Close to $4 billion, coming from both national and international sources, had been invested by the beginning of the 1990s.

Population Change

Dynamic economic processes are transforming the Amazon Basin from a sparsely populated, humid, tropical rain forest into a settled agricultural and resource-extraction region with a firmly established city system and clear urban hierarchy. Simultaneously, contemporary social conditions reflect the dramatic economic and cultural conflicts that are a product of the region's development. For hundreds of years, the Amazon's indigenous peoples have paid a high price for the region's economic exploitation and incorporation into Brazilian national life. As early as the 1700s, the *bandeirantes*, explorers and adventurers from São Paulo, penetrated parts of the Amazon Basin in Mato Grosso and Goiás, capturing native peoples and sending them to the south as slaves. During the rubber boom in the second half of the 19th century, *seringueiros* settled along river courses throughout the region. In an effort to secure hard-to-find labor for rubber tapping, entrepreneurs and their cohorts raided Indian villages and enslaved those they could find to work as rubber tappers and processers. Many were killed or enslaved, some resisted and retreated to more remote rain forest regions in terra firme, and others assimilated into the *caboclo* culture. *Caboclos* are detribalized indigenous and mixed-blood peoples who have developed a subsistence economy focused on the rivers and the váreza. Overall, however, indigenous population numbers declined and native cultures suffered.

The 20th century has only hastened the pace of the assault on indigenous peoples and their cultures as the rain forest lands they inhabited have became the object of both extensive and intensive economic development. Between 1900 and 1960 the indigenous population of the Brazilian Amazon dropped from about 1 million to about 200,000, and some 80 tribes simply ceased to exist. While some 140 tribal groups remained intact in the early 1960s, many have been decimated by the steady onslaught of resource exploitation and colonization in subsequent years. Government officials, ranchers, gold miners, surveyors, road builders, and colonists have all had a hand in the disappearance of indigenous peoples from the region. Indians have been murdered and their lands have been usurped. They have died from introduced diseases (smallpox, influenza, measles, and tuberculosis), and have been subjected to every manner of abuse in order to push them from their ancestral lands and remove them as legitimate claimants of the region's patrimony.

The Yanomamo, living in the northern part of the basin along the Venezuelan border, are the most numerous of the extant indigenous groups, numbering between 20,000 and 30,000. They face constant encroachment from settlers, loggers, and most recently from independent gold miners, or *garimpeiros*.

In 2000 the total population in Amazonia was approximately 12–13 million and was growing rapidly. It continues to do so. Over 10 million reside in the Brazilian portion of the region, with the remaining 2–3 million distributed in Bolivia, Peru, Ecuador, Colombia, and Venezuela. Despite the widespread impression that rural settlers account for most of the region's population, urban dwellers are actually the most numerous.

Urbanization and Urban Places

There is a widespread perception that Amazonia is essentially wild forests, frontier hardships, and farmers and ranchers. This is only part of the story. Cities and urban places are equally important today. By 1980, in the Brazilian portion of the basin, urban dwellers had surpassed rural residents in number, accounting for just over half of the total population of what Brazil called "Legal Amazonia." Twenty-five years later, nearly 60 percent of all residents are classified as urban, and the percentage exceeds 70 percent in the Brazilian states of Amapá and Amazonas.

Urban primacy characterizes the hierarchy of cities in Amazonia, as it does in many other national and regional contexts throughout Latin America. Belém, at the mouth of the Amazon on the Atlantic Ocean, and Manaus, sited some 1,200 km upstream at the confluence of the river's biggest tributary, the Río Negro, dominate the urban hierarchy of the Brazilian Amazon and are the region's megacities. The populations of both cities exceed 1 million inhabitants and combined they account for just about half of all urban residents in the Brazilian portion of the Amazon Basin. Further upstream at the confluence of the Marañón and Ucayali Rivers, Iquitos, Peru, is the Amazon's third largest city, with a population of approximately 250,000.

Situated near the mouth of the Amazon River, Belém was established by the Portuguese to defend against the territorial incursions of other European powers at the beginning of the 17th century. The city's strategic location on the river and its access to a vast interior hinterland contributed to its subsequent development as a key trading center and one of the principal ports for the Amazon region. The Amazon's rubber boom at the end of the 19th century brought the city unprecedented, but short-lived, prosperity. Today, with a population of approximately 1.2 million, the city continues to serve as the principal port of the Amazon region, handling a range of agricultural and forest products (Brazil nuts, pepper, and cassava) and natural resources (aluminum). Railroad and highway connections integrate it with the land transportation network of the rest of Brazil. Industrial activity focuses on consumer goods and agro-industrial processing.

Manaus, nearly 1,200 km upstream from Belém, dominates the urban hierarchy of the central Amazon region. Founded in the late 1600s, it grew slowly until the rubber boom in the late 19th century. Its current population numbers about 1.2 million and it is a major port and industrial center. The Brazilian government has given the city both free port and free trade status; this has promoted considerable commercial and industrial development. Its international airport is the second most active air cargo terminal in the nation, and oceangoing vessels dock regularly at its modern port facilities. Industries include chemicals, electronics, shipbuilding, and production of a wide range of consumer goods.

It is important to note that besides the large urban centers of Belém and Manaus, a wide range of smaller urban places attract unsuccessful agricultural colonists as well as other immigrants. Small boomtowns, populated largely by a landless, rural laborer class that provides the labor for a range of extractive activities—logging, ranching, and mining—typify the Amazonian frontier economy. The economy of these frontier towns is unstable; many wither and some die after the frontier settlement boom passes. In addition to these small boomtowns, local service centers and market towns, with populations ranging well into the tens of thousands and diversified agro-industrial, service, and retail economies, attract considerable population numbers. Finally, a handful of regional centers with diversified economies and populations ranging from 100,000 to 250,000, like Iquitos, Peru, are found in Amazonia (Figure 19.5).

The Guiana Highlands: On the Periphery of the Periphery

The Guiana Highlands is less extensive than the other two highland regions of South America, the Andes and the Brazilian Highlands. At the beginning of the 21st century they are of little economic or social consequence, but eventually may yield abundant natural resources. The region is more remote and inaccessible than any other in Latin America. It is no exaggeration to say that the Guiana Highlands are a region on the periphery of the periphery.

FIGURE 19.5. Riverboats dock along Iquitos' waterfront. Passengers and freight are transported along the Amazon and its tributaries on these small vessels, 1993.

Environment

These ancient and highly eroded highlands and their associated outcrops and piedmonts sit in the north-central portion of the continent (Figure 19.6). The region is bounded on the west by the Orinoco River and the Llanos, on the south by the Rio Negro and the Amazon Basin, and on the north and east by a 100-km-wide coastal plain bordering the Caribbean Sea and the Atlantic Ocean. About half of the Guiana Highlands lie outside of the Latin American culture region, falling in the non-Latin countries of Guyana, Suriname, and French Guiana. The remainder falls principally within the boundaries of Venezuela and Brazil, although isolated outcrops are evident in eastern Colombia.

Structurally, the highlands are more akin to a plateau region than a single mountain chain. A sharp escarpment, coincident with the modern political boundaries between Brazil and its northern neighbors Venezuela, Guyana, Suriname, and French Guiana, rises along most of the plateau's southern extent. Elevations along most of the escarpment do not exceed 1,000 m, but isolated peaks like Mt. Roraima and Neblina Peak do reach considerable altitudes, about 2,800 m and 3,000 m, respectively. A broad piedmont region stretches southward from the escarpment, extending as far south as the Río Negro in the west and to within 100 km of the Amazon River in the east. Northward the plateau grades slowly toward the Llanos and the coastal plain. Capped by resistant sandstone, an unusual landscape characterizes portions of the region where mesas, buttes, and tabular mountains rise precipitously from the surrounding terrain. Angel Falls (979 m), the highest falls in the world, drops from one of these features. The geographical isolation imposed by these features, similar to that of isolated islands like the Galápagos (Ecuador), has encouraged the evolution of unique flora and fauna. The popular science fiction movie *Jurassic Park* (1993) directed by Steven Spielberg, utilized views from the Gran Sabana, a broad plains region within the highlands dotted with over 100 flat-topped

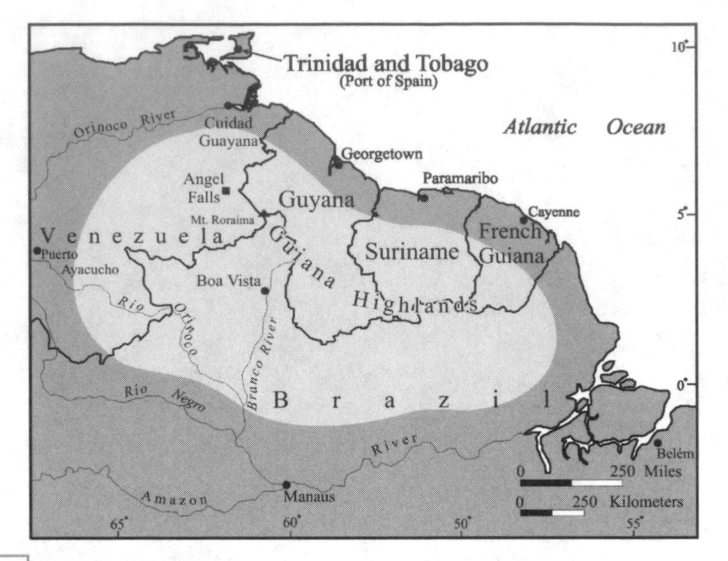

FIGURE 19.6. The Guiana Highlands.

topped mountains, or *tepuis*, to create some of the movie's dramatic special effects.

Its equatorial location between 1°S and 8°N latitudes ensures that the Guiana Highlands have a tropical climate (Figure 19.7). Nevertheless, the geographic patterns of temperature and precipitation are complicated by the region's topography. Weather stations are few and the available data are limited. Mean annual temperatures in the lowlands range between 26° and 28°C and there is little seasonal variation, typically 2°s–3°sC. Daytime highs reach into the mid-30°sC, while temperature minima may drop into the low 20°sC. Temperatures moderate in upland areas.

Precipitation presents a somewhat more complicated panorama. Rainfall totals range from 1,500 mm to over 3,000 mm annually. Local variation in relief, aspect, and the prevailing winds cause much of this variance. The seasonal pattern of precipitation also varies over the region. In the north, wet-and-dry seasons are more clearly pronounced and mirror the seasonal pattern of the Llanos, while in the south precipitation occurs throughout the year in a pattern more characteristic of the Amazon Basin. Although tropical rain forest occupies much of the Guiana Highlands, open savanna landscapes, notably the Gran Sabana, do occur.

Contemporary Conditions

The remoteness and isolation of the Guiana Highlands are such that much of this region remains terra incognita even today. The spectacular Angel Falls, for instance, was not discovered until 1937 and it was not visited for a second time until 1949! The region is largely devoid of permanent settlements. Most colonization and settlement, even on the region's more hospitable periphery, has occurred during the second half of the 20th century. Both the Brazilian and Venezuelan governments have demarcated several extensive indigenous reserves in the region. The economy of the region's 50,000–60,000 native inhabitants has traditionally been limited to subsistence agriculture, hunting, and gathering. The increasing incursions of gold miners, military personnel, and ecotourists into this region present serious challenges to the continued survival of these groups as they succumb to the forces of acculturation and the introduction of European diseases.

Forest products, including tropical hardwoods, vanilla, *chicle*, and medicinal plants, are important in the region's formal economy, while ecotourism is beginning to become more commonplace. The establishment of five national parks in the Venezuelan portion of the Guiana Highlands and four national parks and ecological areas in the Brazilian sector has promoted this incipient industry. Mining, however, is the region's most profitable and dynamic economic sector. Iron ore and bauxite from the Guiana Highlands have fueled industrial growth in Venezuela's Ciudad Guayana. Diamonds and gold are also mined.

Despite the region's considerable geographical extent, its population is miniscule. Only between 250,000 and 300,000 people inhabit the Guiana Highlands. At least half are concentrated in a handful of cities. Boa Vista, Brazil, and Puerto Ayacucho, Venezuela, are the most populous. Both cities are sited on major rivers that skirt the highlands, the Branco and the Orinoco Rivers, and each is the capital city of its territory, Roraima and Amazonas, respectively. The fact that both are territories, rather than full administrative units, reflects the frontier nature of this region.

Boa Vista is one of the handful of planned urban centers built in Latin America in the 20th century. The city's plan suggests grandeur. Set on high ground about half a kilometer from the Rio Branco, its street network focuses on the government palace and civic center. Wide diagonal boulevards radiate from this node, with a series of semicircular streets connecting them. A more traditional grid street pattern parallels the river. Besides serving as a regional service center for the surrounding territory, the city manufactures metal products, construction materials, and petroleum and alcohol fuel from sugarcane and cassava, albeit all on a modest scale. Less grandiose, Puerto Ayacucho provides most of the same commercial and service functions for its hinterland in the Venezuelan sector of the Guiana Highlands.

Summary

The Amazon Basin, a resource-rich mega-region, includes large portions of Brazil, Peru, Colombia, Ecuador, and Bolivia. The floodplains of the Amazon River offer significant resources, but limited access kept outside forces from this region until well into the 19th cen-

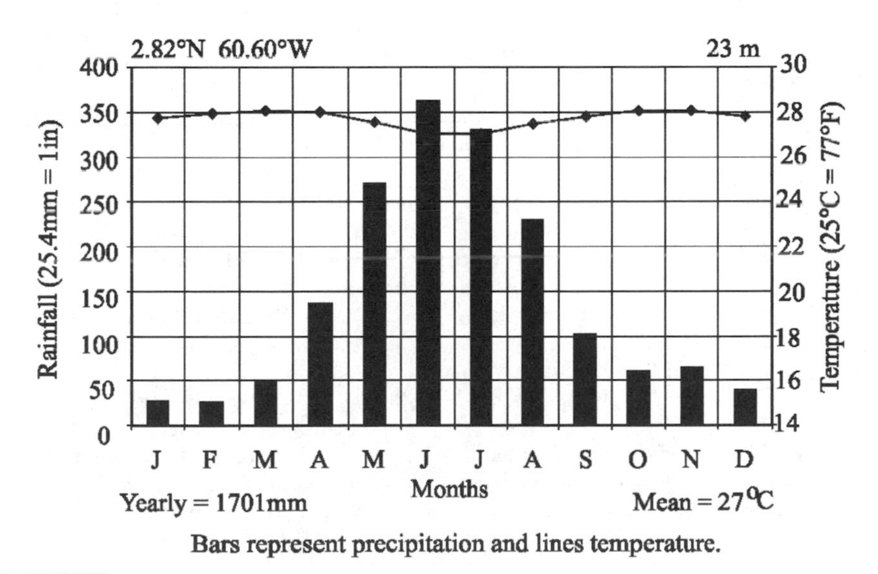

FIGURE 19.7. Climograph of Boa Vista, Brazil.

tury. Early exportable resources included nuts, quinine, and rubber. The rubber boom created wealth and culture. With the latter part of the 20th century came the development of agriculture and continued resource extraction that has resulted in serious land degradation. Approximately 25 percent of the rain forest has been deforested, resulting in the loss of plant and animal species. These developments have also been destructive to the indigenous people. Although for a long time sparsely populated, the Amazon region is developing an urban hierarchy and gaining in population. Belém and Manaus, both with populations in excess of 1 million inhabitants, are the region's major urban centers.

The Guiana Highlands, a remote and sparsely populated region, sit on the periphery of the Amazon Basin. Gold prospectors, loggers, military personnel, and even tourists have begun to reach this area in significant numbers, encroaching on the lands and lifeways of the few indigenous groups that survive in this previously inaccessible region. Unless aggressive measures are taken soon to protect these peoples and the region's environment, it too will face unremitting and destructive environmental assaults.

Further Reading

Browder, J. O., and Godfrey, B. J. (1990). Frontier urbanization in the Brazilian Amazon: A theoretical framework for urban transition. *Yearbook, Conference of Latin Americanist Geographers, 16*, 56–66.

Browder, J. O., and Godfrey, B. J. (1997). *Rainforest cities: Urbanization, development, and globalization of the Brazilian Amazon*. New York: Columbia University Press.

Brown, J. C. (2001). Responding to deforestation: Productive conservation, the World Bank, and beekeeping in Rondonia, Brazil. *Professional Geographer, 53*, 106–118.

Cleary, D. (2001). Towards an environmental history of the Amazon: From prehistory to the nineteenth century. *Latin American Research Review, 36*(1), 65–96.

Coomes, O. T., and Barham, B. L. (1994). The Amazon rubber boom: Labor control, resistance, and failed plantation development revisited. *Hispanic American Historical Review, 72*, 231–257.

Davis, S. H. (1977). *The victims of the miracle*. Cambridge, UK: Cambridge University Press.

Dean, W. (1987). *Brazil and the struggle for rubber: A study in environmental history*. Cambridge, UK: Cambridge University Press.

Denevan, W. M. (1992). *The aboriginal population of Amazonia*. Madison: University of Wisconsin Press.

Foresta, R. A. (1992). Amazonia and the politics of geopolitics. *Geographical Review, 82*, 128–142.

Godfrey, B. J. (1990). Boom towns of the Amazon. *Geographical Review, 80*, 103—117.

Goodland, R. J. A., and Irwin, H. S. (1975). *Amazon jungle: Green hell or red desert?: An ecological discussion of the environmental impact of the highway construction program in the Amazon Basin*. New York: Elsevier.

Hall, A. L. (1989). *Developing Amazonia*. Manchester, UK: Manchester University Press.

Hecht, S., and Cockburn, A. (1990). *Fate of the forest: Developers, destroyers, and defenders of the Amazon*. New York: HarperCollins.

Kaimowitz, D. (2002). Amazon deforestation revisted. *Latin American Research Review, 37*(2), 221–235.

LeBreton, B. (2003). *Trapped: Modern-day slavery in the Brazilian Amazon*. London: Kumarian Press.

Mahar, D. J. (1989). *Government policies and deforestation in Brazil's Amazon region*. Washington, DC: World Bank.

Naughton-Treves, L. (2002). Wild animals in the garden: Conserving wildlife in Amazonia agroecosystems. *Annals of the Association of American Geographers, 92*, 488–506.

Page, J. A. (1995). *The Brazilians*. Reading, MA: Addison-Wesley.

Raffles, H., and WinklerPrins, A. (2003). Further reflections on Amazonian environmental history: Transformations of rivers and streams. *Latin American Research Review, 38*(3), 165–187.

Sierra, R. (2000). Dynamics and patterns of deforestation in the western Amazon: The Nape deforestation front, 1986–1996. *Applied Geography, 20*(1), 1–16.

Simmons, C. S. (2004). The political economy of land conflict in the eastern Brazilian Amazon. *Annals of the Association of American Geographers, 94*(1), 183–206.

Slinger, V. A. V. (2000). Peri-urban agroforestry in the Brazilian Amazon. *Geographical Review, 90*, 177–190.

Smith, N. J. H., et al. (1995). *Amazonia: Resiliency and dynamism of the land and its people*. Tokyo: United Nations University Press.

20 The Latin American Diaspora

Latin Americans have been a people on the move since the middle of the 20th century. Miners, loggers, peasant farmers, ranchers, and entrepreneurs have colonized vast portions of the Amazon rain forest, as well as those of Central America and southern Mexico. Rural-to-urban migration grew almost exponentially in many countries, as poor peasants, young men and women, and others have moved off the land and out of villages and small towns in droves to urban centers and national capitals seeking a better life. The same pressures and attractions that have led some to colonize the rain forest and others to move to urban centers have induced still others to leave their homelands altogether. Migration to neighboring Latin American nations has figured prominently in this movement, but migration to destinations outside of the region has become increasingly consequential in the final decades of the 20th century.

The United States is the principal destination of Latin American migrants. Since the 1950s millions of Latin Americans have migrated to the United States. Their numbers, as well as their proportion of the U.S. population, have grown immensely. Dense concentrations of Latin Americans in some metropolitan centers and regions have transformed these places, extending at least in some manner the cultural boundaries of Latin America into the United States.

Secondary destinations include Western Europe and Japan. In Europe, Latin Americans favor Spain. Since the early 1990s the number of Latin Americans immigrating to that country has increased substantially. In 2001 over 200,000 Latin Americans immigrated to Spain, with Ecuadorians and Colombians being the most numerous. Important secondary flows come from Cuba, Venezuela, and Argentina, countries in which many Spaniards settled in the late 19th and early 20th centuries (Figure 20.1). The current pattern of the country of origin of Latin American residents in Spain closely mirrors this immigration pattern (Figure 20.2).

Latin Americans also settle in Italy, France, Germany, and the United Kingdom in modest numbers. Special provisions in Italian law permit the grandchildren of Italians to obtain Italian passports and to return to Italy as citizens. Millions of Italians settled in Argentina in the late 19th and early 20th centuries. In the early years of the 21st century, as a severe economic crisis gripped Argentina, many thousands of their descendants obtained Italian passports and returned to Europe. Nowhere in Europe, however, do Latin Americans comprise an appreciable proportion of any nation's population. Nevertheless, in some parts of Spain, like the downtown district of Barcelona, Latin Americans are a numerous and important component of the population.

In the last decade, tens of thousands of Japanese Brazilians have immigrated to Japan

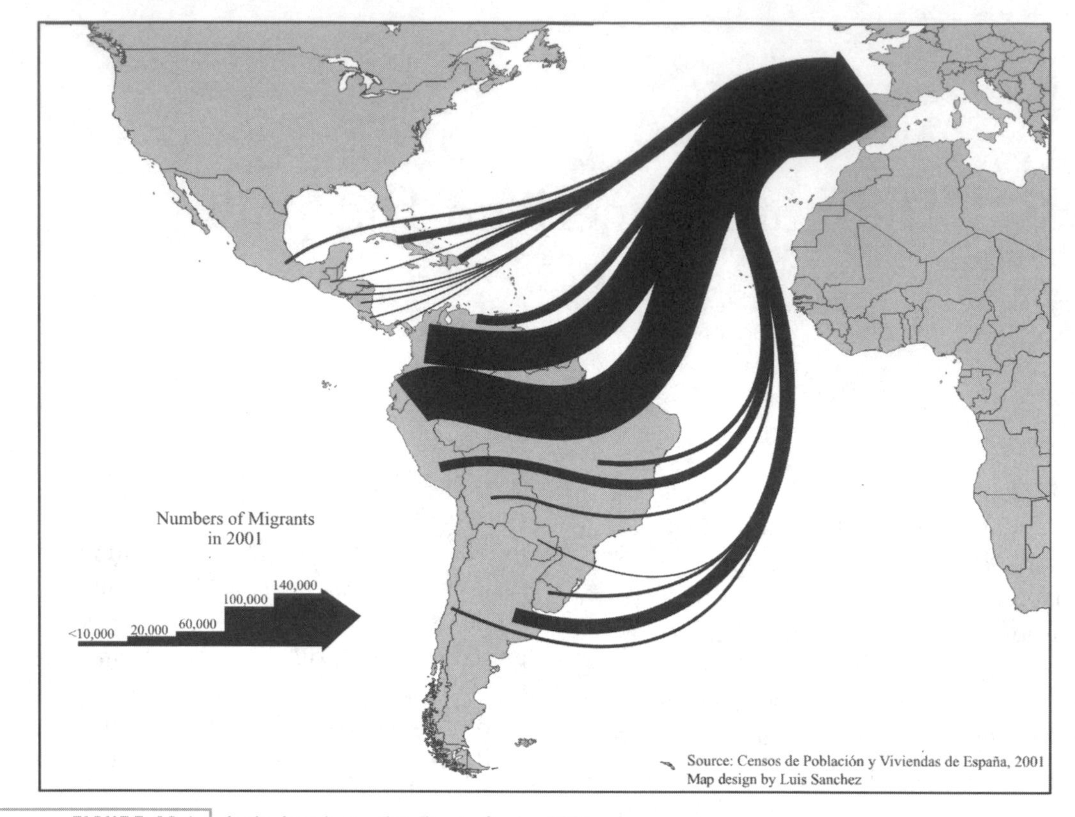

FIGURE 20.1. Latin American migration to Spain, 2001.

as that country has loosened its immigration laws to permit the legal entry of unskilled foreign workers. Because of their number and the historical and cultural linkages between Brazil's community of Japanese descendants and Japan, their immigration has been especially encouraged by the Japanese. Not surprisingly however, most Japanese Brazilians experience extreme culture shock living in Japan, and most return to Brazil (Vignette 20.1).

The fact that Latin Americans have transformed the cultural milieu in some areas of the United States, one of the world's largest and most populous nations, makes it appropriate to consider the United States in a book about Latin America. One might reasonably ask what precisely the "transformation of the cultural milieu" means in this case. Is it simply the appearance of thousands of Mexican restaurants across the suburban landscape or the availability of salsa and tortillas in almost every grocery store in the United States (Figure 20.4)? Or does it have to do with the transformation of the visible landscape, as Latin Americans alter the appearance of residential neighborhoods and establish businesses and business districts catering specifically to Latin American tastes in food, music, and clothing? Alternatively, is it something more profound, like the use of Spanish as a language of instruction in public schools, on voting ballots, and on driver's license examinations? Or is it the appearance of Spanish-language newspapers, radio stations, and cable television channels in major communication markets across the country?

In parts of the United States, it is all of these, as large numbers of Latin Americans concentrate in certain geographic regions, metropolitan centers, towns, and neighborhoods, often becoming numerically dominant. The extent to which these numbers translate into social, economic, and political power varies widely. Cuban Americans wield immense power in Miami, while the political and economic strength of Mexican Americans in Los Angeles is nowhere near proportional to their numbers. Nevertheless, even where their political or economic power is less than their numbers suggest it might be, the cultural impact of Latin Americans in these places is clear and distinct.

In an insightful book, *The Nine Nations of North America*, published in 1981, Joel Garreau, a correspondent for *The Washington Post*, proposes a new regionalization of the United States and Canada. Garreau identifies two North American regions where Latin Americans have profoundly modified the regional culture: the Southwest and south Florida. He calls these regions "Mex-America" and "The Islands," respectively. Mex-America encompasses a broad swath across the southwestern United States, running from south Texas to southern California, including the northern portions of the Mexican states adjacent to the U.S. border. While much of Mex-America lies within the territorial limits of the United States, this is not the case with the Islands. Only a small portion of the Islands' region falls in the United States, the extreme southern tip of Florida, corresponding roughly with the Miami metropolitan area. The remainder of the Islands region, for which Garreau identifies Miami as the capital, includes the entire sweep of the Latin American and other islands of the Caribbean. While Garreau's treatment is laced with humor, and at times his positions are clearly provocative, his argument is simple and clearly supportable: Latin Americans are transforming the United States and leaving their own particular stamp on its evolving human geography.

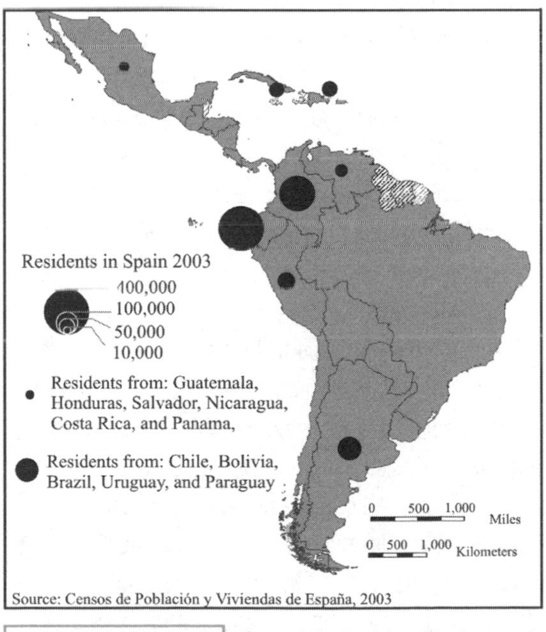

FIGURE 20.2. Countries of origin of Latin Americans for Latin Americans living in Spain, 2003.

Origins

The origins of the United States's Latin Americans are diverse, both historically and geographically. Spaniards colonized frontier regions and established towns as part of what is now the United States even before the Pilgrims arrived at Plymouth Rock in 1620 (Figure 20.5). Their descendants in California and New Mexico were forcefully incorporated into the United States when it emerged victorious in the Mexican-American War in 1848. Puerto Ricans were incorporated into the United States in the late 19th century when Spain lost the Spanish-American War in 1898 and was forced to cede extensive territories to the United States. Still others, millions in fact, have arrived in the last few decades. Many have come legally, as political refugees, students, or family members of U.S. citizens and residents. Many have come illegally, crossing the Florida Straits between Cuba and Florida and the Mona Passage between the Dominican Republic and Puerto Rico on boats and flimsy rafts. Others have made dangerous

VIGNETTE 20.1. JAPAN AND THE LATIN AMERICAN DIASPORA

The changing fortunes of nations' economic health can turn historic migration patterns on their heads. During the early years of the 20th century hundreds of thousands of Japanese emigrated from the island archipelago because population pressure and desperate economic conditions there offered them only a bleak future. The West Coast of the United States was their preferred destination, but rising prejudice, overt discrimination, and eventually restrictive immigration legislation in the 1920s choked off Japanese access to the United States. Brazil replaced the United States as the destination of choice for Japanese as demand for manual labor during the early 1900s brought the immigration of hundreds of thousands from Japan. The coffee boom produced an intense demand for manual labor to tend and harvest the crops on the immense coffee plantations that developed in the cool uplands of São Paulo. The first Japanese immigrants arrived in 1908, when nearly 800 disembarked on Brazil's shores, beginning a process that would eventually bring nearly 1 million Japanese immigrants to the southern highlands of the country.

Hard work, industry, and thrift characterized the Japanese experience in Brazil over the next century. Working initially as agricultural laborers, Japanese immigrants eventually established themselves as agricultural entrepreneurs in the hinterland of São Paulo, where they began raising vegetables and fruits for the city's burgeoning urban market. Japanese farmers exploited the abandonment of coffee plantations on the urban fringe, the fertile soils, and the ready access to urban markets to grow high-value produce. Eventually, these farmers and their offspring migrated to urban areas, especially the city of São Paulo, where a clearly defined Japanese district, Libertade, developed and remains a vibrant ethnic neighborhood today (Figure 20.3). These rural-to-urban migrants firmly established themselves as part of the urban middle class, especially in the southern states of São Paulo and Paraná In 2005 Brazilians of Japanese descent constituted the most numerous expatriate Japanese community in the world, with a population of about 1.3 million, easily surpassing that of the United States, where approximately 1 million reside.

But times change. The transformation of Japan from a defeated nation into an industrial superpower in the decades after World War II and the aging of its labor force have created unfilled labor demands in Japan. Nevertheless, Japan has always been reluctant about allowing foreign labor to enter the country and strict laws have kept out most foreign laborers. Koreans and Chinese have traditionally filled these needs, but beginning in the early 1990s Japan eased its laws to permit people of Japanese origin and their descendants to return to the island nation to work. A stagnating economy, limited economic opportunities, and rampant inflation in Brazil encouraged many Japanese Brazilians to consider returning to Japan to work, albeit for only a short sojourn.

By the end of the the 1990s nearly 250,000 Japanese Brazilians had returned to Japan to enter the labor force. Many of these migrants are young men and women looking to amass enough savings to return to Brazil to start a business or buy a home. Nevertheless, many middle-aged Japanese Brazilians with professional jobs have also migrated, lured by wages that are more than twice what can be earned at home: as much as $2,500 per month in Japan versus $1,000 a month in Brazil. Like most migrants to industrialized countries in the West, the jobs that they have contracted to fill are for unskilled laborers, positions that native Japanese are unwilling to fill. One immigrant described them as "hard, dirty, and dangerous." These jobs have provided an important source of foreign exchange for Brazil. During the mid-1990s remittances aver-

(cont.)

aged nearly $1 billion per year. Recent data from the Central Bank of Brazil show that these numbers have declined in recent years, suggesting that as Japanese Brazilians begin to settle in Japan they are sending fewer yen back home.

The process of integration into Japanese society has not been without difficulties. Known in Japanese as *dekasegi*, or "sojourners," people who have left the island nation and then returned, many Japanese Brazilians feel like second-class citizens and report feelings of discrimination. Japanese law makes it very difficult to achieve citizenship and many feel alienated from Japanese society at large. Small Japanese Brazilian enclaves have developed in about 14 Japanese cities, and private Brazilian schools have sprung up in many of these. It is reported that nearly 40 of these schools operate in Japan, along with four weekly Brazilian newspapers and two Brazilian television stations. Despite these difficulties, it seems likely that Japanese Brazilian immigration will continue. Indeed, other Latin Americans of Japanese origin have also found their way back to the island nation to work during the last decade. Nearly 50,000 have migrated from Peru, Bolivia, Argentina, and Paraguay.

desert crossings in the Southwest or have simply remained by overstaying legal tourist- or student-visa time limits.

Geographically, the countries and regions of origin are also varied, although distance and accessibility have clearly played a role in determining the magnitude of migration flows to the United States. Mexico, from both a historical and a contemporary perspective, has been and remains the single most important source of Latin Americans reaching the United States. Puerto Ricans and Cubans rank a distant second and third, respectively, in contributing migrants to the U.S. mainland. Both of these groups are something of special cases. Puerto Ricans occupy a unique position among Latin Americans because they are citizens of the United States and can move freely between Puerto Rico and the U.S. mainland. Cubans are "special" too because since the 1960s they have received preferential treatment as political refugees from a communist regime, often automatically qualifying for permanent resident status and a host of govern-

FIGURE 20.3. A street scene in Libertade, a São Paulo neighborhood, populated largely by Brazilians of Japanese descent, 2004.

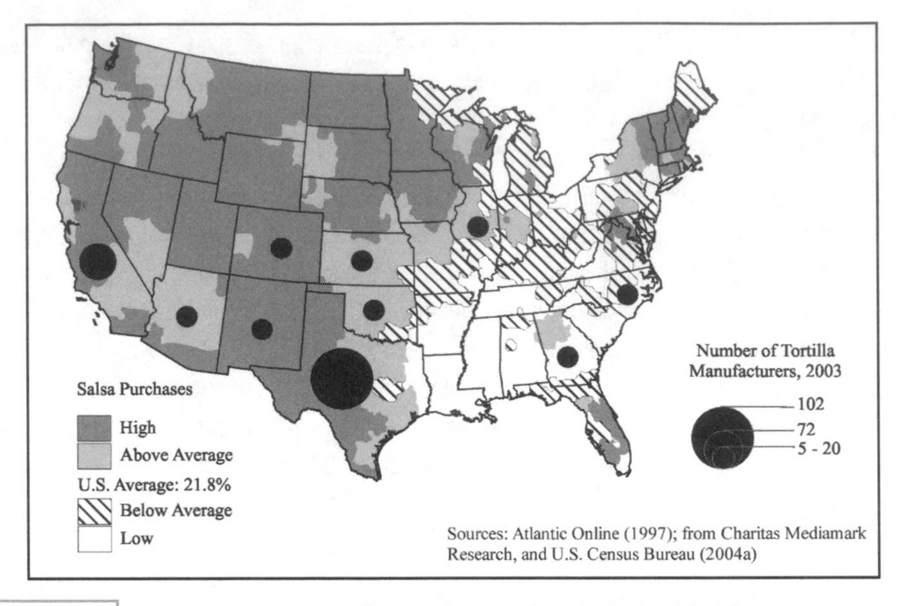

FIGURE 20.4. Salsa purchases and tortilla manufacturers in 2003 in the United States.

ment benefits and social programs, including Social Security. The movie *Papi* (1969), directed by Arthur Hiller, takes a comic look at the special treatment afforded to Cubans when a struggling Puerto Rican waiter attempts to pass off his children as Cuban refugees so they can take advantage of the cornucopia of benefits available to Cuban emigrés. A wide range of other nationality groups, including many from Central America, comprises the balance of the Latin Americans in the United States.

Identification of Latin Americans in the United States.

In the United States, the identification of just who is a "Latin American" is not necessarily clear-cut. Official government population data, collected by the Bureau of the Census, make no specific reference to "Latin Americans." Rather, the census enumerates those individuals who identify themselves as of "Hispanic" origin or descent—where this encompasses nationality, lineage, or the country of origin of one's parents, grandparents, or even distant ancestors. In terms of the U.S. Census, one is either Hispanic or non-Hispanic; the term is a cultural or ethnic identifier. Race is irrelevant, and a Hispanic may belong to any one of the Census Bureau's officially recognized racial groups: American Indian/Native Alaskan, Pacific Islander/Asian, White, and Black. Specifically, respondents are asked to categorize themselves as either of Mexican, Puerto Rican, Cuban, or other Spanish origin or culture.

The system of identification regarding who is Hispanic is filled with contradictions.

FIGURE 20.5. U.S. postage stamp commemorating Spanish settlement in the U.S. Southwest, 1598.

Somewhat curiously, a Spaniard, that is, a person from Spain or of Spanish ancestry or from one of Spain's minor African colonies, is legitimately Hispanic and included in official population statistics for that group. But Brazilians, whose culture is in large part also a product of Iberian exploration, conquest, and colonization, but by the Portuguese rather than by the Spanish, are not enumerated as Hispanics. Nor, of course, are persons of Portuguese origin or descent, of whom there are about 1.1 million in the United States. While the definition of who precisely is Hispanic in the United States seems to be based on arcane academic hair splitting, the results of the census's enumeration of Hispanics have immense political and economic significance. The U.S. federal government uses these data to implement a wide range of social programs; they are especially critical in making decisions about the distribution of public monies, as well as for legislative redistricting, affirmative action programs, and voting rights.

Historically, in the United States, the enumeration of people of Latin American ancestry has not been consistent. Significant changes have been introduced at several junctures, making comparisons over time awkward. For instance, census enumerators classified Mexicans as "other non-white" in the 1930 census, while in the 1950 and 1960 censuses, in states with high percentages of Mexicans, they were enumerated as "white persons of Spanish surname." During these decades, data gathered on place of birth and language use provided the best estimates of the numbers of Latin Americans in the United States. In the 1970 census, the U.S. Census Bureau used a 5 percent sample and a set of objective criteria to identify what was to eventually become known as the nation's Hispanic population. Thus, any individual with Spanish as his or her mother tongue, a Spanish surname, a Hispanic origin, or a Spanish heritage was defined as Hispanic. Five subcategories divided this group: Mexican, Puerto Rican, Cuban, Central and South American, and Other Spanish. The 1980 census marked major changes in the politics of enumeration and nomenclature. The U.S. Census Bureau introduced the term "Spanish-Hispanic origin" and refined the subcategories to just four: Mexican, Mexican-American, Chicano; Puerto Rican; Cuban; and Other Spanish-Hispanic. In addition, all census forms included this question, and more significantly, the categorization was based on each individual's self-classification of his or her ancestry or cultural heritage rather than the use of any objective criteria or the classification of the enumerator.

The Dialectics of Terminology

While the U.S. Census Bureau has settled on the term *Hispanic*, there is less agreement in other quarters about just what the proper term is. Some object to the use of the term because the Census Bureau essentially coined it. Latin Americans from Spanish-speaking countries do not view themselves as "Hispanics," and thus the term has no real meaning outside of the United States.

Alternatively, use of the term *Latino/a* is promoted by others. The term is derived from the regional label, Latin America, and was coined in Paris in the mid-19th century by exiled South American intellectuals. Originally, it was their intention to create a term to distinguish between Anglo-America and Spanish America. However, their anti-Spanish sentiment precluded the use of the term "Spanish America," so they employed the term "Latin America." The meaning of the term has since grown to incorporate Brazil, and in some quarters even Haiti.

In Latin America, few individuals identify with the regional term "Latin America," or consider themselves as Latinos. Such sentiments, when they are held, are most common among intellectuals. In the United States, usage of the terms appears to vary regionally, with the term "Hispanic" being more widely used in the East, while "Latino" enjoys greater acceptance in the West and the Midwest.

However, the vast majority of people of Latin American origin or descent attach little importance to the use of one or another of these terms, often using them interchangeably. Almost invariably, people of Latin American origin or ancestry in the United States identify themselves by their country of origin or by the country of origin of their parents or forebears. In this identification process, the use of compound names is also common, especially for the second and subsequent generations; thus one self-identifies as a "Mexican American" or a "Cuban American." Of course, there are exceptions. The Spanish-speaking natives of the upper Rio Grande Valley in New Mexico and Colorado, whose settlement dates from the 1600s, refer to themselves as "Hispano" as distinct from "Hispanics." For many years, Puerto Ricans in the Northeast identified themselves as "Spanish American." Politically active Mexican Americans have promoted and employed terms like *chicano/a* or *la Raza* to describe themselves.

Latin American Origins in the United States

The use of the term "diaspora" to describe the Latin American population in the United States is, in some ways, a misnomer. While it is true that many Latin Americans have left their homes and immigrated to the United States, the origins of Latin American presence in the country are more complex. The presence of all three major Latin American nationalities represented in the United States—Mexican, Puerto Rican, and Cuban—stems at least in some measure from wars between the United States and its hemispheric neighbors Mexico and Spain.

The admission of Texas as a state into the United States in 1845 provided the catalyst for the Mexican-American War (1846–1848) and the first large-scale incorporation of Latin Americans into the United States. Ostensibly, a dispute over the boundary between the newly admitted state of Texas and Mexico precipitated the war. The Texans, supported by the U.S. Congress, claimed that the Rio Grande River formed the state's southern limit. The well-documented Mexican claim placed the southern limit of Texas along the Nueces River some 200 km to the north. The U.S. president, James Polk, thinking far beyond Texas, saw the dispute as a means of vastly expanding U.S. territory to include California and other western territories. When the Mexicans refused to capitulate on the Texas border issue and then refused the United States's offer of $25 million for the California territory, Polk provoked a military confrontation with Mexico and declared war. The United States opened military fronts with Mexico in the border regions of the Mexican north and west and in the Mexican heartland, laying siege to Veracruz, taking Puebla unopposed, and defeating General Santa Anna's armies decisively in Mexico City in September 1847.

The United States imposed harsh terms on Mexico in the treaty that ended the war. The Treaty of Guadalupe Hidalgo, named for the village outside Mexico City where the treaty was signed, forced Mexico to cede about half of its national territory to the United States (Figure 20.6). The Mexicans recognized the U.S. claim to Texas, and ceded their New Mexico and California territories to the United States as well. In addition to Texas, the United States acquired all or part of what were to become the states of California, Arizona, Nevada, New Mexico, Utah, Colorado, Wyoming, and Oklahoma. In return, the United States recognized the Rio Grande as the definitive border between the two nations.

However, it was not just territory that was incorporated into the United States. About 80,000 Mexicans, many of them descendants of early Spanish colonists, lived in Texas, in the upper Rio Grande Valley, and in California. The

FIGURE 20.6. Mexican territories ceded to the United States in the Treaty of Guadalupe de Hidalgo in 1848.

treaty recognized their property rights and to some extent their cultural rights as well; most Mexicans remained, becoming "Mexican Americans." They were the first significant population group of Latin American origin in the United States. In the succeeding 150 years, immigration from Mexico has swelled these initial tens of thousands to many millions today.

The Spanish-American War (1898) offered another opportunity for the United States to incorporate both Latin American territory and populations into the nation. The explosion of the battleship *U.S.S. Maine* and its sinking in Havana harbor on 15 February 1898 were attributed to Spanish agents. This provided the prime rationale for the declaration of war that soon followed. But other factors contributed to the U.S. declaration of war. First, some Americans had long coveted Cuba, so imperialistic sentiments and commercial interests weighed heavily in support for the war. Second, during the late 19th century, Cubans seeking political independence from Spain had waged a long, bloody, and ultimately unsuccessful guerrilla war against Spanish forces on the island. Support for Cuba's independence from Spain, whose cause was promoted by Cuban exiles like Jose Martí in New York City, helped push the United States into war as well. Third, fanned by the popular doctrine of manifest destiny, which had also provided some of the inspiration for the Mexican-American War, and a fiercely competitive U.S. press that sensationalized and distorted news coverage in Cuba, the United States declared war on Spain in April 1898.

From the perspective of the United States, the war was short and its outcome lucrative. U.S. fatalities were few, less than 6,000, with only about 400 dying in combat. The remainder died from tropical diseases and other illnesses. The Spanish, on the other hand, suffered a stunning and demoralizing defeat, and in the process lost most of their remaining overseas colonies, keeping only a few insignificant colonies in Africa. Spain ceded Cuba, Puerto Rico, the Philippines, Guam, and Midway Island to the United States. In the 100 years that followed, Puerto Ricans and Cubans, respectively, became the second and third largest groups of Latin Americans resident in the United States.

Cuba, Spain's last true colonial jewel, has endured a turbulent relationship with the United States in the 100 years that followed its independence from Spain. The island remained a U.S. protectorate, occupied by U.S. Army troops, until 1903. U.S. corporations moved quickly to establish dominant positions in many sectors of the Cuban economy. These included agriculture (especially sugarcane plantations), refineries, communications, and transportation. The United States granted Cuba independence through the Platt Amendment (1901) but stipulated a range of conditions that effectively established the country as a U.S. protectorate. Key among these were the right of the United States to determine Cuba's foreign policy, to intervene militarily on the island to "restore order" if deemed necessary, to hold title to land, and to establish and maintain naval stations on the island.

In the three decades that followed, the United States exercised these privileges on several occasions. American troops reoccupied Cuba in 1906, 1912, and 1917 when internal unrest threatened U.S. business interests.

In 1903 the United States established a naval base at Guantánamo Bay on the southeastern end of the island, appropriating the entire bay and extensive tracts of adjacent land. While the Platt Amendment was abrogated in 1934, the United States has continued to maintain a permanent military presence on the island at Guantánamo. Sited on one of the best deepwater harbors in the Western Hemisphere and strategically situated adjacent to the Windward Passage between Cuba and Hispaniola, it is one of the largest naval complexes operated by the U.S. Navy. Its infrastructure includes extensive port facilities, an air station, and a vast refugee holding camp capable of receiving tens of thousands. Since 2001, the U.S. military has used the base as a prison for alleged terrorists and enemy combatants, thereby keeping Guantánamo in the news.

The conditions of the treaty that ended the Spanish-American War paved the way for the incorporation of the island of Puerto Rico into the United States as a colony, although this was not necessarily a foregone conclusion at the time. The island's residents were "granted" U.S. citizenship by the Jones Act of 1917. Or, from another perspective, it can be said that they had it imposed upon them. Since then, Puerto Ricans have been able to move to and reside in the continental United States without restrictions.

The island's status remains ambiguous. In the early 1950s the island was allowed greater political autonomy. It become a commonwealth territory of the United States with the right to elect its own governor, write its own constitution, and pass legally binding laws, much like any state, as long as those laws do not contravene the U.S. Constitution or federal legislation. Puerto Rico cannot elect representatives to the U.S. Congress, nor does it participate in the election of the U.S. president. Nevertheless, any Puerto Rican, as a U.S. citizen, who resides permanently in any U.S. state, has full voting rights in local, state, and federal elections!

Even the enumeration of Puerto Ricans is somewhat obscure. Those who reside in any of the 50 states of the United States are enumerated as Hispanics by the U.S. Census Bureau and figure in the bureau's calculation of the nation's population totals and its percentage of Hispanic residents. Puerto Ricans residing in Puerto Rico, on the other hand, do not figure in U.S. population totals, nor does the U.S. Census Bureau enumerate them among the nation's Hispanic population.

Like Mexicans, Puerto Ricans, and Cubans, many other Latin American immigrants to the United States have homelands that have been occupied by U.S. Marines or that have experienced overt and covert intervention in their internal affairs by the U.S. government and its military. In the 1990s, after Cubans, the most numerous Latin Americans in the United States were, in rank order, El Salvadorans, Dominicans, Colombians, Guatemalans, and Nicaraguans. With the lone exception of Colombia, every other country has had one or more major U.S. military interventions during the last 100 years.

The U.S. Marines occupied the Dominican Republic in 1905, between 1916 and 1924, and again in 1965. Marines occupied Nicaragua for even longer, between 1910 and 1933, while the country operated as virtually a U.S. protectorate. During the 1980s, the United States played an essential overt and covert role in supporting the Contra insurgency against the legitimate Nicaraguan government. Covert U.S. assistance helped topple the elected president of Guatemala, Jacobo Arbenz, in 1954, and from the 1970s to the 1990s U.S. military assistance supported the Guatemalan army's deadly counterinsurgency campaigns. A small contingent of U.S. military advisors and close to $10 billion of military assistance aided the El Salvadoran Army in holding the guerrilla fighters of the Faribundo Martí Liberation Front (FMLN) to a stalemate in 1990 after a 10-year insurgency. Since 2000 Colombia has been one of the largest recipients of U.S. foreign aid after Israel, Egypt, and Iraq. In Colombia's case, almost all U.S. aid is for military equipment and operations.

Numbers and Migration Histories

The U.S. Census Bureau enumerated more than 35 million Hispanics living in the United States in 2000. This is roughly equivalent to the population of the entire country of Argentina, the fourth most populous country in Latin America (see Table 10.6). Hispanics now represent 12.5 percent of the population of the United States, displacing blacks (African Americans) as the largest minority population in the nation, and their numbers and percentage of the total population continue to increase. Puerto Rico's population of nearly 3.8 million is not included in these figures. If it were, the percentage of Hispanics in the U.S. population would rise by almost 2 percent. But only those Puerto Ricans living on the U.S. mainland are counted in U.S. population totals. In strictly numerical terms, people of Mexican origin or ancestry dominate the Hispanic population in the United States, accounting for about 59 percent, or 20.5 million people (Table 20.1). Puerto Ricans account for almost 10 percent (about 3.4 million), and Cubans follow with just 3.5 percent (1.2 million). No other national groups yet exceed 1 million according to the U.S. Census, but immigrants from the Dominican Republic and El Salvador now number over half a million and immigration from these countries continues to be strong. Central Americans and South Americans combined total about 9 percent (3 million) of the country's Hispanics; note, however, that their numbers in the U.S. population are the result of migration flows from about 16 different Hispanic countries. The category "Other Hispanics" is a vague category that is difficult to interpret, but accounts for fully 18 percent (6.2 million) of the Hispanic population. It includes less than a million Spaniards and Spanish Americans as well as about 5.3 million whose responses were not classified elsewhere!

Mexican immigration to the United States began in the early decades of the 19th century. It is difficult to state with any real certainty the numbers of Mexican immigrants who came to the United States over the course of the nation's history. This difficulty is attributable to several factors. Significantly,

TABLE 20.1. Hispanic and Latino Population in the United States, by Region and by Country of Origin, 2000

Hispanic or Latino by type	Number	Percent
Mexican	20,640,711	58.5
Puerto Rican	3,406,178	9.6
Cuban	1,241,685	3.5
Other Hispanic or Latino	10,017,244	28.4
Dominican (Dominican Republic)	764,945	2.2
Central American (excludes Mexican)	1,686,937	4.8
Costa Rican	68,588	0.2
Guatemalan	372,487	1.1
Honduran	217,569	0.6
Nicaraguan	177,684	0.5
Panamanian	91,723	0.3
Salvadoran	655,165	1.9
Other Central American	103,721	0.3
South American	1,353,562	3.8
Argentinean	100,864	0.3
Bolivian	42,068	0.1
Chilean	68,849	0.2
Colombian	470,684	1.3
Ecuadorian	260,559	0.7
Paraguayan	8,769	0
Peruvian	233,926	0.7
Uruguayan	18,804	0.1
Venezuelan	91,507	0.3
Other South American	57,532	0.2
All other Hispanic or Latino	6,211,800	17.6
Spaniard	100,135	0.3
Spanish	686,004	1.9
Spanish American	75,772	0.2
Not elsewhere classified	5,349,889	15.2
	35,305,818	**100**
Hispanic or Latino population	35,305,818	12.5
Non-Hispanic or non-Latino population	246,116,088	87.5
Total U.S. Population	**281,421,906**	**100**

The population of Puerto Rico, approximately 3.8 million, is not included in U.S. Census population totals.

Source: U.S. Census Bureau (2001a).

before 1924, Mexicans were not subject to immigration controls and could enter the United States freely. Beginning in that year, proof of identity was required for legal entry. The sheer geographical immensity of a border running some 3,000 km, much of it through unpopulated or thinly populated deserts and mountain ranges, provided ample opportunities for determined immigrants to cross the frontier. By the 1960s, illegal immigration by Mexican nationals began to have a significant demographic impact on the United States.

Despite the difficulties in presenting any precise tabulation of Mexican immigrants to the United States, data collected by the U.S. Immigration and Naturalization Service on the nation's immigration history provides a framework within which to understand the magnitude of legal immigration of Mexicans to the United States. In the decade between 1820 and 1830, Mexican immigrants to the United

States numbered about 5,000 and represented just 3 percent of all immigrants to the country. The number of Mexican immigrants remained stable throughout the rest of the 19th century, never exceeding 7,000 in any single decade. In comparison to the flood of European immigration that washed over the United States in the last half of the 19th century, however, Mexican immigration was slight. Between 1891 and 1900, over 3.5 million immigrants reached the United States, and Mexicans accounted for less than 1,000 of that total.

This pattern changed completely in the decades that followed as settlement, agricultural colonization, population growth, and economic development rapidly transformed the Southwest as well as other parts of the United States. Agriculture, railroad building and maintenance, construction work, and industrial development spawned an intense demand for labor—the cheaper the better—that could not be easily filled. European immigrants were quickly absorbed into the labor force in the East and the Midwest. In the West, immigration by Chinese and Japanese filled some of these labor niches, especially in railroad construction and agriculture, until Asian entry into the United States was barred in the 1880s by a series of restrictive immigration laws known as the "Exclusion Acts." Mexicans filled this void. The country's geographical proximity to the United States, an absence of numerical quotas on the admission of Mexican immigrants until the 1960s, and federal government policies generally favorable to Mexican immigration all contributed to the intense levels of immigration that followed (Figure 20.7). Other factors contributed to this movement as well, most notably the Mexican Revolution (1910–1920), when widespread violence and social unrest led many to abandon their homeland in favor of the United States.

The settlement of Mexicans was most concentrated in the Southwest, stretching northward from the border in a broad band from Texas to California. However, Mexicans workers and immigrants also reached the Midwest in small numbers in the early decades of the 20th century. Mexican communities, albeit small, appeared in a few northeastern cities like Lorain, Ohio, in the 1920s where the railroad and manufacturing offered employment

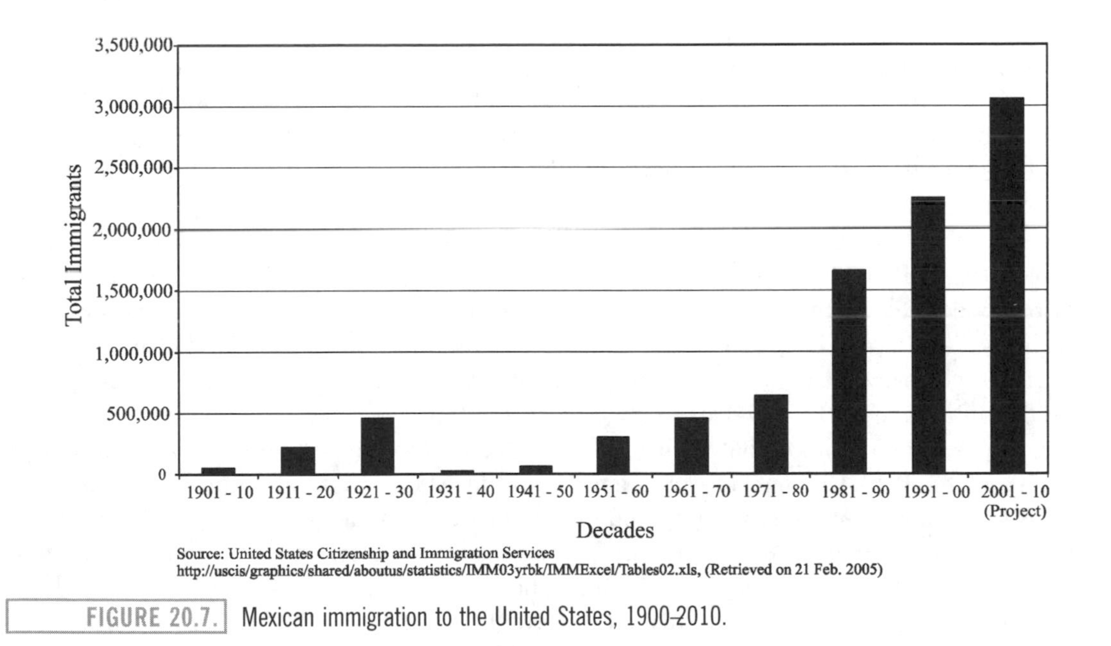

FIGURE 20.7. Mexican immigration to the United States, 1900–2010.

opportunities. In Chicago, the principal migration destination for Mexicans outside of the Southwest, the city's Mexican American population reached 20,000 by 1930.

The trickle of Mexican immigrants that arrived in the United States in the last years of the 19th century became a torrent during the early decades of the 20th century. During the first decade, 1901–1910, almost 50,000 Mexicans arrived in the United States, and between 1921 and 1930 this number increased tenfold to almost 500,000! The Great Depression and widespread unemployment gripped the United States throughout the 1930s. Thus, the northward movement of Mexicans diminished rapidly, with less than 25,000 arriving in the 1930s. Indeed, the 1930s marked one of the saddest episodes in the history of Mexican immigration to the United States, as widespread joblessness fueled prejudice and discrimination against Mexican nationals as well as Mexican American residents and citizens. A repatriation movement clamored for the expulsion of Mexicans, arguing that they contributed to the widespread unemployment in the nation. Forced repatriation during the 1930s sent nearly 500,000 people back to Mexico, most coming from Texas, California, and New Mexico, but others coming from cities deeper in the interior of the country, notably Denver, Chicago, and Detroit.

While World War II created some demand for Mexican labor, and immigration did rise again ever so slightly during the 1940s, it was not until the 1950s that the rate of arrivals again reached levels comparable with the decades before the Great Depression. To some extent, U.S. government policies, specifically the Bracero Program (1942–1964), encouraged the high rates of immigration that occurred in the 1950s and 1960s. Although the program's goal was to promote temporary immigration to fill critical labor shortages in the agricultural sector, many temporary immigrants either stayed permanently or returned permanently later. The number of Mexican immigrants has risen steadily in every decade since 1950. Some 300,000 arrived in the 1950s, 450,000 in the 1960s, 650,000 in the 1970s, and 1.7 million in the 1980s. The pace of Mexican migration quickened again during the decade of the 1990s, when about 2.3 million Mexicans settled in the United States. In the first years of the 2000s, between 200,000 and 300,000 arrived annually. Some projections suggest that over 2.5 million will arrive in the first decade of the 21st century.

Although often overlooked, Mexican immigration to the United States has been one of the most significant demographic trends of the 20th century in both countries. Over 5 million legal immigrants arrived from Mexico during the century. Estimates vary widely, but it is likely that several million additional Mexican immigrants have arrived illegally during this period as well. All indications suggest that this flow will continue largely unabated as the attraction of employment in the United States continues to draw Mexicans northward, particularly if unscrupulous businesspeople continue to go largely unpunished for knowingly employing undocumented workers who work for low wages and no employment benefits. Ineffective immigration enforcement by the U.S. government and its apparent unwillingness to offend business and agricultural interests who profit from the labor of illegal immigrants indirectly encourage the flow. According to official U.S. government statistics, only Germany, Italy, and the United Kingdom have sent more legal immigrants to the U.S. than Mexico. Rapid natural increase and intermarriage have swelled these numbers; in 2000 about 20.5 million people in the United States traced their origin, or that of their ancestors, to Mexico.

Puerto Ricans came to the continental United States in small numbers during the first quarter of the 20th century. This occurred despite the incorporation of Puerto Rico into the United States as a colony in 1898 and the granting of U.S. citizenship to Puerto Ricans in 1917. By 1920, only 12,000 had settled on the mainland. Between 1920 and 1940, their numbers grew slowly as about 40,000 more Puerto Ricans came north, encouraged by labor recruiters and drawn by employment opportunities in agriculture and industry.

Migration accelerated rapidly during the 1940s and 1950s, when almost 700,000 Puerto Ricans left the island. A variety of factors encouraged this movement. Nearly 100,000 Puerto Rican men served in the U.S. armed forces during World War II. Their experiences, including exposure to the U.S. mainland, familiarity with life "stateside," and command of the English language, led many to establish themselves in the United States. A rapidly expanding postwar economy fueled a demand for labor that Puerto Ricans, often encouraged by labor recruiters, moved in to fill. New York City proved to be an early and enduring focus of Puerto Rican settlement. Puerto Rican women found employment in the apparel industry (in the Garment District) and in low-paid service-sector jobs, while Puerto Rican men found unskilled and semiskilled labor and industrial jobs. Urban employment opportunities dominated, although seasonal agricultural employment in the Northeast played a minor role in encouraging Puerto Rican migration.

Since the 1960s Puerto Ricans have transformed the urban landscape of New York City, creating "new" Hispanic neighborhoods. "East Harlem" became "Spanish Harlem" and the "Lower East Side" metamorphosed into "*Loisaida*." Although New York City and its metropolitan region served as the primary destination for island migrants, secondary nuclei of Puerto Rican settlement developed in other eastern and midwestern cities, notably Boston, Cleveland, and Chicago. As Puerto Ricans have moved out of the city proper in increasing numbers, other Hispanic groups have moved into those traditionally Puerto Rican spaces—for example, Mexicans dominate Spanish Harlem today and Dominicans now contest the Lower East Side with wealthy upwardly mobile residents intent on gentrifying the neighborhood!

Employment opportunities in Puerto Rico improved and diversified in the 1960s and 1970s. This affected migration patterns. Puerto Rican and federal government programs and tax incentives encouraged industrial development, creating more manufacturing jobs, while closer contacts with the United States fueled a growing tourist industry and an explosion in service-sector employment. Migration slowed, with some 140,000 Puerto Rican migrants to the United States during the 1960s, and less than one-half that number during the 1970s—just 56,000. A slowing of economic growth on the island in the 1980s propelled some 275,000 Puerto Ricans to the mainland in this decade. This brisk rate of northward movement from the island continued throughout the 1990s. By the end of the decade, nearly 250,000 more had settled "stateside." Return migration to the island is an important element in Puerto Rican migration patterns, but all told, during the 20th century, nearly 1.5 million Puerto Ricans have left the balmy island climate for the mainland. Natural increase has swelled these numbers; in 2000 some 3.4 million people in the continental United States traced their origin, or that of their ancestors, to Puerto Rico. The population of the island itself was only marginally greater, standing at around 3.8 million.

Immigration from Cuba to the United States dates from the mid-19th century. Cuban entrepreneurs began producing cigars in several locations in the United States after it imposed tariffs on imported cigars. Key West, Florida, was an early center of this cigar industry, but it was displaced by Tampa. By the end of the century in Tampa, a thriving Cuban American suburb, Ybor City, had been established around the cigar industry. Another node of Cuban immigration focused on New York City and its suburban areas—for example, Elizabeth, New Jersey, where cigar makers also settled. In addition pro-independence

Cuban refugees and exiles settled in New York City in the wake of their defeat by Spanish government forces in the 1870s. Although Cuban immigrants were not enumerated separately until the 1920s, immigration data suggest that, by the end of the 19th century, Cuban immigrants to the United States totaled between 50,000 and 100,000. The rate of Cuban immigration to the United States remained low throughout the first half of the 20th century; only 50,000 immigrated in the 30 years between 1921 and 1950.

A successful revolution in 1959 presaged major changes in Cuban immigration to the United States. Riding a broad wave of popular support, Fidel Castro led a guerrilla war against the Cuban army and the nation's dictator, Fulgencio Batista. The nationalization of U.S. property in Cuba, strong socialist rhetoric and practice, and a military alliance with the Soviet Union combined to touch off a cold war between Cuba and the United States that has continued ever since. When Cuba's moneyed elite, and much of its upper, middle, and professional classes, began to fear the long-term ramifications of a socialist economy, many sought refuge in the United States. This set in motion one of the most unusual set of circumstances in the immigration history of the United States.

The United States viewed Castro's government as communist and as an extreme threat to the security of the Western Hemisphere. It instituted an economic embargo that has been kept in place since the 1960s and it provided clandestine support for a counterinsurgency movement that attempted to invade Cuba at the Bay of Pigs in 1961. However, the United States also passed a broad range of legislation that granted Cuban emigrés special immigration status and almost immediate legal residence upon arrival. These ardent anti-Castro immigrants also benefited from a host of federal government programs, ranging from social services to business loans and assistance.

By 1961 Cuba's political orientation had pushed tens of thousands of Cubans to immigrate to the United States. Most settled in southern Florida in the city of Miami and surrounding communities in Dade County. But the New York metropolitan area was a secondary focus for Cuban immigrants. Over 200,000 Cubans arrived in the 1960s, another 265,000 in the 1970s, and an additional 140,000 in the 1980s. In 1980, nearly 125,000 Cubans, including a small percentage of common criminals, were permitted by Castro to leave the island. Cuban Americans from Miami rushed to their aid at the port of Mariel, Cuba, in all manner of watercraft, providing one of the more dramatic episodes in the history of Cuban immigration to the United States. Immigration from Cuba waned in the 1990s as U.S. government policies became less favorable to unlimited Cuban immigration. Less than 100,000 Cubans immigrated to the United States during this decade. All told, some 850,000 Cubans immigrated to the United States during the 20th century. People of Cuban origin and their descendants in the United States totaled nearly 1 million at the end of the 20th century.

Distribution in the United States

Paradoxically, the geographical distribution of Hispanics in the United States is both highly concentrated and widely dispersed (Figure 20.8). The "Hispanic heartland," if such a term can even be used, runs across the Southwest of the United States, anchored in the east by Texas and in the west by California, and includes Arizona and New Mexico. Approximately 45 percent of the nation's Hispanics live in these four states. In Arizona, Hispanics account for slightly more than 20 percent of the population, while this swells to fully one-third in both California and Texas, and reaches over 40 percent in New Mexico (Figure 20.9).

Mexican Americans are the dominant

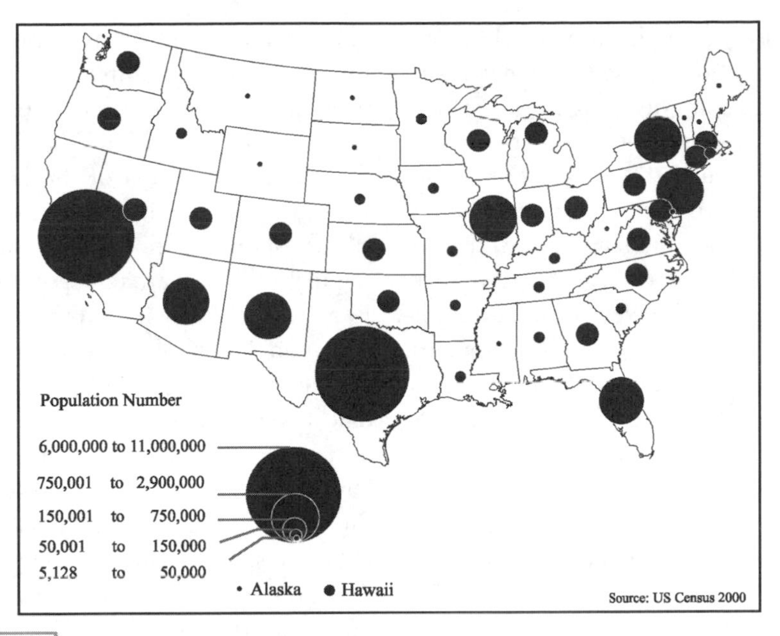

FIGURE 20.8. Hispanic population by state, 2000.

Hispanic population across this region, but there are important variations in this pattern. In New Mexico, for instance, many Hispanics trace their origins in the state back 300–400 years when the region was part of Spain's colonial empire. They often call themselves "Hispanos" as opposed to "Mexicans," "Mexican Americans," or "Chicanos." In the last 20 years, a more variegated pattern of Hispanic immigration, especially in California, has brought appreciable numbers of Central Americans (particularly El Salvadorans, Guatemalans, and Nicaraguans) many of whom were fleeing civil war and political violence, to Los Angeles, San Francisco, Miami, and Washington, D.C. The film *El Norte* (1984), directed

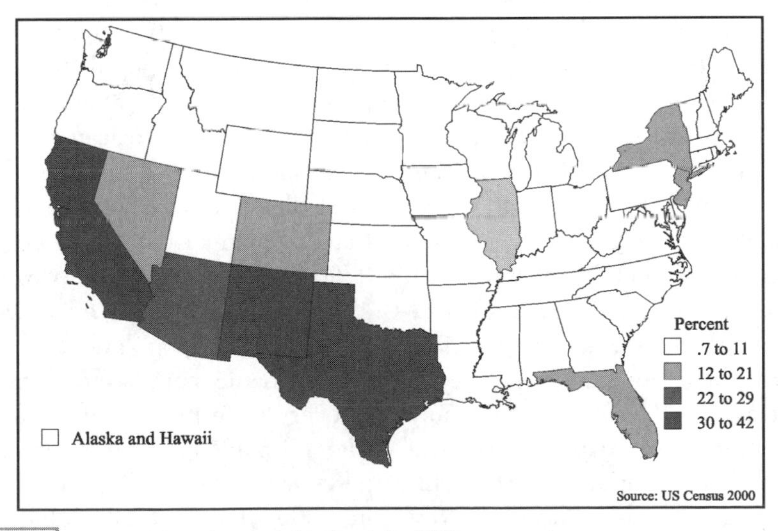

FIGURE 20.9. Percent Hispanic population by state, 2000.

by Gregory Nava, provides a compelling account of the migration of two Guatemalan siblings who make their way through Mexico to escape their war-torn country and seek a better life in Los Angeles.

Dense concentrations of Hispanics are also found in several states far removed from the southwestern United States: New York, Florida, and Illinois. The New York metropolitan area is home to about 3 million Hispanics. Known as the New York Consolidated Metropolitan Statistical Area (CMSA) by the U.S. Census Bureau, it includes New York City and vast surrounding suburban areas in the states of New York, New Jersey, and Connecticut. At the state level, over 10 percent of the population of both New York and New Jersey is Hispanic, but most of this population is concentrated in the New York City metropolitan area. A diverse mix of Latin American nationalities contributes to the region's Hispanic population, but those of Caribbean origin are most numerous, led by Puerto Ricans and followed by Dominicans and Cubans. New York City and Miami are the only places in the United States where South American nationalities reach appreciable numbers. In New York City, the total numbers of Colombian, Ecuadorian, and Peruvian immigrants reach many tens of thousands. Many of the region's Hispanics occupy blighted inner-city neighborhoods—for example, Puerto Ricans in Spanish Harlem, the Lower East Side, and the Bronx—but increasingly other Hispanics, like Mexicans and Dominicans, are replacing the Puerto Ricans who have moved out. Dominicans also have moved into transitional neighborhoods in and around Washington Heights, contributing in part to a commercial and residential revitalization of some areas. Mexicans and Mexican Americans have become increasingly common. Today nearly 200,000 live in New York City. Smaller numbers of Hispanics, particularly Cubans, have concentrated in staunchly middle-class neighborhoods of suburban New Jersey. Many Hispanic professionals, of all national origin groups, have chosen to settle in suburban environments, reflecting a pattern of residential choice similar to that made by others in the upper and middle classes.

Over 2.5 million of the Hispanics in the United States live in Florida, where they constitute about 17 percent of the state's total population. Most are concentrated in southern Florida, especially in and around Miami. Metropolitan Miami alone, known in U.S. Census Bureau parlance as the "Miami–Fort Lauderdale CMSA," has a total Hispanic population of just over 1 million. Historically, Cubans have been the most numerous Hispanics in the state, but this primacy has changed. The 2000 census identified "Other Hispanics and Latinos" as the most numerous census category, with over 1 million. This group includes significant numbers of El Salvadorans, Guatemalans, Nicaraguans, and Colombians. People of Cuban origin and ancestry total about 900,000 and represent about one-third of Florida's Hispanics, followed by Puerto Ricans and Mexicans.

Deep in the U.S. Midwest and far removed from traditional ports of entry for Latin American immigrants, Illinois seems an unlikely destination for Latin Americans. Yet Illinois boasts over 1.5 million Hispanics. Almost all are concentrated in the Chicago CMSA, where they constitute fully one-quarter of the city of Chicago's population. Work in meatpacking plants and other industrial activities initially attracted Mexican and Mexican American immigration, beginning in the 1920s. Almost all of the area's Hispanics are concentrated in inner-city neighborhoods on the South Side of Chicago. Mexicans and Mexican Americans account for over two-thirds of the city's Hispanic population.

Hispanic population numbers increased in every state in the union during the 1990s. The Hispanic population jumped nearly 60 percent from 22 million in 1990 to 35 million in 2000 (Table 20.2)! California and Texas posted gains of nearly 3.2 million and 2.3 mil-

TABLE 20.2. Hispanic Population in the United States, 1990 and 2000, by State

	2000	1990	Total change (1990–2000)	Percent change (1990–2000)	Total % 2000
California	10,966,556	7,687,938	3,278,618	42.6	32.4
Texas	6,669,666	4,339,905	2,329,761	53.7	32.0
New York	2,867,583	2,214,026	653,557	29.5	15.1
Florida	2,682,715	1,574,143	1,108,572	70.4	16.8
Illinois	1,530,262	904,446	625,816	69.2	12.3
Arizona	1,295,617	688,338	607,279	88.2	25.3
New Jersey	1,117,191	739,861	377,330	51.0	13.3
New Mexico	765,386	579,224	186,162	32.1	42.1
Colorado	735,601	424,302	311,299	73.4	17.1
Washington	441,509	214,570	226,939	105.8	7.5
Georgia	435,227	108,922	326,305	299.6	5.3
Massachusetts	428,729	287,549	141,180	49.1	6.8
Pennsylvania	394,088	232,262	161,826	69.7	3.2
Nevada	393,970	124,419	269,551	216.6	19.7
North Carolina	378,963	76,726	302,237	393.9	4.7
Virginia	329,540	160,288	169,252	105.6	4.7
Michigan	323,877	201,596	122,281	60.7	3.3
Connecticut	320,323	213,116	107,207	50.3	9.4
Oregon	275,314	112,707	162,607	144.3	8.0
Maryland	227,916	125,102	102,814	82.2	4.3
Ohio	217,123	139,696	77,427	55.4	1.9
Indiana	214,536	98,788	115,748	117.2	3.5
Utah	201,559	84,597	116,962	138.3	9.0
Wisconsin	192,921	93,194	99,727	107.0	3.6
Kansas	188,252	93,670	94,582	101.0	7.0
Oklahoma	179,304	86,160	93,144	108.1	5.2
Minnesota	143,382	53,884	89,498	166.1	2.9
Tennessee	123,838	32,741	91,097	278.2	2.2
Missouri	118,592	61,702	56,890	92.2	2.1
Louisiana	107,738	93,044	14,694	15.8	2.4
Idaho	101,690	52,927	48,763	92.1	7.9
South Carolina	95,076	30,551	64,525	211.2	2.4
Nebraska	94,425	36,969	57,456	155.4	5.5
Rhode Island	90,820	45,752	45,068	98.5	8.7
Hawaii	87,699	81,390	6,309	7.8	7.2
Arkansas	86,866	19,876	66,990	337.0	3.2
Iowa	82,473	32,647	49,826	152.6	2.8
Alabama	75,830	24,629	51,201	207.9	1.7
Kentucky	59,939	21,984	37,955	172.6	1.5
District of Columbia	44,953	32,710	12,243	37.4	7.9
Mississippi	39,569	15,931	23,638	148.4	1.4
Delaware	37,277	15,820	21,457	135.6	4.8
Wyoming	31,669	25,751	5,918	23.0	6.4
Alaska	25,852	17,803	8,049	45.2	4.1
New Hampshire	20,489	11,333	9,156	80.8	1.7
Montana	18,081	12,174	5,907	48.5	2.0
West Virginia	12,279	8,489	3,790	44.6	0.7
South Dakota	10,903	5,252	5,651	107.6	1.4
Maine	9,360	6,829	2,531	37.1	0.7
North Dakota	7,786	4,665	3,121	66.9	1.2
Vermont	5,504	3,661	1,843	50.3	0.9
United States	35,305,818	22,354,059	12,951,759	57.9	12.5

Source: U.S. Census Bureau (2001c).

lion, respectively, representing increases of over 40 percent in the total number of Hispanics in each state. Nevertheless, the magnitude of these percentage increases pales when compared to the increases in a band of states across the center of the South where Hispanic population growth exceeded 200 percent (Figure 20.10). With an increase of nearly 400 percent, North Carolina posted the largest percentage gain in the nation. Substantial percentage gains also occurred in most states in the Midwest, the Great Plains, and the Mountain West. It is true that the total number of Hispanics in some states remained quite small. Nonetheless, these changes represent a significant shift in Hispanic population distribution in the United States and suggest that soon the country's Hispanic population will be even more widely distributed over the national territory than previously.

Demographic Characteristics and Future Trends

Population statistics portray Hispanics as one of the fastest growing population groups in the United States. In 2000 the Hispanic population surpassed that of African Americans, making Hispanics the largest minority population in the United States, representing almost 13 percent of the total population. This growth has been fueled by the rapid increase in the immigration of Latin Americans to the United States during the decades since 1950. Equally significant and potentially more important in the long run has been the high rates of natural increase among many of the Latin American nationality groups present in the United States. Hispanic population in the United States stood at just about 9.1 million in 1970, but by 2000 this number had more than tripled. Hispanic population totaled about 35 million in 2000—and this does not include the nearly 4 million Hispanics who live in Puerto Rico. During the same period, Hispanics had increased their share of the total U.S. population from just 4.5 percent to nearly 13 percent. Indeed, by 2030, the U.S. Census Bureau projects that Hispanics will account for almost 20 percent of the nation's population, and conceivably will rise to fully 25 percent by 2050. During the same period the percentage of the nation's population attributed to blacks/African

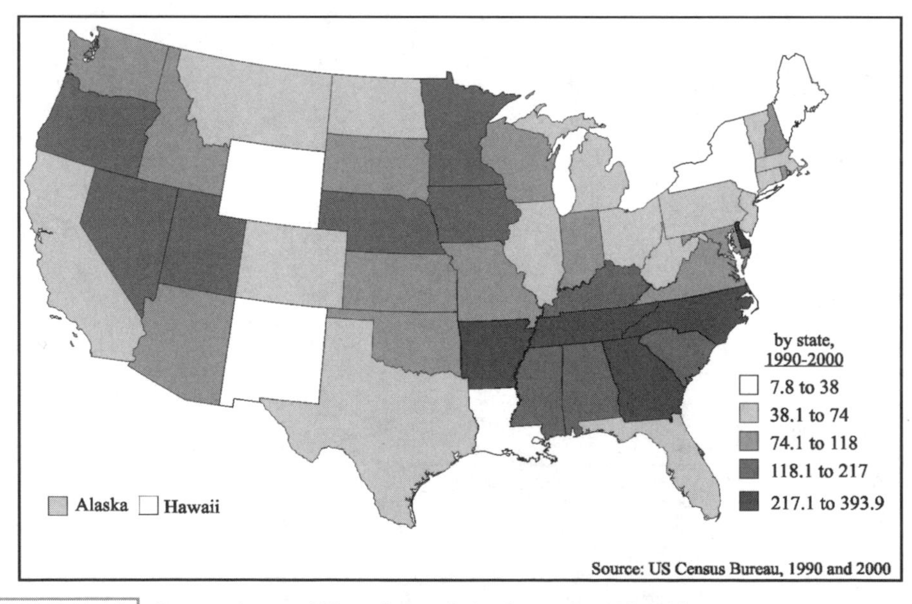

FIGURE 20.10. Percent change of Hispanic population by state, 1990-2000.

Americans is expected to remain stable at about 12 percent. In California and New Mexico, it is possible that Hispanics will constitute a majority of each state's population by 2050.

While Hispanics as a group are the nation's fastest growing minority population, the characteristics of each major subpopulation illustrate the diversity among them. For instance, by the end of the 1990s, continued immigration from Cuba had effectively stopped. Puerto Rican migration to the mainland had slowed. Meanwhile, the number of Mexican immigrants, both legal and illegal, had increased and was in the hundreds of thousands annually. Birthrates also play a critical role. The fertility rates among Mexicans, Puerto Ricans, and Cubans vary considerably. Mexican and Mexican American women in the United States average about 3.3 births over their lifetimes. The rate for Cuban women is just half that, 1.7 births. Puerto Rican women average 2.2 births. The average for the United States as a whole is 2.0. Since the fertility rates for almost all other ethnic groups in the United States are at or below replacement level, the high rates of Mexican fertility in the United States have significant demographic implications for the country's future.

Broadly speaking, Hispanics in the United States are marked by socioeconomic characteristics that place them among the nation's poorer and least educated citizens. In 2002 median annual income for Hispanic households was $34,000, while that for non-Hispanic white households was $48,000. Blacks' and Native Americans' median household incomes in the same year stood at $30,000 and $33,000, respectively. Over 20 percent of all Hispanics in the United States lived in poverty in 2002, in stark contrast to non-Hispanic whites, whose poverty rate was less than 8 percent. Surprisingly, recent studies show that income levels among Hispanics fell nearly 15 percent during the decade of the 1990s, a period during which the U.S. economy was experiencing unprecedented prosperity.

High rates of legal and illegal immigration from Mexico may help hold down wage rates for Hispanics generally, but lower levels of education attainment also seriously handicap their progress in the United States. In 2002 over 40 percent of Hispanics over 25 years old had not completed high school, while only 13 percent of non-Hispanic whites had failed to do so. A similar pattern pertains with respect to higher education, where slightly more than 11 percent of Hispanics held college degrees in 2002, while the average for non-Hispanic whites was 29 percent. On average, English-language fluency among Hispanic immigrants is much lower than it is for immigrants from Asia and Europe, and remains low even after immigration to the United States. Some social scientists and policymakers view these data with alarm, suggesting that Hispanics may be excluded from the pattern of economic advancement so typical of the "immigrant experience" apparent previously in the United States.

Statistical averages, however, can often be misleading, and the socioeconomic characteristics of individuals, families, and national origin groups can vary significantly from the norm. This is clearly the case with Hispanics in the United States. These differences are underscored when comparing a range of labor market characteristics of different segments of the Hispanic population (Table 20.3). In terms of education, stark contrasts exist between Mexicans and other Hispanic groups. Nearly one-third of all Mexicans in the labor force had not finished the ninth grade, yet among Puerto Ricans and Cubans this percentage was only half that for Mexicans. Similarly, barely 8 percent of the Mexican population held a bachelor's degree or higher, while among Cubans this figure stands at about 20 percent! Male and female professional employment follows similar patterns, with Mexicans trailing other Hispanic groups by significant margins. Measures of economic well-being demonstrate other differences. About one-quarter of all

TABLE 20.3. Labor Market Characteristics of Hispanic and Hispanic-Origin Subpopulations in the United States, 2000

Labor characteristics	Mexican	Puerto Rican	Cuban	Central and South American	Other Hispanic	Hispanic	Non-Hispanic	US Total
Bachelor's degree or more	7.8	12.9	19.3	17.4	18	11.1	27.9	26.2
Less than 9th-grade education	32.1	17.9	17.7	22.2	14	27.2	4.5	6.8
Male unemployment	6.9	8.3	5.3	4	7.2	6.5	4.8	5
Female unemployment	8.5	4.8	3.9	6	5.5	7.3	3.9	4.3
Male professional occupation	8.5	19.5	19.2	11.5	18.2	10.7	31.8	29
Female professional occupation	16	23.1	27	19	22	18.3	34	32.8
Male: service occupations, operators, and laborers	46	42	37	50	40	45.7	27	29
Female: service occupations, operators, and laborers	42	29	26	47	39	40.4	22	24
Family below poverty level	21	24.9	8	13.8	17	19.2	7	8.7
Female-headed family	20.7	35.5	17.8	22.8	26.2	22.8	16.8	17.5

Source: U.S. Census Bureau (2001b).

Puerto Rican families live below the poverty level, yet only 8 percent of Cubans families do. Family income data for Hispanic groups also shows similar trends. The proportion of Cuban households earning over $50,000 (43 percent) is nearly double that for Mexican households (23 percent) (Table 20.4).

Most Latin American migrants and immigrants to the United States face substantial legal, social, and economic difficulties in establishing themselves in the country. Low levels of educational attainment, limited financial resources, illegal immigration status, racial/ethnic prejudice and discrimination, and marginal employment opportunities are common problems. Nevertheless, such is not the case for all Latin American immigrants. While the proportion of total immigrants they represent is small, perhaps no more than 10 percent, substantial numbers of Latin Americans reach the United States and settle there under extremely favorable conditions. Typically these immigrants enter the country legally, are well educated, find well-paying employment quickly, and experience few difficulties integrating into U.S. society. In a sense, these are elite immigrants and bear no resemblance to the "tired huddled masses yearning to be free" whom the

TABLE 20.4. Total Money Income of Hispanic and Hispanic-Origin Subpopulations in the United States, 2001

Total with money/ income	Mexican	Puerto Rican	Cuban	Central and South American	Other Hispanic	Non-Hispanic white	U.S. Total
$1–$2,499	2.6	3.7	1.2	1.2	2.9	1.4	1.5
$2,500–$4,999	1.3	2.7	—	1.1	1.2	0.8	0.8
$5,000–$9,999	4.7	9.6	3.4	3.1	6.5	2.6	2.8
$10,000–$14,999	8.1	9.7	7.9	7.5	7.5	4.1	4.5
$15,000–$19,999	937	1.0	5.9	8.4	8.4	5.0	5.5
$20,000–$24,999	9.1	7.1	8.4	7.6	6.5	5.5	5.8
$25,000–$34,999	16.8	14.1	13.2	16.2	13.1	11.5	12.0
$35,000–$49,999	18.3	14.4	15.1	20.4	15.1	16.0	16.2
$50,000–$74,999	16.8	13.7	23.1	17.9	20.9	21.7	21.2
$75,000 and over	12.7	14.9	21.9	16.6	17.9	31.4	29.6

Source: U.S. Census Bureau (1999).

Statue of Liberty beckons from New York Harbor.

During much of the 20th century, and especially during the last half of the century, U.S. immigration policy often sought to encourage the immigration of talented individuals with specific technical or professional skills. These have included nurses, doctors, university professors, engineers, researchers, technicians, and those in other specialized occupations. Talented Latin American nationals, drawn by both professional opportunities and greater financial compensation, have taken advantage of these policies and are frequently found on the professional staff of hospitals, universities, corporations, and research centers. There is perhaps no other area or professional venue in which the contributions of Latin American nationals is more publicly visible than in professional sports—specifically the "All American" game of baseball (see Vignette 20.2).

Summary

In the latter part of the 20th century Latin America has undergone major changes as international out-migration has become increasingly common. While some Latin Americans have migrated to Europe and even Japan, most, numbering in the millions, have chosen the United States as their destination. The growing Hispanic population has influenced U.S. culture, especially in regions with large concentrations such as the Southwest and South Florida. The largest numbers come from Mexico, followed by Puerto Rico and Cuba. The first large Hispanic population in the United States was the result of the Mexican-American War in the 1840s. Mexico was forced to cede approximately half its territory to the United States, including a population of 80,000. Following the Spanish-American War in 1898, Puerto Rico became in essence a U.S. colony, eventually receiving commonwealth status. Cuba became a U.S. protectorate, although officially independent. The United States dominated the political and economic life of Cuba through the first half of the 20th century and established a military base at Guantánamo Bay that still exists. As of 2000, 12.5 percent, or about 38 million, of the U.S. population was Hispanic—this number is roughly equivalent to the population of Argentina. The population continues to grow, with the U.S. Census Bureau projecting that Hispanics will make up 25 percent of the U.S. population by 2050. The socioeconomic characteristics of the Hispanic population place them among the poorest and least educated residents of the United States. Approximately 20 percent live in poverty.

Further Reading

Arreola, D. D. (Ed.). (2004). *Hispanic spaces, Latino places: Community and cultural diversity in contemporary America*. Austin: University of Texas Press.

Bean, F. D., and Tienda, M. (1987). *The Hispanic population of the United States*. New York: Russell Sage Foundation.

Berry, K. A., and Henderson, M. L. (Eds.). (2002). *Geographical identities of ethnic America: Race, space, and place*. Reno: University of Nevada Press.

Brewer, C. A., and Suchan, T. A. (2001). *Mapping Census 2000: The geography of U.S. diversity*. Redlands, CA: ESRI Press.

Carlson, A. W. (1990). *The Spanish-American homeland: Four centuries in New Mexico's Río Arriba*. Baltimore: Johns Hopkins University Press.

Durand, J., Massey, D. S., and Zenteno, R. M. (2001). Mexican immigration to the United States: Continuities and changes. *Latin American Research Review, 36*(1), 107–127.

Foulkes, M., and Newbold, K. B. (2000). Migration propensities, patterns, and the role of human capital: Comparing Mexican, Cuban, and Puerto Rican interstate migration, 1985–1990. *Professional Geographer, 52*, 133–144.

Fox, G. (1996). *Hispanic nation: Culture, politics, and the constructing of identity*. Tucson: University of Arizona Press.

Gann, L. H., and Duignan, P. J. (1986). *The Hispanics of the United States: A history*. Boulder, CO: Westview Press.

VIGNETTE 20.2. ELITE MIGRATION, BASEBALL, AND THE LATIN AMERICAN DIASPORA

Baseball is a popular sport in Latin America. After soccer (*futbol)*, baseball is the second most popular participatory and spectator sport, although overall its popularity runs a very distant second to the former. In addition, avid interest in the sport is focused in only a handful of Caribbean and circum-Caribbean countries, notably Cuba, the Dominican Republic, and Puerto Rico. Still, baseball is an important sport in Mexico, Venezuela, and Nicaragua as well. With the exception of Mexico, the popularity of the sport is essentially ubiquitous within each country. In Mexico, baseball exhibits its greatest acceptance in the states that border the Caribbean on the Gulf of Mexico, Veracruz, Campeche, and Yucatan, as well as in northern states near the United States–Mexico border. Geographical proximity is a clear common denominator in the diffusion of the sport from its origin in the United States to Latin American countries.

Besides proximity to the United States, there are other common factors. Cuba, Puerto Rico, Nicaragua, the Dominican Republic, and Mexico all share a history of U.S. military intervention and occupation, while U.S. oil companies and personnel maintained a long-term presence in both Mexico (until the 1930s) and Venezuela (until the 1960s). These long-term contacts between U.S. servicemen and U.S. overseas employees and local citizens have contributed to the popularization of baseball in these countries. In the Dominican Republic and Cuba, political considerations have led to official government encouragement of the sport. Rafael Trujillo, the Dominican Republic's long-time dictator, encouraged the popularization of the sport in his efforts to forge a unique Dominican identity, while Fidel Castro, again for political reasons, has helped push Cuban baseball to a professional level comparable to that played in the U.S. "Big Leagues." While the United States won the first Olympic Gold medal for baseball in 1988, Cuban teams have won the gold in 1992, 1996, and 2004. In what some have dubbed a "historic encounter" in March 1999, a Big League team from the United States played in Cuba for the first time in 40 years. In the exhibition game, the Baltimore Orioles defeated a team of Cuban all-stars in a hotly contested game that ended with the Orioles victorious 3 to 2 after 11 innings.

Latin American baseball players have been playing in the U.S. Big Leagues since the 1940s. Mexicans, Dominicans, Venezuelans, Puerto Ricans, and Cubans all played in the United States during the 1950s and 1960s, but Cubans were most numerous. Since then, restrictive immigration policies in Cuba have cut off the flow of its players to the north, but others quickly filled the void, with the number of Venezuelans, and particularly Dominicans, growing rapidly in the 1980s and 1990s. In the fall of 1997, about 10 percent of all players on the rosters of Major League teams hailed from the Dominican Republic, and Latin Americans constituted 20 percent of the leagues' players (Table 20.5). In the spring of 2004 Dominicans continued to number approximately 10 percent of players, but the overall percentage of Latin Americans had grown to 25 percent of Major League Baseball's approximately 800 players. A talented pool of novices, financial economies, and less restrictive rules and laws concerning player recruitment make it likely that professional teams in the United States will continue to scout and recruit Latin American players vigorously in the future.

Hamilton, N., and Stoltz Chinchilla, N. (2001). *Seeking community in a global city: Guatemalans and Salvadorans in Los Angeles*. Philadelphia: Temple University Press.

Haverluk, T. W. (1997). The changing geography of U.S. Hispanics, 1850–1990. *Journal of Geography, 96*, 134–145.

Jokisch, B., and Pribilsky, J. (2002). The panic to leave: Economic crisis and the "new emigration" from Ecuador. *International Migration, 40*(4), 75–99.

Kent, R. B., and Huntz, M. (1996). Spanish-language newspapers in the United States. *Geographical Review, 86*(3), 446–456.

TABLE 20.5. U.S. State and Country Origins of Major League Baseball Players, 1997

Ranking	Location	Percent
1	California	15.26
2	Dominican Republic	9.74
3	Florida	9.56
4	Texas	5.79
5	Illinois	3.58
6	Venezuela	3.42
7	North Carolina	3.07
8	Alabama	2.54
8	Puerto Rico	2.54
10	Louisiana	2.46
10	Arizona	2.40

Source: Chass (1998), pp. 1 and 23.

Marcelli, E. A., and Cornelius, W. A. (2001). The changing profile of Mexican migrants to the United States: New evidence from California and Mexico. *Latin American Research Review, 36*(3), 105–131.

McHugh, K. (1989). Hispanic migration and population redistribution in the United States. *Professional Geographer, 41*(4), 429–439.

McKee, J. O. (Ed.). (2000). *Ethnicity in contemporary America: A geographical appraisal* (2nd ed.). Lanham, MD: Rowman & Littlefield.

Nostrand, R. L. (1992). *The Hispano homeland*. Norman: University of Oklahoma Press.

Novas, H. (1994). *Everything you need to know about Latino history*. New York: Plume Books.

Oleksak, M., and Oleksak, M. A. (1991). *Bâsbol: Latin Americans and the grand old game*. Grand Rapids, MI: Master's Press.

Portes, A., and Stepick, A. (1994). *City on the edge: The transformation of Miami*. Berkeley and Los Angeles: University of California Press.

Shorris, E. (2001). *Latinos: A biography of a people*. New York: Norton.

Suárez-Orozco, M. M., and Páez, M. M. (Eds.). (2002). *Latinos: Remaking America*. Berkeley and Los Angeles: University of California Press.

Valdés, M. I., and Seoane, M. H. (1997). *Hispanic market handbook: A definitive source for reaching this lucrative segment of the American consumers*. New York: Gale Research.

21 Conclusion

This book has examined the geography of Latin America and to do so has relied heavily on cultural historical and regional perspectives. Tacitly it has argued that these approaches are central to understanding the contemporary geography of Latin America as well as the region's future. And, at the conclusion of such a book, one has to ask . . . so, what of that future?

Predicting the future is a tricky business. But this fact rarely deters most of us from attempting it anyway. What forces and processes will dictate the directions of geographical change in Latin America in the coming 20, 50, or even 100 years? How will they transform the region's landscapes? Who can really say? No one, of course, but some closing comments are still appropriate.

Place will always matter. The unique characteristics of Latin America as a culture region as well as the peculiar nature of its principal subregions and localities will color the direction and nature of its future geographies. The congruencies and disarticulations of location will exert powerful influences on the region's prospects for economic development. These may be linked to accessibility and the cost of transportation, the presence and accessibility of natural resources, agglomeration economies, market demand, and the availability and character of human resources. But place matters in a myriad of other ways as well, influencing how and where new ideas, innovations, and technologies will or will not take hold and in turn change those same places.

Places exist in context. Understanding context is essential. An appreciation of the cultural and historical context of Latin America is key to comprehending its evolving environmental and human geographies. A solid grounding in the region's culture and historical development provides essential insights that allow one to envision with greater clarity the possible directions of future change. Geographical change does not occur on a blank page. It plays out on the face of an existing landscape and human culture, and the process of geographical change is guided and transformed by both of them.

Globalization and Rapid Change

Globalization is nothing new to Latin America. But its manifestations and their frequency in everyday life have grown tremendously in the last half of the 20th century. Rapid globalization and technological change will continue well into the 21st century. The transformations in the use of money and credit, the integration of banking systems, and the communications revolution are all excellent examples of globalization and the rapid rate of change experienced in recent decades. All portend events in the coming decades.

For example, in the 1970s travelers who visited Latin America or traveled from one country to another within Latin America often had few options other than to carry cash or traveler's checks. Many traveled with wads of cash hidden in money belts or bras. Credit cards were available to only the smallest minority of Latin Americans and often accepted by only a few businesses, usually located in principal cities. Foreign currency transactions or the cashing of traveler's checks often involved lengthy waits and tedious bureaucratic processes, especially if one used the banks that conducted such exchanges.

By the first decade of the 21st century, access to money in Latin America is almost as easy as making a withdrawal from your account in your hometown bank. As a matter of fact, if you have a bank account in the United States, Canada, Western Europe, Asia, or just about any country in Latin America, you can probably go the plaza of any large city and withdraw money from your account from an automatic teller machine (ATM). The use of credit cards is now widespread in Latin America. Once difficult, complicated, and costly, international money transfers from the United States and Europe have become commonplace as overseas remittances feed families and fuel local economies in Mexico, in Central America, in the Caribbean, and increasingly in some South American countries.

A communication revolution that began in the 1990s continues to transform the fabric of human interaction across Latin America. In the 1980s it took 7–10 days for airmail letters posted from Latin American capitals to reach the United States and Europe; letters sent from provincial and departmental capitals might take 2 weeks. Even telegrams could take 24 hours to make it door to door. But conditions in the first decade of the 21st century are vastly different. Need to send a message to a business contact or a friend? And it needs to get there fast? No problem. Just drop into an Internet café and send an e-mail that will arrive in their e-mail box within a few minutes. Perhaps that person will be online. If so, you might just be able to send him or her an instant message (IM) for an immediate real-time text conversation over the web. Or better yet, just use a cellphone.

Telephones have been used in Latin America for over 100 years, but even in the 1980s and 1990s making a call across town could be difficult. Making intercity and interprovincial calls often proved problematic. Call quality was almost always abysmal. An international call from a provincial capital was a big deal—and it might take a whole day to get a line. But satellite and cellphone technology have revolutionized and improved many aspects of telecommunications in Latin America. Dependability and call quality have improved vastly. Moreover, the advent of cellphones has brought communications accessibility to many more people. Today, cellphones are abundant throughout Latin America. They are not cheap and not every one can afford one. But many middle-class folks own them, in part because their cost is often far less than that charged for traditional land lines (which may take up to several years to obtain).

Latin Americans will continue to adopt new technologies. The limited availability of computers and high-speed telecommunications networks in the first decade of the 21st century leaves the tantalizing benefits of the Internet beyond the reach of many Latin Americans. But this will change with rising incomes and increased public access to computer technology and networks. This increased access in turn will reduce the friction of distance within the region and create new opportunities and directions of change.

Movement and Migration

Latin Americans are geographically mobile and promise to be so for the foreseeable future. Since the 1930s rural-to-urban migration has been a major consequence of the transformation of the region from a predominantly rural agricultural/extractive economy to a predominantly urban industrial/service economy. Rural-to-urban migration will persist as an important process for several decades into the future in those Latin American countries that still have predominantly rural economies and a high proportion of rural inhabitants.

International migration, hardly significant in the 1950s, grew appreciably in the final decades of the 20th century. It will continue to be important in the early decades of the 21st century. Some of this international migration will be from poor countries to their wealthier and more dynamic neighbors within the region. But most will be oriented toward the urban industrial and postindustrial economies outside the region. The United States will maintain its position as the foremost recipient of Latin American immigrants. Millions of Mexicans now constitute the vast majority of these immigrants, but these migration flows have become—and promise to remain—more variegated as Central Americans and people from the Hispanic Caribbean have made their way—and will continue to make their way—to many cities across North America.

If present trends continue, Latin Americans and those of Latin American ancestry will account for 25 percent of the U.S. population by 2050. How will they be transformed by the immigrant experience? How will the United States change as a result of exposure to their values and culture? Perhaps after a century of U.S. economic and political hegemony over Latin America, Latin Americans will return the favor and transform their new homeland in profound and unexpected ways in the 21st century.

The 21st century will bring new patterns of international migration. Since the 1990s Latin Americans have been immigrating in appreciable numbers to Europe, and especially to Spain. As populations continue to age, Latin Americans will figure prominently in the immigrant population in Spain, where language and cultural barriers are minimal. The immigration history of Italians in Brazil and Argentina suggests that Italy too may receive appreciable Latin American immigration. A strong demand for both skilled and unskilled labor throughout Europe may encourage notable Latin American immigration to other wealthy economies in the European Union.

Latin American migrants are transforming those countries to which they immigrate, but surprisingly also the homelands they have left behind. One of their most immediate impacts is through financial remittances. Immigrants send billions of dollars each year to Latin America. In some countries, notably Mexico, Cuba, the Dominican Republic, and El Salvador, these remittances now account for a major component of all foreign exchange earnings and comprise an appreciable portion of national income. Economic power begets political power. Political power shapes the directions of social and cultural change. The transformational power of dollar remittances and the political vision of those that send them promise to have a profound effect in some Latin American countries for the next several decades.

Democracy and Democratic Institutions

Democracy holds great promise for Latin America. The region's countries have espoused democratic principles in their constitutions and political structures since the end of the colonial period. Nevertheless, Latin American governance was often long on democratic rhetoric and very short on democratic practice. Military governance and dictatorships were usually the rule. Few functioning democracies existed before the 1980s. However, the 1980s and 1990s brought an increasing commitment to democratic principles and democratic practice throughout the entire region. Citizens demanded greater accountability from their governments, while international organizations and outside powers often tied financial, developmental, and even military assistance to greater government commitments to democracy, democratic institutions, and protection of human rights.

By the first decade of the 21st century, democratic traditions had been strengthened throughout the region at both the national and the local levels. All of the region's countries, except Cuba, could boast democratically elected governments. Many democratic institutions are only in their fledgling stages and the rule of law may be precarious in some places, but democracy and stable legal systems have taken firmer holds across the region than in the past and Latin Americans today seem committed to enhancing democratic practices and traditions.

Democracy and participatory government will dominate the political processes of the region in the coming decades. While there will certainly be political upheaval, violent conflict, and even revolutions, there is little chance of a proliferation of undemocratic governments lead by the military or autocratic dictators in the future. The communications revolution will continue to produce greater transparency in the political process and support for participatory democracy. The migration of Latin Americans to industrialized democracies will contribute to strengthening democratic traditions: the transnational character of many immigrants in the 21st century suggests that many will bring their own exposure to democratic ideas and practices back to their home countries.

Changing Political Hegemonies

The political and economic hegemony of the United States in Latin America will erode in the early decades of the 21st century. The United States will remain the dominant geopolitical force in the region in 2050, but its role will be reduced and it will share its political and economic clout with a broader range of countries and regions. A key component in this change will be a substantial increase in the importance of Asia. That region offers a lucrative international market for Latin America's natural resources and agricultural products. China plays a leading role in generating that demand and is poised to play an increasingly visible diplomatic role as well.

Trade will figure prominently in the unfolding economic geography of Latin America in the 21st century. The United States has pushed a free trade agenda in the region centered on the gradual elimination of customs duties, tariffs, and investment barriers among countries. This strategy capitalizes on each country or region's comparative economic advantages and leads to a more efficient use of resources and greater wealth generation. Policymakers reason that while smaller economies may initially suffer considerable hardships, in the long run these policies will raise the standard of living of all member countries. This and other neoliberal economic reforms are central elements in a formula for regional economic development. The establishment of the North American Free Trade Agreement (NAFTA) in 1994 between the United States, Canada, and Mexico initiated a long-term plan to create a hemispheric free trade zone. The United States envisions a Free Trade Area of the Americas (FTAA) by the early decades of the 21st century.

A monolithic free trade area for Latin America dominated by the United States is no certainty. There are competing visions. The Mercado Común del Sur (MERCOSUR), comprised of Brazil, Argentina, Uruguay, and Paraguay, offers a strong counterweight to U.S. hegemony. These nations may forge critical trade agreements with the European Union (EU) or major Asian countries before FTAA can become a reality. In such a scenario, the United States might end up leading a trade block that includes NAFTA as well as Central America, the Caribbean Basin, and one or two other countries, while MERCOSUR would dominate South America.

Environment and Economic Development

Degradation of the region's natural resource base and environment will continue and will be a pressing concern for decades to come. Population growth, rapid urbanization, and strong demand on world markets for Latin America's agricultural products and natural resources will all contribute to a depletion of the region's nonrenewable and renewable natural resources. Peasant farmers will till steep hillsides and their brethren in the rain forests will clear virgin forests. Rich topsoil will be eroded and rivers will be choked with sediments. Oil companies will drill and miners will claw tons of ore from the mountainsides. Cattle will graze in tropical pastures where tropical hardwood trees once stood. Environmentalists will howl and indigenous people will protest. In some areas citizens and environmentalists will succeed in mitigating the most destructive forms of environmental degradation and some parks and natural preserves will be established and protected. But in other areas the harvest of economic development will include a spoiled environmental resource base. Sustainable development, while an attractive concept, is a long way from being a realistic developmental strategy for most of Latin America.

Since the 1980s neoliberal economic reforms and a booming world economy have brought significant macroeconomic success to many of Latin America's countries. The privatization of government enterprises, a reduction in government payrolls, tax and administrative reform, and the promotion of open transparent markets and investment tools have all contributed to a more dynamic private sector and an increase in wealth and income. Unfortunately, while economic reforms have created more wealth in almost all countries, a highly skewed system of income distribution has meant that the benefits of this income growth have been concentrated in the hands of a few. A rising economic tide in Latin America has failed to raise all boats. Perhaps one-third of the region's population lives in wrenching poverty amid a sea of plenty.

A highly concentrated pattern of wealth and wealth generation will persist in Latin America. Neoliberal economic reforms have invigorated capitalism in Latin America. However, despite growth in gross national income and substantial economic progress for many in the region's upper and middle classes, the working class and the poor face difficult futures. Food, shelter, and basic human services will be in short supply for millions of Latin Americans well into the 21st century and many will continue to live in abject poverty. Adequate medical care, education, and social security will remain elusive for many Latin Americans. Fiscal and social policies that could ensure an equitable tax system and the provision of a basic social safety net have been slow to appear. Increasing democratization of political and social institutions will bring some improvement in socioeconomic conditions. But slow economic progress for the region's poorest citizens will contribute to powerful social conflict in some places and probably even to violent confrontations like those that have plagued Colombia for decades.

Optimistic Perspectives

Despite its problems, there are many reasons to harbor a guarded optimism for Latin America's future. The region is well endowed with agricultural and natural resources. Although population growth has been rapid, it shows signs of abating. In addition, the region's population has not outpaced the ability of its resource base to support it. Latin America's population is youthful, and, by world standards, it is reasonably educated and enjoys an adequate standard of living in many parts of the region. Democracy, democratic processes, and a respect for human rights have become increasingly entrenched in the social traditions of the region in the last half-century and this bodes well. Ongoing political, economic, and administrative reforms initiated during the closing decades of the 20th century have invigorated the region's economy, generated new wealth, and created additional promise for its future.

In 1950 Latin America remained a remote and vaguely exotic land for most outsiders. This is no longer the case. By the first decade of the 21st century, the region and its peoples had become increasingly integrated into the global community and more politically and economically assertive. This trend shows no sign of abating.

Chapter Notes and Additional Sources

Chapter 2

Adas, M., and Adas, S. (2002). *Panorama geogrfico do Brasil: Contradiçes, impasses e disafios socioespaciais* 3rd ed. São Paulo, Brazil: Editora Moderna.

Anonymous. (1995). Is hard to watch. Mexico: A Mennonite community battles the lure of drug money. *Newsweek*, 17 July, p. 33.

Anonymous. (1997). Bolivia: McDonald's opens restaurant in capital. *Akron Beacon Journal* 25 October, p. A11.

Augelli, J. (1986). The Panama Canal area. *Focus*, 36, pp. 21–29.

Berger, J. (2005). What is on TV?: Loved ones from afar: Videoconferences link immigrants and families in homeland. *New York Times*, 8 January, p. B12.

Brenner, L., and Aguilar, A. G. (2002). Luxury tourism and regional economic development in Mexico. *Professional Geographer*, 54 (4)., pp. 500–520.

Brooke, J. (1994). Blumenau journal: A city of blondes battles migrants. *New York Times*, 4 May, p. A4.

Caviedes, C., and Knapp, G. (1995). *South America* 1st ed. Englewood Cliffs, NJ: Prentice-Hall.

Coeyman, M. (1996). Over there: Business is booming overseas and surprise, not just for the biggest guns. *Restaurant Business*, 1 September, pp. 69–79.

Collier, S., Skidmore, T. E., and Blakemore, H. (Eds.). (1992). *The Cambridge encyclopedia of Latin America and the Caribbean* 2nd ed. Cambridge, UK: Cambridge University Press.

Curtis, J. (1995). Mexicali's Chinatown. *Geographical Review*, 85, pp. 335–348.

DePalma, A. (1994). It takes more than a visa to do business in Mexico. *New York Times*, 26 June, p. F5.

DePalma, A. (1994). Mexican attitudes shift in flood of U.S. foods. *New York Times*, 23 November, pp. D1 and D17.

French, H. (1993). Now takeover of canal makes Panama uneasy. *New York Times*, 23 July, p. A4.

García, B. (1996). Fueling growth: Legal framework facilitates franchise expansion. *Business Mexico*, 6, pp. 22–23.

Hamstra, M. (1996). Operators trek into South America with bold expansion plans. *Nation's Restaurant News*, 26 August, pp. 31–32, 36, and 47.

Instituto Nacional de Estadística (Bolivia). (1982). *Atlas censal de Bolivia*. La Paz, Bolivia: Author.

Kent, R. B. (2003). A diaspora of Chinese settlement in Latin America and the Caribbean. In *The Chinese diaspora: Space, place, mobility, and identity*, L. J. C. Ma and C. Cartier (Eds.)., pp. 117–138. Latham, MD: Rowan & Littlefield.

Knapp, G. (1991). *Geografía Quichua de la sierra del Ecuador* 3rd ed. Quito, Ecuador: ABYA-YALA.

Lombardi, C. L., Lombardi, J. V., and Stoner, K. L. (1983). *Latin American history: A teaching atlas*. Madison: The University of Wisconsin Press.

Love, J. L. (1971). *Rio Grande do Sul and Brazilian regionalism, 1882–1930* . Stanford, CA: Stanford University Press.

Marcus, D. L. (1995). Brazil Mcnuts for fast food. *Akron Beacon Journal*, 1 July, p. B5.

McDonalds Brasil. (2005). *Numeros*. Available online at *http://www.mcdonalds.com.br/mc_brasil/numeros.shtml*. Retrieved 26 January 2005.

Morris, W. (Ed.). (1971). *American Heritage dictionary of the English language*. New York: American Heritage Publishing Co.

Nash, N. C. (1992). Latin American Indians: Old ills, new politics. *New York Times*, 24 August, p. A1 and A6.

O'Conner, M. (1994). A new U.S. import in El Salvador: Street gangs. *New York Times*, 3 July, p. 3.

Phelan, J. L. (1968). Pan-Latinism, French intervention in Mexico (1861–1867). and the genesis of the idea of Latin America. In *Conciencia y autenticidad historicas*, J. A. Ortega & Medina (Ed.)., pp. 279–298. Mexico City, Mexico: Universidad Autónoma de Mexico.

Preston, J. (1997). A woman's shooting of attacker rivets Mexico. *New York Times*. 5 February, p. A3.

Rohter, L. (2001). Salvador da Bahia journal: Simmering over who can cook a favorite fritter. *New York Times*. 30 November, p. A4.

Rojas Mix, M. (1986). Bilbao y el hallazgo de America latina: Union continental, socialista, y libertaria. *C.M.H.L.B Caravelle*, 46, pp. 35–47.

Subway. (2005). *Subway around the World*. Available online at *http://www.subway.com/subwayroot/AroundTheWorld/index.aspx*. Retrieved 26 January 2005.

Thompson, G. (2004). Dominicans cast ballots in presidential vote. *New York Times*, 17 May, p. A3.

Toledo, W., and Zapater, E. (1989). *Geografía general y regional de Chile*. Santiago de Chile, Chile: Editorial Universitaria.

United Nations, Department of Economics and Social Affairs, Population Division. (1998). *World urbanization prospects, 1996: Estimates and projections of urban and rural populations and of urban agglomerations*, New York: Author.

Whitaker, A. (1964). *Argentina*. New York: Prentice-Hall.

World radio and TV handbook: The directory of international broadcasting. (2001). Vol. 55. The Netherlands: Billboard Books, 2001.

Chapter 3

American Museum of Natural History. (2003). Mangroves, the roots of the sea: Mangrove threats and solutions. *BioBulletin* (Spring/Summer). Available online at *http: //biobulletin.amnh.org/D/1/3/*. Retrieved 7 February 2005.

Kent, R. B. (1988). The introduction and diffusion of the African honeybee in South America. *Yearbook of the Association of Pacific Coast Geographers*, 50, pp. 21–43.

La Torre-Cuadros, M. D., and Islebe, G. A. (2003). Traditional ecological knowledge and use of vegetation in southeastern Mexico: A case study from Solferino, Quintana Roo. *Biodiversity and Conservation*, 12 (12)., pp. 2455–2476.

Lugo, A. E. (2002). Conserving Latin American and Caribbean mangroves: Issues and challenges. *Madera y Bosques, Número Especial* , pp. 5–25.

National Geographic Society. (1992). *The grand exchange* (Scale 1:50,550,000).

National Research Council, Ad Hoc Panel of the Advisory Committee on Technology Innovation, Board on Science and Technology for International Development. (1989). *Lost crops of the Incas: Little-known plants of the Andes with promise for worldwide cultivation*. Washington, DC: Author.

Chapter 4

Academia de Ciencias de Cuba. (1989). *Nuevo atlas nacional de Cuba*. Havana, Cuba: Author.

Aguado, E., and Bart, J. E. (2001). *Understanding weather and climate*. Upper Saddle River, NJ: Prentice-Hall.

Baker, C. P. (1997). *Cuba handbook*. Chico, CA: Moon Publications.

Bauzá, V., and Marrero, D. (2004). $100 million remitted to Cuba. *South Florida Sun-Sentinel*, 31 March, pp. A1 and A12.

Cruz Báez, A. D., and Boswell, T. D. (1997). *Atlas Puerto Rico*. Miami, FL: Cuban American National Council Inc.

Elliott, A. (2004). A New York vote, but a Dominican contest. *New York Times*, 17 May, p. A21.

Hatchwell, E., and Calder, S. (1995). *Cuba: A guide to the people, politics and culture*. London: Latin America Bureau Research and Action.

Mauricio, V. (1998). Vuelve el Biltmore. *El Pais* (Madrid)., 10 October, p. 45.

Scarpaci, J. L. (2001). Reshaping Habana vieja: Revitalization, historic preservation, and restructuring in the socialist city. *Urban Geography*, 21, pp. 724–744.

Stanley, D. (1997). *Cuba*. Melbourne, Australia: Lonely Planet.

Chapter 5

Bakewell, P. (1997). *A history of Latin America. Empires and sequels*. Oxford, UK: Blackwell.

Burkhholder, M. A. (1998). *Colonial Latin America* 3rd ed. New York: Oxford University Press.

Burns, E. B. (1993). *A history of Brazil* 3rd ed. New York: Columbia University Press.

Collier, S., Skidmore, T. E., and Blakemore, H. (Eds.). (1992). *The Cambridge encyclopedia of Latin America and the Caribbean* 2nd ed. Cambridge, UK: Cambridge University Press.

Crosby, A. (1972). *The Columbian exchange. Biological and cultural consequences of 1492*. Westport, CT: Greenwood Press.

Crow, J. A. (1992). *The epic of Latin America* 4th ed. Berkeley and Los Angeles: University of California Press.

Curtin, P. D. (1969). *The Atlantic slave trade: A census*. Madison: University of Wisconsin Press.

Denevan, W. (Ed.). (1992). *The native population of the Americas in 1492* 2nd ed. Madison: University of Wisconsin Press.

Greenfield, G. M. (1994). Brazil. In *Latin American urbanization: Historic profiles of major cities*, G. M. Greenfield (Ed.)., pp. 62–105. Westport, CT: Greenwood Press.

Handler, J. S., and Tuite, M. L. Jr. (2005). *The Atlantic*

slave trade and slave life in the Americas: A visual record. The Virginia Foundation for the Humanities. Available online at *http://hitchcock.itc.virginia.edu/Slavery/*. Retrieved 14 February 2005.

Hecht, J. (2002). Earliest New World writing revealed. *NewScientist.com*, 5 December. Available online at *http://www.newscientist.com/article.ns?id=dn3151*. Retrieved 16 February 2005.

Helms, M. W. (1975). *Middle America: A culture history of heartland and frontiers*. Englewood Cliffs, NJ: Prentice-Hall.

Hyslop, J. (1990). *Inca settlement planning*. Austin: University of Texas Press.

Jenkins, K. D. (1968). Sources of an insect fat used in Meso-American lacquer. *Yearbook of the American Philosophical Society*, pp. 607–610.

Kent, R. B. (1984). Mesoamerican stingless beekeeping. *Journal of Cultural Geography*, 4, pp. 14–28.

Kraay, H. (2004). Transatlantic ties: Recent works on the slave trade, slavery, and abolition. *Latin American Research Review*, 39 (2)., pp. 178–195.

Lombardi, C. L., Lombardi, J. V., and Stoner, K. L. (1983). *Latin American history: A teaching atlas*. Madison: University of Wisconsin Press for the Conference on Latin American History.

Mörner, M. (1985). *The Andean past: Land, societies, and conflicts*. New York: Columbia University Press.

Motolinía, T. (1950). *Motolinía's history of the Indians of New Spain*. Translated by E. A. Foster. Berkeley, CA: The Cortés Society.

National Geographic Society. (1992). *Spain in the Americas* (Scale 1:15,040,000).

National Geographic Society. (1997). *Ancient Mesoamerica*. (1:2,400,000).

National Geographic Society. (2002). *The Inca: An empire and its ancestors* (Scale 1:5,450,000).

Noble, J. W. (2005). "Mother culture, or only a sister?" *New York Times*, 15 March, p. D1–4.

Pohl, J. (2005). Art and writing in ancient Mesoamerica: Olmec and Zapotec writing. *Foundation for the Advancement of Mesoamerican Studies Inc. John Pohl's Mesoamerica*. Available online at *http://www.famsi.org/research/pohl/pohl_olmec_zapotec_script.html*. Retrieved 16 February 2005.

Wade, N., and Wilford, J. N. (2003). New World ancestors lose 12,000 years. *New York Times*, 25 July, p. A19.

Wagley, C. (1971). *An introduction to Brazil*, Rev. ed. New York: Columbia University Press.

Chapter 6

Bruman, H. (1940).. *Aboriginal drink areas in New Spain*. Unpublished PhD dissertation, University of California, Berkeley.

Consejo Regulador de Tequila (Mexico). (2003). *Geografía: El territorio de denominación, TDO de tequila* . Available online at *http://www.crt.org.mx/*. 22 Retrieved October 2003.

Crowley, W. K., and Griffin, E. C. (1989). Culture areas in Mexico. In *Middle America: Its lands and peoples* 3rd ed., R. C. West and J. P. Augelli (Eds.)., pp. 339–364. Englewood Cliffs, NJ: Prentice-Hall.

Gilbert, A. (1994). *The Latin American city*. London: Latin American Bureau.

Griffin, E. C., and Crowley, W. K. (1989). The people and economy of modern Mexico. In *Middle America: Its lands and peoples* 3rd ed., R. C. West and J. P. Augelli (Eds.)., pp. 284–338. Englewood Cliffs, NJ: Prentice-Hall.

Instituto Nacional de Estadística, Geografía, e Informática (Mexico). (2000). *Censos econónicos, 1999*. Available online at *http://www.inegi.gob.mx/est/default.asp?c=702*. Retrieved 24 September 2003.

Lezama, J. L. (1994). Mexico. In *Latin American urbanization: Historical profiles of major cities*, G. M. Greenfield (Ed.)., pp. 350–395. Westport, CT: Greenwood Press.

Meyer, M. C., and Sherman, W. L. (1995). *The course of Mexican history* 5th ed. Oxford, UK: Oxford University Press.

National Geographic Society. (1994). *Traveler's map of Mexico* (Scale, 1:4,358,000).

Pick, J. B., and Butler, E. W. (2000). *Mexico megacity*. Boulder, CO: Westview Press.

Silver Institute. (2004). Silver production. *World silver survey, 2004*. Washington, DC: Author. Available online at *http://www.silverinstitute.org/supply/production.php*. Retrieved 22 February 2005.

World Gazetteer. (2003). *Current population for cities and towns of Mexico, 2000*. Available online at *http://www.world-gazetteer.com*. Retrieved 9 December 2003.

Wright, J. B. (2001). Mormon *colonias* of Chihuahua. *Geographical Review*, 91, pp. 586–596.

Chapter 7

Bakewell, P. (1997). *A history of Latin America: Empires and sequels*. Oxford, UK: Blackwell.

Collier, S., Skidmore, T. C., and Blakemore, H. (Eds.). (1992). *The Cambridge encyclopedia of Latin America and the Caribbean* 2nd ed. Cambridge, UK: Cambridge University Press.

Gilbert, A. (1994). *The Latin American city*. London: Latin American Bureau.

Godfrey, B. (1991). Modernizing the Brazilian city. *Geographical Review*, 81 (1)., pp. 18–34.

Greenfield, G. M. (1994). *Latin American urbanization: Historical profiles of major cities*. Westport, CT: Greenwood Press.

Hardoy, J. E. (1975). Two thousand years of Latin American urbanization. In *Urbanization in Latin America: Approaches and issues*, J. E. Hardoy (Ed.)., pp. 3–55. Garden City, NY: Anchor Books.

Kent, R. B. (1994). Peru. In *Latin American urbanization: Historical profiles of major cities*, G. M. Greenfield (Ed.)., pp. 446–467. Westport, CT: Greenwood Press.

Lezama, J.L. (1994). Mexico. In *Latin American urbanization: Historical profiles of major cities*, G. M. Greenfield (Ed.)., pp. 350–395. Westport, CT: Greenwood Press.

Lombardi, C. L., Lombardi, J. V., and Stoner, K. L. (1983). *Latin American history: A teaching atlas*. Madison: University of Wisconsin Press for the Conference on Latin American History.

Morse, R. E. (Ed.). (1971). *The urban development of Latin America, 1750–1920* . Stanford, CA: Center for Latin American Studies, Stanford University.

National Geographic Society. (1980). *Visitors guide to the Aztec world: Valley of Mexico* (Scale 1:500,000). and *Mexico City* (Scale 1:50,000). Washington, DC: Author.

National Geographic Society. (1997). *Ancient Mesoamerica* (Scale 1:2,400,000). Washington, DC: Author.

Nutall, Z. (1922). Royal ordinances concerning the laying out of new towns. *Hispanic American Historical Review*, 5, pp. 249–254.

Sargent, C. S. (1994). Argentina. In G. M. Greenfield (Ed.), *Latin American urbanization: Historical profiles of major cities* (pp. 1–38). Westport, CT: Greenwood Press.

Williamson, E. (1992). *The Penguin history of Latin America* London: Penguin Books.

Chapter 8

Brawer, M. (1991). *Atlas of South America*. New York: Simon & Schuster.

Brooke, J. (1995). Colombia's rebels grow rich from banditry. *New York Times*, 2 July, pp. A1 and A6.

Collier, S., Skidmore, T. E., and Blakemore, H. (Eds.). (1992). *The Cambridge encyclopedia of Latin America and the Caribbean* 2nd ed. Cambridge, UK: Cambridge University Press.

DePalma, A. (2002). In trade issue, the pressure is on flowers. *New York Times*, 24 January, p. W1.

Executive Intelligence Review. (1995). FARC: A sort of Wallenstein's army. *Executive Intelligence Review*, November 10. Available online at *http://www.larouchepub.com/other/1995/2245_farc.html*. Retrieved 4 January 2002.

Forero, J. (2001). In the war on coca, Colombian growers simply move along. *New York Times*, 17 March, pp. A1 and A5.

Forero, J. (2002). Farmers in Peru are turning again to coca crop. *New York Times*, 14 February, p. A3.

Gubetich, H. F. (1985). *Geografia del Paraguay*. Asunción, Paraguay: Orbis S.A.C.I.

Hernández Bermejo, J. E., and León, J. (1994). *Neglected crops: 1492 from a different perspective*. Rome: Food and Agricultural Organization of the United Nations. Available online at *http://www.fao.org/docrep/T0646E/T0646E00.htm#Contents*. Retrieved 30 January 2005.

Indiana State University. (2005). The Chaco War, 1932–1935. *Latin American Studies Resources*. Available online at *http://www.latinamericanstudies.org/chaco-war.htm*. 22 Retrieved February 2005.

Lindsay, R. (1998). The Chaco War. *Inventory of conflict and environment*, Case No. 48. Available online at *http://www.american.edu/ted/ice/chaco.htm*. Retrieved 22 February 2005.

Monge, Y. (2001). La inteligencia de Colombia busca asilo en España. *El Pais* (Madrid), 24 June, p. 8.

Nash, N. C. (1991). Volcanic ash is smothering a vast area of Argentina. *New York Times*, 21 October, p. A1.

National Geographic Society. (1964). *Northwestern South America*, (Scale 1:6,652,800).

Pulgar Vidal, J. (1981). *Geografía de Peru: Las ocho regiones naturales del Peru* 8th ed.. Lima, Peru: Editorial Universo.

Rabasa, A., and Chalk, P. (2001). *Colombian labyrinth: The synergy of drugs and insurgency and its implications for regional stability*. Santa Monica, CA: Rand Corporation. Summary in "Drugs and insurgents in Colombia: A regional conundrum," *RAND Research Brief*. Available online at *http://www.rand.org/publications/RB/RB69*. 4 Retrieved January 2002.

Smith, G. H. (1963). *Physiographic diagram of South America*. Maplewood, NJ: Hammond.

Sweeney, J. P. (1999). Tread cautiously in Colombia's civil war. *Backgrounder No. 1264. Heritage Foundation*. Available online at *http://www.heritage.org/Research/LatinAmerica/images/bg1264map2.jpg*. Retrieved 18 February 2005.

Thompson, G. (2003). Behind roses' beauty, poor and ill workers. *New York Times*, 13 February, pp. A1 and A27.

U.S. Geological Survey, Volcanic Hazards Program. (1985, November 13). *Deadly lahars from Nevado del Ruiz, Colombia*. Available online at *http://volcanoes.usgs.gov/Hazards/What/Lahars/RuizLahars.html*. Retrieved 4 February 2005.

Wilson, J. (2002). Colombian rebels will not give up easily. *Financial Times* (London), 1 November, p. 3.

Wilson, S. (2002). Bogotá: An island in a land at war. *Washington Post*, 6 September, p. A16.

Chapter 9

Anonymous. (1995). Is hard to watch. Mexico: A Mennonite community battles the lure of drug money. *Newsweek*, 17 July, p. 33.

Bernhardson, W., and Massolo, M. (1992). *Argentina, Uruguay, and Paraguay: A travel survival kit*. Berkeley, CA: Lonely Planet.

Brooke, E. (1994). Latin American abortions may outpace those in U.S. *New York Times*, 12 April, p. B10.

Brooke, J. (1993). Conversations: Joao Jorge Santos Rodrigues: The new beat of black Brazil sets the pace for self-affirmation. *New York Times*, 11 April, Section 4, p. 7.

Brooke, J. (1993). The godly make Paraguay's "green hell" bloom. *New York Times*, 22 June, p. A4.

Brooke, J. (1994). A city of blonds battles migrants. *New York Times*, 4 May, p. A4.

Centro Latinamerican de Demografia (CELADE). (1981 and 1985). *Boletín Demográfico*, Nos. 27 and 36. Santiago, Chile: CELADE.

Collier, S., Skidmore, T. E., and Blakemore, H. (Eds.). (1992). *The Cambridge encyclopedia of Latin America and the Caribbean* 2nd ed. Cambridge, UK: Cambridge University Press.

Curtin, P. D. (1969). *The Atlantic slave trade: A census*. Madison: University of Wisconsin Press.

DellaPergola, S. (2002). World Jewish population 2002. In *American Jewish Year Book*, Vol. 102. New York: Jewish Agency for Israel. Available online at *http://www.jafi.org.il/education/100/concepts/demography/demjpop.html*. Retrieved 9 February 2005.

Epstein, J. (1995). Mennonites join political fray to save their land: Paraguayan group sees threats to their lifestyle and autonomy. *Christian Science Monitor*, 12 July, p. 14.

Fergusom, J. (1994). *Venezuela in focus*. New York: Monthly Review Press.

Forero, J. (2004). La Paz journal: Even the upscale wear Indian dress, but not to the office. *New York Times*, 14 October, p. A4.

Instituto Brasileiro de Geografia e Estatística. (2002). *Atlas geográfico escolar*. Rio de Janeiro, Brazil: Author.

Jordan, M. (2000). Marketers discover black Brazil. *Wall Street Journal*, 24 November, p. A11.

Kent, R. B. (2003). A diaspora of Chinese settlement in Latin America and the Caribbean. In L. J. C. Ma and C. Cartier (Eds.), *The Chinese diaspora: Space, place, mobility, and identity* (pp. 117–138). Latham, MA: Rowan & Littlefield.

Lombardi, C. L., Lombardi, J. V., and Stoner, K. L. (1983). *Latin American history: A teaching atlas*. Madison: Universiy of Wisconsin Press.

Nathan, J. (1999). An Argentine Passover, then and now. *New York Times*, 31 March, p. F11.

Population Reference Bureau. (2004). World Population Data Sheet 2004. Available online at *http://www.prb.org/pdf04/04WorldDataSheet_Eng.pdf*. Retrieved 12 June 2005.

Rohter, L. (2001). Multiracial Brazil planning quotas for blacks. *New York Times*, 2 October, p. A3.

United Nations, Department of Economic and Social Affairs. (1966). *World Population prospects as assessed in 1963*. New York: Author.

United Nations, Department of Economics and Social Affairs, Population Division. (1998). *World urbanization prospects, 1996: Estimates and projections of urban and rural populations and of urban agglomerations*, New York: Author.

United Nations, Department of Economic and Social Affairs. (2003). *World Population prospects: The 2002 revision*. New York: Author.

United Nations, Department of International, Economic, and Social Affairs. (1986). *World Population prospects: Estimates and projections as assessed in 1984*. New York: Author.

University of California, Los Angeles. Latin American Center. (2000). *Statistical Abstract of Latin America*, Vol. 36.

Chapter 10

Blakemore, H. (1983). Chile. In H. Blakemore and C. T. Smith (Eds.), *Latin America: Geographical perspectives* (pp. 457–531, 2nd ed.). London: Methuen.

Instituto Geográfico Militar (Chile). (1983). *Atlas de la república de Chile*. Santiago, Chile: Author.

Latin America News Digest. (2004). Chile foods exports seen at $7.3 billion in 2004. *Latin American News Digest*, 11 October.

Latin American News Digest. (2004). Chile wine: Exports seen over $800 million in 2004. *Latin American News Digest*, 29 November.

Loveman, B. (2001). *Chile: The legacy of Hispanic capitalism* 3rd ed. New York: Oxford University Press.

Sánchez, A., and Morales, R. (1993). *Las regiones de Chile: Espacio físico y humano-económico*. Santiago, Chile: Editorial Universitaria.

Sims, C. (1994). Making the affordable Chilean wines even finer. *New York Times*, 25 November, p. D8.

Toledo O. X., and Zapater A. E. (1989). *Geografia general y regional de Chile*. Santiago, Chile: Editorial Universitaria.

Chapter 11

Constance, P. (1999). Anatomy of a high-risk location: Why Central America is so prone to disasters. *IDB América*, 26 (7–8), pp. 20–21.

Donkin, R. A. (1977). *Spanish red: An ethnogeographical study of the cochineal and the Opuntia cactus. Transactions of the American Philosophical Society*, Vol. 67, Part 5.

Dugger, C. W. (2004). Supermarket chains crush Central American farmers. *New York Times*, 28 December, pp. A1 and A8.

Elbow, G. S. (1989). Main characteristics of present-day Central America. In R. C. West and J. P. Augelli (Eds.), *Middle America: Its lands and peoples* (pp. 365–386, 3rd ed.). Englewood Cliffs, NJ: Prentice-Hall.

Food and Agriculture Organization. (FAO). (2005). *FAOSTAT.* Available online at *http://faostat.fao.org/faostat/collections?subset=agriculture.* Retrieved 28 September 2005.

Grochembake, J. A. (2003). Central America unites against earthquakes. Available online at *http://www.tierramerica.net/2003/1025/iacentos.shtm.l.* Retrieved 11 November 2004.

Murphy, E. (2002–2003). Guatemala's false dawn: Five years of "peace." *Harvard International Review.* Available online at *http://hir.harvard.edu/articles/?id=946.* Retrieved 20 February 2005.

National Geographic Society. (1986). *Central America* (Scale 1:2,534,000).

National Geographic Society. (1989). *Land of the Maya: A traveler's map* (Scale 1:609,000).

Rohter, L. (1995). Panama changes tune to "Yankee don't go home." *New York Times*, 4 December, p. A3.

Ross, J. (1996). *Mexico: In focus.* London: Latin American Bureau.

Stahler-Sholk, R. (2001). The battle for Guatemala: Rebels, death squads, and U.S. power. *Latin American Perspectives*, 28 (5), pp. 84–86.

Thompson, G. (2004). Panama Canal thinks big, bringing fear to the jungle. *New York Times*, 27 May, p. A4.

United Nations, Population Division of the Department of Economic and Social Affairs of the United Nations Secretariat. (2004). *World population prospects: The 2002 revision*, and *World urbanization prospects: The 2003 revision.* Available online at *http://esa.un.org/unup.* Retrieved 19 November 2004.

U.S. Department of Homeland Security, U.S. Citizenship and Immigration Services, Office of Immigration Statistics. (2004). *Table 2. Immigration by region and selected country of last residence: Fiscal years 1820–2003*. Available online at *http://uscis.gov/graphics/shared/aboutus/statistics/IMM03yrbk/IMMExcel/Table02.xls.* Retrieved 6 February, 2005.

U.S. Department of Homeland Security, U.S. Citizenship and Immigration Services, Office of Immigration Statistics. (2004). *Table 3. Immigrants admitted by region and selected country of birth: Fiscal years 1820–2003*. Available online at *http://uscis.gov/graphics/shared/aboutus/statistics/IMM03yrbk/IMMExcel/Table03.xls.* Retrieved 6 February 2005.

U. S. Energy Information Administration. (2004). *Panama country analysis brief.* Available online at *http://www.eia.doe.gov/emeu/cabs/panama.html.* Retrieved 20 February 2005.

U. S. Geological Survey, Earthquake Hazards Program. (2005). *Earthquake facts and lists-~~E~~arthquakes with 1,000 or more deaths from 1900.* Available online at *http://neic.usgs.gov/neis/eqlists/eqsmajr.html.* Retrieved 6 February 2005.

Zingarelli, D., et al. (2001). *Central America on a shoestring* 4th ed. Victoria, Australia: Lonely Planet.

Chapter 12

Arias, J. (1999). El gobierno de Brasil emprende la lucha contra la esclavitud en las plantaciones. *El Pais* (Madrid), 2 April, p. 22.

Arias, J. (1999). Trabajar como animales con un arma en la sien. *El Pais* (Madrid), 2 April, p. 22.

Barraclough, S., and Domike, A. (1966). Agrarian structure in seven Latin American countries. *Land Economics*, 42 (4), pp. 391–424.

Bobrow-Strain, A. (2004). (Dis)accords: The politics of market-assisted reforms in Chiapas, Mexico. *World Development*, 32 (6), pp. 887–903.

Brockett, C. (1988). *Land, power, and poverty: Agrarian transformation and political conflict in Central America.* Boston: Unwin Hyman.

Burns, E. B. (1993). *A history of Brazil* 3rd ed. New York: Columbia University Press.

Chevalier, F. (1965). *Land and society in colonial Mexico.* Berkeleyand Los Angeles: University of California Press.

Delavaud, C. C. (1980). Agrarian reform in Peru. In *Environment, society, and rural change in Latin America*, D. A. Preston (Ed.), pp. 37–52. New York: Wiley.

Dobyns, H. (Ed.) (1971). *Peasants, power, and applied social change: Vicos as a model.* Beverly Hills, CA: Sage.

Forero, J. (2001). Venezuela preparing to give land to its peasants. *New York Times*, 15 October, p. A3.

Foster, N. (1989). Cuban agricultural productivity. In *Cuban communism* 7th ed., I. L. Horowitz (Ed.), pp. 235–255. New Brunswick, NJ: Transaction.

Gilbert, A. (1974). *Latin American development: A geographical perspective.* Harmondsworth, UK: Penguin Books.

González, M. (1965). Mexico: The lop-sided revolution. In *Obstacles to change in Latin America*, C. Veliz (Ed.), pp. 206–229. London: Oxford University Press.

Hecht, S., and Cockburn, A. (1990). *Fate of the forest: Developers, destroyers, and defenders of the Amazon.* New York: HarperCollins.

Helweg-Larsen, S. (2003). The peace of the oligarchs:

Land distribution and the Guatemalan peace process. *Canadian Journal of Development Studies*, 24 (4), pp. 617–632.

Hill, G. W., and Gollas, M. (1968). *The minifundia economy and society of the Guatemalan highland Indian*. Research Paper No. 30. Madison: Land Tenure Center, University of Wisconsin.

Kraus, C. (1998). This year in Argentina two brothers build an empire. *New York Times*, 14 April, pp. C1 and C4.

Prem, H. J. (1992). Spanish colonization and Indian property in Central Mexico. *Annals of the Association of American Geographers*, 82, pp. 444–459.

Rohter, L. (2002). Brazil's prized exports rely on slaves and scorched earth. *New York Times*, 25 March, pp. A1 and A6.

Rohter, L. (2005). Brazil promises to crack down after nun's shooting death. *New York Times*, 14 February, p. A3.

Schemo, D. J. (1998). In Brazil's tobacco country, conglomerates rule. *New York Times*, 2 April, p. 124.

Seligson, M. A. (1995). Thirty years of transformation in the agrarian structure of El Salvador, 1961–1991. *Latin American Research Review*, 30, pp. 43–74.

Sempat, C. (1992). The colonial economy: The transfer of the European system of production to New Spain and Peru. *Journal of Latin American Studies*, 24, pp. 55–68.

Simmons, C. S. (2002). The local articulation of policy conflict: Land use, environment, and Amerindian rights in eastern Amazonia. *Professional Geographer*, 54 (2), pp. 241–258.

Soustelle, J. (1964). *Daily life of the Aztecs*. Harmondsworth, UK: Penguin Books.

Stavenhagen, R. (1970). *Agrarian problems and peasant movements in Latin America*. New York: Doubleday.

Thiesenhusen, W. C. (Ed.). (1989). *Searching for agrarian reform in Latin America*. Boston: Unwin Hyman.

Todaro, M. (1985). *Economic development in the third world*. London: Longmans.

Whitbeck, R., and Williams, F. (1940). *Economic geography of South America* 3rd ed. New York: McGraw-Hill.

Whitmore, T. M., and Turner, B. L., II. (2001). *Cultivated landscapes of Middle America on the eve of conquest*. New York: Oxford University Press.

Wolf, E., and Hansen, E. C. (1972). *The human condition in Latin America*. New York: Oxford University Press.

Chapter 13

Brooke, J. (1995). Where cocoa was king, the weeds take over. *New York Times*, 25 August, p. A4.

Editora Universo. (1988). *Larouse cultural: Brasil A\Z*. São Paulo, Brazil: Editora Universo.

Magnoli, D., and Araujo, R. (1992). *A nova geografia: Estudios de geografia do Brasil*. São Paulo, Brazil: Editora Moderna.

Noble, J., et al. (2002). *Brazil* 5th ed. Melbourne, Australia: Lonely Planet.

Chapter 14

Arreola, D. D., and Curtis, J. R. (1993). *The Mexican border cities: Landscape anatomy and place personality*. Tucson: University of Arizona Press.

Baker, M. W. (1970). *Land use transition in Mexican cities*. Unpublished PhD dissertation, Syracuse University, Syracuse, NY.

Bromley, R. D. F., and Bromley, R. (1988). *South American development: A geographical introduction* 2nd ed. Cambridge, UK: Cambridge University Press.

Browning, C. E. (1990). Urban primacy in Latin America. *Yearbook, Conference of Latin Americanist Geographers*, 15, pp. 71–78.

Crowley, W. K. (1995). Order and disorder: A model of Latin American urban land use. *Association of Pacific Coast Geographers*, 57, pp. 9–31.

Crowley, W. K. (1998). Modeling the Latin American city. *Geographical Review*, 88, pp. 127–131.

Davis, K. (1969). *World urbanization 1950-70*, vol. 1. Berkeley: University of California, Berkeley, Institute for International Studies.

Eckstein, S. (1990). Urbanization revisited: Inner-city slum of hope and squatter settlement of despair. *World Development*, 18, pp. 165–181.

Ford, L. R. (1996). A new and improved model of Latin American city structure. *Geographical Review*, 86, pp. 437–440.

Ford, L. R. (1999). Latin American city models revisited. *Geographical Review*, 89, pp. 29–31.

Gilbert, A. (1994). *The Latin American city*. London: Latin American Bureau.

Gilbert, A. (1996). *The mega-city in Latin America*. Tokyo: United Nations University Press.

Griffin, E. C., and Ford, L. R. (1980). A model of Latin American city structure. *Geographical Review*, 70, pp. 397–422.

Hardoy, J. E. (Ed.). (1975). *Urbanization in Latin American: Approaches and issues*. Garden City, NY: Anchor Books.

Howell, D. C. (1989). A model of Argentine city structure. *Revista Geografica*, 109, pp. 129–139.

Inter-American Development Bank. (1997). A new future for Rio's *favelas*." *IDB Extra*, supplement to *The IDB*.

Keeling, D. J. (1996). *Buenos Aires: Global dreams, local crises*. Chichester, UK: Wiley.

Kent, R. B. (1994). Peru. In *Latin American urbanization: Historical profiles of major cities*, G. M. Greenfield (Ed.), pp. 446–467. Westport, CT: Greenwood Press.

Lezama, J. L. (1994). Mexico. In *Latin American urbanization: Historical profiles of major cities*, G. M. Greenfield, (Ed.), pp. 350–395. Westport, CT: Greenwood Press.

Mangin, W. (1967). Latin American squatter settlements: A problem and a solution. *Latin American Research Review*, 2, pp. 65–98.

Noble, J., et al. (1998). *Mexico* 6th ed. Melbourne, Australia: Lonely Planet.

Riding, A. (1984). *Distant neighbors: A portrait of the Mexicans*. New York: Vintage Books.

Santos, M. (1979). *The shared space: The two circuits of the urban economy in underdeveloped countries*. London: Methuen.

Scobie, J. (1971). *Argentina: A city and a nation* 2nd ed. New York: Oxford University Press.

United Nations, Department of International Economic and Social Affairs. (1987). *The prospects for world urbanization*. New York: Author.

United Nations, Economic and Social Development Department, Statistics Division. (2002). *Personal computers in use per 100 population and Internet users per 100 population (ITU). Target 18*. Available online at *http://unstats.un.org/unsd/mi/mi_goals.asp*. Retrieved 6 February 2005.

United Nations, Population Division, Department of Economic and Social Affairs. (2002). *World urbanization prospects: The 2001 revision*. New York: United Nations.

Violich, F. (1944). *Cities of Latin America: Housing and planning to the south*. New York: Reinhold.

Wilkie, J. W., Aleman, E., and Guadalupe, J. (Eds.). (2001). *Statistical abstract of Latin America*, Vol. 37. Los Angeles: UCLA Latin American Center Publications.

Chapter 15

Adas, M., and Adas, S. (2003). *Panorama geogrđfico do Brasil* 3rd ed. São Paulo, Brazil: Editora Moderna.

Benson, T. (2004). No streets of gold in São Paulo. *New York Times*, 2 December, p. W1.

Buechler, S. (2004). Sweating it out in the Brazilian garment industry: Korean and Bolivian immigrants and global economic forces in Sao Paulo. *Latin American Perspectives*, 31 (3)., pp. 99–119.

Cohen, R. (2000) Audis and cell phones, poverty and fear. (São Paulo) *New York Times*, 1 January, p. A28.

Instituto Brasileiro de Geografia e Estadística. (2002). *Atlas geogrđfico escolar*. Rio de Janeiro, Brazil: Author.

Rodrigues, J. A. (1977). *Atlas para estudos sociais*. Rio de Janeiro, Brazil: Ao Livro Técnico.

Smith, T. (2003). Difficult times for the coffee industry. *New York Times*, 25 November, pp. W1 and W7.

Chapter 16

Freedom House. (2000). *Map of freedom, 2000*. Available online at *http://www.freedomhouse.org/pdf_docs/research/freeworld/2000/map2000.pdf*. Retrieved 22 October 2003.

Freedom House. (2002). *Map of freedom, 2002*. Available online at *http://www.freedomhouse.org/pdf_docs/research/freeworld/2002/map2002.pdf*. Retrieved 22 October 2003.

Freedom House. (2003). *Annual freedom in the world country scores, 1972 through 2003*. Available online at *http://www.freedomhouse.org/ratings/allscore04.xls*. Retrieved 22 October 2003.

Instituto Nacional de Estadística, Geografía, e Informática (Mexico). (2001). *XII censo general de población y vivienda 2000. Base de datos. Tabulados bẚsicos. Empleo. 12. Población ocupada por municipio, sexo y sector de actividad, y su distribución segŭn ingreso por trabajo en salario mínimo*. Available online at *http://www.inegi.gob.mx/prod_serv/contenidos/espanol/bvinegi/productos/censos/poblacion/2000/definitivos/Nal/tabulados/00em12.pdf*. Retrieved 6 February 2005.

Instituto Nacional de Estadística, Geografía, e Informática (Mexico). (2003a) *Anuario de estadísticas por Entidad federativa. Cuadro 2.18. Defunciones generales por ẚmbito geogrđfico de residencia habitual del fallecido segŭn grupos de edad y sexo 1995, 2000, 2001*. Available online at *http://www.inegi.gob.mx/prod_serv/contenidos/espanol/bvinegi/productos/integracion/pais/aepef/2003/AEPEF%202003%20(parte%201).pdf*. Retrieved 6 February 2005.

Instituto Nacional de Estadística, Geografía, e Informática (Mexico). (2003b) *XII censo general de población y vivienda 2000 Estadísticas sociodemogrđficas. Educación. Resultados educativos. Población de 15 aŭos y mẚs y porcentaje de la misma que es alfabeta por entidad federativa segŭn sexo, 2000*. Available online at *http://www.inegi.gob.mx/est/contenidos/espanol/tematicos/mediano/ent.asp?t=medu16&c=3284*. Retrieved 6 February 2005.

Population Reference Bureau. (2004). *World population data sheet 2003*. Available online at *http://www.prb.org/pdf03/03WorldDataSheet_Eng.pdf*. Retrieved 6 February 2005.

Somini, S., and Rohter, L. (2003). Where faith grows, fired by Pentecostalism. *New York Times*, 14 October, pp. A1 and A10.

Transparency International. (2003). *Corruption Perceptions Index 2003, Table 1*. Available online at *http://www.transparency.org/pressreleases_archive/2003/2003.10.07.cpi.en.html*. Retrieved 22 October 2003.

United Nations. (2002). *World urbanization prospects: The 2001 revision*. New York: Author.

World Bank. (2002). *World development indicators database 2001*. Available online at *http://devdata.worldbank.org*. Retrieved 6 February 2005.

World Bank. (2003). *World development indicators database. GNI per capita 2001, Atlas method and PP.* Available online at *http://www.gotranslators.com/Engl/GNIPC.pdf*. Retrieved 6 February 2005.

Chapter 17

Bernhardson, W., and Massolo, M. (1992). *Argentina, Uruguay, and Paraguay: A travel survival kit*. Berkeley, CA: Lonely Planet.

Carlevari, I. J. F. (1993). *La Argentina 1993: Estructura económica.* Buenos Aires, Argentina: Ediciones Macchi.

Daus, F. A. (1983). *Geografia de la Argentina: Parte humana*. Buenos Aires, Argentina: Angel Estrada y CIA.

Instituto Geografico Militar (Argentina). (1983). *Atlas de la Republica de Argentina*. Buenos Aires, Argentina: Author.

Lewis, C. M. (1983). *British railways in Argentina, 1857–1914*. London: University of London Press.

Lorenzini, H. N., et al. (1993). *Geografia de la Argentina* 2nd ed. Buenos Aires, Argentina: A–Z Editora.

Mandel-Campbell, A. (1997). A stampede on the Pampas: George Soros leads a horde of investors buying Argentine land. *Business Week*, 25 August.

Nash, N. C. (1992). Old Patagonian Express puff, puff, puffs to its end. *New York Times*, 29 June, p. A4

Roccatagliata, J. A. (Ed.) (1992). *La Argentina: Geografía general y los marcos regionales* 2nd ed. Buenos Aires, Argentina: Editorial Planeta.

Rohter, L. (2002). Argentina paying heavily for squandering blessings. *New York Times*, 8 February, pp. A1 and A6.

Winthrop, W. R. (1974). *British owned railways in Argentina: Their effect on economic nationalism, 1854–1948*. London: Institute of Latin American Studies/University of Texas Press.

Zimmerman, K. (1989). Across Argentina on the Patagonian Express. *Washington Post*, 5 March, pp. E1, E8, E9, and E10.

Chapter 18

Becker, E. (2004). Latin migrants to US send billions home. *New York Times*, 18 May, p. C4.

Benson, T. (2004). China fuels Brazil's dream of being a steel power. *New York Times*, 21 May, p. W1.

Benson, T. (2004). New setback for Europe–Latin trade talks. *New York Times*, 28 October, pp. W1 and W7.

Bulmer-Thomas, V. (1994). *The economic history of Latin America since independence*. Cambridge, UK: Cambridge University Press.

Crossley, J. C. (1983). The River Plate countries. In *Latin America: Geographical perspectives* 2nd ed., H. Blakemore and C. T. Smith (Eds.), pp. 383–456. London: Methuen.

Dugger, C. W. (2003). Report finds few benefits for Mexico in NAFTA. *New York Times*, 19 November, p. A9.

EMBRAER. (2004). The company profile. Available online at *http://www.embraer.com/*. Retrieved 27 October 2004.

Ferris, S. (2003). Mexico losing jobs to China. *Plain Dealer*, 16 November, pp. G1 and G4.

Forero, J. (2003). As China gallops, Mexico sees factory jobs slip away. *New York Times*, 3 September, p. A3.

Gonzalez, D. (2002). Fried chicken takes flight, happily nesting in the U.S. *New York Times*, 20 September, p. A4.

Gwynne, R. N. (1999). Globalization, neoliberalism and economic change in South America and Mexico. In *Latin America transformed: Globalization and modernity*, R. N. Gwynne, R. N. Kay, and C. Kay (Eds.), pp. 68–97. London: Arnold.

Gwynne, R. N., et al. (2003). *Alternative capitalisms*. London: Arnold.

Inter-American Development Bank. (2004). *Sending money home: Remittances to Latin America from the U.S. 2004*. Washington, DC: Author.

Klak, T. (1999). Globalization, neoliberalism, and economic change in Central America and the Caribbean. In *Latin America transformed: Globalization and modernity*, R. N. Gwynne, R. N. Kay, and C. Kay (Eds.), pp. 98–127. London: Arnold.

Kleinpenning, J. M. G. (1995). *Peopling the purple land: A historical geography of rural Uruguay, 1500–1915*. Amsterdam, The Netherlands: Center for Latin American Research and Documentation.

Kooster, D. (2003). Regional forest transitions in highland Mexico?: The importance of local institutions in a globalized countryside. *Professional Geographer*, 55 (2), pp. 227–327.

Lombardi, C. L., et al. (1983). *Latin American history: A teaching atlas*. Madison: University of Wisconsin Press.

Lord, M. J. (Ed.). (1998). *The handbook of Latin American trade in manufactures*. Cheltenham, UK: Elgar.

Martins, M. D. (2002). Globalization and development in Brazil. *Latin American Perspectives*, 29 (6), pp. 94–99.

Murray, W. E. (1999). Natural resources, the global economy and sustainability. In *Latin America transformed: Globalization and modernity*, R. N. Gwynne, R. N. Kay, and C. Kay (Eds.), pp. 128–152. London: Arnold.

Rich, J. L. (2001). Maker of small jets sees a silver lining. *New York Times*, 24 October, pp. W1 and W7.

Stiglitz, J.E. (2004). New trade pacts betray the poorest partners. *New York Times*, 10 July, p. A17.

Thompson, G. (2001). Fallout of U.S. recession drifts south to Mexico. *New York Times*, 26 December, pp. C1 and C2.

Thompson, G. (2002). Big Mexican breadwinner: The migrant worker. *New York Times*, 25 March, p. A3.

Thompson, G. (2003). Money sent home by Mexicans is booming. *New York Times*, 28 October, p. A12.

United Nations, Comisión Económica para América Latina y el Caribe (CEPAL). (2002). *Panorama social de América Latina 2001-2002* . Cuadro 14, América Latina (18 Países): Líneas De Indigencia (LI) y de Pobreza (LP), pp. 211–212. Available online at *http://www.eclac.cl/publicaciones/DesarrolloSocial/3/LCG2183PI/Anexos_2002_Ing.pdf.* Retrieved 6 February 2005.

United Nations, Comisión Económica para América Latina y el Caribe (CEPAL). (2004). *Anuario estadístico de América Latina y el Caribe, 2003* . Cuadro 351. Producción de automotores para pasajeros. Cuadro 352. Producción de automotores comerciales, pp. 492–493. Available online at *http://www.cepal.org/publicaciones/Estadisticas/4/LCG2224PB/p2_4.pdf.* Retrieved 7 February 2005.

Weiner, T. (2003). Wal-Mart invades, and Mexico gladly surrenders. *New York Times*, 6 December, p. A9.

West, R. C., and Augelli, J. P. (1989). *Middle America: Its lands and peoples*, Englewood Cliffs, NY: Prentice-Hall.

Chapter 19

Adas, M., and Adas, S. (2002). *Panorama geográfico do Brasil: Contradições, impasses e disafios socioespaciais* 3rd ed. São Paulo, Brazil: Editora Moderna.

Browder, J. O., and Godfrey, B. J. (1990). Frontier urbanization in the Brazilian Amazon: A theoretical framework for urban transition. *Yearbook, Conference of Latin Americanist Geographers*, 16, pp. 56–66.

Browder, J. O., and Godfrey, B. J. (1997). *Rainforest cities: Urbanization, development, and globalization of the Brazilian Amazon*. New York: Columbia University Press.

Center for International Forestry Research. (2004). Beef exports fuel loss of Amazonian Forest. Available online at *http://www.cifor.cgiar.org/docs/_ref/publications/newsonline/36/beef_exports.htm*. Retrieved 10 February 2005.

Collier, S., Skidmore, T. E., and Blakemore, H. (Eds.). (1992). *The Cambridge encyclopedia of Latin America and the Caribbean* 2nd ed. Cambridge, UK: Cambridge University Press.

Correio Braziliense.(Website, Brasília, in Portuguese). (2004). Amazon deforestation causing global warming, Brazilian government says. Posted in English by British Broadcasting Corporation BBC Monitoring Latin America, 10 December 2004.

Ellis, M. (1965). The *bandeiras* in the geographical expansion of Brazil. In *The bandeirantes: The historical role of the Brazilian pathfinders*, R. M. Morse (Ed.), pp. 48–63. New York: Knopf.

Fergusom, J. (1994). *Venezuela in focus*. New York: Monthly Review Press.

Galloway, J. H. (1983). Brazil in *Latin America: Geographical perspectives* 2nd ed., H. Blakemore and C. T. Smith (Eds.), pp. 325–382. London: Methuen.

Godfrey, B. J. (1990). Boom towns of the Amazon *Geographical Review*, 80, pp. 103–117.

Greenfield, G. M. (1994). Brazil. In *Latin American urbanization: Historic profiles of major cities*, G. M. Greenfield (Ed.), pp. 62–105. Westport, CT: Greenwood Press.

Hecht, S., and Cockburn, A. (1990). *Fate of the forest: Developers, destroyers, and defenders of the Amazon*. New York: HarperCollins.

Inter-American Development Bank. (1997). Power project named "deal of the year." *The IDB*, 24 (3), p. 9.

Inter-American Development Bank. (2004). *Sending money home: Remittances to Latin America from the U.S., 2004.* Washington, DC: Author.

James, P. (1938). Changing patterns of population in São Paulo state, Brazil. *Geographical Review*, 28, pp. 353–362.

Lobb, G. G. (2002). Brazil. In *Latin America and the Caribbean: A systematic and regional survey* 4rd ed., B. W. Blouet and O. M. Blouet (Eds.), pp. 401–436. New York: Wiley.

National Geographic Society. (1992). *Amazonia: A world resource at risk* (Scale 1:10,650,000). Washington, DC: Author.

Noble, J., et al. (2002). *Brazil* 5th ed. Melbourne, Australia: Lonely Planet.

Page, J. A. (1995). *The Brazilians*. Reading, MA: Addison-Wesley.

Parker, E. P. (1989). A neglected resource in Amazonia: The Amazon caboclo. In *Resource management in Amazonia: Indigenous and folk strategies*, D. A. Posey and W. Balee (Eds.), pp. 249–259. New York: New York Botanical Garden.

Rohter, L. (1999). A mirage of Amazonian size: Delusions of economic grandeur deep in Brazil's interior. *New York Times*, 9 November, pp. C1 and C2.

Rohter, L. (2001). Amazon populist's killing exposes bitter conflicts. *New York Times*, 12 October, p. A3.

Rohter, L. (2003). Relentless foe of the Amazon jungle: Soybeans. *New York Times*, 17 September, p. A3.

Rohter, L. (2004). Marabá journal: Brazil's problem in a nutshell: Bolivia grows nuts best. *New York Times*, 26 August, p. A4.

Rohter, L. (2004). Brazilians battle Indians: "This is our land." *New York Times*, 15 October, p. A8.

Soili, H. (1985). The effects of deforestation in Amazonia. *Geographical Journal*, 151 (2), pp. 197–203.

U.S. Central Intelligence Agency. (1977). *Brazil. Natural vegetation*, Map No. 503241. Washington, DC: Author.

U.S. Central Intelligence Agency. (1977). *Brazil. Economic activity*, Map No. 503241. Washington, DC: Author

Chapter 20

Atlantic Online. (1997). Salsa consumption in the United States. Available online at *http://www.theatlantic.com/issues/97may/count.htm*. Retrieved 31 January 2005.

Benson, T. (2005). Brazil's Japanese preserve sumo and share it with others. *New York Times*, 27 January, p. A4.

Brooke, J. (2001). Sons and daughters of Japan, back from Brazil." *New York Times*, 27 November, p. A4

Chass, M. (1998). A new baseball strategy: Latin American bargains. *New York Times*, 22 March, pp. 1 and 23.

Collier, S., Skidmore, T. C., and Blakemore, H. (Eds.). (1992). *The Cambridge encyclopedia of Latin America and the Caribbean* 2nd ed. Cambridge, UK: Cambridge University Press.

Deutschmann, D., and Ricardo, R. (2004). *Guantánamo: A critical history of the U.S. base in Cuba* 2nd ed. Melbourne, Australia: Ocean Press.

Duran, J., and Massey, D.J. (1992). Mexican migration to the United States: A critical review. *Latin American Research Review*, 27, pp. 3–42.

Fox, G. (1996). *Hispanic nation: Culture, politics, and the constructing of identity*. Tucson: University of Arizona Press.

Goldberg, C. (1997). Hispanic households struggle as poorest of the poor in the US. *New York Times*, 30 January, pp. A1 and A12.

Holmes, S. A. (1995). Census report finds Hispanic Americans lagging further in college degrees. *New York Times*, 27 July, p. A10.

Holmes, S. A. (1996). Census sees a profound ethnic shift in US. *New York Times*, 14 March, pp. A8 and A14.

Holmes, S. A. (1998). Hispanic births in the US reach record high: Higher fertility rate of Mexicans is cited. *New York Times*, 13 February, p. A12.

Intituto Nacional de Estadística, Censos de Poblacíon y Viviendas (Spain). (2001). *Censos de población y viviendas, 2001*. Available online at *http://www.ine.es/censo2001/censo2001.htm*. Retrieved 15 June 2005.

Instituto Nacional de Estadística, Censos de Población y Viviendas (Spain). (2003). *Revisión del padrón mu - nicipal 2003. Datos a nivel nacional, comunidad autónoma y provincia. Población por país de nacimiento, nacionalidad y sexo*. Available online at *http://www.ine.es/censo2001/censo2001.htm*. Retrieved 6 February 2005.

Jokisch, B. D. (2001). Desde Nueva York a Madrid: Tendencias en la migración ecuatoriana. *Ecuador Debate*, 54, pp. 59–84.

Krauss, C. (2000). With no hope for economy, many Argentines are leaving. *New York Times*, 24 November, p. A3.

Major League Baseball. (2004). 27.3 percent of major league baseball players born outside the United States. *MLB Press Release*, 8 April 8. Available online at *http://mlb.mlb.com/NASApp/mlb/mlb/news/mlb_press_release.jsp?ymd=20040408&content_id=697610&vkey=pr_mlb&fext=.jsp*. Retrieved 21 February 2005.

Meyer, M. C., and Sherman, W. L. (1995). *The course of Mexican history* 5th ed. Oxford, UK: Oxford University Press.

Miyares, I. M., and Gowen, K. J. (1998). Recreating borders?: The geography of Latin Americans in New York City. *Yearbook, Conference of Latin Americanist Geographers*, 24, pp. 31–44.

National Association of Hispanic Publications. (1993). *Hispanics-Latinos: Diverse people in a multicultural society* 2nd ed. Washington, DC: National Association of Hispanic Publications/U.S. Census Bureau.

Navarro, M. (2003). Ads reach out to bicultural Hispanics. *Plain Dealer* (Cleveland), 27 May, p. C2.

Novas, H. (1994). *Everything you need to know about Latino history*. New York: Plume Books.

Portes, A., and Grosfoguel, R. (1994). Caribbean diasporas: Migration and ethnic communities. *Annals of the American Academy of Political and Social Science*, 533, pp. 48–69.

Ramirez, R. R., and de la Cruz, G. P. (2002). The Hispanic population in the United States: March 2002. *Current Population Reports, P20-545*. Washington, DC: U.S. Census Bureau.

Romero, S. (1999). A Japanese exodus in reverse: Brazilians work their way back to the ancestral home. *New York Times*, 16 October, pp. B1 and B4.

Skidmore, T. E., and Smith, P. H. (1992). *Modern Latin America* 3rd ed. Oxford, UK: Oxford University Press.

Swarns, R. L. (2004). Children of Hispanic immigrants continue to favor English, study of census finds. *New York Times*, 8 December, p. A18.

U.S. Census Bureau. (1993). *Hispanic Americans today*. Washington, DC: Government Printing Office.

U.S. Census Bureau. (1998). *Profile of the foreign-born population in the United States, 1997: Detailed tables for P23-195*. Available online at *http://www.census.gov/population/www/socdemo/foreign/foreign98*. Retrieved 29 January 2005.

U.S. Census Bureau. (1999). *Table 10.2. Householders in family households by total money income in 1998, family type, Hispanic origin and race: March 1999*. Available online at *http://www.census.gov/population/socdemo/hispanic/cps99/tab10-2.txt* . Retrieved 29 January 2005.

U.S. Census Bureau. (2001a) *Census of population 2000 QT-P9. Hispanic or Latino by type: 2000 data set: Census 2000 summary file 1 (SF 1) 100-percent data* Geographic Area: United States. Available online at *http://factfinder.census.gov/servlet/QTTable?_bm=y&-geo_id=D&-qr_name=DEC_2000_SF1_U_QTP9&-ds_name=D&-_lang=en*. Retrieved 29 January 2005.

U.S. Census Bureau, Current Population Survey, Ethnic and Hispanic Statistics Branch, Population Division. (2001b) *The Hispanic population in the United States: March 2001 revised detailed tables—Weighted to Census 2000 (PPL-172)*. Available online at *http://www.census.gov/population/www/socdemo/hispanic/ppl-172.html*. Retrieved 21 February 2005.

U.S. Census Bureau. (2001c) *Census 2000 summary file 1, Matrix P8. GCT-P6. Race and Hispanic or Latino: 2000. Data set: Census 2000 summary file 1 (SF 1) 100-percent data geographic area: United States—State; and Puerto Rico.* Available online at *http://factfinder.census.gov/servlet/GCTTable?_bm=y&-geo_id=01000US&-_box_head_nbr=GCT- P6&-ds_name=DEC_2000_SF1_U&-_lang=en&-format=US-9&-_sse=on*. Retrieved 29 January 2005.

U.S. Census Bureau. (2004a). 2003 County Business Patterns. Available online at *http://censtats.census.gov/cgi-bin/cbpnaic/cbpcmp.pl.* Retrieved 28 September 2005.

U.S. Census Bureau. (2004). *Income of households by race and Hispanic origin using 2- and 3-year averages: 2000-2002. Median income and confidence intervals (C.I) in 2002 dollars*. Available online at *http://www.census.gov/hhes/income/income02/3yr_avg_race.html*. Retrieved 22 February 2005.

U.S. Census Bureau, Ethnicity and Ancestry Branch. (2003). *U.S. Hispanic population, 2002*. Available online at *http://www.census.gov/population/socdemo/hispanic/ppl-165/slideshow/*. Retrieved 21 February 2005.

U.S. Citizenship and Immigration Services. (2005). Available online at *http://uscis.gov/graphics/shared/aboutus/statistics/IMM03yrbk/IMMExcel?Table02.xls*. Retrieved 21 February 2005.

U.S. Immigration and Naturalization Service. (1994). *Statistical yearbook of the Immigration and Naturalization Service*. Washington, DC: Government Printing Office.

Index

Abortion, 167
Acapulco, 323
African Brazilians, 20, 85, 154, 170, 241, 242, 243, 342
African cultural influences, 20
African immigration, 154, 156
African slavery, 18, 19, 59, 60
Agave, 102, 103
Agglomeration economies, 101
Agrarian reform. *See also* Land reform
  Mexico, 99
  Latin America, 227
Agrarian transformation, 231–234
Agreste, 287, 290
Agricultural colonization, 352, 355
Agriculture
  agroforestry, 73
  Bajio, 100
  banana plantations, 82, 155, 195, 196, 203, 215, 216, 326
  Central Valley, 182
  Guaraní, 273
  irrigation, 71–73, 76, 91, 93–95, 100–101, 109, 116–117, 181, 223, 229, 288, 319
  Maya, 74
  Mesa Central, 96, 101
  Mesa del Norte, 109
  permanent field, 68
  shifting cultivation (slash and burn agriculture), 67, 68, 71, 206
  subsistence crops, 18, 47, 51, 69, 93–94, 151–152, 181, 183, 193, 197, 217, 289, 341
  terracing, 70, 72, 74, 77, 94, 136
Agro–export economy, 128
Agrovilas, 355
Aircraft, 332
Algorrobo, 151
Allende, Salvador, 183
Alliance for Progress, 225–226, 227
Alpaca, 70, 135
Altiplano, 76, 88, 133, 137
Altitudinal zonation, 44–48, 134
Amazon Basin
  climate, 349–350
  deforestation, 356
  economic development
    19th century, 350–352
    20th century, 352–356
  highway transportation, 355
  large-scale development projects, 357–359
  physiography, 348–350
  population change, 359
  spontaneous settlement, 356–357
  urbanization, 359–360
Amazon River, 87, 88
Amerindians, 15, 75, 136, 359, 362; *see also* Indigenous populations
Andean Community (CAN), 334, 341
Andean Group, 334
Andes, 41, 75, 79, 80
  climate, 133
  coffee, 138, 139
  geology, 131
  indigenous population, 136
  mining, 138, 139
  physiography, 131–133
  plant and animal domestication, 70, 135
  population 20th century, 143
  transportation network, 140
Angel Falls, 361
Antigua (Guatemala), 193
Antioqueños , 33
Antioquia (Colombia), 139
Aquaculture, 46
Araucanian, Indians, 16, 52, 68, 182, 312, 319; *see also* Mapuche
Araucaria (trees), 274–275
Arawak, 16, 58

Arbenz, Jacobo, 196, 225, 375
Argentina, 156, 337, 338
Armero (Colombia), 131
Aspect, 42, 44
Asunción, 151
Atacama Desert, 328
Atahualpa, 76, 77, 136
Audiencia, 125
Automobile, 29, 103, 251, 259, 316, 331, 333, 340
Aymara, 15, 136–137
Aztec, 6, 16, 17, 18, 51, 77, 78, 81, 89, 91, 95, 108, 111, 113, 114, 116, 119, 206, 207, 208
  land tenure, 207, 208

B

Bahia, 123, 124; *see also* Salvador
Bajio, 100, 101
Bañados , 150
Bandierantes, 87, 275–276, 359
Barrio de emergencia, 253, 316
Baseball, major league, 387–388
Batista, Fulgencio, 61
Bauxite, 148, 363
Bay Islands, 84
Beans, 69, 94
Belém, 123–124, 360
Belize, 10
Belo Horizonte, 269, 284
Beni (Bolivia), 293
Birth control, 167, 168
Black Carib, 58, 84, 203; *see also* Garifuna
Blacks, English-speaking, 155, 189, 203
Bluefields (Nicaragua), 203
Blumenau, 23, 157, 274
Boa Vista, 269, 363
Bogotá, 120, 125, 137, 143
Bolas, 310
Bolivia, 140
  land reform, 224
Bolsones (playas), 106
Boom–bust economic cycles, 326–328
BOVESPA, 280
Bracero Program, 378
Brazil's Atlantic Coastal Plain-climate, 236–237
Brazil's Atlantic Coastal Plain-population, 241–242
Brazil's Atlantic Coastal Plain-settlement, 237–238
Brazil
  immigration to
  19th century, 279
  independence, 240
  land ownership, 219
  military and development, 352, 354–355
  modern slavery, 233
  plant and animal domestication, 70
  race, 171
  racial distribution, 174
Brasília, 268–270, 280–285, 291
Brazilian dance, 20
Brazilian Highlands, 41, 271
  geology, 271
  physiography, 271–272
Brazil nuts, 351, 353
Brazilwood, 237
British, immigration, 158
Buenos Aires, 88, 120, 125, 128, 157, 310, 311, 314–316, 323

C

Caatinga, 286–287
Caballero, 211
Cabildo, 121
Cacao, 50, 70, 203, 240, 328
CAFTA (Central American Free Trade Agreement), 341
Cajamarca (Peru), 76, 78, 136, 263
Cali, 144
Callampas, 184
Campesino, 18; *see also Peon*
Campiña Grande, 290
Campo cerrado, 282
Cancún, 26, 100, 269
Candomblé, 20
Canudos, 288–289
Canudos-War, 289
Captaincy General of Guatemala, 12
Capulli, 207, 208
Caracas (Venezuela), 125, 135, 144
Carib, 58, 80
Caribbean, 79, 85
Caribbean Lowlands of Central America, 201
  climate, 201
  physiography, 201–202
  settlement, 202–204
Caribbean National Forest, 57
Carioca, 14, 34, 242
Carnauba wax, 289
Carnival *(Carnaval)*, 20
Casas, Bartholome de las, 208
Castro, Fidel, 61, 380
Catholic Church, 126, 167, 211, 293–294
Catholicism, as a syncretic religion, 20
Cattle raising. *See* Livestock raising
Caudillos, 126, 127

Central America, 12, 79, 85, 96
  immigration, 155
  poverty rates, 198
  rural life, 197–198
Central America, United Provinces of, 125
Central American Common Market (CACM), 334
Central American Highlands, 187–201
  climate, 190–192
  coffee, 191, 194–195, 196, 197
  colonial settlement, 194
  manufacturing, 201
  physiography, 187, 188, 190
  population growth, 196
  pre-Columbian settlement, 193
  volcanism, 190
Central Americans, immigration to U.S., 200–201, 381
Central Andes, urban development, 115
Central Intelligence Agency (CIA), 225
Central Plateau (Brazil), 271, 281
  climate, 280–281
  mining, 282–283, 284
  population, 284
  20th-century economy, 283–284
  vegetation, 282
Central Valley (Chile), 40, 88, 179–185, 336
  agriculture, 182
  climate, 179
  manufacturing, 184
  settlement, 181
  vegetation, 180
Centrales, 214
Cerrado, 275, 282, 350
Cerro de Pasco, 139
Chaco, 149
  climate, 149
  contemporary population, 152
  geology, 149
  vegetation, 150
Chalco, Lake, 93
Chan Chan, 76, 116, 117
Chapala, Lake, 104
Chaqueños , 33
Chavez, Hugo, 228
Chavin culture, 76
Chibcha, 68, 69, 136
Chicago, 382
Chicano/a, 372
Chichimecs, 75, 108
Chico Mendes, 353
Chiefdoms, 68, 207
Chihuahua Desert, 107
Chilango, 172
Chile peppers, 70, 94
Chile, 336, 338
Chimborazo, 133
Chimú culture, 76, 116
Chinampa, 73, 75, 93, 115, 116
Chinatown, 24, 159
Chinese, 22, 24, 61, 158, 214
  immigration, 158
    Cuba, 158, 159
    Mexico, 159
    Peru, 158
Chocolate, 50; *see* Cacao
Christianity, 293–294
Christians
  Catholic; *see* Catholic Church
  Evangelical, 8, 294, 295
  population worldwide, 294
Cinchona, 351
Cities
  Brazil, upper city and lower city, 123
  ports, 120, 128
    Spanish colonial, 120
Ciudad Bolivar, 148
Ciudad Guayana, 148, 269
Civil liberties, 297, 299, 301
Civil war and political violence
  Colombia, 141–142, 143
  El Salvador, 200
  Guatemala, 199–200
  Mexico, 98–99
  Nicaragua, 198–199
  Peru, 142
Civilizations, advanced societies, 69–77, 207
Classic period, Mesoamerica, 114
Clientelism, 9
Climate
  Amazon Basin, 349–350
  Andes, 133
  Atlantic Coastal Plain of Brazil, 236–237
  Caribbean Lowlands of Central America, 201
  Central American Highlands, 190–192
  Central Plateau (Brazil), 280–281
  Central Valley, 179
  Chaco, 149
  Guiana Highlands, 362
  Hispanic Caribbean, 57–58
  Llanos, 146
  Mesa Central, 93–94
  Mesa del Norte, 107
  Pampa, 309
  Paraná Plateau, 273
  Patagonia, 318
  Sertão, 286
  subtropical, 37, 39
  temperate mid-latitude, 40

tropical, 37, 38
tropical wet and dry, 39
Coahuila Desert, 107
Coal, 110, 320
Coca, 32, 46, 54, 70, 71, 136, 141, 153
Cocaine, 46, 71, 136, 141, 144, 153
Cochineal, 70, 194, 324
Coffee plantations, 215
Coffee, 138, 139, 328, 368
Andes, 138, 139
Central American Highlands, 191, 194–195, 196, 197
Paraná Plateau, 276–278
Colca Valley, 77
Colombia, 139, 141
Colonia (Uruguay), 324
Colonial economy, 322
Colonial period, 292–295
Colonial ports, 323
Colonial settlement and economy, Hispanic Caribbean, 58–61
Colonial taxes, 325
Colonias, 103
Columbian Exchange, 49, 50, 81, 82
Compadrazco, 9
Computer, personal, 280
Concepción (Chile), 184, 337
Conselheiro, Antonio, 288–289
Conservatives, 19th century, 126
Contemporary conditions, Guiana Highlands, 362–363
Contras, 199
Conventillos, 251, 316
Convoy system, 323
Copán, 74, 114, 115
Coparenthood, 9
Copper, 138–139, 182, 327, 329, 336, 358
Córdoba (Argentina), 317
Corn, 18, 19, 50, 51, 69–70, 94, 101
Corruption perception index (CPI), 297, 299, 301
Cortés Hernán, 77, 78, 95
Costeño , 14, 33, 172
Cotton, 152, 240
Creoles, 86
Criollo cattle, 151
Crops, 48
monoculture, 60
Cuba, 55, 374
land reform, 225
race, 177
Cuban-Americans, 64, 367, 369–375, 379–380, 382, 385
Cuban immigration to U.S., 379–380
Cuban Revolution, 62, 63
Cubans, 372, 373
Cuidad Bolivar (Venezuela), 147
Cuidad Lázaro Cárdenas (Mexico), 269
Cultural influences
African, 20
Latin Americans in the United States, 366–367
mass media, 27, 28
Mesa del Norte, 110
non-Iberian European, 21
transnational, 27
Cultural regions, 32
Culture
Chimú, 116
Huari, 115
Maya, 15, 16, 74
Moche, 115, 116
Tiahuanaco, 76, 115
Curitiba, 280
Cut flowers, non-traditional exports, 47, 183
Cutervo (Cajamarca), 217
Cuzco, 76, 77, 116, 118, 124, 138

Defeño , 172
Deforestation, 49, 57, 94, 107, 151, 157, 165, 192, 229, 232, 275, 279, 287, 351, 353
Amazon Basin, 356
Democracy, 392–393
Demographic transition model, 166, 168
Dependent development, 128
Development indicators
Latin America, 301,302, 304
Western countries, 303
World Regions, 299
Diamantia, 283
Diamonds, 86, 87, 123, 124, 276, 363
Díaz, Porfirio, 97, 108, 126, 222
Disease, 16, 50, 53, 78–81, 96, 177, 189, 194, 209, 215, 239, 247, 273, 319, 359
Domestication-plants and animals, 50–52, 69–71, 94, 102, 135–136, 194
Dominican Republic, 27, 55, 64, 375
immigration to Puerto Rico, 164
race, 177
Donatarios, 238
Drug trafficking, 141, 144
Dual citizenship, 27, 65
Dualism, economic and social, 140, 141

E

Earthquakes, 92, 192–193
Economic reforms, neoliberal, 31, 101, 142, 183, 185, 224, 232, 296, 336, 337–338, 393–394

Economy
  20th century, 329–338
  Amazon Basin
    19th century, 350–352
    20th century, 352–356
  Central Plateau (Brazil), 20th century, 283–284
  colonial, 322
  export oriented, 328–329
  formal sector, 262–265
  Hispanic Caribbean, 20th century, 62–65
  informal sector, 262–265
  Pampa, 20th century, 314
  Paraná Plateau, 20th century, 278–281
  Sertão
    19th century, 288
    20th century, 289
Ejidatarios, 223, 224
Ejido, 99, 101, 223
El Niño, 40,041, 286
El Salvador, 27, 346, 375
  civil war and political violence, 200
  land reform, 227–228
ELN (*Ejercitio de Liberación Nacional*), 141
EMBRAER, 332
Encomienda, 120, 136, 194, 198
Engenho, 239
English-speaking Blacks, 155, 189, 201, 203
ENSO (El Niño Southern Oscillation), 286
Environmental change, 49
Environmental lapse rate, 134
Estancia, 210, 312, 319
Esteros, 150
Ethnicity, country profiles, 20th century, 170
European conquest
  disease, 78
  Inca, 77
European immigration. *See* Immigration
European Union, 338, 393
Export commodity concentration, 327
Export economy
  Pampa, 19th century, 313–314
  wool, 140
Export-oriented economy, 139, 183, 328–329
Exports, non-traditional, 141, 197
Extractive forest reserves, 353

F

FARC (Fuerzas Armadas Revolucionarias de Colombia), 141
Farm sizes, Peru, 219
Fast food
  in Latin America, 30–32, 344, 345
  Latin American cuisine in U.S., 346
Favelas, 242, 243 257
Fazenda, 210, 233, 277
Feijada, 20
Feira de Santana, 290
Fertility rates, Hispanics in U.S., 385
Film industries, 28
Films, on Latin America
  Romero, 200
  A Man, A Plan, A Canal, 189
  Boca del Lobo, 142
  Central Station, 288
  El Norte, 381
  Fitzcarraldo, 351
  Papi, 370
  Rodrigo D: No Futuro, 144
  The Burning Season, 353
  The Mission, 276
  Xica, 283
Financial remittances, 64, 65, 343, 344, 392
Flagelados, 288
Fleet system, 323, 325
FMLN (*Frente Farabundo Martí de Liberación Nacional*) (El Salvador), 200
Food
  dining customs, 32
  fast food, 31, 32, 344, 345, 346
  Mesoamerica, 69
Foraker Act, 61
Foreign investment, 126
Forest products, 363
Formal sector, 262–265
Fortaleza, 290
Fox, Vicente, 110
Franchise businesses, 30
Free trade, 334
Freedom House, 297
Freedom, 297, 299
Fresnillo, 110
Fruit, 46
  deciduous, 182, 183, 191, 320, 336
  trees, 47
FSLN (*Frente Sandinista de Liberación Nacional*) (Nicaragua), 198–199
FTAA (Free Trade Area of the Americas), 340, 393
Fundos, 181, 210

G

Garifuna, 203; *see also* Black Caribs
Garimpeiros, 359
Gauchos, 52, 53, 310, 311
GDP (Gross Domestic Product), 295
Gente de razón , 176

Gente decente, 176
Geology
  Andes, 131
  Brazilian plateau, 271
  Chaco, 149
  karst, 56, 57
  immigration, 21, 23, 157, 278
  language, 162
Globalization, 342–345, 390–391
GNI (Gross National Income), 295, 296, 297
  PPP (Purchasing Power Parity), 296–297, 298, 299
Goiânia, 269, 285–286
Gold, 86, 87, 123, 124, 282, 328, 363
Golondrinas, 156, 313–314
Gran Carajás (Brazil), 358
Gran Chaco, 149; *see also* Chaco
Gran Colombia (Republic of), 125
Gran Sabana, 362
Greater Antillies, 69, 81, 82
Growth poles, 358
Guadalajara, 91, 93, 96, 98, 99, 103, 104, 120, 124
Guano, 328
Guantánamo Bay, 61, 374
Guaraní, 15, 276, 312
Guatemala City, 125, 193
  climate, 192
Guatemala, 346, 375
  civil war and political violence, 199–200
  land reform, 225
Guayaquil, 88, 120
Guerrilla war, 141, 142
Guevara, Ernesto Che, 64, 141
Guiana Highlands, 41
  climate, 362
  contemporary conditions, 362–363
  physiography, 361–362
Guianas, 10
Guinea pig, 52, 70, 135
Guyana, 10

Hacendados, 127, 212
Hacienda, 52, 53, 99, 101, 127, 138, 183, 211, 212, 213, 222, 223
Hatos, 146
Havana, 65, 128, 249, 253
Hegemony, 126, 393–394
  U.S. in Central America, 195–196
Henequen, 102, 214, 215
Hidalgo, Miguel, 96
Hispanic America, race, 176
Hispanic Caribbean
  climate, 57–58
  colonial settlement and economy, 58–61
  economy and social conditions 20th century, 62–65
  physiography, 55–56
  U.S. military intervention, 61–62
Hispanic, definition in U.S., 370–372
Hispanics
  distribution in the U.S., 380–384
  fertility rates, 385
  in major league baseball, 387–388
  labor force characteristics, 385–386
  population and future trends in U.S., 384–387
  population history in U.S., 375–380
  poverty, 385
Hispaniola, 55, 56
Hispanos, 372, 381
Honduras, land reform, 228
Honeybees, African, 53
Huari culture, 76, 115
Huáscar, 77
Huascarán, 48, 133
Hunting and gathering, 67, 68, 151, 193, 309–310, 318, 362
Hurricanes, 58–60
  Mitch, 59, 60
Hydroelectric power, 279, 284, 358

I

Iberoamerica, 11
Iguaçu, Falls, 272
IMF, 336
Immigration
  20th century, 160
    Venezuela, 161
  African, 154; *see also* Slavery
  Brazil, 19th century, 279
  Brazilians to Japan, 365–366, 368–369
  British, 158
  Central America, 155
  Central American to U.S., 200–201, 381
  Chinese, 22, 24, 61
  Cuban to the U.S., 379–380
  Dominican Republic, 64
  Dominican to Puerto Rico, 164
  European, 21, 86, 155
  German, 23–24, 278
    to Brazil, 157
    to Chile, 157
  illegal Mexican to U.S., 385
  Italian, 155, 156, 278
  Japanese, 22
  Jewish, 25, 157, 158
  Latin Americans to Europe, 365
  Latin Americans to Spain, 365, 366, 367
  Latin Americans to U.S., 365

Immigration (*continued*)
Levantine, 24
Mennonite, 25, 161
Mexican to the U.S., 376–378
Middle Eastern, 158
Portuguese, 86, 157, 283
Spanish, 86, 156, 157
transnational, 26, 27
Venezuela in the 20th century, 161
Welsh, 157
Import substitution industrialization, 331–333
Inca, 69, 76, 136, 137
cities, 116, 117
construction, 77, 77
European Conquest, 77
planned cities, 117
society, 117
INCRA (National Institute for Colonization and Agrarian Reform) (Brazil), 355
Indentured servants, 60
Indigo, 194, 214, 324
Industrial Revolution, 139, 140, 351
Industry. *See* Manufacturing
Infant mortality rate, 298, 301–303, 304
Inflation, 335, 338
Informal sector, 251, 262–265
Iquitos (Peru), 254, 328, 352, 360, 361
Iron ore, 110, 148, 284, 358, 363
Irrigation. See Agriculture, irrigation
Itaipu, Dam, 272
Italians, 21, 51, 155, 156, 157, 161, 178, 215, 278, 279, 314, 365, 392
immigration, 155, 278
ITCZ (Inter-Tropical Convergence Zone), 38, 39, 57, 134, 192

J

Japanese, 22
Brazilians, 368–369
immigration
Brazil, 160
Peru, 159
Peruvians, 159
Jari Project, 357–358
Jesuit missions, 273, 276
Jewish, immigration, 157, 158
Jews, 25
Jones Act, 61, 374
Juárez, Benito, 97, 222
Junin (Peru), 221
Jurquiza, Justo José de, 126

K

Karst, geology, 56, 57

L

La Paz (Bolivia), 120, 135, 136, 145, 146
La Raza, 372
Ladino, 176
Ladinos, 194
LAFTA (Latin American Free Trade Association), 340
Lahars, 131
Land alienation, Amerindian, 209–210
Land bridge, Bering Straits, 67
Land colonization, 229–231
Amazon Basin, 232
directed, 230
semi-directed, 230–231
spontaneous, 229–230
tropical, 231
Land invasion, 253–255
Land reform, 183, 219, 222, 225–229
Bolivia, 224
Catholic Church, 228
Cuba, 225
El Salvador, 227–228
Guatemala, 225
Honduras, 228
Mexico, 99
Nicaragua, 228
Venezuela, 228
Land tenure, 206
Aztec, 207
El Salvador, 220
European, 207
Landless Workers Movement (Brazil), 228–229
Landslides, 131, 132
Language, 8, 10
Aymara, 15, 16
Brazilian Portuguese, 14
Maya, 15, 16
Nahuatl, 16, 18
Quechua, 15, 16
Spanish, 8, 13–15, 18, 59
Tupi-Guaraní family, 15, 16
Lapse rate, environmental, 44, 45
Latifundia, 210, 211, 220, 221, 224
Latin America, films on. *See* Films, on Latin America
Latin American origins in U.S., 367–370, 372–375
Latin American terminology in U.S., 371–372
Latino/a, 371, 372

Latitudinal variation, 37
Legal systems, 8
Lesser Antilles, 69, 81, 84
Liberals, 19th century, 126
Life expectancy, 298, 303, 304
Lima, 87, 120, 124, 249, 255, 257, 323
Limón, 203
Line of Tordesillas, 276
Literacy rates, 298, 303, 305
Livestock raising, 46, 48, 52, 56, 59, 106, 108, 109, 220, 275, 282, 287, 289, 313, 319, 355, 356, 357
  hacienda, 211
  sheep, 311–312
  Mesa del Norte, 108
Llama, 70, 135
Llaneros, 33
Llanos, 145
  climate, 146
  contemporary population, 148
Local government, 294–295
Lunfardo, 21

Machismo, 9
Machu Picchu, 79
Macumba, 20
Mameluco, 275, 289
Managua, 193
Manaus, 328, 350, 351, 360
Mangrove, 46, 237
Manioc (*yuca*, cassava), 50, 71
Manufacturing, 182, 273, 279, 284, 316, 325, 329, 330, 332, 341, 360, 379
  Central American Highlands, 201
  Central Valley, 184
  iron and steel, 185
  Mesa Central, 100, 101, 102, 105, 106
  Mesa del Norte, 109, 110
  Mexico, 102, 104
  Paraná Plateau, 279
Mapuche, 16, 317; *see also* Araucanian
Maquiladoras, 100, 340–341
Marianismo, 9
Marriage, 9
Marron, 84
Mass media, 27, 28
Mata Atlantica, 237, 238, 239, 275
Matorral, 180
Maximilian I, 97
Maya, 17, 49, 73, 74, 75, 113, 114, 115, 178, 187, 193, 194, 197, 200, 204, 206, 303
  calendars, 73
  ceremonial centers, 73
  culture, 15, 16, 73, 74
  language, 15, 16
  writing, glyphs, 73
Medellín, 144
Mennonites, 110, 161, 162
  Bolivia, 25, 161, 163
  Mesa del Norte, 163
  Mexico, 25, 161, 163
  Paraguay, 25, 161, 162
Mercantile system (mercantilism), 59, 322, 326
Mercedes, 181, 208
MERCOSUR (*Mercado Común del Sur*), 338, 340, 341, 393, 394
Mérida, 215
Mesa Central (Mexico), 91–112, 333, 341
  agriculture, 101
  climate, 93
  climate, 93–94
  manufacturing, 100, 101, 102, 105, 106
  physiography, 91–93
  vegetation, 94
Mesa del Norte (Mexico), 75, 93, 97, 98, 100, 106
  agriculture, 109
  climate, 107
  cultural differences, 110
  manufacturing, 109, 110
Mescal, 102
Meseta Central, Costa Rica, 215
Mesoamerica, 17, 69, 73, 79, 80, 194, 207
  Classic Period, 72
  crop triad, 69
  plant and animal domestication, 70
  post-Classic, 74
  pre-Classic, 71
Mestizo, 15, 83, 87–89, 99–100, 122, 147, 170–172, 176, 194, 211, 216, 275
Mexican-American War, 96, 367, 372, 375
Mexican Americans, 367, 369, 371–373, 376–378, 380–382, 385–386
Mexican Constitution, 222, 224
Mexican Flota, 323
Mexican illegal immigration to U.S., 385
Mexican immigration to U.S., 376–378
Mexican Plateau, 91, 92
  geology, 92
Mexican Revolution, 98, 99, 222, 377
Mexico
  colonial transportation network, 96
  industry, 102, 104
  infant mortality rates, 305
  land reform, 99
  literacy rates, 306
  Mesa Central, 91–106, 333, 341
  Mesa del Norte, 93, 98, 106–110
  railroads, 97, 98

Mexico (*continued*)
tourism, 25–26, 33–34, 95, 100, 105, 269
urban hierarchy, 104, 105
Mexico City, 87, 93, 95, 96, 98, 101,103, 120, 124, 128, 249, 255, 372
air pollution, 102
climate, 93
pollution, 102, 103, 104
primate city, 260
public services, 103
shantytowns, 103
transportation, 103
Mexico, Valley of, 72, 73, 74, 114, 208, 345
urban development, 114
Middle America, 13
Middle Eastern, immigration, 158
Migrant labor, 216
Migration, 391–392
first Americans, Amerindians, 67
Puerto Rican to U.S. mainland, 61, 378–379
regional patterns, 163
Argentina, 163
Puerto Rico, 164
Venezuela, 164
Military intervention, 388; *see also* U.S. military intervention
Military strongmen, 127
Minas Gerais, 283
Minifundia, 197, 216–221, 225, 355, 357
living conditions in Guatemala, 217–218
Mining, 100, 109–110, 123, 182, 183,194, 276, 282, 328, 358, 363
Andes, 138, 139
Central Plateau (Brazil), 282–283, 284
copper, 139
Guiana Highlands, 148
iron ore, 109
mercury, 138
Mesa del Norte, 109
nitrates, 328
silver, 108, 138
tin, 139, 140
Miscegenation, 82, 84, 85
Miskito Coast, 202
Miskito Indians, 202
Mission, settlement, 293; *see also Reducciones*
Mita, 136, 208
Mixtecs, 17
Moche culture, 76, 115, 116
Model of the Latin America city structure, 266–268
Monclova, 109
Monroe Doctrine, 195
Monte Albán, 73, 74
Monterrey, 100, 108, 109
Montevideo, 120, 128, 314, 316
Montezuma, 78
Mormons, 110
Mulattos, 85, 170–172, 176
Multinational corporations, 232, 331, 346
Muscovy ducks, 70

N

NAFTA (North American Free Trade Association), 10, 224, 340–341, 393, 394
Nahuatl, 16
Napoleonic Code, 8
Natural gas, 320
Nazca Plate, 130
Neocolonialism, 325, 326, 330, 346
Neo-liberal economic reforms, 31, 101, 142, 183, 185, 224, 232, 296, 336, 337–338, 393–394
Neo-volcanic Transverse Range, 92
Neuquén (Argentina), 318, 320
Nevado de Ruiz, 131
New Laws of the Indies, 122, 208
New York City, 379, 382
Nicaragua, 195, 375
civil war and political violence, 198–199
land reform, 228
Nicaragua, Lake, 188
Nicoya Peninsula, 17, 19, 26, 193
Nitrates, 328
Non-traditional crop exports, 197
Noriega, Manuel, 196
Northern Perimeter Highway (Brazil), 355

O

Oaxaca, 103
Oaxaca, Valley of, 71, 72, 209
Olmec, 72
Operation Bootstrap, 61
Orellana, Francisco de, 350
Other Hispanic Americans, 375
Ouro Prêto, 87, 123, 283

P

Palenque, 73, 114
Palmares, 85
Pampa, 308
climate, 309
economy, 20th century, 314
export economy, 19th century, 313–314

physiography, 308–309
settlement, 310–313
PAN (Partido de Accion Nacional) (Mexico), 110
Panama, 195
Panama Canal Treaty, 25, 82, 189, 190, 195, 203, 215
Panama Galleon, 323
Panama, Isthmus of, 88
Paracas culture, 76
Paraguay, 151
Paraíba Valley, 332
Páramo , 48
Paraná Plateau, 271, 368
19th-century economy, 276–278
20th-century economy, 278–281
Climate, 273
coffee, 276–278
physiography, 273–274
vegetation, 274
Paraná, 333
Paraná River, 272, 273
Pardo, 84
Parícutin, 92
Patagonia, 40, 317
climate, 318
physiography, 317–318
settlement, 19th century, 319
Patrones, 211, 212
Paulistas, 33
Peasant Leagues, 289
Peninsulares, 86
Peones, 211, 212
Perón, Juan Domingo, 126
Peru Current, 43
Peru, 139, 142
colonial city populations, 125
farm sizes, 219
Petroleum, 320, 336, 343
Petrolina (Brazil), 287
Physiography
Amazon Basin, 348–350
Andes, 131–133
Brazilian Highlands, 271–273
Caribbean Lowlands of Central America, 201–202
Central American Highlands, 187, 188, 190
Guiana Highlands, 361–362
Hispanic Caribbean, 55–56
Mesa Central, 91–93
Pampa, 308–309
Paraná Plateau, 273–274
Patagonia, 317–318
Pinochet, Augusto, 232, 336
Pirates, 323
Pisac, 80
Pizarro, Francisco, 76, 78, 136
Planalto Central, 281; *see also* Central Plateau
Planned cities, 268–270
Belo Horizonte, 269
Boa Vista, 269
Brasília, 268
Cancún, 269
Central Plateau (Brazil), 284
Ciudad Guayana, 269
Cuidad Lázaro Cárdenas, 269
Goiânia, 269
Inca, 117
Volta Redonda, 242
Plantations, 212–213, 214–216
banana, 155, 195, 196, 203, 204, 215, 216, 326–328
sugar, 59–61, 63, 81, 163–164, 216, 225, 239, 244, 276–277, 374
Plate tectonics, 130, 192
Platt Amendment, 61, 374
Plaza, 121, 122, 265, 266
Plaza de armas, 121
Pleistocene, 49, 114, 93, 318
Political hegemony, 126, 195–196, 393–394
Political independence, 292
Political rights, 297, 299, 301
Political violence
Colombia, 141 142
Peru, 142
Pollo Campero, 345, 346
Pollution, 184
Mexico City, 102, 104
Population
19th century, 86, 88
20th century, 160
Amazon Basin, 359
Amerindians, 170
Andes, 143
Atlantic Coastal Plain of Brazil, 241–242
birth control and abortion, 167
Central American Highlands, 196
Central Plateau, 284
Chaco, 152
colonial Brazil, 87
country totals, 2000–2050, 169
European, 170
growth in 20th century, 165
growth rates 20th century, 167, 168, 169
growth rates of cities in 20th century, 248–249
Hispanic distribution in the U.S., 380–384
Hispanics future trends in U.S., 384–387
Hispanics in Chicago, 382
Hispanics in Florida, 382
Hispanics in New York City, 382
history of Hispanics in U.S., 375–380

Population (*continued*)
indigenous, 78, 136, 170
indigenous groups, 15, 74–76, 80, 82, 136, 359, 362; *see also* Amerindians
Latin America, 1930–2000, 165
Latin America, urban in 2000, 247
Llanos, 148
Mesa Central, 21st century, 99
mestizo, 83, 170
miscegenation and change, 82
mulattos, 170
rural, 231
sugarcane and change, 80
urban, 1950s, 159
Port cities, 128
Porteño , 14, 34
Pôrto Alegre, 280
Portobelo, 323
Portugal, 8, 10, 11, 34, 82, 86, 87, 124, 157, 210, 238, 240, 273, 276, 283 292, 305, 322, 323, 325, 326, 342
Portuguese
colonial city founding, 118, 123
immigration, 157, 283
settlement, 237–238
Potato, 47–51, 136, 181, 183, 284
Potosí, 138, 264
Poverty, 104, 187, 198, 200–201, 229, 290–291, 394
Poverty rates, 198, 251, 285, 296, 316, 337–339, 385–387
Pre-Columbian urban development, 113
PRI (Partido Revolucionario Institucional) (Mexico), 110
Primate city, 128, 259–262, 314
Protectionism, 331
Public debt, 335
Public health, 166, 247
Public transportation, 98, 254, 258, 263, 270, 314, 336; *see also* Subway
Puebla, 91, 93, 96, 97, 98,102, 103, 104, 105, 120, 124, 372
Pueblo de indios, 122
Pueblos jovenes, 255
Puerto Ayacucho, 363
Puerto Rican migration, to U.S. mainland, 369–372, 374–375, 378–379, 382, 385–386
Puerto Ricans, 372, 375
political status, 374
Puerto Rico, 55, 56, 64, 374
political status, 374
race, 177
Pulque, 102, 103
Puna, 48, 134
Punata (Bolivia), 296
Pupusas, 193
Purchasing power parity (PPP), 295–296

Q

Quality of life measures, 298
Quebracho, 150, 151
Quechua, 15, 16, 130, 136, 303, 325
Querandí Indians, 310
Quetzalcoatl, 78
Quilombos, 85
Quinine, 351
Quito (Ecuador), 135, 117, 120, 125, 138, 144, 145

R

Race, 169
Brazil, racial categories, 173, 174
Cuba, 177
discrimination, Brazil and U.S. compared, 171, 175
distribution in Brazil, 174
Dominican Republic, 177
Hispanic America, 176
personal identity, 172, 173
Puerto Rico, 177
Radio ownership, 28, 29
Railroad construction, 128, 139–140, 158, 215, 329, 358, 377
Railroads
Argentina, 260, 313, 319, 329–330
Mexico, 97, 98, 108
Rain forests, 43–45, 49, 57, 73, 79–80, 192, 202, 229–231, 350, 352–354, 356, 359, 362, 394
Rain shadow, 42, 44, 58, 318
Reagan, Ronald, 199
Recife, 243, 323
Reducciones, 208, 293
Regional integration, 333–334
Religion, 8, 9, 19, 20, 53, 113, 167, 293, 294
Christianity, 293–294
Evangelical Christians, 8, 294, 295
syncretic Catholicism, 20
Remittances, 64, 65, 343, 344
Repartimiento, 120, 136, 208
Resettlement, Amerindian, 209
Residential segregation, 122
Return migration, 379
Rio de Janeiro, 87, 123, 124, 128, 238, 239, 240, 242, 243, 323
Rio de la Plata, Viceroyalty of, 324
Roatán, 84

Romero, Oscar, 200
Rosas, Juan Manuel de, 126
Rubber, 288, 328, 352
Rubber boom, 360
Rubber tappers (*seringueiros*), 352, 353, 354, 359

S

Sacsayhuaman, 79
Saladeros, 311, 313
Salsa purchases in U.S., 370
Salvador (Bahia), 87, 124, 128, 175,240, 243, 323
Sambo (zambo), 83
San José (Costa Rica), 125, 191
San José (Uruguay), 310
San Juan (Argentina), 252, 253
San Juan (Puerto Rico), 65
San Juan River, 188
Sandinistas, 198–199
Sandino, César, 198
Santa Anna, (General) Antonio López de, 126, 372
Santeria, 20
Santiago (Chile), 55, 65, 119, 120, 125, 128, 180,184, 185
Santiago de Estero (Argentina), 150
Santo Domingo, 58
São Francisco, River, 272, 287, 290
Sao Paulo, 123, 128, 263, 279–280, 368
Sapote, 49
Scott, Winfield, 97
Self-help housing, 256
Seringueiros, 352, 353, 354, 359; *see also* Rubber tappers
Serrano, 14, 71
Sertão, 87, 240, 286–290, 351
  climate, 286
  economy, 288–290
  economy, 20th century, 289
  vegetation, 286–287
Settlement
  Atlantic Coastal Plain of Brazil, 237–238
  Caribbean Lowlands of Central America, 202–204
  Central American Highlands, 193–194
  Central Valley, 181
  Llanos, 146
  Pampa, 310–313
  Patagonia, 19th century, 319
  patterns, 19th century, 88, 124
  Portuguese, 237–238
  pre-Columbian, 113
Shantytowns (*favelas, pueblos jovenes, barrios de emergencia*), 184, 242, 252–257, 316
  Mexico City, 103
Sheep raising, 48, 134, 138, 140, 210, 289, 311, 319; *see also* Livestock raising
Shining Path (Sendero Luminoso), 142
Sierra Madre Occidential, 106
Sierra Madre Oriental, 91, 106
Silver Belt, 108
Silver, 106, 109, 138, 182, 194, 325, 327–328
Silver mining, 194
Slavery
  abolition, 154, 213, 240, 288
    effects in Brazil, 278
    impacts on plantations, 213–214
  African, 18, 19, 59, 60, 81, 154, 156, 239
    numbers and destinations, 83
  Amerindian, 81, 239
  Atlantic trade, 81, 84, 155, 156
  in Brazil, modern, 233
  in Cuba, 154
  in Puerto Rico, 154
  Spanish America, 82
Smallpox, 78, 81, 194, 359
Soil erosion, 44
Somoza, Anastasio, 198
South American Plate, 130
Southern Brazilian Highlands, 271
Southern Cone, 149, 154, 155, 165, 246, 300, 308
Soybeans, 152, 273, 275, 279, 308, 314, 356, 357
Spain, 8, 10, 11, 18, 52, 55, 58, 59, 60, 61, 65, 82, 85, 86, 87, 89, 91, 96, 97, 118, 119, 120, 121, 125, 126, 127, 128, 129, 138, 142, 154, 156, 161, 172, 178, 181, 194, 208, 210, 276, 292, 301, 303, 311, 313, 322, 323, 324, 325, 326, 334, 342, 345, 346, 365, 366, 367, 371, 372, 373, 374, 381, 392
Spanish
  colonial city founding, 118, 119, 120
  colonial port cities, 120
  colonial urban development, 118, 119,120
  conquistadores, 69, 75, 77, 118, 119 180, 207
  immigration, 156
  settlement in U.S., 370
  urban form of colonial cities, 121
Spanish-American War, 61, 367, 373
Spanish Civil War, 161
Spontaneous settlement, Amazon Basin, 356–357
Squatter settlements. *See* Shantytowns
Standard Fruit Company, 203
Staple crops, 45, 47
State-owned industry, 332, 336
Steppe, 317, 320
Stroessner, Alfredo, 126
Subduction zone, 130, 192
Subsistence faros, 216–221; *see also Minifundia*
Subway, 103, 144, 184, 257, 314, 315

Subway (food company), 31–32
SUDENE (Superintendencia para o Desenvolvimiento do Nordeste), 288
Sugarcane, 55, 56, 59, 60, 80, 81, 159, 238, 240
Suriname, 10
Sweet potatoes, 46, 71, 193, 273
Syphilis, 80

T

Taíno, 16, 58, 80, 177
Talcahuano, 184, 185
Tarascans, 17, 95
Taxes, colonial, 325
Tegucigalpa, 267
Tehuelche, 319
Teixeira, Pedro, 350
Telecommunications revolution, 342
Television ownership, 28, 30
Tenochtitlán, 73, 75, 78, 81, 95, 114, 115, 116
Tequila, 102, 103
Terra firme, 349
Terra rosa, 276
Texcoco, Lake, 115
Tiahuanaco culture, 76, 115
Tiahuanaco-culture, 115
Tierra caliente, 45
Tierra del Fuego, 321
Tierra fria, 47–48, 134–135, 191
Tierra helada, 48
Tierra nevada, 48
Tierra templada, 46
Tikal, 73, 114
Tin, 326, 327, 329
Tlaxcalans, 75, 78, 95
Tobacco, 50, 86, 232
Toltecs, 74–75, 95, 114
Tordesillas, Line of, 276
Tortillas, 19, 370
Tourism, 48, 63, 64, 65, 159, 184, 203, 204, 241–243, 290, 343, 362–364, 379
  Central American Highlands, 201
  enclave, 25, 26
  Mexico, 105
Trade groups, 334
Trade winds, 42, 43, 56, 58, 134
Trading companies, colonial, 323
Trans-Amazon Highway (Brazil), 354, 355
Trans-Chaco Highway, 162
Transnational, 342
Transportation
  Andes, 140
  colonial Mexico, 96
  export, Argentina, 313
  highways in the Amazon, 354–355
Treaty of Guadalupe Hidalgo, 97, 372, 373
Tribal societies, 67, 206
Trigueño , 177
Tropical rain forest, 49, 57, 73; *see also* Rain forests
Tropical wet and dry forests, 192
Truck farming, 279
Trujillo (Peru), 121
Trujillo, Rafael, 62, 126 177
Tula, 95, 114
Tupians, 273
Tupi-Guaraní, 273
Turkeys, 70

U

U.S. hegemony in Central America, 195–196
U.S. military intervention, 388
  Dominican Republic, 375
  El Salvador, 375
  Guatemala, 375
  Hispanic Caribbean, 61–62
  Nicaragua, 375
U.S.S. Maine, 61, 373
United Fruit Company, 203, 225
United Provinces of Central America (*Provincias Unidas de Centro América*), 13, 125, 194
Urban development
  19th century, 125
  20th century, 246
  Amazon Basin, 359–360
  Central Andes, 115
  Inca cities, 116, 117
  Mesoamerica, 73, 114, 115, 117
  Mexico 19th century, 98
  Portuguese colonial, 118
  Pre-Columbian, 113–114
  primate cities, 128
  Spanish colonial, 118, 119,120–121
Urban form, 257–258, 265–268
  Portuguese colonial cities, 123
  Pre-Columbian, 73, 114
  Spanish colonial cities, 121, 265–266
Urban hierarchy, Mexico, 104
Urban living conditions, contemporary Latin America, 250–252
Urban primacy, 201, 259–262, 314, 360; *see also* Primate city
Urban settlement policy, 256–257
Urban systems, early 19th century, 124
Urban transportation, 257–259
URNG (Unidad Revolucionario Nacional Guatemalteco), 200

Urubamba Valley, 77, 80
Ushuaia, 321

V

Valparaíso, 182, 184
Várezas , 349
Vecindades, 252
Vegetation
  Central Plateau (Brazil), 282
  Central Valley, 180
  Chaco, 150
  Mesa Central, 94
  Paraná Plateau, 274
  Sertão, 286–287
  xerophytic, 42, 57, 94, 107, 180, 286–287
Venezuela
  immigration in the 20th century, 161
  land reform, 228
Veracruz, 323
Viceroyalty of New Granada, 124, 125
Viceroyalty of New Spain, 96, 124
Viceroyalty of Peru, 124
Viceroyalty of Rio de la Plata, 89, 125, 311
Villa, Pancho, 98, 222
Villavicencio, 148
Viña del Mar, 184
Volcanism, 92, 131, 190, 192–193, 318
Volta Redonda, 332

W

Wal-Mart, 344–345
War of the Chaco, 151
War of the Desert, 312, 319
The War of the End of the World, 289
War of the Pacific, 182
War on Drugs, 141
Welsh, 157, 319
Westerlies, 40, 44, 134
Wheat, 96, 101, 110, 183, 313, 315
Wool exports, 106, 138, 140, 210, 311, 316, 317, 319, 327; *see also* Sheep raising
Wool, llama, 70, 136
World War I, 329, 330, 331
World War II, 331, 379

Xango, 20
Xerophytes, 42, 57, 94, 107, 180, 286–287
Xochimilco, Lake, 93, 116

Y

Yanomamo (Yanomami), 78, 359
Yerba mate, 311
Yuca, 20, 45, 51, 71, 151, 193, 273, 284, 289, 360; *see also* Manioc
Yungas, 133

Z

Zapata, Emiliano, 98, 222
Zapotecs, 17, 73
Zona de mata, 236
Zonda, 44

About the Author

Robert B. Kent, PhD, is Chair and Professor of Geography and Planning at the University of Akron. His research and publications have focused broadly on the human geography of Latin America, especially the Andes and Central America, and on the evolving geography of Latin Americans in the United States. He has traveled, lived, and worked throughout Latin America including 3 years spent as an urban and regional planner in Peru for the United States Agency for International Development (USAID). He has worked as a consultant for USAID in Bolivia on numerous occasions. He was a Fulbright Scholar at the Universidad Nacional de San Juan in San Juan, Argentina, and at the Universidad Pedagógica Nacional in Tegucigalpa, Honduras, and also spent a year as a visiting scholar at the Universitat Jaume I in Castellón de la Plana, Spain. He is a past chair of the Conference of Latin Americanist Geographers and a past councillor and treasurer of the Association of American Geographers. Dr. Kent is the editor of *Region in Transition: An Economic and Social Atlas of Northeast Ohio*, a coeditor of *Regional Development and Planning for the 21st Century*, and he has published numerous articles and book chapters on regional development, urban planning, cartography, GIS, and cultural geography. He lives in Akron, Ohio, with his wife, Marialena, and their two children, Robert and Anika.